USEFUL INFORMATION FOR THE CONCHOLOGIST

1 inch (English)
1 pouce (French)
1 pollex (Latin) $\Big\}$ = 12 lines
1 Zoll (German)
1 Ligne (French) = 2.25 mm.
1 line (English) = 2.11 mm.
1 linie (German) = 2.18 mm.
3' 4" 6''' means 3 ft., 4 inches, 6 lines
1 inch = 25.37 mm.

♂ = male

♀ = female

☿ = hermaphrodite

1 fathom = 6 feet
1 meter = 39.37 inches
1 nautical mile = 6,080 feet
1 statute mile = 5,280 feet
1 degree latitude = 60 nautical miles or
about 69 statute miles

Temperature conversion (Fahrenheit-Centigrade)

$$F = \frac{9}{5} C + 32$$

$$C = \frac{5}{9} (F - 32)$$

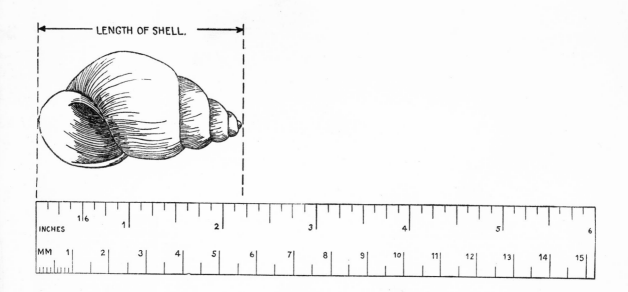

AMERICAN SEASHELLS

THE NEW ILLUSTRATED NATURALIST

AMERICAN SEASHELLS

BY

R. TUCKER ABBOTT, M.S.

Pilsbry Chair of Malacology
Academy of Natural Sciences of Philadelphia

WITH PHOTOGRAPHS BY

FREDERICK M. BAYER, B.S.

United States National Museum
Smithsonian Institution

D. VAN NOSTRAND COMPANY, INC.

PRINCETON, NEW JERSEY

TORONTO   LONDON

NEW YORK

D. VAN NOSTRAND COMPANY, INC.
120 Alexander St., Princeton, New Jersey (*Principal office*)
24 West 40 Street, New York 18, New York

D. VAN NOSTRAND COMPANY, LTD.
358, Kensington High Street, London, W.14, England

D. VAN NOSTRAND COMPANY (Canada), LTD.
25 Hollinger Road, Toronto 16, Canada

First Printing, March 1954
Second (Prepublication) Printing, March 1954
Third Printing, May 1955
Fourth Printing, March 1958
Fifth Printing, December 1960
Sixth Printing, May 1963
Seventh Printing, June 1965

PRINTED IN THE UNITED STATES OF AMERICA

To my children
BOBBY, CAROL, AND CINDY

Preface

This book wrote itself in response to the many hundreds of inquiries on seashells and other mollusks that have been sent to such museums as the Smithsonian Institution. Our natural heritage of seashore treasures has always been of keen interest to Americans, and in recent years there has been such an increase in shell collecting and biological investigations of mollusks that the need for a book like this has become apparent.

American Seashells belongs to the amateurs, for it is their enthusiasm in searching beaches and bays and their limitless curiosity into the ways of molluscan life that have dictated the contents of this book. How do shells grow? How do they form their color patterns? How do they breed and what do they eat? are the kind of questions asked. But the greatest demand has been for a reliable and up-to-date identification work. This need has been felt not only by private collectors, but particularly by students of marine biology and those undertaking research in fisheries and ecology. In meeting these requirements, there has been an attempt to strike a balance between the palatable, popular accounts and the more technical material. The illustrations, the standardization of popular names and the natural history accounts will be of particular interest to the beginner, and it is hoped that the monographic reviews, identification keys and the bibliographies will adequately serve the serious student.

There are over 6000 species of mollusks living in North American marine waters, and a thorough treatment of them all would call for a book many times the size of this. The conchologist will find that the 1500 species discussed or illustrated within these pages include every kind of shell likely to be found in shallow waters, whether collecting is done in Labrador, Florida or along the western shores from Alaska to Lower California.

While considerable original research went into many parts of this book, it should be kept in mind that a popular book covering such a vast fauna is merely an expression of the present state of knowledge of our science and that time and research by others will inevitably render sections of it obsolete.

I would like to express my thanks to Dr. Leonard Carmichael, Secretary of the Smithsonian Institution, for granting permission to publish on and illustrate specimens housed in the United States National Museum. Although the efforts involved in this project did not encroach upon official time, I would like to record my good fortune in being able to consult the National Museum collections on holidays and during after hours. Austin H. Clark stands foremost as spiritual guide and counselor in the many intricacies of preparing a book for the public. Dr. Harald A. Rehder, curator of the Division of Mollusks, with whom I have been pleasantly associated for several years, has kept a weather eye on this project and in not a few instances has made valuable suggestions. I have gratefully and heavily leaned on the Minutes of the Conchological Club of Southern California which represents the work of John Q. Burch, A. Myra Keen, A. M. Strong, S. Stillman Berry and many others. Mr. Gilbert Voss kindly helped me with the section on squid and octopus. This is also true of *Johnsonia*, a magnificent work produced by William J. Clench, Joseph Bequaert, Ruth D. Turner and others. I would also like to thank my friends in the National Museum for constant encouragement.

The heaviest debt is to the countless amateur collectors of American mollusks. Were it not for their enthusiastic pursuit of shells and their unselfish desire to share their treasures with our leading museums, our scientific collections undoubtedly would be half their present size. It is my sincere hope that this book, by its usefulness, will measure up to their kindnesses and friendships.

Illustrations make the book, and *American Seashells* could not have been successfully completed without the aid of Frederick M. Bayer, Associate Curator in the National Museum, who is responsible for the colored plates, including the lovely paintings of western Atlantic nudibranchs. Most of the other photographs were also taken by him. Special thanks are due William J. Clench who made available all the photographs and drawings that have appeared in *Johnsonia*. The colored paintings of Pacific coast nudibranchs are taken from F. M. MacFarland's "The Opisthobranchiate Mollusks from Monterey Bay, California, and Vicinity," which appeared in 1906 in the U. S. Bureau of Fisheries Bulletin 25. All of the exquisite pen drawings of shells, unless otherwise noted, were executed by the U. S. Army Surgeon, John C. McConnell, in connection with researches done by William H. Dall of the U. S. Geological Survey. Our photographs of Florida Thorny Oysters are from specimens kindly sent on loan by Leo L. Burry of Pompano Beach, Florida.

Notable credit is due Chanticleer Company of New York City which took such pains in the preparation of the colored plates, and to the printer who retained with such remarkable fidelity the beauty of the original photographs.

R. T. A.

Washington, D. C.
September 15, 1953.

Contents

List of Plates

PLATE

Foreword

Shell collecting is now taking its place as one of the major outdoor diversions. It has advantages over such pursuits as bird watching or fishing, for you may have even more pleasure in studying your catch at home than in the time spent afield. The thrill of finding a shell new to you, or of watching some rare snail going about its watery affairs, is ample reward for the sunburn and stiff neck you may have from wading around too long with a water-glass. Hours sieving dredgings are counted well spent if a fine volute or turrid turns up in the seaweed and rubbish.

American Seashells gives a comprehensive and well-rounded view of the Mollusca in nontechnical language. It is easy reading for the beginner, but it contains also material indispensable to the advanced malacologist. The chapters on nudibranchs and pteropods are especially welcome, for these beautiful animals have always been slighted in American books. In chapters on the life of the snail and the clam, with the author we "listen in" to the current of molluscan life. The shells become living things, moving and breathing, feeding and mating.

One perplexity of the novice is that different books may give different names for the same shell. The causes of this diversity are explained on a later page. With the facilities of the largest museum in America, the author has been able to speak with authority in those matters of nomenclature. When the problem is zoological and still to be solved by further collections, or by the study of living mollusks, then the cooperation of the keen collector may give the answer sought. Professional malacologists are few. Their work is largely in museums with dead animals. The interesting but long task of collecting from a thousand miles of coast, and observing mollusks alive, has always been in large part a labor of love by private naturalists. Our science owes nearly as much to them as to the work of professional zoologists.

The author belongs to the younger group of malacologists, but he has cultivated the society of mollusks in many lands, from East Africa, the

Philippines and Guam to our Atlantic coasts. His original research covers a wide range, and is at its best in dealing with some neglected or little-known shellfish. This book will do much to increase the knowledge and enjoyment of all of us who hunt the elusive mollusk.

HENRY A. PILSBRY
Curator of Mollusks
Academy of Natural Sciences
of Philadelphia

PART I

The Natural History
of Seashells

CHAPTER I

Man and Mollusks

SEASHELLS and man were closely associated even before the dawn of civilization when primitive man gathered snails, oysters, and other kinds of mollusks along the seashore for food, implements, ornaments, and money. The many kitchen-middens and burial sites in nearly every corner of the world reveal the great extent to which early peoples were dependent upon mollusks. On some coral islands, as, for instance, Barbados, where there was no available stone, nearly all domestic utensils, including knives and axes, were made from seashells. As civilization became more complex, specialization in the use of mollusks increased. From them were obtained dyes, inks, textiles and windowpanes. In the Mediterranean region there was a long period when an entire commercial empire owed its origin and continued success to the Tyrian purple obtained from a seashell. Later, in Roman times, the farming of oysters and edible snails became a major enterprise.

Today the uses of molluscan shells are legion. Jewelers, artists and button manufacturers; biologists, geologists and archaeologists; bird and aquarium dealers; all daily use mollusks or their products. In recent years there has flourished in Florida a five-million-dollar-a-year seashell industry. Throughout the country, the hobby of shell collecting is enjoyed by countless thousands, and it now rivals the popularity of coin collecting. Local and federal agencies are investing millions in research directed toward the more efficient cultivation and utilization of commercially important mollusks.

From another standpoint of perhaps even greater importance mollusks have influenced the activities and welfare of man. Some are extremely destructive to wooden structures in the sea, and others are a serious menace to health, mostly as intermediate hosts to dangerous parasites or as carriers of

3

poisonous micro-organisms. Prior to the advent of ships with metal hulls no vessel on the seas was safe from the borings of molluscan "shipworms." Many ships have disappeared at sea as a result of being weakened by the attacks of these creatures. Even today damage to the extent of millions of dollars is done every year to wharf pilings, small craft, and hemp lines by these bivalves. In many parts of the world the health of millions is seriously menaced by mollusks. It was not until the turn of the century when modern research was directed toward tropical diseases that the full importance of snails as carriers was appreciated. Six major parasitic diseases have been shown to be transmitted by fresh-water mollusks. Thousands of people die each year in China and Egypt from the blood-fluke disease alone. No fatal snail-borne disease is present in North America proper, but visitors to the West Indies and northern South America are warned to keep out of ponds and flooded ditches in these regions.

In other respects, mollusks are of minor medical importance. A number of parasitic diseases of sea birds and fish are carried by marine shells, such as the periwinkles *Littorina* and *Tectarius* and other shore species. During certain seasons of the year, usually in late summer, these snail hosts shed thousands of microscopic larval worms into the sea water. Although normally destined to penetrate the skin of birds, these tiny creatures sometimes attack man and cause an uncomfortable rash or "swimmer's itch" which is often mistaken for jellyfish sting.

Among the most dangerous inhabitants of the coral reefs in the tropical

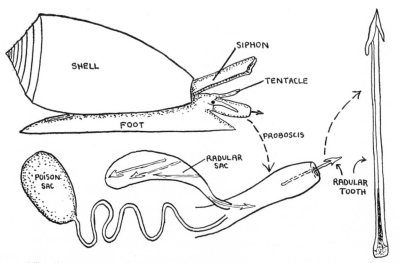

FIGURE 1. The large cone shells of the Indo-Pacific, and possibly those of the Atlantic, can inflict a serious, and at times fatal, sting. The venom leaves the poison sac and, together with the tiny, harpoon-like tooth, is ejected from the snail's proboscis and stabbed into the skin of the victim.

Indo-Pacific are the cone shells (*Conus*), the sting of which is as powerful as the bite of a rattlesnake. Although the beautiful cone shells are among the commonest of Indo-Pacific mollusks, the total number of authentic cases of death from their sting is surprisingly small. No American species have been recorded as harmful to man but, because all cones possess the necessary apparatus, it would be wise to be careful in handling American specimens over two inches in size.

The number of cone stings is few because of the shy nature of the animal. Invariably a snail will withdraw into its shell when disturbed and, unless the cone is held quietly in the palm of the hand for some minutes, there is little likelihood of the collector being stung. The apparatus for the injection of the venom into the skin of the victim is contained in the head of the animal. Bite, rather than sting, is perhaps more descriptive of the operation. The long, fleshy proboscis or snout is extended from the head and jabbed against the skin. Within this tube are a number of hard, hollow stingers, as long and slender as needles. These are actually modified radular teeth, commonly used in other snails to rasp their food. Under a high-powered lens the teeth of the cone shell resemble miniature harpoons. As the teeth are thrust into the skin, a highly toxic venom flows from a large poison gland located farther back in the head, out through the mouth, and into the wound through the hollow tube of the tooth. In some cases, death has taken place in four to five hours after the patient was stung. Not all cases are serious. Andrew Garrett, a famous shell collector of the latter half of the nineteenth century, reported that he was stung by a tulip cone that caused a "sharp pain not unlike the sting of a wasp."

While in recent years the cone shells have received perhaps an undue amount of notoriety as dangerous creatures, they are best known as an aristocratic family of beautiful shells which have been favorites for years among the most discriminating of collectors. For hundreds of years the sound of the auctioneer's gavel has been heard at the sale of valuable collections of seashells, but no shell has created such fevered interest as the Glory-of-the-Seas cone. Its present-day value is in the neighborhood of $400 to $600. This species seems to possess the ideal combination of features which brings high prices—beauty, size, rarity and, above all, mystery or legend. Although the legends connected with the Glory-of-the-Seas are for the most part untrue, the mere mention of its name will invariably cause the blood pressure of shell collectors to rise.

The first published reference to the Glory-of-the-Seas was in 1757. Today the whereabouts of each of the twenty-three specimens is known. The most famous finding was made by the renowned shell collector, Hugh Cuming, in 1838 when he found three specimens at low tide on the reefs at Jacna on Bohol Island in the Philippines. The myth has often been repeated that Cuming returned for more only to find that the reef had sunk during an earthquake, and that since then no other specimens have been found. However, the species is apparently widespread throughout the East Indian region. Specimens have turned up since Cuming's day at Cebu in the Philippines, Amboina Island, and Piru Bay in the Dutch East Indies. A four-inch specimen was found on the shore at Wahaai, Ceram Island, after a storm in 1896. In addition to the existing twenty-three specimens, three were destroyed during World War II and eight, formerly known to exist, are missing. A search in grandmother's attic or along some East Indian beach will doubtless bring others to light.

Collectors of fancy seashells are constantly in search of specimens of outstanding qualities, and although a number of species are well-known for their high value or unusual beauty, the standards by which we judge their rarity and attractiveness are considerably varied. The differences in our appreciation of beauty are natural enough, for the colors, forms and textures of seashells are numerous enough to offer appeal to almost any type of aesthetic appreciation. The man who covets a brilliantly patterned Olive shell of rich golden-red colors may see little in a tiny white shell which another collector treasures for its intricate snow-flake sculpturings.

For many conchologists rarity is gauged by the top price that a specimen may bring; for others the important judging point is the scarcity of the species in nature or perhaps the rarity of specimens in collections. Left-handed, double-mouthed or distorted specimens, like misprints in stamps, are highly valued by many veteran collectors. There are literally hundreds of truly rare species, but most of these are deep-sea shells, some of which are known only from a single specimen. Most of these are small and not particularly attractive. The high-priced shells are found among the showy genera, like the cones, Pleurotomaria slit-shells, volutes, murex shells, scallops and cowries. The Golden Cowrie is the most popular among the so-called rarities, the present-day price ranging from $20 to $60. Some species may be considered rarities for years and command very high prices, until they are collected in large quantities. The Goliath Conch (*Strombus goliath*) is worth about $200 today, but collecting in northern Brazil would undoubedly bring them to light in great quantities and hence would lower the price to a few dollars. The Precious Wentletrap Shell (*Epitonium scalare* or *pretiosum*) of the western Pacific was in such demand years ago that Chinese found it profitable to make counterfeits out of rice paste. The species is now considered reasonably common and is low-priced, but genuine rice counterfeits are now rare and equal in value to the price of the first-known shell specimens.

Some of the most interesting threads of man's early history have been woven around the trade routes of primitive peoples and their dispersal of shells. The discovery by archaeologists in 1895 of the Red Helmet Shell

(*Cypraecassis rufa*) in a grave of the prehistoric Cro-Magnon man in the caves of France was of considerable importance in tracing former trade routes. This species is found only in the Indian and Pacific oceans. Its presence substantiated other archaeological evidence that extensive trade routes for great distances existed among early European man. The Tiger Cowrie (*Cypraea tigris*), another Indo-Pacific species, has been found in a prehistoric pit-dwelling at St. Mary Bourne at Hants, England, and the Panther Cowrie (*Cypraea pantherina*), a Red Sea species, has been found in Saxon women's graves, excavated in several localities in Kent, England.

The seashell with perhaps the widest dispersal by the ancients and modern man is the small, yellow Money Cowrie (*Cypraea moneta*) which was for many centuries the accepted currency in many parts of the world. Although its natural biological distribution is limited to the vast areas of the Indian Ocean, the East Indies and the islands of the tropical Pacific, its use as currency or for ornamentation has been almost worldwide. The three most unusual records are those located in North America.

When the aboriginal sites along the Tennessee River were being investigated at the beginning of this century, five Pacific Money Cowries were unearthed from one of the graves of the Roden Mounds in Alabama. Evidence points to the fact that these burials had been made before the mound makers had any intercourse with white man. The shells were sent to the United States National Museum by their discoverer, and Dr. William H. Dall wrote the following interesting reply:—

> I should incline to the belief that the cowries were imported in or about the time of Columbus' voyages. Bound, as they supposed, for the Indies, where the cowry was formerly (like our wampum) a staple article of barter, the exploring vessels would have undoubtedly carried cowries as well as other articles of trade we know they carried. It would not have taken them long to find out that cowries did not pass as currency with American natives, and reporting this on their return to Spain later traders would not have carried them for barter. The necklace or bracelet you obtained may have passed from hand to hand as a curiosity (as I have known such things to do) until it reached a people who knew nothing of whites 'till much later. In fact your cowries may have come off one of Columbus' own vessels.

If not from one of Columbus' ships, these shells more than likely were brought over from Europe soon afterward by early Spanish explorers. It does not seem so plausible to assume, as some ethnologists do, that these shells were brought by migrating tribes from eastern Asia to America via the Bering Straits long before the time of Columbus.

The Lewis and Clark Expedition brought back in 1805 a handsome dress, possibly of Cree origin, which was adorned with four dozen Money Cowries.

Another Money Cowrie was unearthed near the so-called Onatonabee Serpent Mound of Peterboro County in Ontario, Canada. It is most likely that in both of these cases the shells were the remnants of the Hudson's Bay Company's shell stock which was bartered with the Cree and other Indians well before the time of the Lewis and Clark Expedition.

A lively trade in marine shells took place for centuries among the pre-Columbian peoples of southwestern United States. Archaeological studies in that area have been able to confirm the existence of trade routes which then existed from three principal geographical areas, one along the coast of southern California, a second from the Gulf of California, and the third on the Atlantic side from the coast of the Gulf of Mexico.

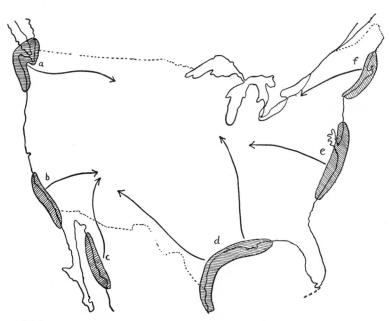

FIGURE 2. Major sources of trade shells used by the early American Indians. **a**, tusk-shells. *Dentalium*, used for money; **b**, abalone shells, *Haliotis*, and the necklace shells, *Olivella*; **c**, *Glycymeris* clams for bracelets; **d**, olive shells, helmet shells, *Cassis*, and many others; **e**, large whelks, *Busycon*, and Venus clams; **f**, wampum from the Venus clam, *Mercenaria mercenaria*.

Marine shells were used primarily as ornaments. Beads of glossy Olivellas and Olive shells were by far the most popular throughout the estimated 1000-year span of trading. Pendants, bracelets, rattles, trumpets and carved shells were popular in that order. Pacific Coast shells were passed on from settlement to settlement to a limited extent by the early Basket Makers (?-500 A.D.) and, with the rise of the late Basket Makers (500-700 A.D.), trading increased from both the Pacific Coast and the Gulf of California.

it was not until Pueblo times that the Atlantic trade reached the southwest when the Pacific trade was also at its zenith.

For years archaeologists were puzzled by the absence in New Mexico of residue shell material which ought to be present wherever bracelets of the *Glycymeris* clam appear. Not until 1930 were the hundreds of ancient manufacturing centers discovered along the Sonora coast of the Gulf of California. There the early Indians sawed out patterns and ground down the clams to a smooth finish. The existence of this industry in the areas where the clams live illustrates one of the fundamental problems of prehistoric trade where beasts of burden were unknown and all goods were carried on men's backs. The finished product was not only much lighter, but also brought a better price.

The Mohaves used a trade route from the Pacific Coast in the vicinity of Los Angeles across the mountains into Nevada and Utah, and they perhaps have the rightful claim to the title of the "Phoenicians of the West." Several routes extended from the Sonora coast of the Gulf of California up to the Gila basin to Pecos in northwestern New Mexico. Around this area there is evidence that the Pacific and Atlantic trade converged during Pueblo times. It is quite likely that Pecos was a trading pueblo between the southwestern peoples and the plains tribes.

In the midwest of the United States an entirely separate trade route existed from the Mound Builders of Illinois (Cahokia group) south to the Gulf of Mexico. Among the mounds of these prehistoric people the Cameo Helmet Shell (*Cassis*), the Fighting Conch (*Strombus*) and the Apple Murex Shell have been discovered—all species from southern Florida or the warmer waters of the Gulf of Mexico.

In recent times, the dispersal of mollusks has been little short of spectacular, particularly so if we may mention in this connection the many large collections that have been assembled in natural history museums. For the last 200 years there has been a steady flow of specimens to these study collections from all lands. Probably the largest mollusk collection in existence, that housed in the United States National Museum in Washington, contains over 9,000,000 specimens and represents about 45,000 kinds. This collection is the result of a century of labor on the part of thousands of ardent enthusiasts who collectively have stooped to pick up mollusks in over 100,000 localities throughout the world.

Added to the scientific traffic of material among dozens of natural history institutions, is the constant and spirited exchange of specimens among thousands of private shell collectors. It is little short of miraculous that in a small Connecticut town one can find in an amateur collection a rare, ivory-like *Thatcheria* shell from 200 fathoms in Japanese waters or a 200-pound valve of the giant clam from the Great Barrier Reef of Australia. In a small cabinet of land mollusks in Boonton, New Jersey, you may find a giant African snail from the Belgian Congo or a tiny ground snail, no larger than a grain of rice, from the Himalaya Mountains of India. The locality labels attached to many of the shells in museums are milestones in recent history—Tobruk, Bizerte, Anzio—Port Moresby, Guadalcanal, Leyte Gulf—Pusan, the Han River, and Peking.

Accidental dispersal of marine shells, even in large quantities, is not uncommon, and many unusual cases have been recorded in newspapers and scientific journals. In the days when the beautiful ear shells or abalones of the Californian coast were used extensively for cabinet inlays, a sailing vessel bound for New York with a cargo of these shells went down in a storm just off Santiago on the south coast of Cuba. For several years, these magnificent shells were being cast ashore on the beaches, much to the delight of local collectors and small children.

A similar case occurred in 1873 when the "Glendowra," a four-masted vessel, homeward-bound from the Philippine Islands on a cowrie expedition, was wrecked off the coast of Cumberland, England. She had on board more than 600 bags of Money Cowries destined for use in the African trade and, during a heavy fog, ran ashore near Seascale. For years these shells were picked up in excellent condition on the nearby beaches. Many collectors, unaware of their history, regarded them as native to the British Isles.

The necessity of taking on ballast to make up for light cargoes on return sailing voyages has been responsible for many introductions of exotic shells to United States ports. The Money Cowrie has been picked up on one of the beaches of Cape Cod and was presumably jettisoned there by a sailing ship returning from the Indian Ocean. Ballast Point in San Diego was years ago a fairly good place to collect Hawaiian shells and, during World War II, a dozen or more species of British marine shells brought in ballast could be found in the vicinity of Long Island, New York.

Wholesale dispersal of marine shells has been carried out purposely by man on several occasions. With malice toward none, it may be said that considerable competition for the tourist trade exists between the Atlantic and Gulf coasts of Florida. Lacking the abundance of attractive seashells which are now considered prime tourist bait, the Atlantic coasters have made up for it by their aggressive ingenuity. It is reported that some Miami hotel owners have sent trucks to the rich beaches of the Gulf Coast, loaded them with molluscan spoils and brought them back to dump on their own relatively shell-less beaches.

Mollusks have been used extensively in art and literature, and throughout history we find numerous uses of shells as symbols. In many parts of the world, and especially along our motor highways, the scallop shell is a familiar trademark to motorists. The "Shell" Transport and Trading Company had its origin in London, England, during the middle of the last century when shell ornaments on boxes, screens and frames were popular in Early Victorian drawing rooms. The founding brothers, Marcus and Samuel Samuel, traded in shells from all parts of the globe, but as a side line they began to deal in the sale of kerosene. With the advent of the new "parafine oil" lamps and, later, the combustion engine, it was not long before they were marketing oil exclusively. Soon afterward their company was merged with the Royal Dutch interests. Until 1904 they used a trademark emblem patterned after the Sun-rayed Tellin (*Tellina radiata* of the West Indies), but this was later replaced by the now world-famous emblem of the European Jacob's Scallop (*Pecten jacobaeus*). The scallop on the letterhead of the company's stationery is a fossil species from California.

The 200-odd oil tankers of the Shell Oil Company are named after various genera of mollusks, the first ship launched being christened the S. S. *Murex*. Aboard each vessel, a specimen of her namesake mollusk is mounted in a glass exhibit case. Naming and securing shells for the first hundred ships was comparatively easy, but recently the choice of new names has resulted in the unfortunate selection of obscure genera based on rare and, in some cases, microscopic species. Some ships bear names based on the same genus —nautical synonyms!

The use of the scallop is a very ancient one. As a source of food and as an eating dish it was used in prehistoric times. It is pictured on the coins of the early Phoenician outpost of Saguntum (now Murviedro, Spain). All through the middle ages the scallop shell was used as a religious symbol, especially in connection with pilgrimages to the shrine of Saint James at Compostella and the crusades to the Holy Land. Three different popes granted a faculty to the Archbishops of Compostella to excommunicate all who sold scallop shells to pilgrims anywhere except in the city of Compostella. Today many of the family shields of England bear scallop shells, indicating that their ancestors made pilgrimages to the Holy Land.

It is interesting to note that one of the earliest shell collections known to us contained a Jacob's Scallop. This was unearthed from the ruins of Pompeii, together with *Conus textile* and the Pearl Oyster of the Indian Ocean, in what appears to have been a natural history collection. It is not beyond the realm of possibility that this was the remains of the Natural History Society of Pompeii, of which the distinguished naturalist, Pliny, was probably a member. It was Pliny who first recorded the swimming activities of the scallop, and he observed that it was able to dart above and skip along the surface of the water.

In our modern age of synthetic dyes and highly mechanized textile industries, we little appreciate the part played by dye-producing mollusks in the history of the ancient civilizations of the Mediterranean. The power and fame of the Phoenicians, who were the great traders, navigators and colonizers of that region as early as 1500 B.C., were largely due to their monopoly of the Tyrian purple dye. The ancient cities of Tyre and Sidon (now Souro and Saidi in Lebanon) became great banking centers and the crossroads of commerce between Asia, Africa and western Europe. Although archaeological findings indicate that purple dye from species of *Murex* was in use in Crete as early as 1600 B.C. and in Egypt by 1400 B.C., these two Phoenician cities had managed to monopolize the industry and to expand their prosperous enterprises by 1000 B.C. The continual search for new beds of *Murex* is probably one of the reasons for their later colonization of Malta, Sicily, Utica, Carthage and Gades (now Cadiz). These ports served as trading stations and, as evidenced by the great piles of unearthed *Murex* shells, as subsidiary purple dye factories. The imperial coins of grateful Tyre bore for many years the imprint of the *Murex* shell. It is interesting to note that the name Phoenicia comes from the Greek *phoenix*, "red," which may well allude to the red or magenta color variations of the molluscan purple.

It is now the general consensus that three species of marine snails were used in the Mediterranean. Although all three were present in many areas, the city of Tyre employed in the main *Murex brandaris*, while the great banks of shells discovered near Sidon in recent times were almost exclusively made up of *Murex trunculus* (see plate 10, figs. i and j). The "buccinum" of the Roman naturalists probably was *Thais haemastoma*.

The high cost of the purple dye was largely due to the long and arduous process of manufacture. A recent experimenter used about 12,000 specimens of *Murex brandaris* before obtaining 1.5 grams of pure dye, and he estimated that one pound of dye in ancient times was worth from $10,000 to $12,000.

The dye-producing fluid is exuded from an elongate gland which is situated on the inner wall of the mantle between the rectum and the gills. The fluid is colorless to milky-white when first produced, but when exposed to direct sunlight, it changes immediately to bright yellow, then passes through shades of pale-green to bluish and finally red-purple. During this photochemical process a strong odor is given off which resembles rotting garlic. The Tyrians collected vast quantities of living snails and ground up the smaller specimens in caldron-shaped holes in the rocky shore. Larger specimens were cracked open and the gland-supporting mantle ripped off and thrown into the holes. Salt was added to this juicy mass to prevent excess rotting, and then the sun was allowed to act on it for two or three days. This material was transferred to vessels of tin or lead and then diluted with five or six times its bulk in water. A ten-day period of moderate boiling followed, during which time the scum was constantly removed. Test pieces of wool were allowed to soak for five hours to ascertain if the desired strength of dye had been reached.

Our modern concept of purple is quite different from that of the ancients. They understood it as several colors ranging from dull crimson and magenta to violet-purple. The most expensively dyed cloth was made in Tyre and was more on the order of a dull red. In Sidon, where *Murex trunculus* was mainly used, the color was closer to our modern idea of purple. The wide range in hues of Tyrian purple was brought about by different strengths of, and varied techniques in making, the dye, including the double-dip system of dibapha in which the first bath consisted of extracts from *Thais* and the second dip taking place in *Murex* dye. The type of cloth and weave also produced wide variations.

There is no question that cloth dyed with Tyrian purple was extremely valuable and at times vied in value even with gold. Hence it was reserved for the use of the wealthy and the hangings of temples. The Babylonians are said to have used it for the dress of their idols. A few of our museums possess small pieces of Egyptian mummy wrappings which were dyed with Tyrian purple. However, it is necessary to make a chemical analysis to prove the presence of this dye, for the ancients were able to produce a similar color by double dyeing with indigo and madder.

The Bible makes several references to this valuable purple. Moses used it for the works of the tabernacle, as well as for the clothing of the high priest. Among the presents which the Israelites made to Gideon were purple raiments that belonged to the kings of Midian. Much later, according to Acts 16, verse 14, a seller of purple from Thyatira was converted by St. Paul at Philippi.

Aristotle and Pliny both gave fairly detailed accounts of the industry widespread throughout Asia Minor. Plutarch records that when Alexander took possession of Susa he found among the treasures of Darius 5000 talents in weight (290,000 pounds) of purple cloth. Athenaeus states that the dye was extensively used as a cosmetic and was applied as a lipstick and rouge in Rome. At the fateful battle of Actium, the ship of Marcus Antonius and Cleopatra was distinguished from the rest of the fleet by having sails solidly dyed in Tyrian purple. It is difficult to believe, as many authorities claim, that the Tyrians kept the process a secret even for a short time, for we find that factories existed throughout most areas in the Mediterranean.

In Rome only senators were allowed to wear a broad purple stripe (*latus clavus*) around the opening of the tunic. Laws were finally introduced by Nero and again by Theodosius (379-395 A.D.) prohibiting the wearing of Tyrian purple except by the Emperor himself. Except for its later use by the Christian church, especially in cardinal cloaks, the crimson color ceased to be worn or manufactured after the fall of the Roman Empire and the conquest of Tyre by the Arabs in 638 A.D. It would scarcely pay to revive the industry except perhaps as a novelty item for tourists. The color is not particularly exciting to the modern eye, and, in addition, it may be synthetically produced at low cost, so that one has no assurance that a souvenir textile is actually dyed with molluscan purple.

The Mediterranean area and the west coast of Africa were not the only regions where mollusks were used for dyeing. In the British Isles the art seems to have been known from very early times. The Celts of England and the Lake Dwellers of Ireland (about 1000 B.C.) used the common *Thais lapillus* which is also abundant on New England shores. As late as the eighteenth century this species was used for marking linen in England, Scotland, France and Norway. The French used molluscan purple to dye the parchment of rare books, some examples of which are still bright after 800 years.

Had the Phoenicians possessed the compass and ventured to the West Indies, they would have marveled at the abundance of our Wide-mouthed Purpura, *Purpura patula*, and its large production of rich violet dye. Collectors who have put live specimens in a cloth bag will recall the bright, durable stains that have appeared soon afterward in the fabric. Many shell collections contain this species in which specimens still retain purple stains on the outside of the shell. The subspecies, *pansa*, was used in prehistoric times for dyeing cotton on the northwest coast of South America and the west coast of Central America. Even today the Tehuantepec Indians of Mexico use the Pansa Purpura for dyeing cotton threads. The natives have put into effect a plan of conservation and, instead of crushing the shells, they carefully "milk" the living specimens by pressing in the animal to squeeze out the juice. They then return the mollusks to the rocks and revisit them at a later date. The cotton threads are individually drawn through the liquid to obtain the fast dye.

In 1711 Reaumur accidentally discovered that the egg capsules of *Thais lapillus* were a simpler and more abundant source for the purple dye. As *Murex* egg capsules mature, they take on a characteristic purplish hue. It is possible that this was the secret, if such existed, that the Tyrians guarded so jealously. Latest experiments indicate that the purple dye is a derivative of indigo containing bromide.

Probably most, if not all, species of *Murex*, *Thais*, *Purpura* and other members of the Muricid family produce this bromide, dye-giving secretion. It has been suggested by some workers that this secretion serves as an anesthetic on various oysters, clams and chitons upon which they prey. However, the presence of purpurase in the egg capsules does not favor this view. In addition, the dye-producing gland is closely associated with the reproductive system and not with the salivary glands or any other organs of the proboscis. Many other carnivorous families which attack other living mollusks in a manner similar to that of the Muricids do not produce this dye.

Inks and dyes are produced by many other mollusks, the *Sepia* cuttlefish being an outstanding example. Purple dye has been recorded in the Purple

Sea Snails, *Janthina*, in the Wentletraps, *Epitonium*, in some of the Mitras and Olive shells and in the sea hares, *Aplysia*. These substances are known to be irritating to fish and other would-be predators, and its purpose as a defense mechanism seems most likely. Probably future experiments will show that the egg capsules of *Murex*, loaded as they are with purpurase, are distasteful to fish and have an unusually high survival value.

CHAPTER II

Life of the Snails

THE private lives of the snails, or gastropods as they are more correctly called, are almost as varied as the different kinds of seashells that are found along our beaches. More than half of the 80,000 species of existing snails live under marine conditions, the remainder being air-breathing land species or inhabitants of fresh water. In their evolutionary struggle for existence, they have shown an amazing diversity in adapting themselves to nearly every condition found in the sea. There are snails that creep, jump, swim, burrow, some that are permanently anchored to rocks and a few that live inside other marine creatures. In a few cases, as in some conchs and top shells (*Trochus*), the snail may play host to small fish and tiny crabs.

Gastropods have experimented in all manner of forms, colors and sizes. In size they vary from the two-foot-long Horse Conch of Florida (*Pleuroploca gigantea*) to the microscopic Vitrinellas that scarcely exceed the size of a grain of sugar. Some species display unusual ornamentation and, as in the *Murex* shells, produce long, delicate spines. There are few objects in nature that can vie in beauty with the glistening sheen found in the shells of the olives and cowries. On the other hand, the beautiful sea slugs or nudibranchs may entirely lack a shell. The Carrier Shell, *Xenophora*, has acquired the strange habit of collecting shells, bits of coral and other hard objects, and cementing them to its own shell.

WHERE THEY LIVE

From the high levels of the coastal cliffs to the canyons of the ocean's bottom, a thousand kinds of habitats have been adopted by marine gastro-

16

pods. A few species of nerites and periwinkles are known to ascend trees near the seashore, although tree-dwelling is best known among certain tropical land snails. In the tropics, the *Tectarius* prickly-winkles habitually live in or near splash pools along the rocky coast where spray from the waves and drenching rains are constantly changing the temperature and salinity. When the pools are dry the snails are often able to withstand weeks of hot sun and parched conditions.

Three kinds of snails in American waters are forever destined to wander at large on the surface of the open ocean. The purple *Janthina* snails are born, live and, in most instances, die at sea. These pelagic snails live upside down and remain at the surface by means of a small raft of bubbles. Small bubbles of air are entrapped in a special mucus secreted by the animal. This clear fluid congeals upon contact with salt water and air, and it adheres to the foot. The entire float has much the appearance of crumpled cellophane. The female attaches her small eggs to the underside of the float where they are partially shaded from the sun's rays. The Janthinas live off the coasts of our southern states, and during certain seasons they are commonly cast ashore in California, Florida and the Gulf States. Specimens have been blown off their Gulf Stream course and been washed ashore in New England and even the British Isles.

As is the case with so many other pelagic creatures, the shell surface of *Janthina* which faces downward (the spire of this upside down shell) is colored a light, milky blue. This is probably a protective coloration which blends with that of the surface of the sea, which to an underwater observer is similarly colored. For some unknown reason Janthinas are completely blind.

Two other groups of gastropods live at the surface of the ocean and, like *Janthina*, live an upside down existence. These are the tiny brown *Litiopa* snails which adhere to floating sargassum seaweed by means of a silken thread of mucus, and the heteropods or fin-footed sea snails which remain afloat by paddling a wide, fin-shaped foot. The latter group includes the rare and highly prized *Carinaria*, the *Atlanta* shells and the shell-less *Firoloida*.

Not all pelagic mollusks live solely at the surface. The transparent, delicate-shelled sea butterflies or pteropods (pronounced tero-pods) remain several fathoms below the surface during the daylight hours but move upward toward the surface at night. In many equatorial areas pteropods exist in great numbers, and the steady rain of the sinking shells of the dead mollusks litter the ocean's bottom many feet deep. Among the sea slugs, one species of nudibranch (*Scyllaea*) is always pelagic, while the small and beautiful Bat Sea Slug, *Gastropteron rubrum*, makes nocturnal trips from the bottom of the shallow bay to the surface. The two pancake-shaped lobes of the foot of this snail are flapped up and down much in the manner of a bat in flight.

However, the pelagic habitat and the ability to swim are the exception among the snails. The intertidal zone which is intermittently flooded and drained by the moving tides is well stocked with many kinds of creeping snails. Many *Nassarius* Mud Snails live exclusively on the warm, flat mud-

FIGURE 3. **a,** The Nassa Mud Snail, *Nassarius*, crawling under the sand with its siphon extended into the water above; **b,** cutaway view of a prosobranch snail showing the direction of water currents (arrows) down the siphon, over the gills and out from the right side of the body. (After Ankel 1936.)

bars of quiet bays. Among the carnivorous snails, we find that their ecologic stations are determined by the location of the worms or bivalves upon which they feed. One or two species of *Terebra* and *Polinices* Moon Shell are found burrowing in the sand of beach slopes where they are able to find their favorite clams, but the majority of these snail genera are found from low-tide mark to a depth of several fathoms. Since most marine gastropods are nocturnal in habit and shun bright sunlight, many species spend their time hidden in crevices under rocks. This affords protection to themselves and their eggs from predators, bright sun and violent wave action.

A great number of species live in deep water, and frequently their vertical distribution is limited to relatively narrow ranges. From some 500 dredging samples taken off southeastern Florida by the late J. B. Henderson's yacht "Eolis," Bayer's Dwarf Olive (*Olivella bayeri*) was found in depths ranging from 25 to 115 fathoms. On the other hand, the Greenland Moon Shell has been found from twelve feet to over two miles in depth.

In their experimental search for new living places, a few gastropods have evolved strange associations with other marine animals. The dwarf Cyphoma (*Simnia*) lives on the latticed blades of seafans, while the root-like bases of the same seafans may be honeycombed with pockets of the *Coralliophila* shells. Some species of *Trivia* cowries not only live with the compound ascidians or sea squirts (*Botryllus*) but also feed upon them. Deep holes are eaten into the ascidian in which the female snail deposits her flask-shaped egg capsules (fig. 9). Among the *Eulima* and tiny Pyram snails there are many species which parasitize sea urchins and certain kinds of clams. Several species of *Stylifer* live embedded in the flesh of starfish, and only a wart-like

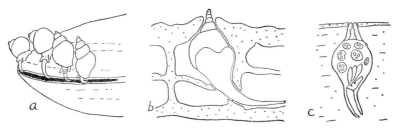

FIGURE 4. Three stages of parasitism. a, the Pyram Snails, *Brachystomia*, make daily visits to tap the body fluids of the mussel, *Mytilus*; **b,** the adult of the *Stylifer* Snail becomes encased in the tissues of the starfish; **c,** the *Entocolax* Snail is embedded in the flesh of a holothurian sea-cucumber and has lost shell, operculum and mouth parts.

swelling and a bit of shell spire projecting above the surface reveal their presence. One species of *Eulima* lives inside the intestinal tract of the sea cucumber and obtains its food by tapping the nutritious juices of its host by means of a modified, syringe-like snout.

HOW THEY GROW

In most cases the shell material in the snails is secreted by special glands located along the edge of the fleshy mantle of the animal. Within the aperture or mouth of the shell a certain amount of reinforcing material may be secreted by the roof of the mantle, especially in the case of the heavy trochid shells which are nacreous within the aperture. The foot is often the source of shell material, not only as the site of the formation of the hard trapdoors or opercula of the turban and natica shells but also as an important addition to the shell itself. The actual formation of calcium carbonate and the formation of the various layers are discussed in more detail in the chapter on clams.

In some groups of gastropods, particularly certain wentletraps and liotias, the mantle edge is capable of producing exquisitely fine filigree or porous shell structure whose intricate designs and overlapping layers can best be seen with the aid of a magnifying glass. In the cowrie shells, the mantle has two large extensions which are spread at will over the entire outside of the shell, This covering mantle continually adds thin paintings of shell material over the entire outer surface of the adult shell. In these groups, where the outside of the shell is protected by the mantle, there is no production of protective, horny periostracum. This is a tough, mat-like and often hairy covering to the shell which prevents acids and marine growths from doing damage.

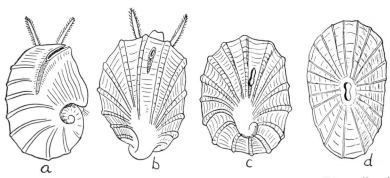

FIGURE 5. Four stages in the growth of the Keyhole Limpet, *Fissurella*, showing how the spire gradually disappears and the marginal slit becomes an apical hole. **a, b, c,** highly magnified; **d,** natural size. (After Boutan 1886.)

In contrast to the bivalves, many gastropods exhibit certain modified shell structures when they become reproductively mature. In many instances adulthood of an individual is accompanied by the formation of a thickened or flaring apertural lip. It is most pronounced in the Strombus conchs, Pelican's Foot (*Aporrhais*), marginellas, cowries and others. Such development is unknown in the cone shells, Busycon whelks, abalones and a host of others.

Growth of the gastropod shell is more rapid in young individuals. Some species apparently continue to grow in size during their entire life span, while others cease once sexual maturity is reached. In the murex shells and frog shells (*Cymatium*) and certain *Cassis* helmet shells a strong varix or thickened rib may be formed at the edge of the shell lip at regular intervals regardless of sexual or seasonal conditions. Each thick varix represents a resting period in growth. Collectors may have noticed that they seldom find murex shells in a growth stage between varices. This is because over ninety percent of the snail's life is spent in the varix stage and because additional growth between varices takes place in less than two days.

The color pattern of shells is a graphic representation in time of the secretory activity of the pigment-producing cells located along the mantle edge. The ground color is produced by the whole line of cells; banding is produced by the special activity of groups of cells, often sharply localized. Where the activity of these groups is cyclical, blotching results; where the active focus moves up and down the mantle edge, or where activity spreads from a focus, there may be formed zigzag, V-shaped or circle patterns. A review of the biochemistry of shell pigments has been made by Alex Comfort (1951).

Rate of growth and span of life in gastropods vary according to the species and ecologic conditions. The maximum age of marine species is very imperfectly known. Undoubtedly many species live for only two, three or four years. The common European periwinkle (*Littorina littorea*) found in New England has been kept alive in captivity for twenty years. Large specimens of the Horse Conch, the Queen Conch (*Strombus gigas*) and the Cameo King Conch (*Cassis*) probably represent ten to twenty-five years of growth. The nudibranch sea slugs are believed to be short-lived, and *Aeolis* and *Goniodoris* have been shown to survive only into the second year. It is quite likely, though, that the *Aplysia* seahares and the *Bulla* Bubble Shells live for at least five years.

The ultimate size of individuals in species in which the sexes are separate may be influenced by the sex of the individual. In many groups, such as the buccinid and Busycon Whelks, the Strombus conchs, periwinkles and others, the shells of the females are always considerably larger. In the Pale Lacuna Periwinkle (*Lacuna pallidula*), the females are from five to ten times as large as the males.

Considerable variation in size results from the diet of mollusks. It has been experimentally shown that the Oyster Drill snails (*Urosalpinx cinerea*) eating *Mya* clam and oyster meat show the greatest increase in growth, while those feeding on barnacles and *Mytilus* mussels show the least amount of growth. It has also been found that snails of this species living in brackish water grow to a larger size than those living in pure sea water. Colonies of snails exhibiting these ecologic characters have been erroneously considered new species by some workers.

HOW THEY FEED

The gastropods are much more imaginative in their selection and manner of acquiring food than the bivalves and other mollusks. Unlike the clams, most snails travel in search of their food. A great proportion of the marine gastropods are carnivorous, but some are detritus feeders, others are vegetarians, and a few, like their bivalve relatives, are suspension feeders.

Among the flesh-eating snails, there have been many modifications in the structure of the mouth parts, including the proboscis and the teeth. In some the snout has remained very simple, and the snail merely pushes the end of its mouth against its food and tears off bits with the tongue-like radula or row of teeth. But in others a remarkably long, tube-like extension is developed which, when not in use, is retracted within the snout or head of the snail. When a living *Melongena* Crown Conch is quickly picked up, one can frequently see the three-inch-long, tubular proboscis being withdrawn into the snail's head. This indicates that a clam or worm, upon which the snail was feeding, is located at that spot one or two inches below the surface of the mud.

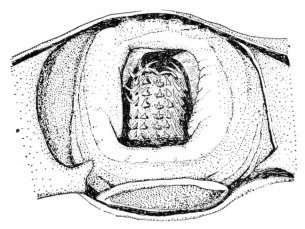

FIGURE 6. The open mouth of the Moon Snail, *Natica*, showing the radula ribbon and its teeth. X3. (Redrawn from Ankel 1936.)

In the *Natica* Moon Shells, there is a muscular disk on the under surface near the end of this extensible proboscis, which serves as a suction disk while the radula is at work on the clam shell. No evidence of the presence of acid has been presented so far. Once the clam is perforated, the long proboscis is wiggled down into the flesh of the clam and the moon shell is able to remove most of the flesh without opening up the valves of the clam. Some *Murex* Snails and the *Busycon* Whelks open their clam victims by applying suction with the sole of the foot and by prying apart the valves with the edge of the outer lip of the shell.

The large group of rachiglossate snails (those having three large teeth in each radular row) are for the most part predators. The Tun Shells and *Cassis* Helmet Shells feed upon live sea urchins. The *Xancus* Chanks, *Busycon* Whelks and others feed upon live clams. The Nassa Mud Snails, however, are purely scavengers, and their ability to detect the odor or taste of spoiled meat in the water is highly developed. Among the toxoglossate snails (those with tiny, needle-like, harpoon-shaped teeth as shown in figure 1), the cones and *Terebra* shells have a highly developed poison gland and duct which are presumably used in quieting their prey.

Vegetarians are found among the more primitive gastropods. All of the limpets, nerites, trochids and turban snails graze on seaweeds. However, many of the "middle-class" snails, among them the ceriths, *Modulus*, and some periwinkles, limit their feeding to swallowing mud detritus on the bottom from which they obtain small algal cells and diatoms. The common Atlantic Slipper Shell feeds in the same manner as the oyster, and its stomach is found to contain the same diatomaceous food. Just as in the oyster, a food current of water is set up in the mantle cavity and the pectinate gill acts as a food sieve. The food particles are entrapped on the gills by a mucus secreted by an endostyle which is located at the base of the gill. Tiny cilia move the food along a groove on the side of the body to a pouch located near the mouth where it is then taken in up through the proboscis. *Turritella communis* of Europe buries itself in mud and has a ciliary feeding habit. This snail remains for days in one spot just below the surface of the mud. An inhalant depression in the mud is made by lateral movements of the foot, and the action of thousands of cilia creates a current which brings food-laden water into the mantle cavity. There is a unique exhalant siphon constituted by two overlapping folds, and through this are expelled water and fecal pellets without disturbing the surrounding mud.

The most extreme modifications in the entire molluscan phylum have occurred in connection with the feeding habits of certain parasitic snails. For years the *Entoconcha* snails found inside the *Synapta* sea cucumbers were thought to be some form of parasitic worm. The "head" of the mollusk is attached in leech-like fashion to a blood vessel of the host, and its worm-like body is embedded in the gonads of the sea cucumber. The adult parasite has no shell, sensory organs, nervous system or radula. It is little more than a tube adapted to absorbing the blood of the host and carrying on self-fertilization. Were it not for the tiny young found inside the adult with their small shell and operculum, it is doubtful if these creatures would ever have been thought to be mollusks.

The passage of food from the buccal cavity, through the esophagus to the stomach is facilitated by muscular contractions of the wall of the alimentary tract and by saliva produced by the two salivary glands. The hind end of the esophagus may be modified into a gizzard, and in many Bubble Shells, especially *Scaphander*, there are several large, cucumber-shaped plates armed with hard corrugations which grind the food into small particles. The stomach proper consists of a simple enlargement of the digestive canal. Its wall may be smooth, furrowed, or lined with spines. As in most bivalves, some snails possess a jelly-like crystalline style which projects into one corner of the stomach and dissolves off digestive enzymes. The so-called "liver" of the snail which forms most of the upper part of the soft, coiled viscera is actually a digestive gland where food material is broken down and absorbed into the blood stream.

HOW THEY BREATHE

Breathing by most aquatic marine snails takes place through the gills where oxygen is obtained from the sea water and where the waste gases are dissolved. The numerous gill leaflets are usually located on the inner side of the mantle. Except in the primitive snails with a pair of gills, water is brought into the mantle cavity through the siphonal canal or through the region to the left of the head. It then bathes the gills and passes out on the right side of the body. The current of water is maintained by thousands of microscopic, lashing, hair-like cilia mostly on the gill leaflets.

Like the bivalves, the snails display a wide variety of types of gills. The most primitive groups, such as some of the Keyhole Limpets, Slit-shells, Pleurotomarias and abalones have two pairs of gills. They are of equal size in the Keyhole Limpets, but in some others the right one is considerably smaller. In the higher groups of snails, the left gill is the only one remaining. In the *Cerithidea* snails, the gills are reduced to mere stumps, and respiration takes place in the mantle skin itself. The sea slugs have lost their ctenidia but have evolved very complicated and beautiful gill-like organs on the sides and back of their bodies. Many of these gills have taken on the shape of miniature shrubs and trees.

HOW THEY REPRODUCE

The subject of reproduction among the gastropods is a fascinating study of many important phases of biology. Our final concepts of the formation of species, our understanding of zoogeography, distributional methods and the basis of sex determination are dependent on a fuller knowledge of reproduction. The manner of assuring fertilization of eggs, the various methods of egg-laying and brooding of young and the interesting types of larval development are horizons of research that are now being expanded.

The gastropods exhibit nearly every possible modification of sexuality. Two of the three orders of snails, the opisthobranchs containing the sea slugs and the land snail pulmonates, combine a complete set of male and female organs in the same individual. The gonad produces both sperm and eggs, but there are separate ducts for the products of each sex. Despite the dual sex life, all mature individuals experience the mating instincts of both sexes, and during copulation there is a mutual exchange of sperm. In some sea slugs, the tectibranchs, several individuals may form rows or a ring of copulating snails. In some fresh-water pulmonates, self-fertilization is sometimes practiced, and some experimenters have bred over ninety generations, extending over twenty years, without cross-fertilization between individuals.

The marine gastropods contain representatives of several categories of sexuality. Dual sexuality or hermaphroditism as found in the pulmonates is also known in some species of *Acmaea* Limpets, *Janthina*, *Odostomia*, *Stilifer*, *Valvata* and the Paper Moon Snail, *Velutina*. The sexuality of this type, however, is more of the consecutive type, in which the gonads at first produce sperm and later in the season only eggs.

Sex reversal is especially characteristic of the Slipper Shell family. The best known examples belong to the Cup-and-saucer Shells, *Calyptraea* and *Crucibulum*, and the true Slipper Shells, *Crepidula*. Individuals function as

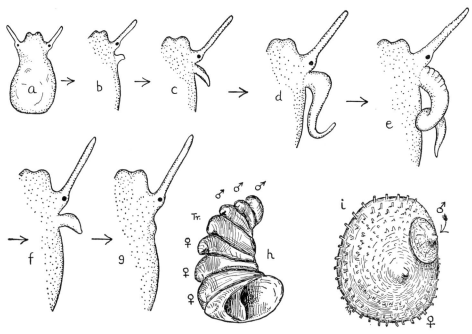

FIGURE 7. Sex reversal in the Slipper Shells, *Crepidula*. **a** to **e**, animal with shell removed to show the development of the verge in the male phase; **f** and **g**, atrophy of the verge and the change to the female phase; **h**, a group of attached *Crepidula fornicata*, showing the smaller males (♂) at the top and the females (♀) below; **i**, *Crucibulum spinosum* with the small male attached to the female. (After W. R. Coe 1943.)

the male sex when young and as females when fully grown. The change-over may be gradual with the individual being ambisexual for a short period, or the male phase may suddenly disappear with the loss of its associated organs, and the female organs may then quickly develop. The males are much smaller than the females. In most species, each young male tends to creep about until it finds an individual of the same species in the female phase, whereupon it attaches itself to the dorsal side of the female's shell in a position adjacent to the female copulatory organs (fig. 7i). In other species the

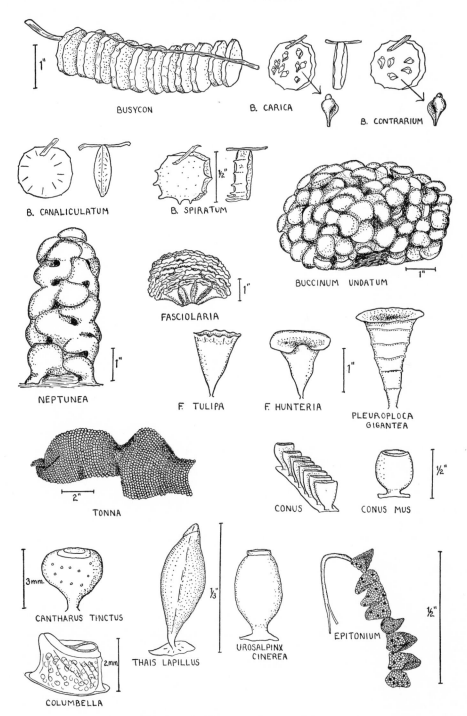

FIGURE 8. Gastropod egg cases.

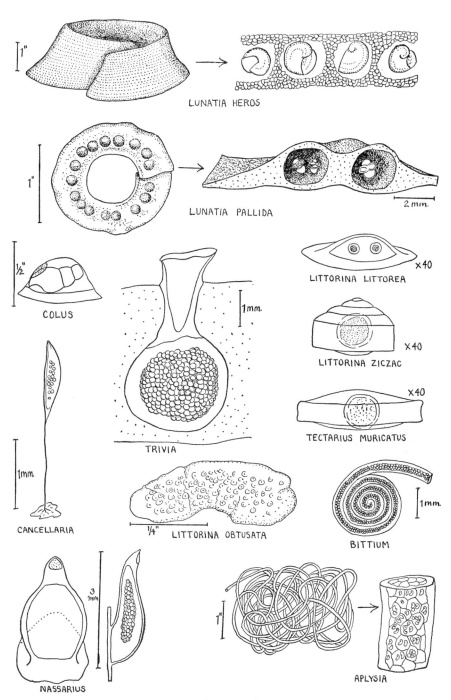

LUNATIA HEROS

LUNATIA PALLIDA

2 mm.

COLUS

LITTORINA LITTOREA ×40

LITTORINA ZICZAC ×40

TECTARIUS MURICATUS ×40

TRIVIA

1mm.

CANCELLARIA

¼" LITTORINA OBTUSATA

BITTIUM

1mm.

NASSARIUS

3 mm

APLYSIA

1"

FIGURE 9. Gastropod egg cases.

males usually occupy positions in the vicinity of the female and move to the mating position at night. Occasionally bachelors are found which either by chance or choice remain solitary throughout the entire male phase.

Soon after hatching from the egg, and in one species (*Crepidula adunca* Sowerby from Panama) even before hatching, a slender copulatory organ, the verge or phallus, grows out from the body behind the right tentacle (fig. 7). As the female phase develops later in life, the verge begins to shrink and is finally absorbed as the female organs take form. Associated with these changes is a marked alteration in behavior, whereby the wandering individual, which was so characteristically masculine when young, now becomes strictly sedentary. She receives her mate, lays her eggs in capsules beneath her foot and broods her young until they are prepared for their own independence.

In our Common Slipper Shell, *Crepidula fornicata*, those individuals which live on muddy bottoms where there are no solid objects to which they can attach themselves, frequently pile up in groups of six to twelve or more. These groups continue from year to year, newly arrived young in the male phase attaching themselves to the top of the pile as the old, female-phase individuals die at the bottom.

Most marine prosobranchs, however, are of separate sexes (dioecious or unisexual). While some species in which the sex products of both sexes are discharged freely into the water have no outward morphological features, there are a great number of gastropods in which the male has an external copulatory organ or verge. The shape and position of the verge are often used in classifying families, genera or species.

Depending upon the species, and sometimes the genus, the females take care of their young in a variety of ways. In some there is no motherly instinct, and the eggs are liberated directly into the water where they float away on the chance of being fertilized by the free-swimming sperm from a nearby male. (See fig. 9 with *Tectarius* and *Littorina*.) In other types the eggs are fertilized and undergo development to the adult-like form in the uterine portion of the oviduct. Others have developed a kangeroo-like pouch in the tissues of their back where the young are allowed to develop to the adult form. Once liberated, however, the young do not return to the pouch. Viviparity or the giving birth to young alive (technically ovoviviparity) is known in *Planaxis*, *Littorina saxatilis* and a number of fresh-water species in several different families.

THE EGG CASES OF SNAILS

Among a large proportion of the marine gastropods, the females form special egg cases or capsules into which the eggs are placed, and where the eggs may develop in an undisturbed, food-laden medium. Very frequently extra eggs (nurse eggs) are added which serve as food for the young that hatch first. The young may emerge from the egg cases as miniature replicas of their parents and commence a life of crawling and feeding, or they may escape as free-swimming larval forms. The latter are known as veligers and possess special organs for swimming. The larval shell is often quite different from the adult shell and, in some species, there may be an extra shell or echinospira encasing the entire veliger.

There are many types of egg cases, and some of these are illustrated in figures 8 and 9; others are briefly described under the generic or family discussions in the identification section. Several types of egg-laying may be found within a single family or even genus.

1. Eggs Laid in Capsules and Attached to the Bottom:
  Rissoidae, Caecum, Epitonium, Thais, Murex, Colus, Neptunea, Busycon, Buccinum, Melongena, Nassarius, Bela, Mangelia, Voluta, Conus, Columbella, Fusinus, Cancellaria, Marginella, Neritidae and others. Of these, some have nonpelagic development: some *Murex, Conus, Natica* and most *Marginella;* others have pelagic, free-swimming young: *Nerita,* some *Murex,* some *Conus* and some *Natica.*

2. Eggs Laid in Gelatinous Masses or Strings:
  Acmaea, Gibbula, Fissurella, Lacuna, Littorina obtusata, some *Turritella, Bittium, Triphora, Cerithium, Capulus, Strombus, Aporrhais, Cassis,* all opisthobranchs and heteropods.

3. Eggs Laid in Capsules and Protected by the Female:
  Crepidula, Calyptraea, Janthina, Cypraea, Hipponix, Vermetus.

4. Eggs Laid in Sandy Collars:
  Polinices and *Natica.*

5. Eggs Shed and Developing Suspended in Water:
  Some *Acmaea,* some *Gibbula, Tectarius,* some *Littorina, Haliotis,* and the heteropods, *Atlanta* and *Oxygyrus.*

In some groups of snails which are more or less sedentary, the egg capsules may be protected by the female. In the cap-shell, *Hipponix,* the underside of the foot of the female has a tough, reinforced ridge of flesh to which she attaches her gelatinous egg sacs. In some worm-shells, *Vermetus,* whose shells are permanently attached to the rocks, the eggs are deposited on the inside of the female's own shell.

The time and length of breeding differs among mollusks depending mainly on the geographical locality, the temperature of the water, phases of the moon and the inherent characteristics of the species. Some species spawn once a year for a few weeks only, while others may produce eggs half of the year as long as the temperature is suitable.

The eggs, larvae and young have been described for many species by famous workers such as Gunnar Thorson, Marie Lebour and others. The common European Periwinkle (*Littorina littorea*) will serve here as an ideal example of the pelagic type of development. The female spawns two to twelve hours after copulation by the male. About 200 single egg capsules are shed during the night. During the entire breeding season of six months, the total number of egg capsules per female is estimated at about 5000, and a half dozen copulations are necessary to ensure fertilization of all the eggs. The helmet-shaped capsules are shed freely and float about in the water. Each contains from one to nine eggs. The free-swimming young, called veligers, hatch on the sixth day and remain afloat for two weeks or more, depending upon temperature conditions, then sink to the bottom and begin an adult snail's life of crawling. They reach maturity on the second or third year and may live for five to ten years.

In contrast to this mode of spawning, the Left-handed Whelk of Florida (*Busycon contrarium*) lays its horny strings of egg capsules during a relatively short period of a few weeks. On the west coast of Florida egg-laying usually takes place in the spring. The female digs down well below the surface of the sand and attaches the first few capsules to a buried rock or broken shell. As the process of extrusion of the egg capsules continues, the female moves toward the surface until its siphon can protrude into the water to allow easy respiration. As more capsules are made, the string may loop out into the water above the hidden adult. From five to fifteen cases may be formed each day, and a completed two-foot-long string may have nearly a hundred capsules. Within each case there may be two to twenty-five eggs which in a few weeks will develop to quarter-inch-long young. These miniature replicas of the adults eat their way out of the case at a special "door" and commence crawling and feeding immediately. The Left-handed Whelk begins spawning at a relatively early age, commonly when no larger than three inches. In such cases the capsules are only a half inch in diameter, while larger females may produce capsules about the size of a half dollar.

CHAPTER III

Life of the Clams

OF THE approximate 15,000 species of existing clams or bivalves, four fifths live in the sea, while the remainder are inhabitants of fresh-water rivers, lakes and ponds. Throughout the seventy or so families of this class, the clams show an amazing diversity of ways of adapting themselves to almost every kind of aquatic environment. There are clams that swim, burrow, dangle by silken threads, others that are permanently cemented to rocks and corals, some that live a sedentary life of attachment to other marine creatures. In size, they vary from the 500-pound giant Tridacna clams of the East Indies, which reach a length of over four feet, to the pinhead-sized Amethyst Gem Shells (*Gemma*), which so heavily populate some of our intertidal flats. In ornamentation and coloration the clams are almost unexcelled in their wide range of beautiful hues and bizarre shapes.

WHERE THEY LIVE

The bivalves have selected a wide variety of ecological stations in life. While many must live in strictly marine waters, a few have adapted themselves to the brackish waters and estuaries and inland bays. One species, the 'Coon Oyster of Florida and the West Indies, has "taken to the trees" and is able to withstand exposure to the air for several hours, or even days, between high tides. In its early, free-swimming stages, the oyster is carried by the rising tide in among the roots, trunks and overhanging boughs of the mangrove trees where it settles and attaches itself. Feeding, growth and reproductive activities take place only during the few short hours of high

31

tide. The Coquina Clam, *Donax,* is faced with much the same problem of making the most of high tides but, in contrast to the sedentary life of its oyster cousin, it leads a very active existence on the sandy beaches along the open ocean. It is an attractive sight when a scouring wave suddenly studs the white beach with dozens of brightly hued clams. The tumbling motion and sudden exposure to light act as a stimulus to the clam which instantly thrusts out its small muscular foot and rapidly pulls itself down into the sand again. During the three or four hours in which the waves are sweeping the middle and upper sections of the beach, the tiny clams may be uncovered and obliged to burrow down again several hundred times.

While many clams prefer clean sand as a habitat, others are habitual mud-dwellers. The handsome Angel Wing, *Barnea costata,* is usually found in mud so soft and deep that Florida collectors find it extremely difficult to reach them. The Angel Wing is usually located one or two feet below the mud surface and maintains its connection with the bay's waters with its long siphon. Because of its popularity as a souvenir and collector's item, methods have been devised to collect them at high tide from a boat or barge. Powerful jets of water are forced through hoses, the mud is swept away from the clams, and then hand-nets are employed to gather them. In more shallow regions where a mixture of sand in the bay bottom permits walking, the exposed Angel Wings are gathered by hand at the next low tide.

The majority of marine clams live in a substrate of sandy mud, but a few have become specialized to the extent of making burrows in exceedingly compact clay, as in the case of the Arctic Saxicave, *Hiatella arctica,* and the False Angel Wing, *Petricola.* A few groups such as the Date Mussels, *Lithophaga,* and the Piddocks, *Pholas,* burrow into corals, other shells or soft rocks such as sandstone and limestone. The shipworms, *Teredo* and *Bankia,* are expert at drilling out their long, tube-like homes in wooden planks of ships, wharf pilings, and manila hemp. So too is the Wood Piddock or *Martesia.*

A large proportion of bivalves are found in shallow water, but many others are typically deep-water dwellers. The bathymetric range for some species may be narrowly defined in the case of certain scallops, Dipper Clams (*Cuspidaria*) and astartes. On the other hand, some species found in a few feet of water may also occur in depths of over two miles. One species of Abra Clam, *Abra profundorum* E. A. Smith, has been dredged in the mid-North Pacific at a depth of 2,900 fathoms—over three miles!

HOW THEY GROW

The shelly valves of clams are the product of the fleshy mantle. This thin, leaf-like organ covers the animal as the flyleaves cover the body of a book and, by its physiological activities, secretes the hard valves of calcium carbonate, which thus come to occupy the position of the covers of the book. In the simplest form of mantle the edges are free except on the back, where the hinge of the shell is located, corresponding to the arrangement of a book. Sea water may enter the cavity enclosed by the mantle at almost any place. In many groups of bivalves, however, the mantle edges may be fused, not only along the back where the valves are joined together but along all or most of the lower margins. Openings are usually present to accommodate the foot and siphons when such organs are developed.

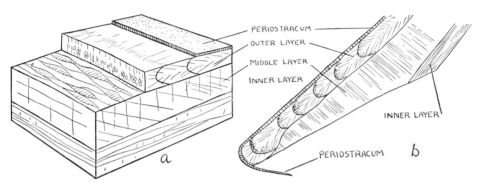

FIGURE 10. Structure and layers of a clam shell (*Tellina tenuis*). a, Diagrammatic representation of a small piece of shell; b, Cross-section of shell showing the loose end of periostracum around the margin of the shell. (After Trueman 1942.)

From its food supply the clam absorbs minerals into its blood system which are then carried to the mantle. A certain amount of shell deposition takes place along the thickened borders of the mantle, although a small amount, including pearly or nacreous material in some species, is laid down by other parts of this organ. The liquid secretion of lime salts becomes crystallized when mixed in a colloidal albumen which is also produced by the mantle. Several types of shelly material are laid down in definite layers, and the structure and composition may vary depending upon the family or genus of mollusks. The structure of a layer may be *prismatic* (made up of tiny, individual, closely packed prisms), *foliated* (layers built up of overlapping leaves), *nacreous* (mother-of-pearl), *granular* (like grains of sugar stuck together), *crossed lamellar* (a common type in which the long lamellae are rectangular), or it may be *homogeneous* with no visible structure. The mineral character of these layers may be calcite (2.7 times as heavy as water) or aragonite (2.9 times as heavy as water), both of which are forms of calcium carbonate.

The shell of the tellin clam (*Tellina*), for instance, is made up of three layers of calcium carbonate and the horny periostracum. The latter consists of a very thin layer of conchiolin, probably not more than 0.003 mm. in thickness. In other clams, such as *Arca*, it may be many times as thick. It is normally secreted from a group of cells situated just under the tip of the mantle. The three shell layers are: (1) The outer layer of shell which consists of elongate radial prisms of calcite. These are arranged in concentric bands which are plainly visible on the outer surface of the shell. (2) The middle layer which is entirely composed of aragonite in the form of "crossed lamella." This specialized structure is peculiar to mollusks. (3) The inner layer which is a homogeneous layer of porcellaneous material.

The Pen Shells of the genus *Pinna* commonly found on the west coast beaches of Florida offer an excellent demonstration of prismatic structure. When the surface is examined with a high-powered lens, it appears to be honeycombed. What you see are the ends of the needle-like prisms of calcite which, although closely packed together, are separated from each other by a thin varnish of conchiolin. By examining the edge of the broken shell you can make out the prisms in side view.

Most clams continue to grow in size during lifetime, but the greatest increase takes place during the first year or two. A species may show considerable variation in its manner of growth under different living conditions at various localities. Thus the Pacific Razor Clam (*Siliqua patula*) in its southern range in California grows much faster and reaches a length of about five inches in three years. In Alaska, however, it grows more slowly, taking five to eight years to reach the same size. Yet the northern colonies continue to grow for a greater length of time, some living for fifteen to eighteen years and eventually reaching a length of over six inches. This is also true of the Pacific Cockle (*Clinocardium nuttalli*) which in ten years grows to three inches in length in California, but in Alaska it survives sixteen years to reach a length of five inches.

The maximum age is known for a few species of clams. It is believed that the giant Tridacna clam of the Indo-Pacific lives for perhaps a hundred years, but this has not been confirmed by experiments or accurate calculations. The average age of the Atlantic Bay Scallop (*Aequipecten irradians*) is about sixteen months, its maximum age only two or three years. The average age of a five-inch Pismo Clam (*Tivela stultorum*) on the Pacific Coast is about eight years, its maximum age twenty-five years. The Common Blue Mussel (*Mytilus edulis*) grows to about two inches the first year, to four inches the second year but, beyond this, it grows very little although it may live for a total of seven or eight years. The Soft Shell Clam (*Mya arenaria*) takes about five years to reach an edible size of three or four inches and may live for ten years. The Washington Clam (*Saxidomus nuttalli*) lives ten to fifteen years or longer, while Nuttall's Gaper Clam (*Schizothaerus nuttalli*) may survive for seventeen years.

HOW THEY FEED

Normally one does not think of clams and oysters as being very active feeders and certainly, in comparison with the voracious methods of fish and squid, the bivalves are rather peaceful eaters. Yet in their characteristic way they are highly efficient and, in proportion to their size, possess a large and varied menu. Most clams feed on minute plants and, in a relatively short time, can filter from the sea water an extraordinary number of living diatoms and dinoflagellates—microscopic, swimming plants—and protozoa of the ocean. A few genera, such as the small *Cuspidaria* and *Poromya* clams, are carnivorous and feed upon small living or dead animals, usually crustaceans and annelid worms.

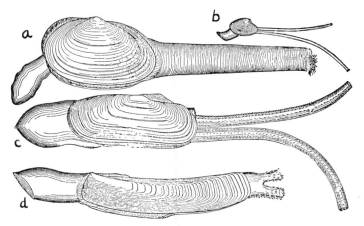

FIGURE 11. Extended animals of some bivalves, showing various types of siphons. **a,** *Mya arenaria* Linné; **b,** *Tellina agilis* Stimpson; **c,** *Tagelus plebeius* Solander; **d,** *Ensis directus* Conrad. (From A. E. Verrill 1873.)

The bivalves fall into two general classes of feeders—*suspension feeders* which merely pump water through their mantle cavity and thus obtain free-swimming or suspended creatures from the water; or *deposit feeders* which suck up food from the muddy bottom with their long, mobile inhalant siphons. Among the *suspension feeders* are the oysters, scallops, venus clams, cockles, the shipworms and many others. They may or may not possess siphons, but when present these are generally short. The *deposit feeders* include such forms as *Tellina, Macoma* and *Abra* which all have long siphons.

Whether food is taken in through the inhalant siphon as in the tellins or through a slit in the mantle as in the scallops, it must pass over the gills. These filament-like organs are covered by a thin sheet of mucus. Food passing through the gills becomes ensnarled in the mucus which is transported by water currents and myriads of tiny, hair-like cilia. Mucus is constantly being secreted and carried to the food grooves bordering the gills, along which the food-laden strands are carried to the mouth.

Our common Atlantic Oyster and those in France are frequently found with green gills. The "green oysters" of Marennes, France, are famous for their supposed medicinal qualities. Americans are inclined to shy from "green oysters," because they fear the color may be a sign of spoilage. Oysters feeding upon the small diatom, *Navicula ostrearia*, digest these single-celled plants and absorb from them large quantities of blue pigment. In the tissues of the oyster's gills the pigment appears in the form of a sickly but quite harmless green. Occasionally, however, our oysters may take on a general greenish tint, not due to diatoms but to an increase in the amount of copper in the tissues. Such oysters have a rather brassy taste.

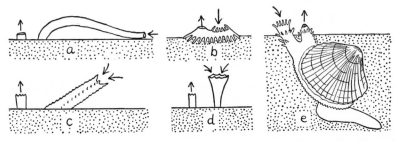

FIGURE 12. Siphons of bivalves projecting above the sand bottom. *Mya* (b) is a suspension feeder, the others deposit feeders. **a**, *Tellina* and *Macoma;* **b**, *Mya;* **c**, *Gari;* **d**, *Donax;* **e**, *Trachycardium.* (After C. M. Yonge 1949.)

The clam has considerable choice in what it wishes to eat, and it can reject undesirable particles of sand or oversized pieces of food. The gills and the two fleshy palps, or flaps guarding the mouth, help in sorting out the right-sized organisms. Acceptable food is taken into the funnel-shaped mouth, passed through a short esophagus and enters the stomach. Inside the stomach, a further selection of food may take place with indigestible matter being passed on immediately through the intestine. The best food passes from the stomach into the digestive gland where it is broken down chemically and absorbed into the blood stream.

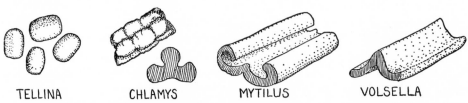

TELLINA   CHLAMYS   MYTILUS   VOLSELLA

FIGURE 13. Fecal pellets of mollusks are characteristic in shape and may aid in identification of genera and species. (After H. B. Moore 1931.)

Lying close to the stomach is a sac which contains a cucumber-shaped, jelly-like crystalline style. The end of this style projects into the stomach. It rotates clockwise and dissolves its enzymes in the stomach which aid in digesting the food, that is, in converting starch into sugar. The style, numerous cilia and the furrows on the stomach wall aid in churning the food.

The fecal pellets of mollusks are often very distinctive for the various genera and species. Some are cylindrical rods, others elongated strings or ribbons, and a few consist of strands wrapped up in round balls. In cross-section, some rods are characteristically bi- or trilobed.

Feeding is not done at all times, although a great part of a bivalve's life is spent in securing food. The oyster, for instance, spends from seventeen to twenty hours of each 24-hour period in taking in water for the purpose of feeding and breathing. Individuals living in the intertidal zone and left dry by receding tides or exposed to water heavily charged with silt spend considerably less time feeding. During cold periods, when the water temperature falls below 40° F, the oyster goes into a state of hibernation, and it ceases to feed because of the lack of coordination of the ciliary motion along the surface of the gills. Under ideal conditions, the Giant Pacific Oyster (*Crassostrea gigas*) filters 5½ quarts of water per hour at 77° F (1 quart at 34° F). In a year, the total amount would fill a 10,000-gallon tank car.

Perhaps the most startling modification of obtaining food nutrients is exhibited in the giant Tridacna clams of the Indo-Pacific reefs. These clams literally "farm" colonies of brown-colored algal plants (*Zooxanthellae*) in their huge, exposed mantle edges. Unlike most clams, the Tridacnas spread their valves open and expose their mantles to as much sunlight as possible for the benefit of these single-celled seaweeds. In addition, small, fleshy tubercles grow on the surface of the mantle in which are located lens-like, clear cells. Sunlight can thus penetrate down into the flesh and be diffused into areas which otherwise would not receive enough light for the algae. Surplus plant cells are engulfed by phagocyte blood cells of the clam and transported to the digestive gland for absorption as food. The giant clams also feed in the conventional gill-to-mouth manner and are therefore not entirely dependent on the algae. The algae, however, must have a clam as a host in order to survive. This peculiar symbiosis is found to a lesser extent in the Bear Paw Clam (*Hippopus*), the Heart Cockle (*Corculum*) and the nudibranch, *Phestilla*. This phenomenon is not to be confused with the pathologic entry of the parasitic blue-green algae in fresh-water mussels, *Anodonta* and *Unio*.

While the gills are the main organs for catching, sorting and transporting food in the majority of clams, they are limited to respiratory functions in a few groups. The smallest and most inefficient gills are found in the primitive protobranchs (*Nucula*, *Nuculana*, etc.) and in the small, highly evolved septibranch clams (*Poromya* and *Cuspidaria*). In order to make up for the loss of efficient food-gathering gills, the palps near the mouth have become

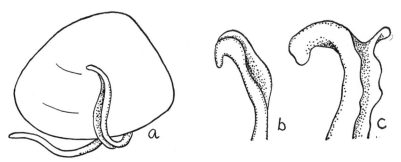

FIGURE 14. The pair of proboscides in the Nut Clams, *Nucula*, sweep up food and transport it to the mouth. The gills are not used in gathering food as in the majority of clams. a, X5; b and c, ends of the proboscides. X10. (After K. Hirasaka 1927.)

very specialized. In the Nucula Nut Clams, a pair of strong, muscular, contractile organs serve as food gatherers. These proboscides are very flexible, moving about freely in all directions. Food material is picked up by the tip and is carried swiftly down a large groove in the proboscis to a palp pouch and then to the stomach by means of minute cilia. E. S. Morse very aptly described the action of these appendages in our Atlantic Nut Clam, *Nucula proxima:*

> Without seeing the behaviour of these appendages it is difficult to appreciate the remarkable action of these feeding organs. The graceful movements of these beautiful and translucent appendages, exceeding the length of the shell, sweeping rapidly the bottom of the dish in which they are confined, or even turned back and feeding on the surface of the shell, are a most curious and interesting sight.

HOW THEY BREATHE

Oddly enough, the gills of the bivalves are not primarily used for respiration, despite their conspicuous size. As has been noted, their main function is in connection with feeding. Some experts deny their role as respiratory organs entirely, claiming that the mantle with its extremely effective blood supply serves as the main place of oxygen and carbon dioxide exchange. It has been found that blood coming from the mantle to the heart is completely charged with oxygen received from the sea water. Undoubtedly, however, the gills do absorb oxygen to some extent. Indirectly, the gills are extremely useful in respiration, since they produce the all-important currents which bring in oxygen and remove carbon dioxide dissolved in the water.

A certain amount of respiration may take place even when the valves of certain bivalves are completely shut during exposure to dryness or to heat from the sun. What little air may be trapped within the mantle cavity of the animal is soon used up. Oxygen is then obtained anerobically (without contact with air) by cleavage of reserve glycogen substances stored in the clam's tissues. Carbon dioxide builds up and is dissolved in the fluid in the mantle cavity, and the resulting increase in acidity may dissolve or etch away portions of the shell. Shells of oysters and the *Patella* limpets which are kept dry on the rocky coast for unusually long periods show considerable etching on the inside. Shells of the Date Mussels (*Lithophaga*), which live in a small volume of water in their rock burrows, are etched in this manner, while those specimens which live in the same volume of well-aerated water are not etched.

Bivalves can be forced to cease respiration for several days without succumbing, but they are very susceptible to polluted waters and excess amounts of silt. The "red tide" caused severe destruction to the marine fauna on the west coast of Florida in 1946, and for several years afterward the "shelling" on famous Sanibel Island was little better than it is on Coney Island Beach, New York. "Red tides" have occurred from time to time in California, Washington State, Japan, Australia and elsewhere. They are caused by an unusual increase in the numbers of single-celled dinoflagellates, *Gonyaulax*. It is believed that billions of these organisms not only deplete the oxygen supply but also clog the gills of fish, mollusks and other animals which die in vast numbers and further befoul the ocean. Fortunately, these "red tides" spend themselves out, and the coastal waters return to normal in a few years.

Another species of *Gonyaulax* (*G. catanella*) may be ingested by mussels and clams and, although it does no harm to the mollusk, it is highly toxic to humans who may eat the infected shellfish. A number of deaths have occurred on both of our coasts from this type of mussel poisoning. There is no way of distinguishing poisonous from sound mussels by their appearance, and heat does not destroy the poison. Mussel poisoning occurs along the California coast from May 15 to October 15. There is another such center in Nova Scotia.

Among the various schemes of classification of the bivalves, the type of gill structure has been used by many students of phylogeny (the study of molluscan ancestral trees), such as Lankester, Pelseneer, Ridewood and others. Opponents to this system, such as Neumayr, Munier-Chalmas, Dall, Cotton and others, have based their classification on the hinges of the shell valves. Neither system is without its weaknesses, and in some modern schemes the two systems are employed together.

There are four main types of gills: (1) *Protobranch*, in which the gills are flat, plate-like, unreflected lamellae and are regarded as the most primitive (*Nucula, Yoldia*, etc.); (2) *Filibranch*, in which the gills are long curtains folded back against themselves and held close to each other by the

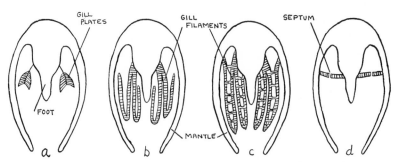

FIGURE 15. Diagrammatic cross-sections of clams showing the major types of gill structure. **a**, protobranch; **b**, filibranch; **c**, eulamellibranch; **d**, septibranch.

interlocking of the tiny cilia on the surface of the gill filaments (arks, mussels, scallops, etc.); (3) *Eulamellibranch*, similar to the filibranchs except that the gill curtains are united by cross-channels (astartes, cardiums, venus clams, tellins and many others); (4) *Septibranch*, which have very degenerate gill structures consisting of two pallial chambers with only gill slits or very reduced gill filaments acting as windows to the chambers (*Cuspidaria* and *Poromya*).

HOW THEY REPRODUCE

The staid bivalve has made his share of contributions to experiments in sex and reproduction, and throughout the class we find varying degrees of sexual differentiation, as well as all manner of ways of insuring proper fertilization, protection of the young and thus the continuation of the species.

The pelecypods have no copulatory organs or other external sexual characteristics, with the exception that in certain species of fresh-water mussels, the marine astartes and a few other genera, the two sexes can be distinguished by the shape of the adult shell. The majority of the bivalves as a group are predominantly of separate sexes, but at least four percent of those adequately studied are known to deviate from the strictly dioecious, or unisexual, condition.

A few species are true hermaphrodites in which the same individual contains both female and male sex organs which may produce eggs and sperm simultaneously. In this group are found certain species of *Pecten, Tridacna* (the Giant Pacific Clam), *Kellia, Dinocardium, Gemma, Tivela* (the Pismo Clam), *Thracia, Poromya*, the shipworm *Teredo diegensis* and the fresh-water genera *Anodonta, Pisidium* and *Sphaerium*. In some of these the eggs are fertilized within the mantle cavity, and the young complete development to the adult form in brood pouches on the gills of the parent. Usually self-fertilization does not occur, for in the majority of these species the sperm is discharged before the eggs are mature in the same individual (protandric hermaphroditism).

Other kinds of bivalves are accustomed to practicing sex reversal in which the early part of their lives is spent as males and their "adulthood" as females. In the Quahog (*Mercenaria mercenaria*), nearly all individuals experience a male phase in which functional sperm is produced while the clam is only a few months old. Following this initial male phase, about half of the population turns female to produce eggs, while the other half remains male. No further sex change takes place.

Sex reversal is apparently very popular among some of the oysters, such as our native Pacific Coast *Ostrea lurida*. In this species there is a series of male and female phases. There may be three changes within a single year. Usually the male phase comes on first. Alternating sexuality also occurs in our Atlantic Oyster (*Crassostrea virginica*), but the early sex organs are capable of turning toward either male- or femaleness. It is not known, at present, to what extent environmental conditions determine the direction of sex change. It has been shown, however, that under unfavorable circumstances, when circulation of water is poor and the food supply low, there are more female oysters in a colony. When conditions improve, the percentage of males increases considerably.

Thorough studies have now been made to show that normally no sex reversal occurs and that the sexes are separate and of equal numbers in a given colony in the following species: *Modiolus demissus, Mytilus californianus, Septifer bifurcatus, Anomia simplex, Mytilus edulis, Petricola pholadiformis, Donax gouldi, Mya arenaria* (Soft-shell Clam) and the Angel Wing, *Barnea costata*.

The number of eggs produced by the female bivalve may vary considerably depending upon the species and environmental conditions. Species which retain the fertilized eggs within their bodies for further development invariably produce fewer eggs than those species which discharge them into the water. The oysters are probably among the greatest molluscan producers of eggs. C. R. Elsey estimates that one female *Crassostrea gigas* of Japan and our northwest Pacific Coast may discharge into the water each year eggs numbering 1000 to the eighth power. If all survived in five generations, the aggregate would be large enough to make eight worlds like ours. Needless to say, enemies and unfavorable conditions kill off most of the young.

In contrast to this prodigious effort on the part of the oyster, the Dwarf Turton Clam (*Turtonia minuta*) deposits only 12 to 20 eggs which are neatly encased in oval egg masses of gelatinous material. While most species of Nucula Nut Clams discharge their eggs freely into the water, one New England species, *N. delphinodonta*, deposits from 20 to 70 tiny, opaque brown eggs in a gelatinous sac which is attached to the posterior end of the valves of the shell. Small bits of debris and mud stick to the outside of this sac, which probably serve as a camouflage. Many bivalves keep the developing young within the mantle cavity or in the meshes of the gills until the tiny shells are quite well advanced in development. With the aid of a high-powered lens one may readily see tiny juvenile clams inside the translucent adult shells of such genera as *Gemma*, *Parastarte*, *Psephidia*, *Transennella*, *Kellia*, *Lepton* and *Lasaea*. The odd Dwarf Milner Clam of California (*Milneria minima*) incubates about 50 young in a peculiar external pouch. The valves are indented on the ventral margins to form a neat exterior pocket. To prevent the young from dropping out, a sheath of periostracum is stretched over the entrance. When the small clam shells have grown sufficiently to fend for themselves, the sheath is "unzipped," and all tumble out into the free world.

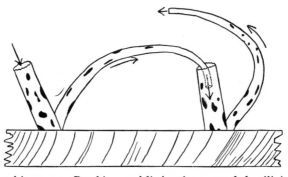

FIGURE 16. The shipworm, *Bankia gouldi*, in the act of fertilizing its neighbor. The spotted siphons are shown projecting from the wood in which these bivalves live. Arrows indicate the direction of water currents. X5. (Redrawn from W. F. Clapp 1951.)

In practically all cases, the sperm from bivalves is liberated into the water where it comes in contact with unfertilized eggs that have been previously released. In cases where eggs are retained by females, the sperm is sucked in through the inhalant siphon of the mother. Only one instance of pseudo-copulation is known. In 1951 workers at the W. F. Clapp Laboratories observed Gould's Shipworm (*Teredo*) placing their exhalant siphons down into their neighbors' inhalant siphons and discharging what is presumed to have been sperm.

HOW THEY SENSE AND SEE

Bivalves are the least "brainy" of the mollusks and, although the central nervous system forms a rather complicated latticework throughout the body, its three pairs of "brains" are merely swellings or ganglia in the larger nerves. The pair of so-called cerebral ganglia control the actions of the lip palps near the mouth, parts of the mantle, and they also receive "nerve notices" from the tiny organs of balance, the otocysts. The second major pair of ganglia are the pedals which supply the foot. This pair is large in the clams that use the foot for digging or burrowing, but it is extremely small or aborted in the oysters in which the foot is not used. The third pair, or visceral ganglia, is usually the largest and supplies the adductor muscles and the visceral mass. The remarkable eyes of the scallops are connected with this pair of visceral ganglia.

Many of the bivalve larvae possess true paired eyes, but in all cases these are lost when the animal transforms into the adult stage. The adults of a number of clams and mussels have developed pigment spots sensitive to changing light, but in the scallops true eyes are well-developed. When the shell of a scallop is open there can be seen just within the margin of each valve a line of small, brilliant, emerald-like dots on the mantle, each of which is a small eye fully equipped with cornea, focusing lens, receptive retina and conducting nerves.

HOW THEY BURROW AND SWIM

There are bivalves that swim, leap, crawl and burrow deeply in mud, sand or clay, and some that bore into wood, rock and even lead casings of submarine cables. Even the rock-bound oyster and the stuck-in-the-mud clam have their days of wandering about as free-swimming larvae before they settle down to a life of permanent attachment or clumsy crawling.

The habit of swimming among adult bivalves is rare. The scallops and the *Lima* File Clams not infrequently swim. Only under the abnormal condition of finding themselves "unearthed" do the *Ensis* Razor Clams and the *Solemya* Veiled Clams practice jet propulsion through the water. The Razor Clam swims backward in quick, short jerks by first extending its long cylindrical foot out from the shell and then suddenly withdrawing it with great force. This action, together with the closing of the shell valves, quickly forces the water within the mantle cavity out through the openings at the anterior or foot end. Thus the razor clam darts through the water with its pumping foot to the rear. In *Solemya*, the foot is in front of the animal as it swims. In this case the water is admitted around the foot but is expelled from the opposite end through the siphons.

The highly developed swimming ability of the scallops accounts for the migratory powers of the great schools of these active bivalves. One would normally expect the direction of swimming taken by a scallop snapping its valves together to be "backward" in the direction of the hinges. Although that type of movement is on rare occasions used as an escape measure, the typical swimming movement is in the opposite direction with the free edge of the shell going in front, so that the animal appears to be taking a series of bites out of the water. This odd action is made possible by the vertical, curtain-like edges of the muscular mantle. When the valves are snapped shut by the powerful adductor muscle, water is driven out, not past the mantle curtains but through the regions around the hinge or ears of the shell. By manipulating these curtains, which can be extended or withdrawn locally, the scallop is able to vary the amount and position of exodus of water and hence can direct its course. If accidentally turned over onto the wrong valve, the scallop can execute a neat flip and regain its normal position.

While the scallop always swims with its valves in a horizontal plane, the *Lima* File Clams most frequently progress edgewise, that is, with the breadth of the valves vertical or slightly oblique. The long, colorful tentacles of the *Lima* keep the animal momentarily suspended in water while the valves are being opened in preparation for another "bite" forward. The *Lima* is a poor swimmer and, because of its habit of building nests under rocks, apparently has no incentive to undertake migrations as is done by

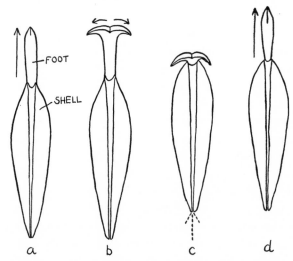

FIGURE 17. The mode of locomotion in the bivalve, *Yoldia limatula*. **a**, the foot is thrust forward; **b**, the muscular flaps are spread apart to form an anchor; **c**, the foot is withdrawn, thus pulling the animal forward; **d**, by closing the flaps together, the foot is made ready for another thrust forward. (After Drew 1900.)

the scallops. Scallops of the genus *Chlamys* are equally poor swimmers and, unlike adult *Pecten*, spin small byssal threads for attachment to the bottom.

Burrowing in sand and mud is accomplished by the foot of the bivalve. The principle is the same in all digging clams. The foot is slowly protruded with the pointed tip wriggling down into the mud. During this extension the end of the foot is kept small, but when it reaches its greatest extension the end is suddenly swelled into a great bulb and the whole foot becomes very rigid. This is accomplished by injecting blood into the foot. The bulbous end serves as an anchor while the clam withdraws the foot and pulls its entire shell deeper into the mud. In the case of the Razor Clam, this action is accompanied by a jet of water against the mud ahead. The dislodged mud is washed up the sides of the shell and out the burrow. The action is similar to the pile driver that opens a way for the pile by a somewhat similar stream of water.

Because of its long, powerful foot the Razor Clam is capable of leaping. Generally, when this clam is lying on the surface of the mud, the foot is bent back under the shell and is then suddenly made rigid with the result that it is straightened out with great rapidity. In some cases the animal may turn itself end over end.

Many types of clams are especially adapted to boring into hard clay, shale, sandstone and concrete. The Date Mussels, *Lithophaga*, possess acid-secreting glands as an aid to penetrating limestone. The shell of the clam would of course be dissolved by this acid were it not for the thick, protective covering of periostracum. The Saxicave Borers, *Hiatella*, may live attached by a byssus on surfaces that they cannot penetrate, or they may bore into soft rock. The boring of the adults is wholly mechanical and is accomplished by rubbing the edges of their shell valves against the rock. *Hiatella* stays near enough to the surface to allow its siphons to protrude just outside its cone-shaped burrow. In three years the burrow is only three fourths of an inch in length and, after eight years of constant grinding and new gowth of shell, it is only one and a half inches in length.

The shipworms burrow long distances into the wood but retain contact with the "outside world" by means of the long fleshy, tube-like extension of the body. Boring is accomplished by the two valves at the anterior end of the tunnel. The denticulated ridges of the shell are the cutting tools, and the foot and muscles aid in rotating the shells back and forth. Burrowing may progress at the rate of as much as four inches a month. In the genus *Bankia*, at the posterior end of the worm-like animal there are two tiny, feathery pallets. These are used to plug the entrance of the burrow, thus giving protection from enemies, changes in salinity or other adverse conditions. When the shipworm is undisturbed, the pallets are drawn inside and the siphons extended into the water for breathing and feeding.

CHAPTER IV

Lives of the
Other Mollusks

IN ADDITION to the bivalves and snail classes, the mollusks include three other groups which are not so frequently seen at the seashore and whose combined number of living species probably does not exceed two thousand. Two of these classes, the *Amphineura* or Chitons and the *Scaphopoda* or Tuskshells, are among the lowliest and most sluggish of the mollusks, but the third class, the well-known *Cephalopoda*, including the squid and octopuses, contains the largest, fastest and most ferocious of all backboneless animals.

THE SQUID AND OCTOPUSES—*CEPHALOPODA*

The octopuses and the giant squid have been spine-chilling characters in adventure tales from the days of the ancient Greeks to the undersea film thrillers of Hollywood. Nothing seems more appropriate for a horror scene then the sudden appearance of a tentacle-lashing, beady-eyed octopus just as the hero-diver finds the long-lost treasure chest. And few authors of strange sailing voyages can resist retelling the numerous instances in which gigantic squid have wrapped their arms about the riggings and dragged ship and hapless crew to the bottom.

But despite the fanciful nature of most, if not all, of these stories, there are enough scientific facts to convince the skeptic of the ferocity, speed and unusual intelligence of these creatures. Canadian and American fishermen have long been familiar with giant squid and have often captured

46

crippled individuals and used them for bait. A number of these giants have been brought into museums, and others, stranded on beaches after storms, have been measured and recorded by reliable observers. *Architeuthis* of the North Atlantic waters is known to reach a total length of 55 feet. The longest arms of this specimen are 35 feet, while the length of the body from tip of tail to the base of the arms is 20 feet. The greatest circumference of the body is 12 feet. Sperm whales which feed upon smaller squid have often been locked in battle with these giants. The skin of these whales is sometimes heavily marked with circular scars caused by the suckers of the squid.

The octopus does not reach a very large size. The largest known species occurs on the west coast of North America where, in Alaska, *Octopus punctatus* attains a length of 16 feet or a radial spread of nearly 28 feet. However, the arms are very small in diameter, and a specimen of such long proportions has a body length of not more than a foot. The octopus occasionally found in the Lower Florida Keys is usually less than three feet in radial spread. A dead specimen cast on a beach near Nassau, Bahama Islands, was reported to have an arm length of five feet, and it was estimated that the entire creature weighed about 200 pounds. This, however, is without verification. Recent reports of octopus holes 100 feet across seen in the Bahamas from the air were made by untrained observers. There is no satisfactory evidence that any of these species of *Octopus* has ever intentionally attacked man, or that any person has ever been seriously injured by one. The octopus is a rather sluggish and timid creature, seeking shelter in holes and crevices among the rocks, and is usually small. It feeds mainly on bivalve mollusks but will also eat snails, fish and crustacea. Its hideouts along the shore can usually be detected by the presence of empty shells.

Locomotion among the cephalopods varies from a slow, "tentacle-walking" pace, both in and out of water, to the rapid, jet-propulsion darts which are so characteristic of the squid. The so-called aerial "flight" of squid, like that of the flyingfish, is actually a gliding operation and largely depends upon the initial speed attained under water. Squid have frequently landed on the decks of ships a dozen or more feet above the surface of the ocean. When a school of squid is alarmed by an approaching ship or by marauding fish, the fleeing squid dart from the water simultaneously and all in one direction rather than individually fanning out in several directions in the manner of flying fish.

The squid darts backward, forward, or in any other direction by means of the reaction of the jet of water which is ejected with great force from the siphon, and direction of movement is controlled by the bending of the siphon. Even when it is confined to a limited space, as in a fishpound, it is not an easy matter to capture it with a dip-net, so rapid is its movement.

When it is darting rapidly, the lobes of the caudal fin are closely wrapped around the body, and the arms are held tightly together to form a streamlined outline. Except when attacking or escaping, the squid swims less strenuously, using the caudal fin as a balancing organ.

There are few sights as interesting as that of squid engaged in capturing and devouring young mackerel. During the summer this chase may be observed from certain wharves in New England. In attacking mackerel the squid darts backward among the fish with the velocity of an arrow, and then turns obliquely to one side and seizes a fish, which is almost instantly killed by a bite in the back of the neck by the squid's sharp beak. The bite is always made in the same place, cutting out a triangular piece of flesh, and is deep enough to penetrate to the spinal cord. The attacks are not always successful and may be repeated a dozen times before one of the wary fish can be caught. Between attacks a squid may suddenly drop to the bottom and, resting on the sand, change its color to that of the sand so perfectly as to be almost invisible. Ordinarily, when swimming, it is thickly spotted with red and brown but, when darting among the mackerel, it appears translucent and pale. The schools of young mackerel often move close to shore where the water is shallow and offers more protection. In their eagerness to capture fish, the squid frequently force themselves up on the beach where they perish by the hundreds. At such times they often discharge their ink in large quantities.

Many species of octopuses and squid possess an ink sac and, in moments of great excitation they may expel a large cloud of black or brown liquid through the siphon. The ink is of a caustic nature and, in addition to its use as a "smoke screen," it is believed to be distasteful to hungry fish. Two sources of sepia ink are a species of squid found along the southeastern coast of China and another found in the Mediterranean Sea.

Many geologic eras ago the cephalopods possessed large and showy shells. Today, however, shells produced by this class are a rarity. The most spectacular shell is found produced by the Indo-Pacific Chambered Nautilus, *Nautilus pompilius*. On our shores, the small, white, spirally coiled shell of *Spirula* is frequently encountered on southern beaches. The three-inch-long Spirula squid which produces this shell is a denizen of deep water. In other squid the internal shell has been reduced to a simple slab of chalky material (the cuttlefish bone fed to canaries) or, in the case of the Loligo squid, to a thin, elongate shaft of transparent, horny material.

By an odd turn of fate, squid are heavily preyed upon by adult cod, mackerel and other fish, and no doubt some young mackerel which have escaped by a tentacle's breadth have lived to devour later their would-be assassins. Squid are taken in large quantities in nets and weirs each year, and they constitute one of the main fish baits on the Grand Banks. They are frequently eaten by peoples of the Mediterranean area and the Orient but to a much lesser extent by Americans.

In contrast to the speedy squid, the octopus is relatively a slow-moving creature, although it can swim away at a fairly rapid rate by using the same water-jet system of propulsion; it lacks the caudal fins of the squid. The underside of the eight arms of the octopus are studded along their entire lengths with cup-like disks or acetabula. When a sucker is pressed against any smooth surface, the center is withdrawn to create a vacuum which ensures a powerful attachment. An octopus can "tentacle" along with remarkable agility and at night may even take to short excursions out of water. I have known of an octopus kept in a small aquarium in Bermuda to push the lid off the top, crawl down the table and off the veranda in an attempt to reach the ocean. It crawled more than a hundred feet toward the sea before it succumbed and was attacked by ants. There have been many authentic accounts of encounters with octopus on exposed tidal reefs, and a few observers state that the octopus can keep up with a man in a brisk walk.

Even more astounding than the locomotive powers of the cephalopods are their amazing displays of bright, glowing lights and color changes. The shallow-water species have embedded in their skin chromatophores whose expansion and contraction are controlled by the nervous system. Emotion, excitation or response to the color of surrounding objects will effect the color changes in the octopus. Among the deep-water squid, many of which are phosphorescent, gorgeous underwater pyrotechnics are frequently displayed which far outshine the brightest of fireflies and glowworms. Specimens of *Lycoteuthis* brought up from considerable depths and kept alive in chilled water have had their photographs taken by their own light. The body looks as if it were adorned by a diadem of brilliant gems. The middle organs of the eyes shine with ultramarine blue, the lateral ones with a pearly sheen. Those toward the front of the lower surface of the body give out a ruby-red light, while those behind are snow-white or pearly, with the exception of the middle organ which is sky-blue. Some squid have astonishingly complex bull's-eye lanterns; others have mirrored searchlights. A species of *Heteroteuthis* is able to spurt out a luminous secretion from its funnel and the jet of water following it draws out the bright globules into long, shining threads.

The sexes in the cephalopods are separate, except for two or three isolated examples. In most of the species females are much more numerous, the ratio of females to males being 100 to 15 in some species of the *Loligo* squid and 100 to 25 in some of the *Octopus*. The most outstanding feature is the morphological differences between the two sexes. In the *Argonauta* or Paper Nautilus, the females are 10 to 15 times as large as the males which completely lack the beautiful shell used by females for storing eggs. The female also differs in having its two dorsal arms enlarged at the end to form a veil or mold with which she secretes a shell. In a vast number of species, the males are characterized by having one of the arms modified to form a copulatory organ. This arm is known as the hectocotylus. In certain octopods, including the *Argonauta*, this arm is broken off and left in the female to fertilize the eggs. In all the other groups the hectocotylus is simply held inside the female until copulation is complete. It is interesting to note that more than 2000 years ago Aristotle recorded the presence of the hectocotylus arm in the octopus and correctly associated it with its sexual purpose.

In the males the sperm is gathered into large sacs or spermatophores of several inches in length. These sacs find their way in some unknown manner into the hectocotylus arm. Each sac contains a tiny, coiled, spring-like filament which spews the ripe sperm out of the sac.

The eggs of the cephalopods are laid in various ways. They may be single and floating in the pelagic species, such as *Oegopsida*, congregated together in a shelly nest as in *Argonauta*, laid in jelly tubes as in the *Loligo* squid, or anchored in grape-like bunches under rock ledges as in the *Octopus*. The embryo emerges from the egg fully developed and does not have a free veliger stage. With the aid of a lens it is possible to see the beautiful splotches of bright chromatophores in the skin of the tiny young even before they hatch. About a hundred eggs are laid at one sitting by the octopus; the squid egg strings from one female may contain over 40,000 eggs. Some species of octopuses take pains to watch over their brood of eggs and from time to time may carefully go over them with their tentacles to remove dirt.

THE CHITONS—*AMPHINEURA*

Amateurs and professionals alike have found the chitons or coat-of-mail shells an extremely interesting and fruitful field of study, and no collection is complete without at least three or four representatives of this strange group of mollusks. The chitons closely resemble the gastropods except that they bear eight shelly plates. For those who wish to excel in a more serious study of a relatively small class of mollusks, no more inviting series of species awaits the collector than our chitons of the rocky shores. They are dealt with in this book in some detail, for no popular shell book has hitherto attempted to open the doors to this supposedly "difficult and poorly known" group.

There are nearly fifty species in our Atlantic waters and perhaps twice that number on the Pacific Coast, and yet this represents fewer species than are found in the single family of Wentletraps or *Epitonium*. Some private collectors, such as the late Dr. R. B. Bales, were able in a few years to make larger and finer collections of Florida chitons than are found even now in our leading museums. Hitherto unknown species await the enthusiastic specialist in chitons.

Except for the more easily recognized and common species, such as the Pacific Coast *Katherina tunicata* and *Amicula stelleri,* most identifications require a simple understanding of the various parts of the shells and the patience to remove one or two of the eight valves for observation under a hand lens. Tourist collectors have little time to devote to the special but simple methods of collecting and preserving chitons, and unless they are willing to take to the shore a bucket, penknife, some thin slabs of wood or small glass plates and some soft twine, it would be best for them to concentrate their searches on the olives and cowries.

No one man has done more for the encouragement of chiton collecting in America than Dr. S. Stillman Berry of Redlands, California, and we cannot do better than to follow his simple directions. Curled-up specimens in collections signify ignorance of methods or lack of time while in the field. The chitons are easily flipped from the rock surfaces by quickly inserting a knife blade beneath the edge of the animal. If the chiton is then quickly transferred to the wet surface of a piece of shingle or glass of the appropriate width, it may be possible to flatten the creature before it curls into a ball. Wood and chiton should be tightly wound with twine or strips of cotton cloth, so that the animal will die in this flattened position. The bound chiton may be soaked in 60 to 80 percent alcohol for an hour or more, or in fresh water, for killing. Unbind, scrape the meat away from the underside, being careful not to damage the outer rim or girdle. Rewrap on the wood and set in the hot sun or oven to dry thoroughly. If scientific study is to be done at a later date, it is best to keep a few specimens permanently in a jar of about 70 percent alcohol. When specimens roll up before they can be straightened against your piece of wood, they may be dropped into the bucket of sea water where they will eventually straighten out and allow a second attempt of transfer.

Habitats of chitons are usually specialized for each species, some being found only on the underside of rocks between the tide levels, others on wave-dashed headlands, a few in tidepools, others only in deeper waters offshore. *Pinna* pen shells recently cast ashore often have tiny chitons attached on the outside of the shell. Those who do not have dredging facilities may acquire the latter species through exchange or by purchase from several of our excellent shell dealers. A watchful eye and variation in collecting localities will soon bring familiarization with the various habitats of most of the species.

IDENTIFICATION OF CHITONS

It is essential for accurate identifications to refer to one or more of the ten technical terms used in describing the various parts of the chiton. A few minutes' study of figure 18 will prepare the reader for the photographs, identification drawings and descriptions of the species. Jumping to conclusions from the photographs instead of ascertaining the family or genus first will lead to discouraging results.

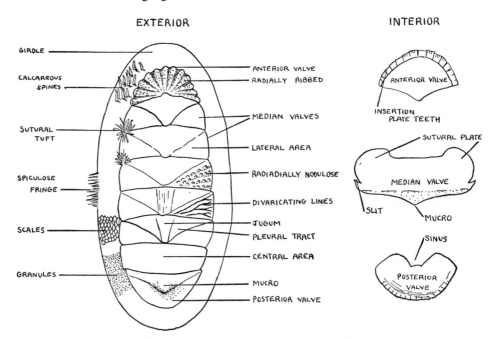

FIGURE 18. Parts of the chiton shell.

All the chitons discussed in this book bear eight shelly valves which cover the body of the creature and are bound together by a leathery girdle. The chitons without valves (*Aplacophora*) are too rare to be conveniently included here. A view of the underside of a living chiton will show the rather small, separate head and its mouth and, behind this, the larger, oblong foot. On each side of the foot is a straight row of closely packed gills. The head bears no tentacles or eyes, although the valves of many chitons bear numerous shell eyes.

Each chiton possesses three types of valves: (1) the anterior valve at the head end, (2) six intermediate valves, and (3) the posterior valve at the hind end. The shape and ornamentation of these valves are used for identification purposes, and for this reason the various areas of the valves have been named. Removal of the last two valves by the soaking of dried specimens in warm water for five or ten minutes will usually afford sufficient information. The upper surface as well as the under surface of each valve has characteristic areas which aid in identification.

In one family of chitons, the *Chitonidae*, the upper surfaces of the valves of some species bear microscopic eyes which consist of an eye capsule, cornea, iris, lens, retina and optic nerve, but they are probably useful only in sensing changes of light intensity and passing shadows.

The girdle is the leathery rim which encircles the eight valves. In some species the girdle entirely or partially covers the valves. The surface of the girdle may be covered with beautiful little scales or with spines, hairs or tufts of bristles. Unfortunately these characters vary among individuals and cannot always be used to separate species, although the general types are fairly reliable in distinguishing genera.

The radula or ribbon of teeth is very long, and is composed of thick and dark amber-colored teeth. There are usually about seventeen teeth in each transverse row, in the following order reckoned from the center: one simple, small central; flanked on each side first by a translucent minor lateral and then by a major lateral which bears a conspicuous black cusp; next, two boss-like uncinal plates; then a twisted spatulate uncinal; and, finally, three scale-like external uncini. The radula of the chitons have not been demonstrated as useful characters in separating species because of their great variability, although some workers claim that the major laterals are useful.

The sexes are separate in the chitons. Some species lay eggs in a glutinous, indistinct mass. There may be a free-swimming veliger stage in some species. In other species the young live under the mantle edge of the mother for protection.

THE TUSK-SHELLS—*SCAPHOPODA*

To our Northwest Pacific Indians and our early pioneers the tusk-shells were a familiar form of wampum, but today few Americans would recognize one on sight. The 200-odd known living species are for the most part inhabitants of deep water, although a few of our American species live in relatively shallow water and are frequently washed ashore. The shells resemble miniature elephant tusks open at both ends, and the sluggish creature lives embedded obliquely in sand and mud, with only the small end of the shell projecting above the surface of the substrate.

Like many gastropods, the scaphopods possess a single shell and a set of radular teeth but, like the bivalves, they have a nonlobed velum in the larval or veliger stage, and in adulthood have a wedge-shaped foot and lack a definite head. They lack gills but absorb oxygen from the sea water through the tissues of the mantle. Water is first taken in through the small posterior end of the tube-like shell. The water slowly builds up inside the mantle cavity of the animal over a period of about ten minutes; then, after a short period of rest, the water is suddenly expelled in the opposite direction. As in the manner of feeding among the bivalves, ciliated ridges within the mantle ensure passage of food particles to the region of the mouth. However, the primary method of feeding is by means of a number of long, cephalic filaments or captacula which are anchored to the two flattened lobes flanking the mouth. The club-shaped ends of these tiny filaments are tactile and prehensile and are capable of capturing Foraminifera and other similar minute organisms. These captacula project out in all directions from the

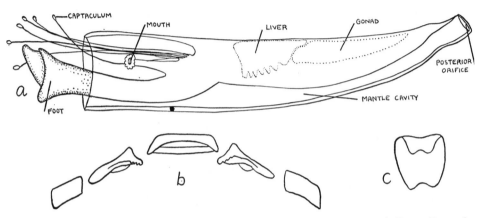

FIGURE 19. **a**, Diagrammatic drawing of the internal anatomy of *Dentalium;* **b**, radular teeth of *Dentalium;* **c**, the central tooth found in the radula of the *Siphonodentaliidae.*

larger, anterior end of the shell. Frequently, they are broken or torn off in the searchings through the sand but are soon regenerated. This accounts for the difference in length of the captacula in many specimens.

The embryonic shell or prodissoconch of the scaphopods is cup-shaped and consists of two shelly valves, which subsequently unite to form a tube. They may still be seen at the initial end in some specimens of *Siphonoden-talium,* but are always absent in adult *Dentalium.* The adult shell is open at both ends. It is added to at the larger, anterior end by the mantle edge, while at the posterior end there may be a gradual loss of shell through wear and absorption. The tiny posterior slits or notches that are characteristic of some species are formed by reabsorption of the previously solid shell wall. The shell wall is made up of three thin layers of calcareous material; this is in contrast to the similar-appearing worm-tubes that have only two layers. In cross-section, the shell may be round, slightly elliptical, octagonal or polygonal in shape, depending upon the species. The presence or absence of microscopic longitudinal or concentric riblets and the nature of the apical slits are often useful for identification. Few of our American species, other than a few pinkish or yellowish forms, can boast of colorful shells; but in the East Indies such forms as *Dentalium elephantinum* Linné are brightly hued in various shades of emerald green and jade. Some species of *Dentalium* have a terminal pipe projecting out of the posterior end.

There are only two families in the class *Scaphopoda*—the *Dentaliidae* and the *Siphonodentaliidae*. Both families are well-represented in our waters, the former by numerous species of *Dentalium*, the latter by members of the genus *Cadulus*.

> *Dentaliidae:* Shell tusk-shaped, increasing in size regularly with the greatest diameter at the mouth end. Foot conical. Central tooth of radula twice as wide as long.
>
> *Siphonodentaliidae:* Shell bulbous near the middle with the mouth end generally contracted. Foot vermiform, capable of expansion into a rosette-like disk at the end. Central tooth of radula almost square.

CHAPTER V

Collecting
American Seashells

Instructions for collecting seashells are much akin to revelations of the secrets of good cooking. Everyone has his favorite methods and, despite the most expert advice one may obtain from books, experience is the surest path to success. There are, however, a number of fundamentals which will help guide the collector in obtaining a representative series of our seashells with the least trouble. The hints offered here are more in the order of how collecting problems may be solved than a revelation of how and where to find rare specimens.

The most successful collectors mix together four ingredients to obtain what appears to most of us "unusual luck" in finding good shells. These are a knowledge of the habits of mollusks, a familiarization with the physical conditions of the ocean and the seashore, a sensible choice of collecting equipment and, perhaps most important, a large proportion of perseverance. The first three of these may be acquired, to some degree at least, from books and from the advice of veteran collectors, but only keen observation in the field and many hours of trial collecting will develop satisfactory techniques. It is true, of course, that strolling the more productive beaches at certain times of year will produce encouraging results, but soon the common species have been collected and more often than not the remaining specimens are far from perfect. The moment a collector ceases to be a beachcomber and begins to search for living mollusks in their natural haunts he has opened unlimited possibilities of acquiring a remarkably beautiful and complete collection.

56

It is most surprising how many treasures within arm's reach are lost to the uninitiated. A waterlogged board if kicked aside may be found to contain three or four kinds of interesting wood-boring clams; a rock unturned at the end of the beach may still shelter a pair of cowries or a nest of orange-tentacled Lima Clams; or the seafan momentarily admired and cast aside may be the holdfast for a colony of rare, purple Simnia snails. All mollusks have their particular ecological niches or favorite haunts, whether a very limited type of locality or more extensive areas such as mud flats, rocky shores or the open ocean. To be forearmed with a knowledge of where our species live will often bring rich rewards from salt marshes, eel-grass flats, mangrove trees, the backs of other marine creatures, the underside of boats or even the stomachs of fish. The tracks made by gastropods on sand or mud bottoms are characteristic for many species and can aid in hunting down live specimens. So, too, holes of certain shapes and sizes in the sand flat are a betrayal of the clam occupant deep below. At times it is worthwhile to know when and where gregarious mollusks gather to breed. Their appearance is often clocked not only by the seasons but often by tidal conditions and the time of day. Most intertidal species reveal themselves more frequently about half an hour after the tide has begun to rise. A great number of species are more active a few hours after dark, while others are content to wait until early morning before starting on their foraging missions.

Attention to tides, seasonal moods of the ocean and the effects of winds and currents is put to good use by the expert collectors. September seems to be the most favorable time, for instance, to gather shells on the Carolina strands. During late April and early May there is more likelihood of the Purple Sea Snail, *Janthina*, being washed ashore on the east coast of Florida. After winter gales, some New England beaches may be strewn with millions of large Surf Clams, *Spisula*.

Low tide is obviously the best time to collect, and most collectors make long-range plans to catch the spring tides. Local newspapers publish the times of low and high tides, but many serious collectors prefer to use the Coast and Geodetic Survey Tide Tables to plan well in advance for the lowest tide of the month. Tide Tables for the Pacific and Atlantic coasts may be obtained for a fraction of a dollar from the U.S. Department of Commerce, Washington 25, D.C.

As you may well know, the rise and fall of tides are caused by the attraction of the moon, and to a lesser extent by the sun. Choose the time of the new and full moon for collecting, for that is when the sun and moon are uniting their forces to give the lowest or spring tides. Low tide lasts for about fifteen minutes, but profitable collecting may be done one hour before or after. It is sometimes useful to know that the tides are about fifty minutes later the following day. Be aware of the dangers of rising tides, especially if you have waded a long distance out to some small isle at low tide. Tidal currents can sometimes be extremely strong at the narrow mouths of inlets, and swimmers are urged to familiarize themselves with local conditions.

If one were to take into the field the collecting equipment which has been recommended by friends and books, one would certainly resemble a busy Christmas shopper in full knightly armor. Crowbars, bilge pumps, shovels, rakes, sets of screens, hammer and chisels, even water wings and miner's caps have been suggested. It is true that these and many other pieces of equipment are ideal for very specific and limited purposes, but for general collecting simplicity and lightness of gear are most essential. If at a later date you wish to collect a certain species which lives in rocks, take along that hammer and chisel.

Streamlined collecting in the intertidal areas when it is calm calls for little more than a pair of canvas shoes, bathing suit and a few small cloth bags. Wear shirt and pants if the sun is bright and your tan still underdeveloped. Two or three cotton bags may be tucked under the belt until ready for use. Most shells may be picked up by hand, and the more fragile ones put in matchboxes or thumb-sized vials. When a breeze is blowing wrinkles on the surface of the water, it is impossible to see the bottom, and many collectors use a glass-bottomed bucket or merely a diving mask floated on the surface to clear a view. A square or oblong bucket about a foot each way and ten or twelve inches high may be made of light wood and the glass set in the bottom and held in place with a thin layer of white lead and strips of molding or quarter-rounds. If the inside is painted dull black reflections on the glass will be held to a minimum. For a clearer view wet the inside of the glass occasionally. The water bucket is useful to those who enjoy diving for shells. It not only serves as a friendly support between dives, but may be used as a collecting receptacle. Diving masks or water goggles are indispensible for collecting many species which are normally found in waters down to twenty feet in depth.

A fine-mesh wire screen bought in any kitchen utensil store can be put to excellent use in sandy or muddy areas where many interesting small shells live. Screening for mollusks is a favorite pastime with many collectors, and many types and sizes of screens have been designed. Copper mesh should be used if you plan to screen over a period of a few months.

Forceps are sometimes useful in getting small shells out of rock crevices, but in general it does not pay to search individually for minute shells. Mass screening or taking a large bagful of bay bottom or beach drift home for leisurely sorting in the evenings brings richer rewards. Shaking clumps of seaweed over the screen often gives encouraging results, for many uncommon species are found nowhere else. Breaking apart coral blocks often reveals interesting rock-boring clams.

DREDGING AND TRAPPING

There are few active shell collectors who have not given serious thought to trying their hand at dredging. This is especially true if one has spent several summers in one locality and acquired a large and representative collection of the littoral and intertidal species. Over half of our American species prefer to live below the low-tide mark and, although storms occasionally cast up samples of this rich fauna on our beaches, trapping and dredging are

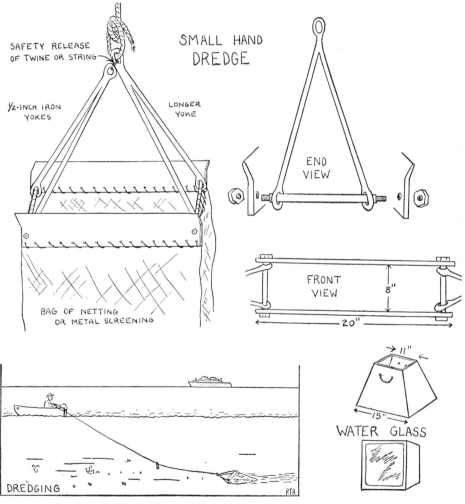

FIGURE 20. Dredging and collecting gear.

the only satisfactory methods of collecting deep-water species. In many instances larger and more perfect specimens of normally shallow-water species are found in moderately deep waters just offshore.

Dredging, like fishing, is a science as well as an art. It requires a basic knowledge of boats, equipment and bottom conditions. Firsthand experience is a necessity before satisfactory results can be obtained. It is also an expensive operation in which costs increase geometrically the deeper one dredges. Those who are financially willing to spend several hundred dollars in elaborate operations are urged to seek the advice of one of several of the Florida or California shell dredgers. However, very profitable collecting in depths less than 100 feet may be undertaken from a rowboat at relatively little expense.

One of the prime prerequisites of safe dredging is a healthy respect for the ocean and her many moods. Limit rowing operations to calm inlets and bays. Sudden squalls, high winds, swift currents and blistering sun on the open ocean are serious adversaries to even the "saltiest" fisherman. Prepare for each trip with care, and back your operations with a knowledge of local tides, currents, the weather and bottom conditions.

There are many types of dredges, and the larger your boat and engine the more elaborate may be your dredge. For rowboat operations the simplest type consists of a triangular or rectangular iron frame with a pair of iron bridles which are tied together. A fine-meshed fishing net is sewn to the frame. The free end of the net is not sewn but merely tied together, so that the contents can be removed from the back. A net of this sort is apt to be ripped on rough bottom, so that a canvas sleeve or tube open at the back end should be sewn to the frame and allowed to cover the outside of the net. The Burches of California, renowned for their west coast dredging, have had better luck with a triangular dredge and copper screen net. The leading edges of these smaller types, which are rarely more than two feet across, should be sharp and flare a little in order to dig moderately deep into the bottom.

Very remarkable results over mud bottoms may be obtained by using a small trawl. This is a modified dredge whose leading edges are of lead-weighted lines. The mouth of the trawl is kept open by a small, slanting board at each end. This type has the advantage of not digging up large quantities of ooze and mud.

In waters less than 150 feet in depth, the tow line may be of ½" or ⅝" manila rope, although the tendency for this to float in deeper waters necessitates the use of lead weights placed at intervals along the line. About 300 feet of line will suffice for hauls not deeper than 100 feet.

Only through trial and error will you learn the many tricks of dredging. The feel of the line will tell you whether the dredge is cutting into mud or gravel or is skipping over the bottom. Sometimes it is next to impossible to dredge downhill, so try in the opposite direction if your dredge is failing to dig in. On dredges with iron bridles it is suggested that one arm be attached to the frame merely by a small cord, so that it will break loose and free the dredge should it snag on rocks or corals.

When hauls are brought aboard they should be screened and washed to remove mud and sand. If this is not convenient, at least the extraneous material may be thrown away and the remainder put in sacks for home sorting. In tropical waters, gloves should be worn to prevent serious stinging by certain kinds of sponges. Be sure to make a record of the depth, location and date of haul.

Fish and lobster men often bring up rare shells in their traps, and this suggests, of course, the possibility of setting one's own traps. Successful traps may be purchased or built with a little ingenuity, if the entrances are made so that snails can easily enter. Dead fish or spoiled meat will attract the carnivorous gastropods, but to date no magical "catnip" has been found to lure the herbivorous species. Even simpler than the trap is the system of weighting a burlap bag of spoiled meat with rocks near the low-water line. *Nassarius* Mud Snails, *Melongena* Crown Shells and a host of other species may be collected nearby the next night.

If you have yet to collect your first live Olive or Terebra shell, wade along the shores of a sandy bay on a quiet, moonlight night, and with the aid of a flashlight follow along the trails in the sand. A dozen daytime visits to the same locality will never compare to one hour of night collecting. Not only are sand-dwelling mollusks on the move, but in rocky regions the cowries, mitras and murex shells are out from under their hiding places and traveling along in full view.

It is perhaps appropriate here to mention the dangers of over-collecting in certain localities. This is to be avoided particularly if certain species have taken several seasons to build up their populations even to a moderate size. By leaving at least most of the immature specimens and perhaps one or two adults, you will assure yourself of good collecting at the same spot at a later date. While it is unreasonable to expect people to roll back the rocks they have overturned, some collectors do this in order to obtain additional specimens on the next visit. Once destroyed by sunlight and air, protective algae and sponges need many months to grow back. However, the blame for extinction of many beautiful mollusks at Lake Worth, Florida, and in many other places rests not with greedy collectors but with super-drainage experiments, city pollution and construction work.

Keeping accurate locality data with specimens you have collected is most essential. Many private collections are eventually left to museums for the enjoyment and use of future generations. Today's crowded museums must rightfully dispose of specimens which have no data and are therefore of no scientific value. Large and beautiful collections representing much time and cost would have been of inestimable value to science had someone only taken the time to record where each specimen was collected. "Australia," "Hawaii" or "California" is not enough. An example of good data would be: "North end of Captiva Island, Lee County, Florida. Leo Burry, collector. July 4, 1952." Many careful collectors add interesting notes concerning the depth of water, type of bottom, abundance, and so forth. A rare shell in perfect condition, correctly identified and with accurate data, is almost worth its weight in gold.

PREPARING SHELLS FOR THE COLLECTION

The beauty and value of a collection depends largely on the manner in which specimens are cleaned and the methods in which the shells are arranged and housed. The majority of snails and clams, whether they be marine, land or fresh-water, may be cleaned of their animal soft parts by merely boiling in fresh or salt water for about five minutes. The meat may be extracted with a bent safety pin or icepick, depending on the size of the specimen. Shells which have a highly glossed or enameled finish, such as the cowries and olives, should never be thrown directly into boiling water. Start them in warm water, bring slowly to a boil, and then let cool gradually. Any rapid change in temperature will crack or check the polished surface. Save the horny operculum or trapdoor of those species that have them. When the shell is dry, a plug of cotton will hold the operculum in the aperture.

Many species are difficult to clean even when the boiling system is used. Usually the tip end of the animal's body remains in the shell of such genera as *Terebra*, *Vasum* and *Xenophora*. Vigorous shaking or syringing with a powerful blast of tap water will get most out. Filling the shell half full of water and setting it out in the shade for a day or so with an occasional syringing will help. If odors still persist a few drops of formaldehyde introduced into the shells, plus a cotton stopper, will eliminate the objections.

In the Pacific Islands most collectors bury their shells alive a few inches under soft, dry sand. In a few weeks the specimens are dug up and washed. The sand must be sifted for smaller shells and the opercula. Some people who do not object to flies set their shells upside down in the sand and allow blowfly larvae or maggots to clean out the meat in a week or so. Vigorous rinsing of the shell is all that is necessary.

Many delicate snails, including most land species and small fragile clams, may be placed in fresh water overnight and then syringed or picked clean. This system works well with *Dentalium*, *Janthina*, *Marginella*, *Olivella*, *Trivia* and *Cyphoma*, although the last four genera may require a two-day soak. Bivalves are usually the easiest to boil and clean. Allow your pairs to dry in the flat, open or "butterfly" position, as this will permit ready inspection of the hinge teeth for identification purposes.

There are many minute species which obviously cannot be boiled and picked clean. Shells less than one third of an inch may be soaked in seventy percent grain alcohol, and then placed in the sun to dry thoroughly. This strength of alcohol is also ideal for pickling squid, octopus or the soft parts of other mollusks. Isopropyl alcohol may be used, but it is best to use this at a fifty percent strength. Never use formaldehyde (or formalin) to preserve mollusks. The shell turns soft, loses color and often crumbles away in a few months.

When a shell has been cleaned of its soft parts, it must next be prepared for the collection. Most shells are ready for display and most attractive in their natural state. However, a large number of gastropods, whose beauty is hidden by coral and algal growths, are in need of a certain amount of "face lifting." A stiff brush, soapy water and diligence will usually suffice. Many collectors soak specimens in a strong chlorine solution for a few hours. This removes a great part of the unsightly growths and will not damage the shell. It will also remove the natural periostracum or thin corneous layer on the outside of the shell. However, when you have several specimens to add to your collection, it is best to keep at least one in its natural state.

Very few expert collectors use acid in treating shells, since this often gives specimens a very unnatural, although colorful, sheen. It is used occasionally to remove limy deposits and to brighten up old specimens. Commercial dealers dip the Pink Queen Conch, for example, for five or ten seconds in a vat of one part muriatic acid to four parts of water and then rinse in fresh water. Shells may be dipped with forceps in full strength oxalic or muriatic acid for two seconds and then immediately put under running cold water. This may be repeated until the desired effect is obtained, but it should be pointed out that any acid treatment ruins most shells for scientific study.

Polishing abalone shells and cutting cross-sections of larger shells require special equipment such as electrically run burring wheels and circular diamond cutters. A visit to a shell factory will be of profit to those wishing to undertake this interesting hobby.

THE SHELL COLLECTION

Although seashells are easy to keep since they do not deteriorate and generally do not fade in color like many insects, they present many special problems in housing because of their many sizes and shapes. There are three general types of collections—the knickknack shelf, the display arrangement and the study collection.

The first of these is usually the result of a summer's random beach collecting by the novice or a living-room auxiliary to the main collection. Many important private collections have started in this manner.

The display collection for museums, libraries, clubs or even the home is limited by the pocketbook and by the type of secondhand display cabinets that can be afforded. Little more is needed than common sense attention to matters of good artificial lighting, attractive but neutral background, neat labeling, choice of specimens and especially the avoidance of overcrowding. The exhibit should be designed for its eye-appeal as well as for its interest. One has a wide choice of themes—a selection of local shells, mollusks of economic or medical interest, shells of odd habits, examples of colors and patterns and a host of others. The labels of exhibits showing classification should bear the scientific and common names and the geographical range. Miniature display boxes with cotton background and glass or cellophane covering are very popular and, if of uniform size, may be neatly stacked in a closet when not in use.

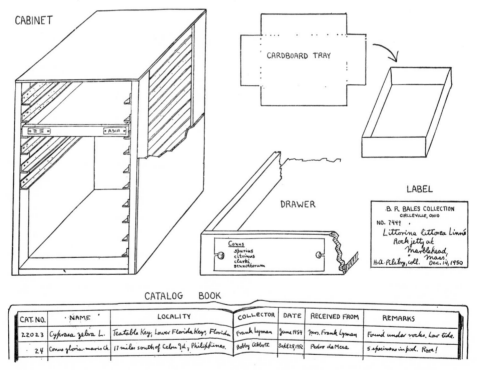

FIGURE 21. The shell collection.

The name "study collection" may sound ominous to some but, if a few simple principles are followed, this type of housing will bring more joy and less work than any other system. It is not only neater, more compact and equally as attractive as the display type, but it also permits the collector to locate any specimen quickly and add new material with a minimum of rearranging. The simplicity, uniformity and mobility of equipment, such as drawers, trays, labels and vials, and the use of the biological or systematic order of arrangement are the essence of a good collection.

The choice of cabinet and style of drawers will be limited, of course, by the collector's pocketbook. The accompanying designs are the result of many years of observing private and institutional cabinets, and they are offered here as an ideal toward which you can strive.

If the cabinet is made in a roughly oblong shape and is about table-height, additional cabinets may some day be set alongside for desk space or set on top of each other without causing the top drawers to be too high to reach. Pine, basswood or any of the whitewoods may be used. It has been reported that certain oaks have a detrimental effect on shells which have been stored away for years. It is best to have a cabinet door which swings open all the way (180 degrees), although so hinged that the drawers may still be pulled out when it is open only 90 degrees. Some students prefer the type of door which lifts off.

The ideal cabinet unit has the following dimensions: outside measurements, height 40″ (or 80″), width 22″, depth 32″. Runners for drawers, 30″ long. If wooden, ½″ × ⅜″ and set 2¼″ apart. If galvanized sheet iron, 2½″ wide and bent along the midline to form an L. Inside measurements, wooden drawers 20″ × 30″ and 1⅝″. No runners or handles are necessary on the drawers.

All cardboard trays to hold specimens should be ¾″ in depth, and all their other outside dimensions should be multiples of the smallest type of tray. This unit may be 1½″ × 2″, the next largest tray 3″ × 2″, then 3″ × 4″, then 4½″ × 6″, and the largest of all 8″ × 9″. It is inadvisable to have more than five sizes of trays, since this complicates curating and the making or ordering of future stocks. Odd-sized trays make neat arrangement impossible. Cardboard trays covered with glossy-white enameled paper may be purchased in any large city, or a simple style may be made by cutting out and folding pieces of shirtboard as shown in our illustration. The corners are held together by adhesive paper or butcher's tape. The various sets, or lots as they are called, of each species should be placed in the trays and arranged in the drawer from left to right, beginning at the front. Many students separate the species or genera by turning over an empty box which may bear a label indicating the genus or species.

Small glass vials without necks are used to hold smaller specimens. Cotton is best for plugging the vials, since corks are expensive, are difficult to obtain for various-sized vials and eventually deteriorate. When a lot consists of a hundred or more small specimens which will not easily go into vials, it is convenient to use a covered box 3″ × 4″ and 2″ deep. The label should be pasted on the lower left corner of the lid. A duplicate label or a slip of card bearing the catalog number should be placed in the box. Some people can afford to have glass-covered boxes.

A catalog is most essential, and its single purpose is to prevent the loss of valuable locality data. If each specimen bears the same number as the label and catalog entry, it can be returned to its proper tray in case of accidental spilling. A thick ledger about 12″ × 8″ may be purchased at a second-hand office equipment store at small expense. Headings may be arranged across both pages as shown in our figure. More space should be given "Locality" than any other section. Run your catalog numbers from 1 on up. Do not experiment with mystical letters indicating the locality, collector or date of cataloging, since all this information will be on your label and in your catalog. A card catalog arranged systematically is useless, time-consuming and a duplication of the information already available from your collection.

Specimens should be numbered in India ink with a fine pen. Shells that are too small to number may be put in vials or covered boxes, but do not fail to add a small slip bearing the catalog number.

The housing of molluscan animals, octopus and other soft-bodied creatures which must be preserved in seventy percent grain alcohol is expensive and generally beyond the scope of the average private collector. It may be mentioned, however, that preserving jars with rubber rings and clip-on glass lids are the best. Vials with necks may be plugged tightly with cotton and set upside down in the jars.

The mollusk collection should be arranged systematically, that is, in biological sequence, with the first drawer containing the primitive abalones, followed by the limpets and on up to the specialized bubble shells (*Bulla*). The small chiton, cephalopod and scaphopod classes may be put at the beginning of the gastropods or between them and the bivalves. You may wish to place your unsorted or unidentified material in the last few drawers. Once you have a species represented in your collection, do not stop there. Add other lots from other collecting regions. You will then learn to appreciate individual, ecological and geographical variations.

Exchanging. An amazing amount of traffic of duplicate material exists throughout the country and in many parts of the world today. Exchanging is an ideal way of sharing your local rich hauls and of obtaining species beyond your collecting sphere. A list of the many hundreds interested in exchanging is published in several directories of conchologists and naturalists. Sound out your prospective exchanger to learn what species or type of material he desires, since some advanced collectors are extremely "choosy." Always give accurate locality data and send as perfect specimens as you can. Some people make up elaborate exchange lists which they send around to other collectors. Exchanging, although worthwhile, is time-consuming, and great care must be taken that the upkeep of your main collection does not suffer.

Excellent specimens with largely reliable locality data may be obtained from a number of dealers. Their prices are often high, but this is justified, at least with regard to locally dredged material, by the high cost of operating boats and replacing dredges. Like antiques and costume jewelry, the prices of shells vary with what people will pay.

Shipping. When sending shells on exchange or to some other collector for identification, always include a fully inscribed label with each lot. Most shells are best protected by loose wrapping in old newspaper. Small or fragile shells should be boxed with cotton. Mail or express shipments up to twenty pounds will travel safely in cardboard cartons obtained from the grocery store. The top and bottom should be padded with two inches of crumpled newspaper. Small lots are conveniently sent in mailing tubes. It is inadvisable to send living snails through the mails, and foreign imports of living land and fresh-water mollusks are prohibited by law except by prior permission from The Surgeon General, U.S. Public Health Service or from The U.S. Department of Agriculture, Washington 25, D.C.

Identification services. Besides popular books and a few professional papers available in public libraries, there are few places where amateurs may turn for expert determinations. Fortunately, not a few private collectors are even more familiar with their local faunas than are the professional workers. Although some charge small fees for their services, most are only too happy to identify your "sticklers." It is customary to name only material which has been sorted and which has accurate and detailed locality data, and to send a sufficient series so that the identifier may retain a sample for his efforts. It is a breach of etiquette to send material before asking if the identifier is willing to undertake the task. Sending photographs is highly unreliable and is tantamount to saying you do not trust the specimens out of your hands. Some museums will identify specimens if you are unable to do so after serious effort, and this, of course, can be done only if the curator or research worker has the time. Never send more than five species at a time. It is surprising how many people abuse this service, purely voluntary on the part of the expert, by sending unsorted, data-less shells. It is more important that the professional spend his time in caring for his vast collections, doing his research and writing for the benefit of all, than in identifying for the few. Medical workers, agriculturalists, archaeologists, fisheries men, ecologists and other professional malacologists already demand a great deal of his time.

OUTSTANDING COLLECTIONS

There are a number of very lovely private collections in the United States, some devoted wholly to marine species, others limited to land or fresh-water types. Many represent years of collecting, others an expenditure of many thousands of dollars. To mention a few would be to slight many another. The best private collections are in California, Florida, Connecticut, the New York area and Massachusetts. As time passes, private collections are either sold, lost or left to some public or university museum, so that today we find the largest collections housed by public or endowed institutions.

The United States National Museum, under the Smithsonian Institution in Washington, D.C., contains what is undoubtedly the largest mollusk collection in the world. Until Dr. Paul Bartsch, now retired, was curator, it was second in size to that of the British Museum in London. Today, this study collection contains over 9,000,000 specimens, 600,000 lots or suites and in the neighborhood of 36,000 species and subspecies. Its curator at present is Dr. Harald A. Rehder, and his associates are Dr. J. P. E. Morrison.

The Museum of Comparative Zoölogy at Harvard College, Cambridge, Massachusetts, has risen to second place in the United States within the last fifteen years. It is famous for its well-kept collection of about 7,000,000 specimens, 300,000 lots and approximately 28,000 species and subspecies. Its present curator is Dr. William J. Clench, noted for his development of students in mollusks. Dr. Ruth D. Turner is assistant curator.

The Academy of Natural Sciences of Philadelphia, Pennsylvania, is third or fourth in size and contains an unusual amount of valuable material. Its present curator, Dr. Henry A. Pilsbry, has been with the institution for over sixty years, and he has contributed more to our science than any other worker. He was preceded by two equally famous curators, George W. Tryon and Thomas Say, America's first malacologist. The author is the present incumbent of the Pilsbry Chair of Malacology.

In the Midwest, one of our largest fresh-water and land collections is located at the Museum of Zoology, University of Michigan, Ann Arbor. Dr. Henry van der Schalie, an expert on fresh-water clams, is the curator. The Chicago Museum of Natural History in Illinois contains a small but adequate collection and is under the care of Dr. Fritz Haas, a scientist well-versed in many phases of malacology.

There are no very large study collections in southeastern United States, although one of the finest exhibit collections is on display at Rollins College in Winter Park, Florida. It is well worth visiting, for the collection is beautifully lighted and arranged and is instructively labeled. Of equal brilliance, the Simon de Marco collection of rarities is housed in the commercial Florida Marine Museum near Fort Myers, Florida.

Among the leading American collections that in the California Academy of Sciences in Golden Gate Park, San Francisco, stands foremost in western United States. It is a large and well-kept collection, supplemented by an excellent library. Drs. G. Dallas Hanna and Leo Hertlein have in the main been responsible for its successful growth. The Paleontology Collection at Stanford University contains a large series of recent and fossil mollusks. It is particularly strong in material from the Pacific northwest. Its present curator, Dr. Myra Keen, is one of our outstanding malacologists who specializes in marine bivalves. A very large collection of marine mollusks is housed at the San Diego Natural History Society, Balboa Park, San Diego. The Museum of Paleontology, University of California, Berkeley, also has a very large collection.

CHAPTER VI

How to Know
American Seashells

The satisfaction of gradually becoming master of a study and the enjoyment of devoting a full interest to one of the many fields of natural history, whether it be wild flowers, butterflies or seashells, are the two strongest motivations among naturalists in their search for new facts and additional specimens. "Knowing seashells" is not so much a state of knowledge, attained after so many years of study, as it is a continuous process of adding to our store of information and experience. Through personal observation, by taking advantage of what others have discovered and recorded, and by increasing our ability to identify species, we gradually become familiar with our mollusks.

What's the name of that shell? Is it rare or common? How does it live? Where can I find more and better specimens? These are four of the most frequently asked questions among shell collectors. Because people who are incurably or only mildly "shell-shocked" are continually asking for the names of shells, over three fourths of this book is devoted to the problem of recognizing and naming our American seashells.

To know the name of a shell is in many ways to know the object itself. What we may gain in observation of the shell, the animal which builds it or the habits of the creature, we can, with its correct name, compare with the findings made by other students. "So this is Sozon's Cone!" transforms the shell in your collection into an object of rarity and opens the door to fascinating accounts of fatal, venomous cone shells or the tales of bygone shell

70

auctions. If *Anachis avara* is swarming over an oyster bed, no one takes particular note, but the mere mention of the Destructive Oyster Drill, *Urosalpinx cinerea*, brings the shell-fishery man to the scene to eliminate the pest.

The identification of one of the 6,000 species found in our waters is not always a simple task. True, by flipping through pages of illustrations we may spot the shell in question or at least a near relative. This method will sometimes bring us close enough so that reference to the text will reveal the correct identity. However, unless it is realized that many species differ only in seemingly slight characters and, conversely, that other species show wide variation in color or shape, misidentifications can result. How hopeless a task it would be to separate into species the various color varieties of the Common Coquina Shell (*Donax variabilis*) or the many shapes and sculptural varieties of the Western Dog Winkle (*Thais lamellosa*). Yet how many would not at first fail to notice the differences between the shell of McGinty's Cyphoma and the Flamingo Tongue (*Cyphoma gibbosa*)? But look at the obvious differences in the color patterns of the animals shown on plate 8.

Marine mollusks are exceedingly responsive to varying ecological conditions. The presence of certain salts and minerals in the mud often dictates the degree to which certain colors are developed or to what extent spines are produced. In highly exposed areas, where surf waves pound against the shore, snail shells are usually devoid of delicate sculpture. These differences caused by environment are often difficult to distinguish from those which are genetic or naturally inherent characters of the species. So, too, there is often great genetic variation within a species, just as we have brunettes, blondes and redheads among humans. It is not an easy problem, even for the professional, to define the limits of a species, nor to say with authority that a certain specimen represents a "form" or is an example of a subspecies or even different species.

What is a species? Volumes have been written in answer to this question, and the subject is one of continuous investigation by many biologists working with all forms of animals and plants. Every population of mollusks is inherently different, and these differences, however minute, are morphological, physiological or genetic. One need only collect a common species in several localities along our coast and carefully examine them in order to reach this conclusion. It is this factor of geographical variation, together with timely isolation and selection, which has been largely responsible for the evolutionary production of species. The development of species is a continuous and very gradual process and, when we settle upon a reasonably homogeneous series of populations and label them as, say, *Melongena corona*, we are merely "snapping a candid camera shot" of a species living today, one whose picture looked quite different several million years ago during the Pliocene period. Within the geographical range of this species we find a series of populations on the west coast of Florida which seem to be attempting a "break-away" from the typical form, and to this geographical race the name *Melongena corona perspectiva* has been given. Perhaps in another million years, through fortuitous isolation (geographical or reproductive) and selection, it will merit recognition as a full species. Elsewhere throughout the range of *corona*, we find minor groups of variants, some that are individuals stunted by ecological conditions, others that are minor genetic variations which seem to crop up at random in all parts of Florida. These ecotypes, aberrations and varieties, although actors in the evolution game, do not warrant subspecific names.

There have been many attempts to define a species. A very excellent summary of the various definitions has been published in Ernst Mayr's interesting book entitled *Systematics and the Origin of Species* (Columbia University Press, 1942). Mayr defines species as groups of actually or potentially interbreeding natural populations, which are reproductively isolated from other such groups by geographical, physiological or ecological barriers. Unfortunately, this biological concept of species cannot as yet be used extensively in the field of mollusks, for malacology is largely in the purely descriptive and cataloging stages, and the majority of species being described today are still based on the old-fashioned morphological species concept.

While the species is considered by some people as an objective entity in nature, nearly everyone agrees that a genus is merely a convenient and arbitrary grouping of closely related species. This is also true of many higher categories such as the subfamily and family which are merely convenient groupings of closely related genera. However artificial, the system is extremely useful, for it permits us to arrange the species in our collections and our scientific reports in a logical, evolutionary and biological sequence.

IDENTIFICATION FEATURES

These are the many morphological features exhibited in mollusks which are used for identifying species and in understanding the evolutionary relationships existing between members of the higher categories, such as genera, families or orders. It must be realized that in some groups of shells certain types of characters, such as number of spines, shape of aperture or color markings are used to distinguish species, while in other groups these will prove useless and reliance may have to be put on the number of folds in the columella, the number of teeth in the aperture or the sculpturing on the operculum. These key features are pointed out in their appropriate places throughout this book.

The verbal tools which are used in the study of mollusks are especially designed to assure a method as accurate as possible for telling apart the

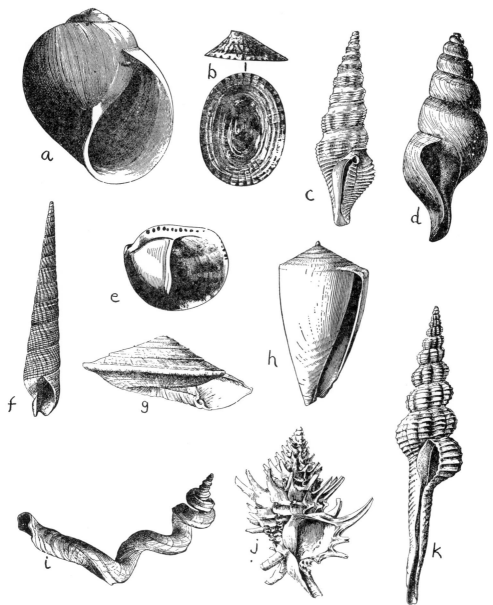

FIGURE 22. Various shapes of gastropods. **a,** globose (*Lunatia heros* Say); **b,** cap-shaped (*Acmaea testudinalis* Müller); **c,** fusiform and with an anal notch on the outer lip (*Cerodrillia acestra* Dall); **d,** sinistral or left-handed (*Antiplanes vinosa* Dall); **e,** slipper-shaped (*Crepidula fornicata* Linné); **f,** high-spired (*Terebra nassula* Dall); **g,** low-spired (*Architectonica peracuta* Dall); **h,** cone-shaped (*Conus*); **i,** whorls freely coiled (*Vermicularia spirata* Philippi); **j,** spinose (*Murex hystricinus* Dall); **k,** spindle-shaped (*Fusinus eucosmius* Dall).

100,000 or more living, and many more fossil, species of mollusks. It is impossible to avoid using technical names for various parts of the shell and its animal, such as apex, spire, whorls, operculum, etc., for most of these words have no counterpart in everyday language. Familiarization with these few terms is gained easily and rapidly as trial identifications and references to the illustrated glossaries are made. Many of the technical terms explained below are not employed in this book, but they are presented for the sake of those readers who intend to use more advanced works.

Gastropod Features

Shape of shell. It is this character that is instinctively used at first when identifying a snail shell, and little would be gained in discussing at length what our photographs so clearly demonstrate. However, the shape of the adult shell in some species may differ radically from its young stages as may be seen in the illustrations of the cowries (pl. 6g) or the American Pelican Foot (*Aporrhais*, pl. 23c). Monstrosities caused by embryological defaults or by injury in early life have always been a source of error in identification, and in certain extreme cases many species have been erroneously described as new.

Parts of the shell. As the typical gastropod mollusk grows, it adds to the spiral shell and produces turns or *whorls*. The first few whorls, or nuclear whorls, are generally formed in the egg of the mollusk and usually differ in texture, color and sculpturing from the postnuclear whorls which are formed after the animal has hatched. When the nuclear whorls are marked off from the remainder of the whorls they are often referred to as the *protoconch*. The last and largest whorl which terminates at the aperture of the shell is known as the *body whorl*. The *periphery* is an imaginary spiral area on the outside of the whorl, usually halfway between the suture and the base or at a point where the whorl has its greatest width. The Giant Atlantic Pyram (pl. 4q) shows a narrow color band on the periphery of the last whorl. The whorl just before the last whorl often has distinctive characters and has been differentiated by the name *penultimate whorl*. Above this the succeedingly smaller earlier whorls in the pointed apex of the shell are known as *apical whorls*. The rate of expansion of the growing whorls and the degree to which the succeeding whorls "drop" determine the shape of the shell. The sides of the whorls may be flat, globose, concave, channeled or ribbed. The juncture of each whorl against the other forms a *suture* at the top or above the shoulder of each whorl. The suture may be very fine—a mere tiny, spiral line—or it may be deeply channeled (see *Busycon canaliculata*, the Channeled Whelk, pl. 23n). Sutures may be wavy, irregular, slightly or deeply indented or impressed.

The *anterior end* of the shell is that end which is in front when the animal is crawling. The aperture, the siphonal canal (when present), the head and the tentacles of the mollusk are at this end. The *posterior end* is the opposite, where the apex and nuclear whorls are located, hence it is sometimes referred to as the apical end. When we speak of the posterior side of a rib or a bar of color we mean the side nearest to the apex or away from the anterior end of the mollusk. The total distance between the two ends of the shell is known as the *length*, although this measurement is often called the height.

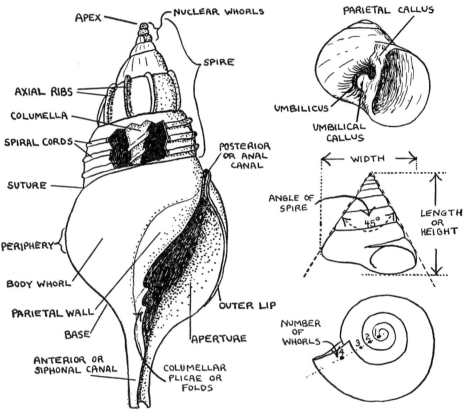

FIGURE 23. Parts of the gastropod shell.

The *aperture* of the shell is the hole or space at the end of the body whorl into which the mollusk can withdraw itself. The edge of the body whorl which borders the aperture is known as the *lip* (sometimes called the *peristome* in technical works). Sometimes the lip is thickened greatly or flaring like an old-fashioned blunderbuss. Any startling development of the lip is generally a sign of adulthood. If the lip thickens into an unusually large, rounded, sharp rib it is known as a *varix*. Varices may be produced at various stages in the growth of a shell, and their number and position are used as identifying characters (see *Bursa*, the frog shells, pl. 9k).

For the sake of convenience, the part of the lip which is away from the center of the shell or is not next to the axis of the shell is known as the *outer lip*. Opposite this on the other side of the aperture is the inner lip or *parietal wall* which may be thickened, armed with teeth (see *Nerita*, pl. 4) or have a parietal shield (see the helmet shells, *Cassis*, pl. 23v). The inner lip is continuous with the thickened axis or *columella* of the shell about which the whorls are developed. In many kinds of marine gastropods, especially the murexes, the columella extends forward and forms the tube-like anterior *siphonal canal*. In a few genera there is a small *posterior canal* formed at the upper or posterior end of the aperture (see *Bursa*, pl. 9k).

The outer lip in a few genera has a very characteristic notch or slit. It is longest in the very rare, large Pleurotomaria shell (pl. 3). The "stromboid notch" in the conchs is weak but distinct. In the abalones, *Haliotis* (pl. 2), the slit is replaced by a series of small, round anal holes. Nearly all the turrids are recognized by their "turrid notch" on the upper portion of the outer lip. The Keyhole Limpets, *Fissurella*, have reduced the slit to a single small hole which is located at the apex of their cap-shaped shells, although in their young stages the slit is well-developed at the edge of the shell (see fig. 5).

The sculpturing on the exterior of the shell—ribs, nodules, cords, threads, indented lines, pits, spines, etc.—are grouped into two basic types: (1) The *axial sculpture*, that is, any markings, ribs or lines which run across the whorl in line with the axis of the shell or from suture to suture. Sometimes it is called longitudinal sculpture. Varices, growth lines and the outer lip are axial features. (2) The *spiral sculpture*, which is spirally arranged in the direction of the suture or in line with the direction of the growth of the whorls. Thus we often speak of spirally arranged color bands (as in the Tulip Shell, *Fasciolaria hunteria*, pl. 13c), or axially arranged color streaks (as in the Lightning Whelk, *Busycon contrarium*, pl. 23-o).

The *umbilicus* is a hole or chink in the shell next to the base of the columella, which is formed because the whorls are not closely wound against each other at their anterior or basal end. The umbilicus may be quite large and deep as in the sundial shells, *Architectonica* (pl. 4m). Commonly there is a spiral cord in the umbilicus which may terminate in a button-like callus. Some species are differentiated by the size, position or color of this *umbilical callus* (see *Polinices duplicatus*, pl. 5k). About a fourth of our marine species are umbilicated to some degree or another.

Teeth (not to be confused with the radular teeth in the animal's proboscis or mouth cavity) are often present in the aperture. The Distorted Shell, *Distorsio* (pl. 25z), is an extreme example, but some shells have teeth on the parietal wall only (*Nerita*) or on the inside of the outer lip (*Cassis*).

The *periostracum* is a horny covering which overlays the exterior of the shell in many species and, like the shell, is secreted and shaped by the fleshy mantle of the animal. The periostracum (erroneously called the epidermis) may be very thin and transparent or only slightly tinted (as in some volutes, moon shells and the smaller conchs); or it may be like a thick coating of shellac which flakes off when dry (as in the Queen Conch, *Strombus gigas*). In a few buccinids, some frog shells (*Lampusia*) and the vase shells (*Vasum*), the periostracum may be very thick and often have clumps which simulate hairs and bristles. It is wholly absent in many groups, including the cowries, olives and marginellas. It is primarily a protective coating and prevents damage from boring sponges and water acids.

When axial and spiral sculpturing are equally prominent and cross each other at right angles, a *cancellate* or decussate sculpture is produced. *Reticulate* sculpture is similar, but the lines do not cross at right angles.

Growth lines are mentioned in many of our descriptions and these refer to the axial lines which run parallel to the edge of the apertural lip. These are irregularities in the shell, usually very small but sometimes coarse, which mark places where growth of the shell was stopped for a relatively long time. Sometimes the lip of the aperture becomes stained or slightly thickened during these brief rest periods (probably a few days apart), and, when additional growth takes place, these blemishes are left as growth lines.

FIGURE 24. Various types of opercula. **a**, calcareous (*Turbo*); **b**, under surface of same showing the paucispiral, corneous layer to which the foot muscle is attached; **c**, calcareous and paucispiral (*Nerita*); **d**, paucispiral and corneous (*Littorina*); **e**, ungulate and corneous (*Busycon* and *Vasum*); **f**, multispiral and corneous (*Livona*); **g**, concentric and corneous (*Buccinum*).

The *operculum* is a horny or calcareous plate firmly attached to the dorsal side of the posterior end of the foot. When the head and foot are withdrawn into the shell, this "trapdoor" is the last part to be pulled in, and it thus serves as a protection against enemies and, in many species, seals the shell from either noxious fluids or the drying effects of the sun and air. When the foot is extended and used in crawling, the operculum serves as a foot-pad on which the heavy shell may rest and rub without injury to the soft foot. The operculum is present in many families of marine mollusks, and it often

serves as a useful identification character. It is absent in adults in the following families: *Marginellidae, Cypraeidae, Tonnidae, Haliotidae, Acmaeidae,
Fissurellidae, Janthinidae,* and nearly all of the sea slugs (opisthobranchiates,
nudibranchs, bullas, etc.). Some genera lack this organ, such as *Oliva* and
Cypraecassis, although their close relatives, *Olivella, Ancilla, Phalium* and
Cassis possess well-developed opercula. Nearly all *Voluta* are without the
operculum, except our West Indian Music Volute. This is also true of the
genera *Conus* and *Mitra* whose various species may or may not possess one.
In the Alaskan volute, *Volutharpa ampullacea* Dall, 15 percent have an operculum, 10 percent only traces of the operculigenous area and 75 percent
without a trace of either. The presence or absence of this part of the animal
is not always a good classificational character.

Many families, genera and species (although not in so many cases as
generally believed) possess a characteristic type of operculum. Calcareous
or hard, shelly opercula are found in the turban shells ("cat's eyes" of
Turbo), the rissoids, the nerites, and the natica moon shells. The color and
sculpturing of these opercula are used for identification purposes. The liotias
(*Liotiidae*) possess a horny operculum which is overlaid by rows of calcareous beads. Among the horny or corneous opercula there are several important and characteristic types which we have illustrated in figure 24.

The radula. The minute teeth or radula (also called the odontophore
or lingual ribbon) located in the mouths of all classes of mollusks, except the
clams, are so very distinctive in the various families, genera and species that
they have been used as a fairly reliable identification criterion. Our present
arrangement of the gastropod families is based largely upon the radula, although many other anatomical characters of the animal and shell are equally
important. The Greek naturalist, Aristotle, mentioned the radula of snails as
early as 350 B.C., but a fuller account was given by the Dutch naturalist,
Swammerdam, in the seventeenth century. The Italian malacologist, Poli,
was the first to figure the radulae of gastropods, cephalopods and chitons.

The radula is attached to the floor of the buccal cavity or inner mouth
and consists of a ribbon-shaped membrane to which are attached many small,
fairly hard teeth. The radula ribbon is maneuvered back and forth in somewhat licking fashion as the animal rasps its food. The teeth are arranged in
transverse rows on the ribbon (see fig. 6). The number of rows may vary
from a dozen (in some nudibranchs) to several hundred. Each transverse
row contains a specific number of teeth, depending on the family or group
to which the snail belongs. In the taenioglossate snails (many families, including *Cypraeidae, Strombidae, Cerithiidae* and *Littorinidae*) there are generally
only seven teeth in each row, but each of these teeth has a distinctive shape
and a specific number of tiny cusps on its edges. The tooth in the center is
called the *rachidian* or central. Flanking this tooth on each side is a *lateral*.

Beyond each lateral there is first an *inner marginal* and finally an *outer marginal*. This makes seven teeth in all. In the rachiglossate snails (*Muricidae, Buccinidae, Olividae,* etc.) there are only three teeth per row—the rachidian and a strongly cusped lateral on each side. The four toxoglossate families (*Conidae, Turridae, Terebridae* and *Cancellariidae*) have lost their rachidians and laterals and have retained only the marginals.

The docoglossate snails (*Acmaeidae* and *Patellidae*) have less than twelve teeth per row but are peculiar in that there are two to four identical rachidians or centrals. In the rhipidoglossate families (*Trochidae, Fissurellidae, Neritidae*) the radula is very complicated, and the very numerous laterals at the end of each row are called *uncini*. Among the gastropods which do not have a radula are the *Pyramidellidae, Eulimidae,* the genus *Coralliophila,* adult *Harpa* and a few genera of nudibranchs.

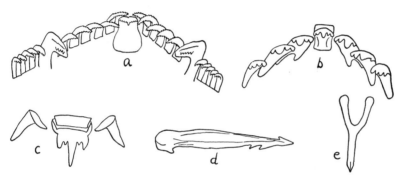

FIGURE 25. Types of radular teeth found in the prosobranch gastropods. **a,** rhipidoglossate (*Calliostoma doliarium* Holten); **b,** taenioglossate (*Littorina irrorata* Say); **c,** rachiglossate (*Purpura patula* Linné); **d,** toxoglossate (*Conus clarki* Rehder and Abbott); **e,** reduced rachiglossate (*Scaphella junonia* Shaw). All greatly magnified and representing only a single transverse row of teeth.

We have figured several main types of gastropod radulae (fig. 25), but other examples have been included in the systematic section when they are of especial use in identification. It is not expected that many amateurs will want to prepare and examine radulae but, because so many serious private collectors and many biology students will find this identification tool indispensable, we have included brief instructions on the preparation of radula slides.

Preparation of the radula. In large specimens, such as the whelks or conchs, the proboscis may be slit open from above and the round buccal mass removed. Occasionally, the proboscis is withdrawn far inside the animal, but it is easily located below the thin skin on the dorsum just posterior to the tentacles. The flesh may be torn away with the aid of small dissecting needles until the glistening, worm-like radula pops out. In order to remove

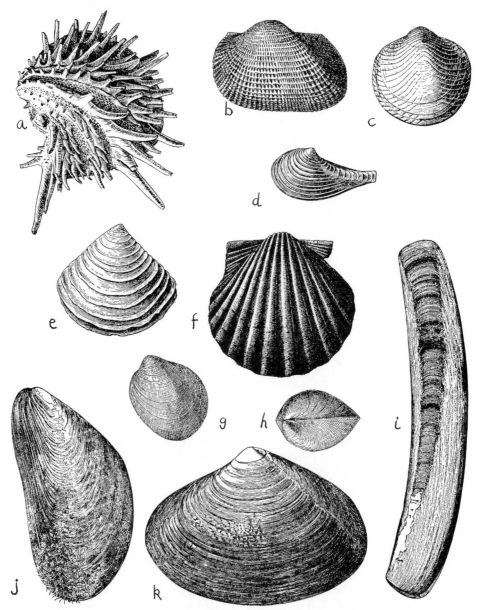

FIGURE 26. Various shapes of bivalves. **a**, spinose (*Echinochama californica* Dall); **b**, quadrate (*Arcopsis adamsi* E. A. Smith); **c**, orbicular or circular (*Divaricella*); **d**, rostrate (*Nuculana hamata* Cpr.); **e**, trigonal (*Crassinella*); **f**, fan-shaped or pectinate (*Aequipecten irradians amplicostata* Dall); **g** and **h**, obliquely ovate and aequivalve (*Crenella columbiana* Dall); **i**, elongate (*Ensis directus* Conrad); **j**, mussel-shaped (*Modiolus modiolus* Linné); **k**, ovate (*Spisula polynyma* Stimpson).

the last traces of flesh, the radula may be soaked in a saturated solution of potassium hydroxide (KOH) for a few minutes. A solution of common lye will do as well. Animals whose flesh has been hardened by a preservative will have to be carefully boiled for a few minutes or soaked overnight in KOH or lye. Small specimens may be dropped whole into this alkaline solution if only the radula is desired. Transfer the radula successively to several watch-glasses of clean water in order to rid it of all traces of KOH. The radula may then be placed in one or two drops of water on a clean, glass microscope slide and, by observation under the dissecting microscope, a few teeth may be teased apart with fine needles. Leave some of the ribbon intact to show the relative position of the teeth. Add a square cover slip for study under the compound microscope. In water mounts such as these, stains are usually unnecessary. This temporary preparation may be permitted to dry for a day, the cover slip gently lifted, a few drops of euporol or mounting medium added, and the cover slip replaced to make a permanent slide. Some workers prefer to go from water to eosin stain to ninety-six percent alcohol and then to euporol, but this is an unnecessary elaboration. There are also excellent, permanent, plastic mounting mediums on the market. Canada balsam and glycerine jelly eventually deteriorate. Keep in mind that KOH or lye will burn flesh and eat holes in clothing.

PELECYPOD FEATURES

Shape of shell. In most families of bivalves, the shape of the shell is extremely important as a species character, and only in a few groups, such as the oysters and mussels, is shape so variable within a species as to be of little taxonomic value. Shape of shell, as a whole, is of little value in determining families or genera, except in a few instances such as *Pecten, Spondylus* and *Pinna.*

Parts of the shell. The two valves of a clam are bound together by a brown, chitinous *ligament,* and usually hooked together by a *hinge* which is furnished with interlocking *teeth.* The valves are kept closed by powerful, internal *adductor muscles* but kept spread open by the action of the ligament when the animal relaxes or after it is dead. Each *valve* is a shallow, hollow cone, with the apex, from which point growth of the valves commences, turned to one side. This apex is termed the *umbo* (plural: umbos, umbones) or *beak.* The hinge and its teeth are usually just below the beak on the inside of the valve. The *prodissoconch* is the embryonic shell of the bivalve, and corresponds to the protoconch or nucleus of the gastropods. It is generally eroded away in adults, but when preserved it serves as a useful identification character, especially in such groups as the oysters.

Right and left valves. It is important to distinguish one valve from the other and to determine which is the anterior or posterior end, for many identification features are used in relation to these orientations. The dorsal or *upper margin* is located on the beak or hinge side; the *ventral margin* is the opposite side. The beaks usually are pointed or curved toward the anterior end which is generally the less pointed end of the shell. The ligament in the great majority of cases is posterior to the beaks. When present, the heart-shaped impression called the *lunule* is anterior to the beaks. When a clam is placed on its ventral margins on the table with the dorsal hinge margin up, and with the anterior end away from the observer, the right valve is on the right, the left valve to the left. Another quick way is to observe the concave, interior of a valve with the hinge margin away from the observer and to locate the U-shaped pallial sinus impression (see below). If the sinus opens toward the left, it is a left valve, and vice versa for the right valve.

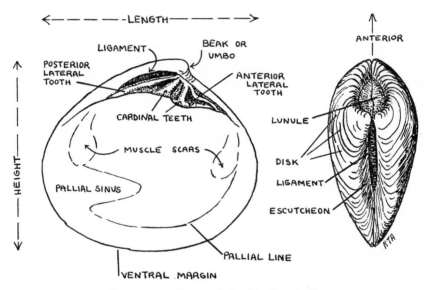

FIGURE 27. Parts of the bivalve shell.

In most bivalves, the two valves are of the same size (*equivalve*), but in some genera one valve is larger and slightly overlaps the other (*inequivalve*). In *Ostrea*, *Pandora* and *Lyonsia*, the left valve is the larger; in *Corbula*, the right valve is the larger. A bivalve is said to be *equilateral* when the beak is midway between the anterior and posterior ends of the valve. Most bivalves, however, are *inequilateral* with the beak placed nearer one end.

In many forms, the margins of the valves do not fit closely together, but have an opening called the *gape* somewhere along the margin. In the Soft-shelled Clam, *Mya*, the gape is posterior and through it protrudes the siphon (siphonal gape); in *Rocellaria* it is anterior and large and serves for the passage of the foot. Some clams, such as *Solen* and *Ensis*, gape at both ends. In *Arca* there is a small notch or opening on the ventral margin for the passage of the anchoring organ, the byssus. This is called the *byssal notch.*

The *ligament* is a brown, horny band located above the hinge, and is generally posterior to the beaks. As a rule, the greater part of the ligament is externally placed on the shell, but in some genera it may be partially or entirely internal. The ligament consists of two distinct parts, which may occur together in the same species or separately in others—the ligament proper and the internal cartilage or *resilium*. In most cases, the two portions are intimately connected with one another, but in some clams, such as *Mya* and *Mactra*, the cartilage is entirely separate (the resilium) and is lodged within the hinge in a spoon-shaped *chondrophore*. The external ligament is inelastic and insoluble in strong alkali (KOH). The cartilage is very elastic, slightly iridescent and soluble in KOH.

Muscle scars or impressions. The interior, concave surface of the valve

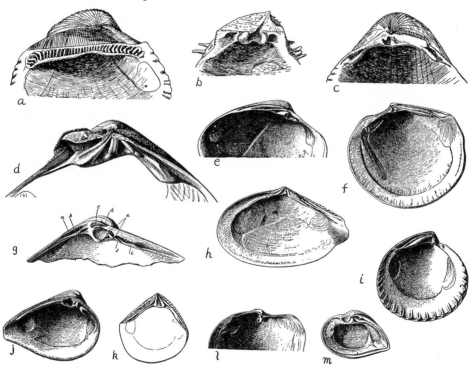

FIGURE 28. Various types of bivalve hinges. **a,** Arcidae (*Noetia ponderosa* Say); **b,** Spondylidae (*Spondylus*); **c,** Cardiidae (*Dinocardium vanhyningi* Clench and Smith); **d,** Veneridae (*Tivela stultorum* Mawe); **e,** Veneridae (*Callocardia texasiana* Dall); **f,** Lucinidae (*Phacoides annulatus* Reeve); **g,** Mactridae (*Mactra alata* Spengler); **h,** Tellinidae (*Tellina idae* Dall); **i,** Carditidae (*Venericardia*); **j,** Mactridae (*Rangia*); **k,** Crassatellidae (*Crassinella lunulata* Conrad); **l,** Periplomatidae (*Periploma discus* Stearns); **m,** Corbulidae (*Corbula*).

possesses a number of useful identification features. The large muscles which serve to close the valves leave round impressions on the surface. When two muscles are present, as in the venus, lucine, tellin and other clams, they are known as the anterior and posterior muscle scars respectively. The fine, single-lined impression produced by the muscular edge of the mantle is known as the pallial line. The pallial line may have a U-shaped notch at the posterior end of the valve indicating the presence of a siphon and its siphonal muscles. This is known as the *pallial sinus*. It is entirely absent in genera possessing no retractile siphons.

The hinge. This is one of the most important identification features in the bivalves, and often many hours of fruitless search can be avoided when the major types of hinges and their various parts are understood. There are many types of hinges from those without teeth (*edentulous*) to those with a complex pattern. We have figured below some of the major types of hinges. The teeth are distinguished as *cardinals*, or those immediately below the umbo, and the *laterals*, or those on either side of the cardinals. In many inequilateral bivalves the teeth have become so distorted or set out of place that it is often difficult to distinguish the cardinals from the laterals or to determine which ones are absent. We have labeled the teeth in several groups in the systematic section of this book to overcome this difficulty. In *Chama*, for instance, the cardinals have been pushed up into the umbo and have become a mere ridge, while the strong anterior lateral has become nearly central and simulates a cardinal.

Sculpture. In many groups, such as the scallops (*Pecten*), sculpture is of paramount importance in determining species. In most other groups it is used in conjunction with other characters. There are two major types of sculpture—*concentric* and *radial*—and both of these may be present in many forms, such as ridges, ribs, nodules, spines, foliaceous processes (leaf-like), threads, beads, indented striae (fine lines), etc. Concentric growth lines of varying degree of development are seen on most bivalves. They are always parallel to the margins of the valves, may be exceedingly fine or very coarse, and they generally indicate former growth and resting stages. Radial sculpture, running from the umbones to the lower or end margins of the valves, is exemplified in the ribs of *Cardium* (pl. 32), *Pecten* (pl. 33) and others. Concentric and radial sculpture may occur together to form a cancellate sculpture as in *Chione cancellata* (pl. 39h). In a few genera, such as *Poromya*, the valve's surface may be granulose, as if finely sugar-coated.

The periostracum or protective chitinous sheath overlaying the exterior of the valves is present in most bivalves. It may be extremely thin and transparent so that it imparts a high gloss to the shell, or it may be thick and matted or even very coarse and stringy so that the valves appear to be bearded, as in *Volsella* and *Arca*.

NAMES AND NOMENCLATURE

In order to discuss the various kinds of mollusks, we must use standardized names which are understood or recognized by students in every part of the world. For this reason, Latin names, or latinized forms, are employed as the official medium for nomenclature. It is not at all necessary to have a knowledge of Latin or Greek in order to label a seashell. Nor is it supposed that one should attempt to remember the names, although it adds to the enjoyment of the study to absorb those of a few commoner species. In fact, it is not difficult to remember such scientific names as *Venus, Mitra, Oliva* and *Conus.* It may be of interest to beginners to know that few professional malacologists can remember more than a hundredth part of the total number of names. They, too, consult books to refresh their memories.

Popular names. Popular or vernacular names in seashells are in great need of standardization and, while their use sometimes has its drawbacks, there is no reason they cannot become as acceptable to the amateur as have the popular names of birds, fishes and wild flowers. It is true that one species may be known by one name in New England and another in Florida, but these are generally names which are in use by local fishermen and not necessarily accepted by amateur shell collectors. In the face of so much name changing in the scientific literature because of legalistic technicalities, the existence of a few provincial popular names seems little enough excuse for not attempting to standardize the common names of seashells. Throughout this book we have presented both scientific and popular names. The latter have been derived from several sources and listed only after careful consideration of the evidence. Private collectors, shell dealers, professionals and, in some cases, many popular books, both recent and old, have contributed to the final choice. In a few instances, alternate popular names which are well-entrenched along wide regions of our coast have been listed. Popularization of patronymic names, such as Clark's cone for *Conus clarki,* has been simple. Direct translations of the Latin have in many but not all cases been advisable. Many obvious direct translations have been avoided in order to avoid confusion with names already used for shells in other regions of the world. It is interesting to note that many popular names in use today were recorded by early eighteenth century writers, and that a few popular generic names are to be found in the writings of Aristotle and Pliny. We have not, of course, employed the rule of using the name first employed as is done in scientific nomenclature (rule of priority). It is hoped that this first listing of 1100 popular names of American seashells will bring fuller enjoyment to the many amateurs who do not desire to "wrestle" with scientific names.

Scientific names. A mollusk is given two parts to its scientific name—the genus, which is akin to a surname, such as Smith or Jones, and the species which is akin to a first name such as William or Julia. The generic name is always capitalized, e.g., *Conus*, *Strombus* or *Arca*, but the specific name which comes after the genus name is not, e.g., *princeps*, *pilsbryi* or *floridensis*. It is also customary to add the name of the person who described and christened the species; thus the Queen Conch of southern Florida and the West Indies is known as *Strombus gigas* Linné. If subspecies or geographical races are recognized, the name may appear, for example, as *Melongena corona perspectiva* Pilsbry or *M. corona corona* Gmelin, the latter being the typical race. We have employed subgenera throughout the book as center headings. They may also be written into the name in parentheses: *Janthina* (*Violetta*) *globosa* Swainson. It is wrong to put a generic synonym in the middle of the name, as *Busycon* (*Fulgur*) *carica* Gmelin.

Some authorities may put the author's name in parentheses, for example, *Modulus modulus* (Linné). This means that the species was first described under another genus, in this case, not *Modulus* but *Trochus*. Unfortunately, as our science becomes more advanced, parentheses must be used in the majority of the species, and their usefulness becomes offset by the tax on one's memory as to whether or not they are to be employed in the various species. Modern workers are attempting to abandon this useless frill of nomenclature, and in this book they are not used. Dates following the author's names refer to the date of publication and serve the useful purpose of tracking down the original reference. It should be noted that the "double i" ending is no longer used in species names (not *smithii*, but *smithi*).

Name changing. There is nothing more annoying than having a well-known and frequently used scientific name changed; and the field of mollusks seems to be having its lion's share of tossing out of old friends for utter strangers. There are two basic kinds of changes—zoological and nomenclatorial. Everyone will condone the former, for it is obvious, as our knowledge increases, that certain genera or even species will be found to be mixtures, and this necessitates separating and applying new names. In this book, for example, *Fasciolaria gigantea* is changed to *Pleuroploca gigantea*. The Horse Conch, *P. gigantea*, does not have characters like those of the tulip shells, and it cannot be put in the genus *Fasciolaria* with such species as *F. tulipa* Linné and *F. hunteria* Perry. For the same reason, what has been called by many workers *Ostrea virginica* is now *Crassostrea virginica*. *Venus mercenaria* is now *Mercenaria mercenaria*.

Nomenclatorial name changing is hardest for everyone to accept. As not infrequently happens, a species may be given several different names inadvertently by various authors. The International Commission for Zoological Nomenclature has set up an extensive set of rules; among these is the rule of priority by which the earliest valid name is chosen if several names are available. Unfortunately, the earliest name may have been overlooked for many years, and its subsequent discovery will "knock out" one which has been in use for a long time. Thus about thirty years ago the whelk genus *Fulgur* Lamarck 1799 was abandoned for *Busycon* Röding 1798. The same fate may be met by well-known species. Thus *Busycon pyrum* Dillwyn 1817 now becomes *B. spirata* Lamarck 1816. It is believed that "rock bottom" will be reached some day, so that few, if any, further changes will occur. Nevertheless, it is with considerable regret that I change a number of familiar names in this book.

Occasionally, certain names are conserved or "frozen" by the Commission if they are well-established and are in danger of being replaced by an earlier but obscurely known name. The following marine genera of mollusks are on the conserved list: *Aplysia, Arca, Argonauta, Buccinum, Bulla, Calyptraea, Columbella, Dentalium, Mactra, Modiolus, Mya, Mytilus, Neritina, Ostrea, Sepia, Spirula, Teredo.* Many others, including very familiar species names, need to be added to this list. There are many technical refinements to nomenclature, and those interested in such matters are referred to *Procedure in Taxonomy* by Schenk and McMasters (Stanford University Press).

Pronunciation of scientific names. There is no official pronunciation established for names, and for certain words it may vary from one county to another. Many pronunciations not based on classical rules have become established and passed on from generation to generation. A few examples, classical or not, are given below:

Oliva (all-eeva), *Eulima* (you-lee-mah), *Chiton* (kite-on), *Chama* (kam-ah), *Chione* (kigh-own-ee), *Cypraea* (sip-ree-ah), *Cyphoma* (sigh-fo-mah), *versicolor* (ver-sik-o-lor said quickly), *Busycon* (boos-eekon), *Janthina* (yan-theena), *Xenophora* (zen-off-fora), *gigas* (rhymes with "jibe gas": ji-gas), *conch* (konk), *radula* (rad-you-lah), *operculum* (oh-perk-you-lum), *smithi* (smith-eye), *ruthae* (rooth-ee). The pronunciations of some of the authors are: Linné (lin-ay) or sometimes Linnaeus (lin-ee-us), Gould (goold), Deshayes (desh-ayz), Orbigny (or-bee-nee), Gmelin (mell-an), Bruguière (broo-gui-air), Kiener (keen-er), Mighels (my-els), Couthouy (koo-thoo-ee).

Common abbreviations of names of well-known authors. Although most popular and scientific books spell in full the names of authors of scientific designations, a large number of articles and most museum labels bear only abbreviations. For this reason, a short list of frequently seen examples is included:

A. Ads.—A. Adams

A. and H.—Alder and Hancock

Ag.—Aguayo

Btsch.—Bartsch

B. and S.—Broderip and Sowerby

Brod.—Broderip

Brug.—Bruguière

C. B. Ad.—C. B. Adams

Cl.—Clench
Con.—Conrad
Coop.—Cooper
Couth.—Couthouy
Cpr.—Carpenter
Dautz.—Dautzenberg
Desh.—Deshayes
Dkr.—Dunker
d'Orb.—Orbigny (d'Orbigny)
Esch.—Eschscholtz
Dill.—Dillwyn
G. and G.—Grant and Gale
Gld.—Gould
Gmel.—Gmelin
Hemp.—Hemphill
Hert.—Hertlein
L. or Linn.—Linné; Linnaeus

Lam. or Lk.—Lamarck
Midff.—Midendorff
Migh.—Mighels
Mts.—E. von Martens
Nutt.—Nuttall
Old.—Oldroyd
Orb.—Orbigny (d'Orbigny)
Pfr.—Pfeiffer
Phil.—Philippi
Pils.—Pilsbry
Q. and G.—Quoy and Gaimard
Raf.—Rafinesque
Röd.—Röding (or Roeding)
Rve.—Reeve
Sby. or Sow.—Sowerby
Val.—Valenciennes
Verr.—A. E. Verrill

PART II

Guide to the American Seashells

[*Systematic Account*]

THE SYSTEMATIC ACCOUNT OF THE AMERICAN SEASHELLS includes the scientific and common names, geographical ranges, descriptions, comparative remarks and habitats of 1500 of the 6000 species of marine mollusks found in North American waters. The areas covered are from Alaska to southern California and from Labrador to the Gulf of Mexico. Consequently a large part of the molluscan fauna of Bermuda, the West Indies, Lower California and the Gulf of California is included. Monographic accounts and more detailed information on the remaining species and their ecology, anatomy and habits may be found by consulting the papers listed by subject matter in the appendix on molluscan literature.

CHAPTER VII

Periwinkles
Conchs and Other Snails

Class GASTROPODA

Subclass PROSOBRANCHIA   *Order ARCHAEOGASTROPODA*
Superfamily PLEUROTOMARIACEA
Family SCISSURELLIDAE
GENUS *Scissurella* Orbigny 1823
Subgenus *Schizotrochus* Monterosato 1884

Scissurella crispata Fleming   Crispate Slit-shell

Massachusetts to eastern Florida and the West Indies. Europe.

3.5 mm. (⅛ inch) in width and 3.0 mm. in length. 4 to 5 whorls. Fragile, frosty-white in color and sculptured by very delicate reticulations. Umbilicus small, round and very deep. Periphery of whorls angulate and with two thin, sharp spiral lamellae. Between these there is an open slit running from the edge of the thin apertural lip back about ⅕ of a whorl. Uncommon from 60 to 500 fathoms.

The Florida Slit-shell, *S. proxima* Dall (fig. 29), from South Carolina to the Lower Florida Keys, differs in being half as large, with a more rounded periphery, and a higher spire, so that the length is about equal to the width of the shell, and in having a weaker pair of peripheral lamellae. Uncommon from 20 to 434 fathoms.

FIGURE 29. Florida Slit-shell, *Scissurella proxima* Dall, ¹⁄₁₆ inch (Massachusetts to Florida, 20 to 430 fathoms).

Family PLEUROTOMARIIDAE
GENUS *Perotrochus* P. Fischer 1885

Perotrochus quoyanus Fischer and Bernardi
                 Quoy's Pleurotomaria

Gulf of Mexico and the West Indies.

1½ to 2 inches in length and width. Umbilicus sealed over. Sculpture of finely beaded, small spiral threads. Characterized by a relatively short but wide slit at the periphery of the body whorl just behind the outer lip. Color dull orange-yellow with darker maculations. Base white. Interior slightly pearly. Dredged from 73 to 130 fathoms. One of our rarest seashells.

Subgenus *Entemnotrochus* P. Fischer 1885

Perotrochus adansonianus Crosse and Fischer   Adanson's Pleurotomaria

Plate 3d

Cuba and the Lesser Antilles.

3 inches in length, and slightly more in width. Umbilicus round, very deep. Sculpture of coarsely beaded, moderately small spiral threads. Slit on periphery of whorl narrow and very long (½ of a whorl). Color cream with a salmon blush and irregular, small patches of red. Base similarly colored. Dredged from 94 to 100 fathoms, but sometimes brought up in fish traps. This is an exceedingly rare species.

Family *HALIOTIDAE*
GENUS *Haliotis* Linné 1758

Haliotis cracherodi Leach   Black Abalone

Plate 2f

Coos Bay, Oregon, to Lower California.

6 inches in length, oval, and fairly deep. Outer surface smoothish, except for coarse growth lines. Usually 5 to 8 holes are open. External color bluish to greenish black. Interior pearly-white. A fairly abundant, edible species although not fished commercially to any great extent. A littoral species which clings to rocks between tide marks. Some shells may lack the holes (unnecessarily named *H. c. holzneri* Hemphill, *H. c. imperforata* Dall and *H. c. lusus* Finlay). A subspecies, *H. c. californiensis* Swainson, occurs on Guadalupe Island and is characterized by 12 to 16 very small holes. *H. c. bonita* Orcutt is the same as this subspecies. *H. c. splendidula* Williamson is the typical *cracherodi*.

Haliotis rufescens Swainson   Red Abalone

Plate 2a

Northern California to Lower California.

10 to 12 inches in length, oval, rather flattened. Outer surface rather rough, dull brick-red with a narrow red border around the edge of the shell. Interior iridescent blues and greens, with a large central muscle scar. 3 to 4

holes are open. Fished commercially below 20 feet, especially between Monterey and Point Conception. The legal minimum size for sportsmen is 7 inches, and the catch is limited to 5 specimens per person per day. This is a popular food and, when polished on the outside, makes an attractive mantel piece.

Haliotis corrugata Gray   Pink Abalone
Plate 2c

Monterey, California, to Lower California.

5 to 7 inches in length, almost round, fairly deep, with a scalloped edge and strong corrugations on the outer surface. 3 to 4 large tubular holes are open. Exterior dull-green to reddish brown. Interior brilliant iridescent. The variety *diegoensis* Orcutt is the same. Abundant in its southern range. The legal minimum collecting size is 6 inches.

Haliotis fulgens Philippi   Green Abalone
Plate 2b

Farallon Islands, California, to the Gulf of California.

7 to 8 inches in length, almost round, moderately deep, and sculptured with 30 or 40 raised, coarse spiral threads. Exterior dull reddish brown; interior iridescent blues and greens. 5 to 6 holes are open. Fished commercially in southern California. The legal minimum size is 6¼ inches. *H. splendens* Reeve, *H. revea* Bartsch, and *H. turveri* Bartsch are the same species in all likelihood.

Haliotis walallensis Stearns   Northern Green Abalone

Westport, Washington, to Point Conception, California.

4 to 5 inches in length, elongate, flattened, with numerous spiral threads. Exterior dark brick-red, mottled with pale bluish green. 5 to 6 holes are open, and their edges are not elevated. This is a small, relatively scarce species.

Haliotis assimilis Dall   Threaded Abalone
Plate 2d

Farallon Islands to San Diego, California.

4 to 5 inches in length, oval, fairly deep, with weak corrugations and weak to strong spiral threads. 4 to 5 holes open, tubular. Outer color mottled with brick-red, greenish blue and gray. *H. aulaea* Bartsch is a little more corrugated than usual, and it may be this species. *H. smithsoni* Bartsch and *H. sorenseni* Bartsch appear to be giant specimens of *assimilis* Dall.

Haliotis kamtschatkana Jonas

Japanese Abalone
Plate 2e

Japan, southern Alaska to Point Conception, California.

4 to 6 inches in length, elongate, with a fairly high spire. 4 to 5 holes open which have raised edges. Outside of shell rudely corrugated, but a few specimens may have weak, spiral cords. This is a small species, uncommon in California, but increasingly abundant northward.

Haliotis pourtalesi Dall

Pourtales' Abalone

Off the Lower Florida Keys.

½ to 1 inch in length, elongate, with 22 to 27 wavy, spiral cords. Outside waxy yellow to light-brown with a few irregular patches of reddish orange. A light-orange band runs from each hole to the edge of the shell. Inside pearly-white. A very rare species, and the only one recorded from our eastern coast. It has been dredged from 65 to 200 fathoms. Beware of young specimens from other oceans labeled as this species.

Family FISSURELLIDAE
(Keyhole Limpets)
Subfamily EMARGINULINAE
Key to the Genera of *Emarginulinae*

a. Apex at the same level as the base of the shell:

    b. With an internal septum . . . . *Zeidora*
    bb. Without internal septum . . . . *Nesta*

aa. Apex above the base of the shell:

    c. Slit at anterior edge . . . . *Emarginula*
    cc. Slit at anterior middle:

        d. Funnel around slit on inside . . . . . . *Puncturella*
        dd. No funnel around slit . . . . *Rimula*

GENUS *Emarginula* Lamarck 1801

Emarginula phrixodes Dall

Ruffled Rimula
Plate 17-o

Off North Carolina to eastern Florida and the West Indies.

⅓ inch in length, thin but strong, and with a small, narrow slit on the anterior slope of the shell near the margin. Base oval. Color translucent-white. Interior glossy. Concentric cords and 20 to 24 radial ribs cross each other to form a knobby, cancellate pattern. Dredged occasionally off the Miami area in 35 to 90 fathoms.

GENUS *Rimula* Defrance 1827

Rimula frenulata Dall   Bridle Rimula

Figure 30d

Off North Carolina to eastern Florida and the West Indies.

⅓ inch in length, thin, very delicate. Anal slit in the middle of the anterior slope of the shell and arrow-shaped. Base elongate-oval. Shell ⅓ high as long. Sculpture of fine cancellations. Margin finely crenulate. Color translucent-white to cream or rust, generally a deeper shade at the apex. The commonest species of American *Rimula*, but rare in collections. Dredged 5 to 150 fathoms, especially off the Miami area.

GENUS *Puncturella* R. T. Lowe 1827

Puncturella noachina Linné   Linné's Puncturella

Circumpolar; south to Cape Cod; south to the Aleutians.

½ inch in length, conical, laterally compressed, with an elliptical base. 21 to 26 primary radial ribs between each of which are added a smaller, secondary rib farther down. Margin crenulate. Tiny slit just anterior to the apex, and internally it is bordered by a funnel-shaped cup on each side of which is a minute, triangular pit. Color uniformly white, internally glossy. May be collected under rocks at lowest tides in its northern range but also occurs in waters over a mile deep. Common.

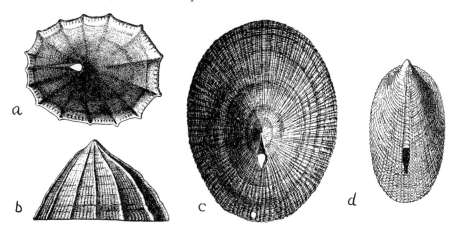

FIGURE 30. Pucturellas. **a** and **b**, *Puncturella cucullata* Gould; ¾ inch (Pacific Coast); **c**, *P. galeata* Gould, form *major* Dall; ¾ inch (Pacific Coast); **d**, *Rimula frenulata* Dall; ⅓ inch (Florida).

Puncturella cucullata Gould   Hooded Puncturella
                                                 Figure 30a, b

Alaska to La Paz, Mexico.

¾ to 1 inch in length, moderately strong. Apex small, elevated and hooked over toward the anterior end. Behind it is a small, elongate slit penetrating through the shell. Internally the slit is separated from the apex by a calcareous, convex shelf. Exterior with 14 to 23 major ribs and with 1 to 5 smaller radial ribs between the main ones. The fewer the ribs, the stronger they are. Shell dull-gray externally, glossy-white inside. The border is crenulated. Found at low tide in Alaska and dredged from 20 to 75 fathoms off southern California.

Puncturella galeata Gould   Helmet Puncturella
                                                 Figure 30c

Aleutian Islands to Redondo Beach, California.

½ to ¾ inch in length, similar to *cucullata*, but with an almost smooth basal edge; with numerous, much finer radial ribs, and with the internal shelf behind the slit reinforced by a second, straight shelf. Commonly dredged in mud from 10 to 75 fathoms.

Subfamily *DIODORINAE*
Genus *Diodora* Gray 1821

Keyhole Limpets with the internal callus of the hole truncated and frequently minutely excavated behind; shell with its basal margin never raised at the ends. Central tooth of the radula wide. Compare with *Fissurella*. *Diadora* is a misspelling.

Diodora cayenensis Lamarck   Cayenne Keyhole Limpet
                                                 Plate 17m

Virginia to south half of Florida and to Brazil.

1 to 2 inches in maximum diameter. Orifice just in front of and slightly lower than the apex. Many radial ribs with each fourth one larger. Color variable from whitish, pinkish to dark gray. Interior white or bluish gray. Just behind the callus of the orifice on the inside there is a deep pit. *D. listeri* is much more coarsely sculptured. A common intertidal to moderately deep water species. It was named by Thomas Say one month after Lamarck's description as *D. alternata*.

Diodora listeri Orbigny   Lister's Keyhole Limpet
                                                 Plate 17l

South half of Florida and the West Indies.

1 to 2 inches in maximum diameter. Similar to *D. cayenensis* but differs in that: (1) every second radial rib is larger; (2) concentric threads are more distinct and, by crossing the ribs, form little squares; (3) radial ribs often have nodules or scales. Color usually white, cream or gray, sometimes with obscure radial bands. Intertidal. Common in the West Indies.

Diodora minuta Lamarck   Dwarf Keyhole Limpet

Southeast Florida and the West Indies.

½ inch in maximum diameter, rather thin, depressed. Apex at anterior third of shell. Base elliptical, raised slightly at the center, so that the shell rests on its ends. Short front slope slightly concave, back slope convex. Orifice narrow and trilobated. Exterior shiny, with numerous, finely beaded radial ribs. Color white, with many of the ribs entirely or partly blackened. Margin very finely crenulate. Internal callus around hole frequently bounded by a black line. Not very common. Dredged 6 to 72 fathoms, but has been picked up on beaches. Do not confuse with *D. dysoni* which is more likely to be encountered, especially at Sanibel Island.

Diodora dysoni Reeve   Dyson's Keyhole Limpet
                                                         Plate 17n
Florida, the Bahamas and West Indies to Brazil.

½ to ¾ inch in maximum diameter, depressed and with straight sides. Base ovate. Apex slightly in front of the middle and characterized by a blunt knob situated behind the posterior wall of the small, almost triangular orifice. Sculpture of 18 strong ribs with three smaller ones between, and with numerous concentric lamellae. Color milky-white or cream with 8 solid, broken or dotted black rays. Margin sharply crenulated with the denticles arranged in groups of four. Distinguished from *cayenensis* by the shape of the orifice. Moderately common, sometimes washed ashore.

Diodora aspera Eschscholtz   Rough Keyhole Limpet
                                                         Plate 18b
Cook's Inlet, Alaska, to Magdalena Bay, Mexico.

1½ to 2½ inches in maximum diameter, slightly less than ⅓ as high. The roundish to slightly oval, flat-sided apical hole is $\frac{1}{11}$ the length of the shell and about ⅓ back from the narrow, anterior end of the shell. Sculpture of coarse radial and weaker concentric threads. Color externally is grayish white with about 12 to 18 irregularly sized, purplish blue, radial color bands. Commonly found clinging to rocks at low tide. In the south, dredged no deeper than 20 fathoms, and often found on the stems of kelp.

Diodora murina Arnold　　　　　　　　　Neat-ribbed Keyhole Limpet

Crescent City, California, to Magdalena Bay, Mexico.

¾ inch in length, similar to *aspera*, but smaller, with a lower, more rounded apex, with convex sides, a narrower shell, and with finer, much neater cancellate sculpturing. Color white or with few, or many, broken radial rays of gray-black. The apical hole is nearer the anterior end. Moderately common on rocks. This is *D. densiclathrata* of authors, not of Reeve.

GENUS *Lucapina* Sowerby 1835

Shell thin, low-conic, with the apex in front of the middle. Orifice rather large, roundish. Margin finely crenulated. Fleshy mantle covers most of the shell; foot larger than shell.

Lucapina sowerbii Sowerby　　　　　　　Sowerby's Fleshy Limpet
Plate 17h

Southeast Florida and the West Indies to Brazil.

¾ inch in length, oblong in outline. With about 60 alternating large and small radiating ribs. Also with 9 to 13 raised, concentric threads. Color white to buff, with 7 to 9 small, splotched rays of pale brown. Inside whitish; callus sometimes bounded by an olive-green streak. Outside of orifice not stained. Uncommon under rocks at low tide zone. It has been erroneously called *L. adspersa* Philippi.

Lucapina suffusa Reeve　　　　　　　　Cancellate Fleshy Limpet
Plate 17k

South half of Florida and the West Indies.

1 to 1½ inches in length, oblong in outline. Much like *L. sowerbii*, but larger, a delicate mauve to pinkish, and with a bluish-black orifice. Inside grayish to dirty-white. Not uncommon under rocks. Formerly called *L. cancellata* Sowerby.

GENUS *Lucapinella* Pilsbry 1890

Shell depressed, conical, less than ¾ inch, with a large orifice and thickened margins.

Lucapinella limatula Reeve　　　　　　　File Fleshy Limpet
Plate 17i

North Carolina to south half of Florida and the West Indies.

⅓ inch in length, resembling *L. sowerbii*, but smaller with a proportionately larger apical hole which is sharp at its top edge and which is nearer the center of the shell. The ends of the shell are slightly turned up and the sides are slightly concave. Sculpture of about 2 dozen heavily scaled, radial ribs and numerous, fine, thread-like concentric ridges. Color whitish with weak mauve or brown discoloring. Commonly dredged off southeastern Florida, from 6 to 60 fathoms.

Lucapinella callomarginata Dall   Hard-Edged Fleshy Limpet
Plate 18e

Bodega Bay, California, to Nicaragua.

¾ to 1 inch in length, narrower at the anterior end, quite flat. Base flat and usually with strong crenulations on the under edge. Sides slightly concave. Apical hole narrowly elongate, slightly nearer the anterior end, about ⅕ the length of the shell and with flat inner sides. Sculpture coarsely cancellate with the radial ribs stronger and often scaled. Color dark-gray with irregular, darker, radial color-rays. Rather rare under rocks in the low tide zone.

Genus *Megathura* Pilsbry 1890

Megathura crenulata Sowerby   Great Keyhole Limpet
Plate 18a

Monterey, California, to Cedros Island, Mexico.

2½ to 4 inches in length, ⅕ as high. Apical hole large, with rounded sides, ⅙ the length of the shell, and bordered externally by a white margin. Interior glossy-white. Basal edge finely crenulate. Exterior finely beaded and light mauve-brown. Animal much larger than the shell, with a massive, yellow foot and a black or brown mantle that nearly covers the entire shell. Common in many low-tide, rocky areas, such as breakwaters.

Genus *Megatebennus* Pilsbry 1890

Megatebennus bimaculatus Dall   Two-Spotted Keyhole Limpet
Plate 18d

Alaska to Tres Marias Islands, Mexico.

¾ to ⅝ inch in length, low, with ends turned slightly up. Apical hole elongate-oval, located at the center of the shell and about ⅓ the length of the shell. Numerous radial and concentric threads give a fine cancellate sculpturing. Color dark-gray to light-brown with a wide, darker ray on each side of the hole, and occasionally at each end. Interior white to grayish. Animal several times as large as the shell, variable in color—red, yellow or white. Common under stones at low tide.

Subfamily FISSURELLINAE
GENUS *Fissurella* Bruguière 1789
Subgenus *Cremides* H. and A. Adams 1854

Fissurella nodosa Born   Knobby Keyhole Limpet

Plate 17d

Lower Florida Keys and the West Indies.

1 to 1½ inches in length. 20 to 22 strongly nodulated, radial ribs. Margin sharply crenulated. Interior pure-white. Orifice oblong. An intertidal rock-dweller. Uncommon in Florida and the Bahamas; abundant in the West Indies.

Fissurella barbadensis Gmelin   Barbados Keyhole Limpet

Plate 17f

Southeast Florida, Bermuda and the West Indies.

1 to 1½ inches in length. With irregular radiating ribs. Orifice almost round. Inside with green and whitish concentric bands. Border of orifice deep-green with a reddish-brown line. Outside grayish white to pinkish buff, generally with purplish lines between the small ribs. Commonly blotched with purple-brown. Lives on wave-dashed rocks. Common.

A similar, rather rare species, *F. angusta* Gmelin, also intertidal and frequently covered with calcareous algae, occurs on the Florida Keys. The shell is flattish, pointed in front, and its internal callus is light-brown to reddish brown, but not bounded by a reddish line as in *barbadensis*.

Fissurella rosea Gmelin   Rosy Keyhole Limpet

Plate 17e

Southeast Florida and the West Indies to Brazil.

1 inch in length, thin, flattish, narrower at the anterior end. Orifice slightly oblong. Many radiating, small, rounded riblets. Color of alternating whitish to pale-straw and pinkish rays. Interior pale-green at the margins, blending to white in the center. Green orifice callus bordered by a pinkish line. Common in beach drift. Do not confuse with the larger, more elevated *F. barbadensis*.

Fissurella volcano Reeve   Volcano Limpet

Plate 18c

Crescent City, California, to Lower California.

¾ to 1 inch in length, ⅓ to almost ½ as high. Orifice at the very top, very slightly nearer the somewhat narrower anterior end, and elongate with deep, flat inner sides. Sculpture of numerous rather large, but low and rounded, radial ribs of varying sizes. Base of shell slightly crenulate and with color blotches. Exterior grayish white to dark-slate with numerous radial rays of mauve-pink. Interior glossy-white, often with a fine pink line around the callus at the apex. Foot of animal yellow; mantle with red stripes. Very common on rocky rubble at low tide. The variety *crucifera* Dall is merely a color form with white radial bands.

Subgenus *Clypidella* Swainson 1840

Fissurella fascicularis Lamarck　　　　　Wobbly Keyhole Limpet
Plate 17g

Southeast Florida and the West Indies.

¾ to 1½ inches in length. Both ends turned up (can be rocked back and forth on a flat table). Orifice toward the anterior end, keyhole in shape. Color a faded magenta. Interior whitish, tinged with pale-green or pink. Inner callus of orifice white with a narrow red line. Uncommon in Florida.

Superfamily *PATELLACEA*
Family *ACMAEIDAE*
Subfamily *ACMAEINAE*
Genus *Lottia* Sowerby 1833

Lottia gigantea Gray　　　　　Giant Owl Limpet
Plate 18j

Crescent City, California, to Lower California.

3 to 4 inches in maximum diameter, oval in outline, low, with the apex close to the front end. Exterior dirty-brown, rough, commonly stained with algal green. Interior glossy, with a wide, dark-brown border. Center bluish with an "owl-shaped" whitish to brownish scar in the very center. Very common at or above high tide line where the sea spray may reach them. In the south they grow to a large size. Frequently polished and used as souvenirs.

Genus *Acmaea* Eschscholtz 1830
Subgenus *Acmaea* s. str.

Acmaea mitra Eschscholtz　　　　　White-Cap Limpet
Plate 18r

Alaska to Lower California in cold water.

1 inch in maximum diameter, thick, pure white, conic in shape, and with an almost round base. Apex pointed and near the center. Often covered with small, knobby nullipore growths. Commonly washed ashore. It lives in cold water below the low tide level.

Subgenus *Collisella* Dall 1871
Pacific Coast Species

Acmaea pelta Eschscholtz   Shield Limpet
                                                             Plate 18n

Alaska to Lower California.

1 to 1½ inches in maximum diameter, elliptical in outline, with a moderately high apex which is placed ⅓ to almost ½ way back from the anterior
end. With about 25 axial, weakly developed, radial ribs. Edge of shell slightly
wavy. External color of strong black radial, often intertwining, stripes on a
whitish cream background. Interior usually faint bluish white, with or without a dark-brown spot. Inner border edged with alternating black and cream
bars. A common rock-dweller.

Acmaea fenestrata Reeve   Fenestrate Limpet
                                                             Plate 18t

Alaska to Lower California.

1 to 1½ inches in maximum diameter, almost round in outline, rather
high, and smoothish. The northern subspecies, *cribraria* Gould (pl. 18w),
found from Alaska to northern California, has interior with various shades of
glossy, chocolate-brown, and with a narrow, solid black border. The exterior
is plain dark-gray. The typical southern *fenestrata* Reeve (pl. 18t), found
from Point Conception south, has an external color pattern of regular dot-
tings of cream on a gray-green background. Its interior has a small, brown
apical spot surrounded by a bluish area and bordered at the margin of the
shell with brown. Intergrades occur near Point Conception. This species is
the only Pacific *Acmaea* which lives among loose boulders that are set in sand.
It only feeds when submerged. Common.

Acmaea conus Test   Test's Limpet
                                                             Plate 18g

Point Conception, California, to Lower California.

¾ inch in maximum diameter. Shell low, and like *A. scabra*, is with
distinct but widely spaced, radial ribs. Distinguished from *scabra* by its
glossy, smooth interior which often has an evenly colored brown center.
A. scabra has a rough interior center and the brown stain looks smeared.
However, this species may be a form of *scabra*. It is very abundant south of
La Jolla and is found with *A. scabra* and *A. digitalis*.

Acmaea limatula Carpenter   File Limpet
                                                             Plate 18-o

Puget Sound to Lower California.

1 to 1¾ inches in maximum diameter, elliptical to almost round in outline, low to quite flat. Characterized by radial rows of small beads which sometimes may be crowded together to form tiny, rough riblets. Exterior greenish black. Interior glossy-white, younger specimens having a blue tint. Patch of brown on inside generally weak or absent. Edge of shell usually with solid, black-brown, narrow band. Occasional albinos are cream-brown or tan on the outside. Compare with *A. scutum* which is smooth and has a barred band of color on its under edge.

Acmaea digitalis Eschscholtz — Fingered Limpet
Plate 18f

Aleutian Islands to Socorro Island, Mexico.

1¼ inch in maximum diameter, elliptical in outline; generally with a moderately high apex which is minutely hooked forward and which is placed ⅓ back from the anterior end of the shell. The 15 to 25 moderately developed, coarse, radiating ribs give the edge of the shell a slightly wavy border. Color grayish with tiny, distinct mottlings of white dots and blackish streaks and lines. Inside white with faint bluish tint and with a large, usually even, patch of dark-brown in the center. Edge of shell with a solid or broken, narrow band of black-brown. Common. Do not confuse this species with *A. scabra* which does not have the "hooked-forward" apex and is not glossy on its internal brown patch. Compare also with *persona*.

Acmaea persona Eschscholtz — Mask Limpet
Plate 18q

Aleutian Islands to Monterey, California.

1 to 1¾ inches in maximum diameter, with characters much the same as those of *digitalis*, but differing in being smoothish, larger, often slightly higher, and in having a strong tint of blue or blue-black inside. I am inclined to believe that Pilsbry is correct in considering *digitalis* as a smaller, ribbed form of *persona*, despite the fact that recent workers place these two species in different subgenera. It is possible that colder waters allow the smooth *persona* form to express itself. The Mask Limpet is very common from Monterey north. It is an intertidal dweller where strong waves flush the rock crevices. It feeds mostly during the ebb tide and is more active during dark hours. The small southern subspecies, *strigatella* Carpenter, is about ½ inch in size, dark gray-blue inside, and externally with a mass of intertwining or joining radial bars of brown on a bluish or gray-white background.

Acmaea scabra Gould — Rough Limpet
Plate 18l

Vancouver, B.C., to Lower California.

1¼ inch in maximum diameter, elliptical in outline, generally with a low apex which is placed ⅓ back from the front end. The 15 to 25 strong, coarse radiating ribs give the edge of the shell a strong crenulation. Color dirty gray-green. Underside of shell whitish, irregularly stained in the center with blackish brown. Edge of shell between the serrations is stained blackish to purplish brown. A common species found clinging to rocks high above the water line but within reach of the ocean spray. *A. spectrum* Nuttall is the same species. Do not confuse with the smaller *A. conus* which is evenly glossed, instead of coarse and dull, on its interior center.

Acmaea testudinalis scutum Eschscholtz   Pacific Plate Limpet

Alaska to Oregon (common) to Lower California (rare).

1 to 2 inches in maximum diameter, almost round in outline, quite flat, with the apex toward the center of the shell. Smoothish, except for very fine radial riblets in young specimens. External color greenish gray with slate-gray radial bands or mottlings. Interior bluish white with faint or darkish brown spot. Inner edge with band of alternating bars of black or brown and bluish white. The name of this species was also known as *tessulata* Müller. The typical *testudinalis* from the Arctic Seas and New England rarely, if ever, exceeds a size of 1½ inches, is not so round, and has a darker, more concentrated brown patch on the inside. Intergrades exist in Alaskan waters. The Pacific race was also named *patina* Esch.

Acmaea asmi Middendorff   Black Limpet

Alaska to Mexico, clinging to the gastropod, *Tegula*.

¼ inch in maximum diameter, high-conic, elliptical in outline, and solid black inside and out. In the northern part of its range, the Black Limpet is found living attached to the common snail, *Tegula funebralis* A. Adams.

Acmaea triangularis Carpenter   Triangular Limpet

Southern California to Gulf of California.

¼ inch in maximum diameter, oblong in outline, side view distinctly triangular. Color whitish with 3 or 4 vertical, rather broad, brown stripes on each side. Found among coralline algae from the shore line down to several fathoms. Uncommon.

Acmaea depicta Hinds   Painted Limpet

Santa Barbara, California, to Lower California.

½ inch in maximum diameter, very narrow, 3 times as long as wide. Sides straight with brown vertical stripes on a whitish background. Smoothish. This species is found on the broad-leaved eel-grass of the estuaries. Abundant in certain localities, such as Mission Bay.

Acmaea instabilis Gould   Unstable Limpet
Plate 18zz

Alaska to San Diego, California.

1 to 1¼ inches in maximum diameter, oblong with a rather high apex. Sides compressed. Lower edge curved so that the shell rocks back and forth if put on a flat surface. Exterior dull, light-brown. Interior whitish with faint brown stain in the center and with a narrow, solid border of brown. Inhabits the stems or holdfasts of large seaweeds. Moderately common.

Acmaea insessa Hinds   Seaweed Limpet
Plate 18z

Alaska to Lower California.

½ to ¾ inch in maximum diameter, narrowly elliptical, with a high apex, and colored a uniform, greasy light-brown. Abundant on the stalks or holdfasts of the large seaweeds, such as *Egregia*.

Acmaea paleacea Gould   Chaffy Limpet

Vancouver, B.C., to Lower California.

¼ inch in maximum diameter, very fragile, translucent-brown, 3 or 4 times as long as wide. Sides straight with fine, raised radial threads. Abundant on the narrow-leaved eel-grass of the open coast.

Atlantic Coast Species

Acmaea testudinalis testudinalis Müller   Atlantic Plate Limpet

Arctic Seas to Long Island Sound, New York.

1 to 1½ inches in maximum diameter, oval in outline, moderately high with the apex nearly at the center of the shell. Smoothish except for a few coarse growth lines and numerous, very fine axial threads. Interior bluish white with a dark- to light-brown center and with short, radial brown bars at the edge. Exterior dull cream-gray with irregular axial bars and streaks of brown. A common littoral species in New England. Formerly referred to as *A. tessulata* Müller. The form *alveus* Conrad is a thin, elongate, heavily mottled ecological variant which lives on eel-grass.

Acmaea antillarum Sowerby

Antillean Limpet
Plate 17a

South half of Florida and the West Indies.

¾ to 1 inch in maximum diameter, usually very flat, rather thin, oval in outline but narrower at the anterior end. Neatly sculptured with numerous radial threads. Color variable: exterior whitish with a few or many narrow or wide radial rays of brownish green. Interior glossy whitish with a dark- or light-brown callus. Borders or sometimes the entire inside marked by numerous radial lines of purple-brown. These are often divided near the edge of the shell. Uncommon in Florida, but abundant in the West Indies. *A. candeana* Orbigny and *A. tenera* C. B. Adams are the same.

Acmaea pustulata Helbling

Spotted Limpet

Southeast Florida, the West Indies and Bermuda.

1 inch in maximum diameter, oval in outline, moderately flat with rounded sides. Shell thick, with coarse axial ribs which are crossed by fine concentric threads. Interior glossy-white, with the central callus yellowish. Exterior chalk-white, dull. Sometimes flecked with red-brown dots and bars. Common. Formerly known as *punctulata* Gmelin. A deep-water form, which is perhaps a young phase, of this species is very thin, light-rose in color, with a tiny, sharp apex and is occasionally flecked with red. It may be called *A. pustulata pulcherrima* Guilding.

Acmaea leucopleura Gmelin

Dwarf Suck-On Limpet
Plate 17b

Southeast Florida and the West Indies.

⅓ to ½ inch in maximum diameter, high-conic, with numerous, alternating black and white rays. The black rays divide into two near the edge of the shell. Radial riblets weak, usually black. Interior white, often stained brown or black on the callus. Frequently found adhering to the underside of large gastropods such as *Livona pica*. Common. *A. cubensis* Reeve and *A. simplex* Pilsbry are probably this species.

Acmaea jamaicensis Gmelin

Jamaica Limpet
Plate 17c

Southeast Florida and the West Indies.

½ inch in maximum diameter, moderately high, with roundish sides, thick, with about 15 to 20 rather large, rounded, white radial ribs on a black-brown background. Sometimes completely white. Interior white, occasionally with a black-spotted edge and with a thickened central callus which is light-brown to black. *A. albicosta* C. B. Adams and *A. fungoides* Röding are the same. Moderately common in the West Indies, occasionally found on the Lower Florida Keys.

Family *LEPETIDAE*
GENUS *Lepeta* Gray 1842

Small, flattish, uncoiled shells which are "hat-shaped," similar to *Acmaea*, but the embryonic nucleus is spiral; the animal has no external gills and the proboscis is produced into a labial process on each side. The radula has a median tooth, which in *Acmaea* is absent.

Lepeta caeca Müller   **Northern Blind Limpet**
                                                       Plate 17j

Arctic Seas to Cape Cod, Massachusetts.

¼ to ½ inch in maximum diameter, moderately conic, with straight sides, oval-elongate in outline. Rather fragile, dull-white to brownish externally and with fine, granulose, crowded, radial threads. Interior white or tinged with pink. Apex usually eroded. A common cold-water species often dredged in shallow water off New England.

Superfamily *TROCHACEA*
Family *TROCHIDAE* (Top Shells)
Subfamily *MARGARITINAE*
GENUS *Margarites* Gray 1847
Subgenus *Margarites* s. str.

Margarites costalis Gould   **Northern Rosy Margarite**
                                                       Plate 17t

Greenland to Cape Cod, Massachusetts. Bering Strait to Port Etches, Alaska.

¼ to ⅜ inch in length, a little wider, with 5 evenly and well-rounded whorls. Narrowly and deeply umbilicate. Angle of spire about 90 degrees. Next to last whorl with 10 to 12 smoothish, raised, spiral threads. Columella and outer lip thin, sharp, the latter finely crenulate. Color rosy to grayish cream. White within the smoothish umbilicus. Aperture pearly-rose. Commonly dredged from 10 to 62 fathoms. *M. groenlandicus* Möller is the same. Formerly known as *M. cinereus* Couthouy.

Margarites groenlandicus Gmelin

Greenland Margarite
Figure 31d

Arctic Seas to Massachusetts Bay.

½ inch in length, ¾ inch in width. Angle of spire 110 degrees. Whorls strongly rounded, aperture round, umbilicus wide and deep. Outer lip and columella very thin. Base smooth; top of whorls with about a dozen smooth spiral lirations or almost entirely smooth (form *umbilicalis* Broderip and Sowerby). Nucleus glassy smooth. Suture finely impressed. Color glossy-cream to tan. Aperture pearly. Commonly dredged from 5 to 150 fathoms.

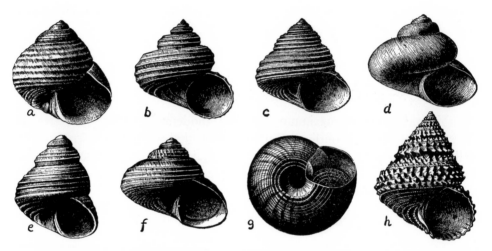

FIGURE 31. American Margarites. **a,** *Margarites succinctus* Cpr., ⅛ inch (Pacific); **b,** *Solariella peramabilis* Cpr., ½ inch (Pacific); **c,** *Margarites pupillus* Gould, ½ inch (Pacific); **d,** *M. groenlandicus* Gmelin, ½ inch (northern Atlantic); **e,** *M. lirulatus* form *parcipictus* Cpr., ¼ inch (Pacific); **f** and **g,** *Solariella obscura* Couthouy, ¼ inch (Atlantic); **h,** *Lischkeia cidaris* Cpr., 1 inch (Pacific).

Margarites lirulatus Carpenter

Lirulate Margarite
Figure 31e

Santa Barbara, California, to the Coronado Islands.

¼ inch in length, 4 to 5 whorls, strong, semi-glossy. Very variable in color (solid purple, whitish with dark-brown variegations and sometimes with a spiral row of dark squares on the periphery), and variable in the number and strength of the small, smooth, spiral cords. Base rounded. Umbilicus narrow but deep. Suture well-impressed. Interior iridescent. Common in shallow water. *M. parcipictus* Cpr. and *obsoletus* Cpr. are forms of this species.

Margarites succinctus Carpenter　　　　　　Tucked Margarite
Figure 31a

Alaska to Lower California.

⅛ inch in length, 4 whorls, slightly wider than long, smoothish except for microscopic, weak threads or incised lines. Umbilicus small, round, deep. Exterior grayish brown, commonly with microscopic, brown, spiral lines. Aperture dark-greenish iridescent. Littoral on algae; common.

Subgenus *Pupillaria* Dall 1909

Margarites pupillus Gould　　　　　　Puppet Margarite
Figure 31c

Bering Sea to San Pedro, California.

⅓ to ½ inch in length, whorls 5 to 6, upper whorls with 5 to 6 smoothish, small, spiral threads, between or over which are microscopic, axial, slanting threads. Umbilicus a minute chink. Exterior dull, chalky whitish to yellowish gray. Aperture rosy to greenish pearl. Apex usually eroded. A common littoral species in the northern half of its range. Also dredged in 50 fathoms.

GENUS *Lischkeia* P. Fischer 1879
Subgenus *Turcicula* Dall 1881

Lischkeia bairdi Dall　　　　　　Baird's Spiny Margarite
Plate 3c

Bering Sea to Coronado Islands, Mexico.

2 inches in length, moderately fragile, sculptured with varying number of spiral rows of fairly large beads. No umbilicus. Shell white with a thin, glossy, yellowish-green periostracum. Interior of aperture pearly-white. This is a choice deep-water species much sought after by collectors. Moderately common in 100 to 600 fathoms.

Subgenus *Cidarina* Dall 1909

Lischkeia cidaris Carpenter　　　　　　Adams' Spiny Margarite
Figure 31h

Alaska to Lower California.

1 to 1½ inches in length, moderately solid, similar to *L. bairdi* but with a higher, flat-sided spire. The suture is usually more impressed. Color gray to grayish white. Moderately common from 20 to 350 fathoms.

Subgenus *Calliotropis* Seguenza 1903

Lischkeia ottoi Philippi   Otto's Spiny Margarite

Nova Scotia to North Carolina.

⅝ to ¾ inch in length, equally wide. Moderately thin with a sharp lip. The round, narrow umbilicus is partially covered by the top of the columella. Color pearly-white. Sculpture of whorls in spire with 3 evenly spaced spiral rows of prickly beads. Suture wavy. Base of shell with 4 to 5 spiral threads which bear smaller, often obscure, beads. Nuclear whorls with axial lamellae. *S. regalis* Verrill and Smith is the same species. Common from 50 to 100 fathoms.

GENUS *Solariella* Wood 1842
(*Machaeroplax* Friele 1877)

Solariella obscura Couthouy   Obscure Solarelle
Figure 31f, g

Labrador to off Chesapeake Bay, Virginia.

¼ inch in length, similar to *Margarites costalis*, but with whorls made more angular by one large, feebly beaded, spiral cord above the periphery. Base smoothish except for microscopic, spiral scratches. Umbilicus narrower and bordered by an angular rim. Color grayish to pinkish tan, often worn to reveal a pearly-golden color. Aperture pearly-white. Some specimens may have weak axial riblets below the strongest spiral cord on the periphery of the whorl. Commonly dredged from 3 to 400 fathoms, especially on the Grand Banks.

Solariella lacunella Dall   Channeled Solarelle
Figure 32b, c

Virginia to Key West, Florida.

⅜ inch in length, equally wide, thick, pure white. Whorls convex. Aperture circular, internally pearly. Suture channeled. Whorls with 6 spiral cords, bottom 3 smooth, the upper ones axially beaded. Nuclear whorls glassy, with microscopic axial ribs. Umbilicus round, narrow, deep, lined with spiral rows of coarse beads. Very commonly dredged from 18 to 100 fathoms.

Solariella lamellosa Verrill and Smith   Lamellose Solarelle
Plate 17x

Massachusetts to Key West, Yucatan and the West Indies.

⅛ inch in length, similar to *S. lacunella*, but with a much deeper channel at the suture below which are numerous, small axial, short lamellar-like ribs.

Middle of whorl with a strong, sharp, smooth or beaded, spiral thread. Base of shell smoothish except for one smooth spiral thread near the periphery and one heavily beaded cord bordering the deep, round umbilicus. Entire shell with numerous microscopic incised lines. Very commonly dredged from 35 to 150 fathoms, but also recorded from 683 fathoms.

Solariella peramabilis Carpenter   Lovely Pacific Solarelle
                                                                   Figure 31b

Alaska to San Diego, California.

½ to ¾ inch in length, equally wide, solid, semi-gloss. Aperture circular. Umbilicus fairly wide, round, very deep. Whorls 7, shouldered just below the suture by a flat shelf. Lower ⅔ of whorl with numerous weak spiral cords that are smoothish in the last whorl but crossed by numerous axial riblets in the early whorls. Color tan with light-mauve stains and mottlings. Interior iridescent. Moderately common from 20 to 339 fathoms.

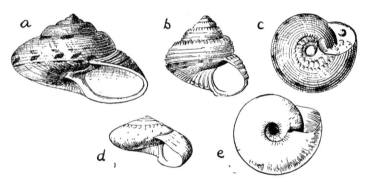

FIGURE 32. **a**, *Gaza watsoni* Dall (nat. size); **b** and **c**, *Solariella lacunella* Dall X4; **d** and **e**, *Microgaza rotella* Dall X4. All from the Atlantic.

GENUS *Microgaza* Dall 1881

Microgaza rotella Dall   Dall's Dwarf Gaza
                                                                   Figure 32d, e
North Carolina to south Florida and the West Indies.

¼ inch in diameter, spire flat, surface smooth except for a spiral row of low pimples just below the suture. Whorls about 5. Umbilicus fairly wide, very deep, its squarish edge bearing numerous, neat, rounded creases. Columella straight. Color whitish gray with a beautiful opalescent sheen, especially inside the aperture. Top of whorls colored with chestnut, zebra-like axial stripes. Very commonly brought up in dredging hauls off Miami from 50 to 100 fathoms. The form *inornata* Dall lacks the pimples just below the suture.

Subfamily *CALLIOSTOMATINAE*
GENUS *Calliostoma* Swainson 1840

Calliostoma euglyptum A. Adams   Sculptured Top-shell
Plate 17w

North Carolina to Florida and Texas.

¾ inch in length, equally wide. Angle of spire about 70 degrees. Sides of whorls slightly concave. Periphery well-rounded. No umbilicus. Whorls with 6 major, well-beaded, spiral cords between each of which is a much smaller, weakly beaded thread. Color dull-rose, sometimes with axial flammules of cream. Nucleus pink or, when worn, dark purple. Moderately common in some localities from low tide mark to 32 fathoms.

Calliostoma zonamestum A. Adams   Chocolate-lined Top-shell
Plate 3n

Lower Florida Keys and the West Indies.

¾ to 1 inch in length, slightly wider. Angle of spire about 70 degrees. Sides of whorls flat; periphery sharp; base flat. Umbilicus deep, smooth-sided, white. Whorls characterized by 10 spiral, beaded threads between each of which there is a dark-chocolate line. Base olive with about 5 to 6 fine, brown, spiral lines. A very beautiful and moderately rare species much sought after by collectors.

Calliostoma roseolum Dall   Dall's Rosy Top-shell

North Carolina to both sides of Florida and Yucatan.

½ inch in length, ¾ as wide. Angle of spire about 50 degrees. Sides of whorls well-rounded, and with 8 to 9 crowded spiral rows of numerous neat beads. Columella upright, strong, with a slight twist. Color of shell light orange-tan to cream, often with arched splotches of darker color running axially across the whorl. No umbilicus. Aperture pearly-rose. Uncommon from 12 to 100 fathoms.

Calliostoma pulchrum C. B. Adams   Beautiful Top-shell

North Carolina to Florida, the Gulf of Mexico, and the West Indies.

⅜ inch in length, ¾ as wide. Angle of spire about 50 degrees. Sides of whorls straight. Characterized by a pair of strong, spiral cords just above the suture which are white with distantly spaced red-brown dots. Rest of whorl pearly-green with 6 to 7 very weak (or sometimes strong) beaded spiral threads. Columella almost upright, its inner side rounded, pearly. No umbilicus. Moderately common from 1 to 40 fathoms.

Calliostoma jujubinum Gmelin   Jubjube Top-shell
Plate 3p

Lower Florida Keys, the Bahamas and the West Indies.

½ to 1¼ inches in length. Characterized by the deep, narrow, smooth-sided umbilicus which is bordered by a spiral, beaded thread, and by the swollen, rounded periphery of each whorl, which in the spire is located just above the suture. Color ranges from brownish cream to reddish and is often maculated with white splotches near the periphery. Typical *jujubinum* has a spire angle of about 50 degrees; the spiral threads on the whorls are weakly beaded, and the umbilicus is almost closed.

C. *jujubinum tampaense* Conrad (North Carolina to both sides of Florida to Yucatan) varies in spire angle from 50 to 65 degrees, is not always so swollen at its periphery, and has 9 to 10 well-beaded spiral threads between each suture.

Calliostoma occidentale Mighels and Adams   North Atlantic Top-shell

Nova Scotia to Cape Cod, Massachusetts. Europe.

½ inch in length, equally wide. Whorls convex and with 3 to 4 strong spiral cords, the 2 lower ones smooth, the upper one beaded. Color pearly-white. No umbilicus. Outer lip fragile. Moderately common from 10 to 365 fathoms.

Calliostoma bairdi Verrill and Smith   Baird's Top-shell
Plate 3–o

Massachusetts to North Carolina (and to Florida).

1 to 1¼ inches in length, about as wide. Angle of spire about 70 degrees. Sides of spire straight to slightly convex. Base rather flat. Periphery angular. Sculpture of 6 to 7 spiral rows of small, neat beads, with those on the topmost row being the largest. Suture difficult to find. No umbilicus. Color brownish cream with faint maculations of light reddish. Not uncommon from 43 to 250 fathoms.

C. *bairdi psyche* Dall (North Carolina to Key West, 30 to 130 fathoms) is usually ¾ inch in length, slightly wider, with a spire angle of about 75 to 80 degrees, and the color is lighter and more pearly. Base with 3 or 4 spiral brown lines. It has a chink-like depression beside the umbilicus. Uncommon. C. *subumbilicatum* Dall is a form of this species whose umbilicus is half open.

Calliostoma tricolor Gabb   Three-colored Top-shell
Figure 33e

Moss Beach, California, to Cape San Lucas, Mexico.

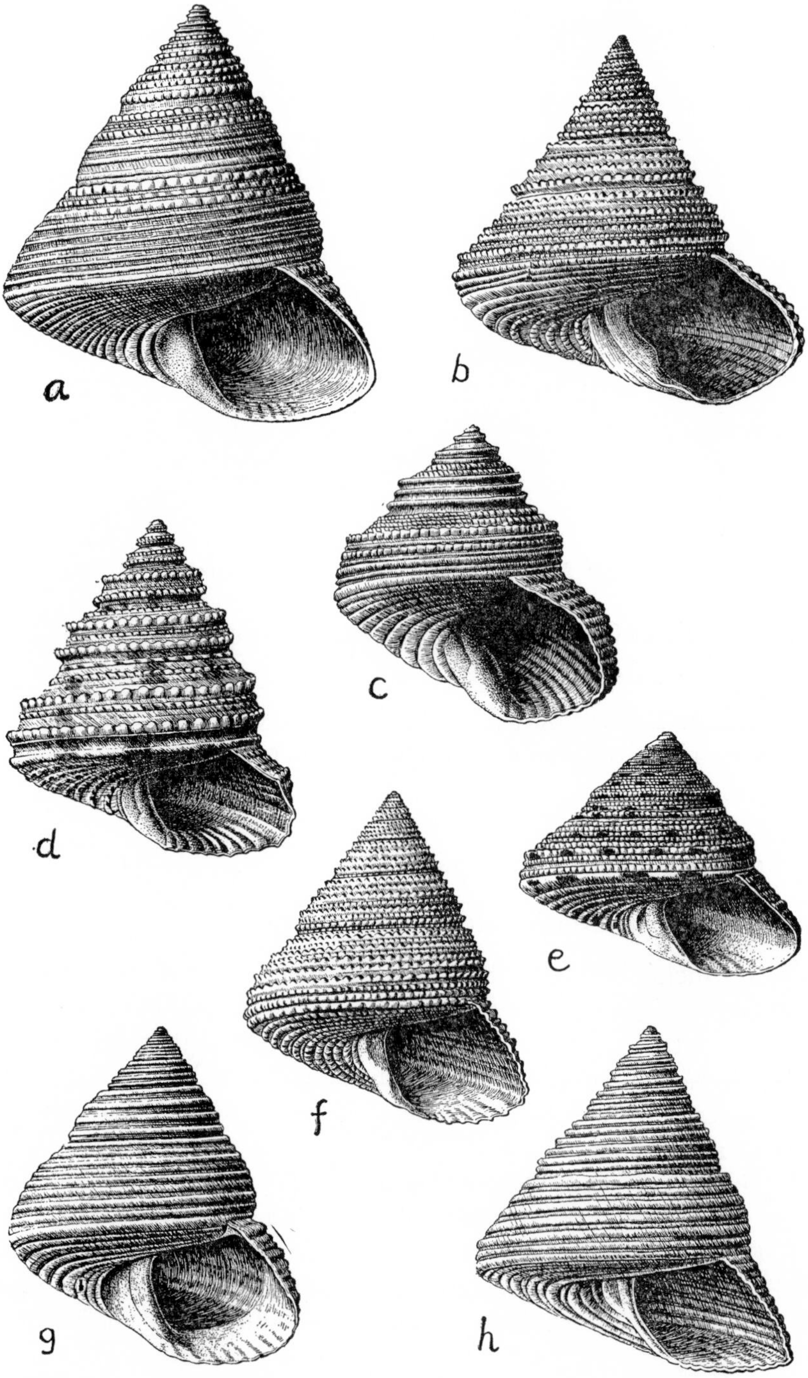

Figure 33. Pacific *Calliostoma*. **a,** *gloriosum* Dall; **b,** *variegatum* Cpr.; **c,** *splendens* Cpr.; **d,** *gemmulatum* Cpr.; **e,** *tricolor* Gabb; **f,** *annulatum* Sol.; **g,** *ligatum* Gld. (*costatum*); **h,** *canaliculatum*. All about natural size. (From Dall 1901.)

PLATE 1 — THE FIVE CLASSES OF MOLLUSKS

GASTROPODA
(a) Turnip Whelk, p. 237.

PELECYPODA
(b) Calico Clam, p. 416.

CEPHALOPODA
(c) Paper Nautilus, p. 485.

AMPHINEURA
(d) West Indian Chiton, p. 324.

SCAPHOPODA
(e) Dall's Pacific Tusk, p. 327.

PLATE 2

PACIFIC COAST ABALONES

a. RED ABALONE, *Haliotis rufescens* Swainson, 10 to 12 inches. Left, artificially polished; right, natural (Northern California to Mexico), p. 92.

b. GREEN ABALONE, *H. fulgens* Philippi, 7 to 8 inches. Left, exterior; right, interior (California to Mexico), p. 93.

c. PINK ABALONE, *H. corrugata* Gray, 5 to 7 inches (Monterey, California to Mexico), p. 93.

d. THREADED ABALONE, *H. assimilis* Dall, 4 to 5 inches (California), p. 93.

e. JAPANESE ABALONE, *H. kamtschatkana* Jonas, 4 to 6 inches (Japan, southern Alaska to northern California), p. 94.

f. BLACK ABALONE, *H. cracherodi* Leach, 4 to 6 inches (California to Mexico), p. 92.

PLATE 3

TURBANS, TOP-SHELLS, STAR-SHELLS

a. CHANNELED TURBAN, *Turbo canaliculatus* Hermann, 2 inches (Southeastern Florida and the West Indies), p. 123.

b. SUPERB GAZA, *Gaza superba* Dall. 1 inch (Gulf of Mexico), p. 118.

c. BAIRD'S SPINY MARGARITE, *Lischkeia bairdi* Dall, 2 inches (Pacific Coast), p. 109.

d. ADANSON'S PLEUROTOMARIA, *Perotrochus adansonianus* Crosse and Fischer, 3 inches (West Indies), p. 92.

e. GREENISH TURBAN, *Astraea olivacea* Wood, 1½ inches (Gulf of California), not in text.

f. QUEEN TEGULA, *Tegula regina* Stearns, 1½ inches (Southern California), p. 120.

g. CHESTNUT TURBAN, *Turbo castaneus* Gmelin, 1½ inches (Southeastern United States and the West Indies), p. 123.

h. CARVED STAR-SHELL, *Astraea caelata* Gmelin, 2 inches (Southeastern United States and the West Indies), p. 124.

i. AMERICAN STAR-SHELL, *Astraea americana* Gmelin, 1½ inches (Southeastern Florida), p. 124.

j. GREEN STAR-SHELL, *Astraea tuber* Linné, 1½ inches (Southeastern United States and the West Indies), p. 124.

k. and m. LONG-SPINED STAR-SHELL, *Astraea longispina* Lamarck, 2 inches (Southeastern United States and the West Indies), p. 123.

l. SHORT-SPINED STAR-SHELL, *Astraea brevispina* Lamarck, 2 inches (West Indies only), p. 123.

n. CHOCOLATE-LINED TOP-SHELL, *Calliostoma zonamestum* A. Adams, 1 inch (Florida Keys and the West Indies), p. 112.

o. BAIRD'S TOP-SHELL, *Calliostoma bairdi* Verrill and Smith, 1 inch (Massachusetts to Florida), p. 113.

p. JUJUBE TOP-SHELL, *Calliostoma jujubinum* Gmelin, 1 inch (Florida Keys and the West Indies), p. 113.

q. CHANNELED TOP-SHELL, *Calliostoma canaliculatum* Solander 1786, 1 inch (Pacific Coast), p. 115.

PLATE 4

NERITES, PURPLE SEA-SNAILS, SUN-DIALS

a. BLEEDING TOOTH, *Nerita peloronta* Linné, 1 inch (Southeastern Florida and the West Indies), p. 128.

b. FOUR-TOOTHED NERITE, *Nerita versicolor* Gmelin, ¾ inch (Florida and the West Indies), p. 128.

c. ANTILLEAN NERITE, *Nerita fulgurans* Gmelin, ¾ inch (Southeastern Florida and the West Indies), p. 129.

d. ROUGH-RIBBED NERITE, *Nerita scabricosta* Lamarck, 1 inch (Pacific side of Central America), not in text.

e. ZEBRA NERITE, *Puperita pupa* Linné, ½ inch (Southeastern Florida and the West Indies), p. 129.

f. TESSELLATE NERITE, *Nerita tessellata* Gmelin, ¾ inch (Southeastern United States and the West Indies), p. 128.

g. OLIVE NERITE, *Neritina reclivata* Say, ½ inch (Southeastern United States and the West Indies), p. 129.

h. EMERALD NERITE, *Smaragdia viridis* Linné, ¼ inch (Southeastern Florida and the West Indies), p. 130.

i. VIRGIN NERITE, *Neritina virginea* Linné, ½ inch, 6 color phases (Southeastern United States and the West Indies), p. 129.

j. COMMON PURPLE SEA-SNAIL, *Janthina janthina* Linné, 1 inch (Pelagic, warm seas), p. 160.

k. GLOBE PURPLE SEA-SNAIL, *Janthina globosa* Swainson, ¾ inch (Pelagic, warm seas), p. 160.

l. DWARF PURPLE SEA-SNAIL, *Janthina exigua* Lamarck, ¼ inch (Pelagic, warm seas), p. 160.

m. COMMON SUN-DIAL, *Architectonica nobilis* Röding, 1½ inches (Southeastern United States and the West Indies), p. 142.

n. KEELED SUN-DIAL, *Architectonica peracuta* Dall, ¾ inch (Southeastern Florida and the West Indies), p. 143.

o. KREBS' SUN-DIAL, *Architectonica krebsi* Mörch, ½ inch (Southeastern United States and the West Indies), p. 143.

p. HENDERSON'S NISO, *Niso hendersoni* Bartsch 1953, 1 inch (Southeastern United States), not in text, Holotype.

q. GIANT ATLANTIC PYRAM, *Pyramidella dolabrata* Lamarck, 1 inch (Bahamas and the West Indies), p. 289.

r. FLAMINGO TONGUE, *Cyphoma gibbosum* Linné, ¾ inch (Southeastern United States and the West Indies), p. 183. Also see plate 8.

s. McGINTY'S CYPHOMA, *Cyphoma mcgintyi* Pilsbry, ¾ inch (Florida), p. 184.

t. FINGERPRINT CYPHOMA, *Cyphoma signatum* Pilsbry and McGinty, 1 inch (Florida), p. 184.

PLATE 5

CONCHS, TRITONS, AND MOON-SHELLS

a. WOOD'S PANAMA CONCH, *Strombus granulatus* Wood, 2½ inches (Gulf of California to Panama), not in text.

b. ATLANTIC CARRIER-SHELL, *Xenophora conchyliophora* Born, 2 inches (Southeastern United States and the West Indies), p. 173.

c. HAWK-WING CONCH, *Strombus raninus* Gmelin, 3 inches (Southeastern Florida and the West Indies), p. 175.

d. ANGULAR TRITON, *Cymatium femorale* Linné, 6 inches (Southeastern Florida and the West Indies), p. 195.

e. ROOSTER-TAIL CONCH, *Strombus gallus* Linné, 5 inches (Southeastern Florida and the West Indies), p. 175.

f. TRUMPET TRITON, *Charonia tritonis nobilis* Conrad, 14 inches (Southeastern Florida and the West Indies), p. 196.

g. WEST INDIAN FIGHTING CONCH, *Strombus pugilis* Linné, 3 inches (Southeastern Florida and the West Indies), p. 173.

h. FLORIDA FIGHTING CONCH, *Strombus alatus* Gmelin, 3 inches (Atlantic Coast from North Carolina to Texas), p. 174.

i. PANAMA FIGHTING CONCH, *Strombus gracilior* Sowerby, 3 inches (Gulf of California to Panama), not in text.

j. BROWN MOON-SHELL, *Polinices brunneus* Link, 1½ inches, (Southeastern Florida and the West Indies), p. 186.

k. SHARK EYE, *Polinices duplicatus* Say, 2 inches (Atlantic Coast), p. 186.

l. COLORFUL ATLANTIC NATICA, *Natica canrena* Linné, 1½ inches (Southeastern United States and the West Indies), p. 191.

m. EXCAVATED NATICA, *Stigmaulax elenae* Recluz, 1½ inches (Pacific side of Panama), not in text.

Plate 7

SEA-WHIP SNAILS

a. COMMON WEST INDIAN SIMNIA, *Neosimnia acicularis* Lam., ½ inch. Yellow phase on Yellow Sea-whip, *Leptogorgia virgulata* Lam., and lavender phase on Purple Sea-whip, *Leptogorgia hebes* Verrill (North Carolina to the West Indies), p. 182.

b. and c. CALIFORNIAN PEDICULARIA, *Pedicularia californica* Newcomb, ½ inch. Attached to the hydrocoralline, *Allopora californica* Verrill. c is the heavy form, *ovuliformis* Berry (Southern California) p. 182.

d. DECUSSATE PEDICULARIA, *Pedicularia decussata* Gould, ½ inch (Georgia to the West Indies), p. 181.

e. SINGLE-TOOTHED SIMNIA, *Neosimnia uniplicata* Sby., ½ inch (Virginia to the West Indies), p. 182.

f. DALL'S TREASURED SIMNIA, *Neosimnia piragua* Dall, 1 inch. Holotype (West Indies), p. 182.

g. WESTERN CHUBBY SIMNIA, *Neosimnia avena* Sby., ½ inch (Monterey, California, to Panama), p. 183.

h. INFLEXED SIMNIA, *Neosimnia inflexa* Sby., ½ inch (Monterey, California, to Panama), p. 183.

i. LOEBBECK'S SIMNIA, *Neosimnia loebbeckeana* Weink., ¾ inch (Monterey, California, to Gulf of California, form *barbarensis* Dall), p. 183.

j. PANAMA CYPHOMA, *Cyphoma emarginata* Sby., (4 specimens), 1 inch (Lower California to Panama), not in text.

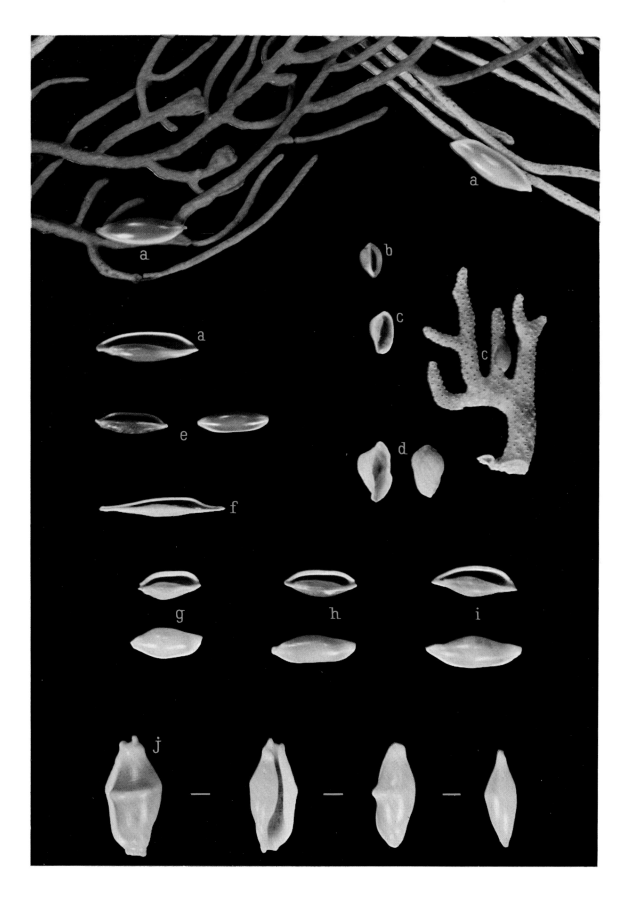

PLATE 8

Living Flamingo Tongues on the Rough Sea-whip, *Muricea muricata* Pallas.

Upper right: McGinty's Cyphoma, *Cyphoma mcgintyi* Pilsbry, 1 inch (Southeastern Florida), p. 184.

Lower Three: Flamingo Tongue, *Cyphoma gibbosum* Linné, 1 inch (North Carolina to West Indies), p. 183.

PLATE 9

BONNETS, TUNS AND FROG-SHELLS

a. ROUGH PANAMA HELMET, *Cypraecassis coarctata* Wood, 3 inches (Gulf of California, south), not in text.

b. LARGE PANAMA HELMET, *Cypraecassis tenuis* Wood, 5 inches (Pacific side of Panama), not in text.

c. RETICUATED COWRIE-HELMET, *Cypraecassis testiculus* L., 3 inches (Florida, south), p. 194.

d. ATLANTIC PARTRIDGE TUN, *Tonna maculosa* Dill., 3 inches (Southeastern Florida and the West Indies), p. 199.

e. SCOTCH BONNET, *Phalium granulatum* Born, 3 inches (North Carolina to the West Indies), p. 192.

f. SMOOTH SCOTCH BONNET, *Phalium cicatricosum* Gmelin, 2 inches (Southeastern Florida and the Caribbean), p. 193.

g. PEAR WHELK, *Busycon spiratum* Lam., 4 inches (North Carolina to Mexico), p. 236.

h. ROYAL BONNET, *Sconsia striata* Lam., 2½ inches (Gulf of Mexico, south), p. 192.

i. COMMON FIG SHELL, *Ficus communis* Röding, 3 inches (North Carolina to Gulf States), p. 200.

j. DOG HEAD TRITON, *Cymatium cynocephalum* Lam., 2 inches (Southeastern Florida and West Indies), p. 196.

k. GAUDY FROG-SHELL, *Bursa corrugata* Perry, 3 inches (Southeastern Florida, south; Lower California to Ecuador), p. 198.

l. ATLANTIC HAIRY TRITON, *Cymatium martinianum* Orb., 2½ inches (North Carolina to the West Indies), p. 195.

PLATE 10

MUREX SHELLS AND ROYAL TYRIAN PURPLE

a. WEST INDIAN MUREX, *Murex brevifrons* Lam., 4 inches (Southeastern Florida and the West Indies), p. 203.

b. GIANT EASTERN MUREX, *Murex fulvescens* Sby., 6 inches (North Carolina to Texas), p. 203.

c. REGAL MUREX, *Murex regius* Swainson, 6 inches (Gulf of California to Panama), not in text.

d. BEAU'S MUREX, *Murex beaui* Fischer and Bern., 4 inches (Gulf of Mexico and the West Indies), p. 202.

e. LACE MUREX, *Murex florifer* Reeve, 3 inches (Florida and the West Indies), p. 203.

f. ANTILLEAN MUREX, *Murex antillarum* Hinds, 3 inches (West Indies) not in text.

g. BEAU'S MUREX, *Murex beaui* F. and B., 5 inches. Deep-water form (Florida and West Indies), p. 202.

h. CABRIT'S MUREX, *Murex cabriti* Bernardi, 2 inches (Florida and the West Indies), p. 201.

i. BANDED DYE MUREX, *Murex trunculus* L., 3 inches (Mediterranean Sea), p. 12.

j. SPINY DYE MUREX, *Murex brandaris* L., 3 inches (Mediterranean Sea and West Africa), p. 12.

k. CABBAGE MUREX, *Murex brassica* Lam., 7 inches (West Mexico), not in text.

l. APPLE MUREX, *Murex pomum* Gmelin, 3 inches (North Carolina to the West Indies), p. 202.

m. PINK-MOUTHED MUREX, *Murex erythrostomus* Swainson, 6 inches (Gulf of California to Panama), not in text.

n. to q. Strips of paper dyed with Royal Tyrian Purple from the Mediterranean Sea and France. Various shades were obtained by the ancients by varying the concentration of snail dye, the number of dips and the species of snail. n. Light dipping from *Murex trunculus*. o. Light dipping from *Murex brandaris*. p. Heavy concentration from *Murex brandaris* and *Thais haemastoma*. q. Frequent dips in a heavy bath from *Murex trunculus* and *Thais lapillus* (see p. 12).

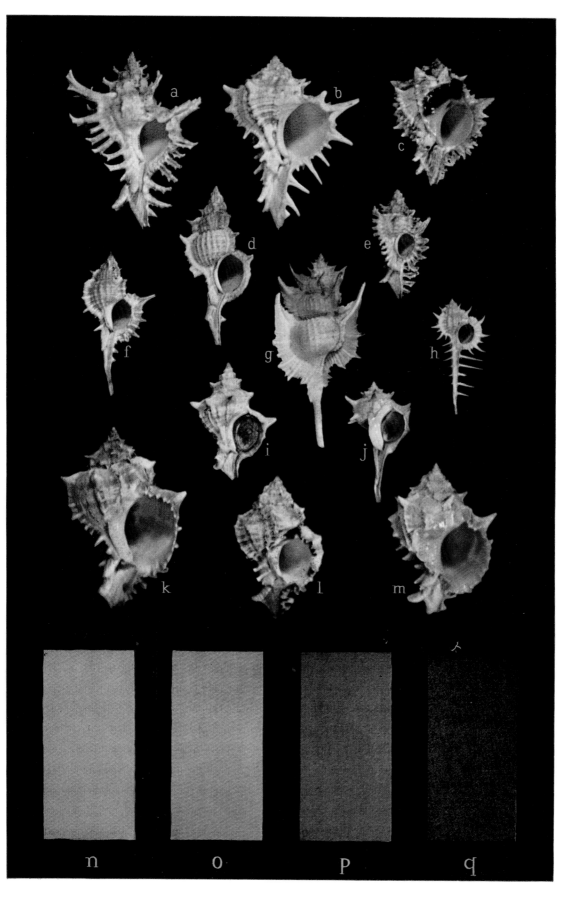

PLATE 11

SPINDLES, DWARF OLIVES AND MARGINELLAS

a. BROWN-LINED LATIRUS, *Latirus infundibulum* Gmelin, 3 inches (Florida, Keys and West Indies), p. 241.

b. MCGINTY'S LATIRUS, *Latirus mcgintyi* Pils., 2 inches (Southeastern Florida), p. 241.

c. ORNAMENTED SPINDLE, *Fusinus eucosmius* Dall., 3 inches (Gulf of Mexico), p. 243.

d. CHESTNUT LATIRUS, *Leucozonia nassa* Gmelin, 1½ inches (Florida to Texas, south), p. 240.

e. WHITE-SPOTTED LATIRUS, *Leucozonia ocellata* Gmelin, 1 inch (West Florida to West Indies), p. 241.

f. SHORT-TAILED LATIRUS, *Latirus brevicaudatus* Reeve, 2 inches (Florida, Keys and West Indies), p. 241.

g. TURNIP SPINDLE, *Fusinus timessus* Dall, 3 inches (Gulf of Mexico), p. 243.

h. and j. WEST INDIAN DWARF OLIVE, *Olivella nivea* Gmelin, ½ inch (Southeastern Florida and West Indies), p. 246.

i. JASPER DWARF OLIVE, *Olivella jaspidea* Gmelin, $^1/_3$ inch (Southeastern Florida and West Indies), p. 246.

k. ORANGE MARGINELLA, *Prunum carneum* Storer, ¾ inch (Southeastern Florida and West Indies), p. 254.

l. ROYAL MARGINELLA, *Prunum labiatum* Kiener, 1 inch (off Yucatan, Mexico, south), p. 256.

m. WHITE-SPOTTED MARGINELLA, *Prunum guttatum* Dill., ¾ inch (Southeastern Florida and West Indies), p. 256.

n. COMMON ATLANTIC MARGINELLA, *Prunum apicinum* Menke, ½ inch (North Carolina to Texas and West Indies), p. 257.

o. ROOSEVELT'S MARGINELLA, *Prunum roosevelti* Btsch. and Rehd., 1 inch, holotype (Bahamas), p. 254.

p. ORANGE-BANDED MARGINELLA, *Hyalina avena* Val., $^1/_3$ inch (North Carolina to West Indies), p. 258.

PLATE 12

OLIVE SHELLS

a. LETTERED OLIVE, *Oliva sayana* Ravenel, 2½ inches (North Carolina to the Gulf States), p. 245.

b. PANAMA FALSE OLIVE, *Agaronia testacea* Lam., 2 inches (West Central America), not in text.

c. NETTED OLIVE, *Oliva reticularis* Lam., 1½ inches (Southeastern Florida and West Indies), p. 245.

d. ANGULATE OLIVE, *Oliva incrassata* Solander, 2 inches (Gulf of California to Peru), not in text.

e. TENT OLIVE, *Oliva porphyria* L., 3 inches (Gulf of California), not in text.

f. POLPAST OLIVE, *Oliva polpasta* Duclos, 1½ inches (West Mexico), not in text.

g. SPLENDID OLIVE, *Oliva splendidula* Sby., 2 inches (Panama), not in text.

h. VEINED OLIVE, *Oliva spicata* Röding, 2 inches (Lower California to Panama), not in text.

i. PURPLE DWARF OLIVE, *Olivella biplicata* Sby., 1 inch (Washington to Lower California), p. 247.

PLATE 13

ATLANTIC TULIPS, SPINDLES AND VOLUTES

a. FLORIDA HORSE CONCH, *Pleuroploca gigantea* Kiener, 3 inches (young) (North Carolina to Florida). For adult see pl. 23y, p. 242.

b. TRUE TULIP, *Fasciolaria tulipa* L., 4 inches (North Carolina to the West Indies), p. 242.

c. BANDED TULIP, *Fasciolaria hunteria* Perry, 3 inches (North Carolina to Gulf States), p. 242.

d. COUE'S SPINDLE, *Fusinus couei* Petit, 4 inches (off Yucatan, Mexico), not in text.

e. SCHMITT'S VOLUTE, *Scaphella schmitti* Bartsch, 5 inches. Holotype (off south Florida), p. 251.

f. JUNONIA, *Scaphella junonia* Shaw, 5 inches (North Carolina to Texas), p. 250.

g. COMMON MUSIC VOLUTE, *Voluta musica* L., 2 inches (West Indies), p. 250.

h. FLAME TEREBRA, *Terebra taurina* Solander, 5 inches (Florida, the Gulf and West Indies), p. 265.

i. ROYAL FLORIDA MITER, *Mitra florida* Gould, 1½ inches (Florida and West Indies), p. 248.

j. DOHRN'S VOLUTE, *Scaphella dohrni* Sby., form *florida* Cl. and Ag., 3 inches (off Florida and Cuba), p. 251.

k. COMMON NUTMEG, *Cancellaria reticulata* L., 1½ inches (North Carolina to Florida), p. 252.

l. WHITE GIANT TURRET, *Polystira albida* Perry, 4 inches (off Florida, the Gulf of Mexico and West Indies), p. 268.

m. DELICATE GIANT TURRET, *Polystira tellea* Dall, 3 inches (off South-eastern Florida), p. 268.

n. LATIRUS-LIKE VASE, *Vasum (Siphovasum) latiriforme* Rehd. and Abb. 1951. 2 inches. Holotype (off Yucatan, Mexico), not in text.

o. MINIATURE TRITON TRUMPET, *Pisania pusio* L., 1½ inches (Florida and West Indies), p. 233.

p. STRIATE BUBBLE, *Bulla striata* Brug., 1 inch (Gulf of Mexico, south), p. 277.

q. BROWN-LINED PAPER-BUBBLE, *Hydatina vesicaria* Solander, 1 inch (Florida and West Indies), p. 276.

PLATE 14

ATLANTIC CONES

a. CARROT CONE, *Conus daucus* Hwass, 1½ inches (Florida and West Indies), p. 260.

b. JULIA'S CONE, *Conus juliae* Clench, 1½ inches (off south Florida), p. 261.

c. SOZON'S CONE, *Conus sozoni* Bartsch, 3 inches (South Carolina to Gulf of Mexico), p. 261.

d. FLORIDA CONE, *Conus floridanus* Gabb, 1½ inches (North Carolina to Florida), p. 261.

e. DARK FLORIDA CONE, *C. floridanus floridensis* Sby (North Carolina to Florida), p. 261.

f. VILLEPIN'S CONE, *Conus villepini* F. and B. 1½ inches (Gulf of Mexico), p. 263.

g. GOLDEN-BANDED CONE, *Conus aureofasciatus* Rehder and Abbot, 3 inches (Gulf of Mexico), p. 260.

h. SENNOTT'S CONE, *Conus sennottorum* Rehd. and Abb., 1 inch (Gulf of Mexico), p. 261.

i. CLARK'S CONE, *Conus clarki* Rehd. and Abb., 1 inch (off Louisiana), p. 264.

j. STIMPSON'S CONE, *Conus stimpsoni* Dall, 1½ inches (Gulf of Mexico), p. 263.

k. MAZE'S CONE, *Conus mazei* Desh., 2 inches (Gulf of Mexico and West Indies), p. 264.

l. GLORY-OF-THE-ATLANTIC CONE, *Conus granulatus* L., 2 inches (Southeastern Florida, south), p. 264.

m. CROWN CONE, *Conus regius* Gmelin, 3 inches (Florida and West Indies), p. 262.

n. JASPER CONE, *Conus jaspideus* Gmelin, ¾ inch (Florida and West Indies), p. 262.

o. MOUSE CONE, *Conus mus* Hwass, 1 inch (Southeastern Florida and West Indies), p. 262.

p. ALPHABET CONE, *Conus spurius atlanticus* Cl., 3 inches (Florida and Gulf of Mexico), p. 260.

PLATE 15

NEW ENGLAND NUDIBRANCHS

a. RED-FINGERED EOLIS, *Coryphella rufibranchialis* Johnston, 1 inch (Arctic to New York), p. 310.

b. PILOSE DORIS, *Acanthodoris pilosa* Abild., 1 inch (Arctic to Connecticut; Alaska), p. 305.

c. DWARF BALLOON EOLIS, *Eubranchus exiguus* A. and H., $1/5$ inch (Arctic to Massachusetts), p. 309.

d. JOHNSTON'S BALLOON EOLIS, *Tergipes despectus* Johnston, $1/3$ inch (Arctic to New York), p. 309.

e. FROND EOLIS, *Dendronotus frondosus* Ascanius, 2 inches (Arctic to Rhode Island; to Washington), p. 307.

f. ATLANTIC ANCULA, *Ancula cristata* Alder, $1/2$ inch (Arctic to Massachusetts), p. 306.

g. PAPILLOSE EOLIS, *Aeolis papillosa* L., 2 inches (Arctic to Rhode Island; to California), p. 308.

h. PAYNTED BALLOON EOLIS, *Eubranchus pallidus* A. and H., $1/2$ inch (Arctic to Boston), p. 310.

i. YELLOW FALSE DORIS, *Adalaria proxima* A. and H., $1/2$ inch (Arctic to Maine), p. 306.

j. ORANGE-TIPPED EOLIS, *Catriona aurantia* A. and H., $1/2$ inch (Arctic to Connecticut), p. 308.

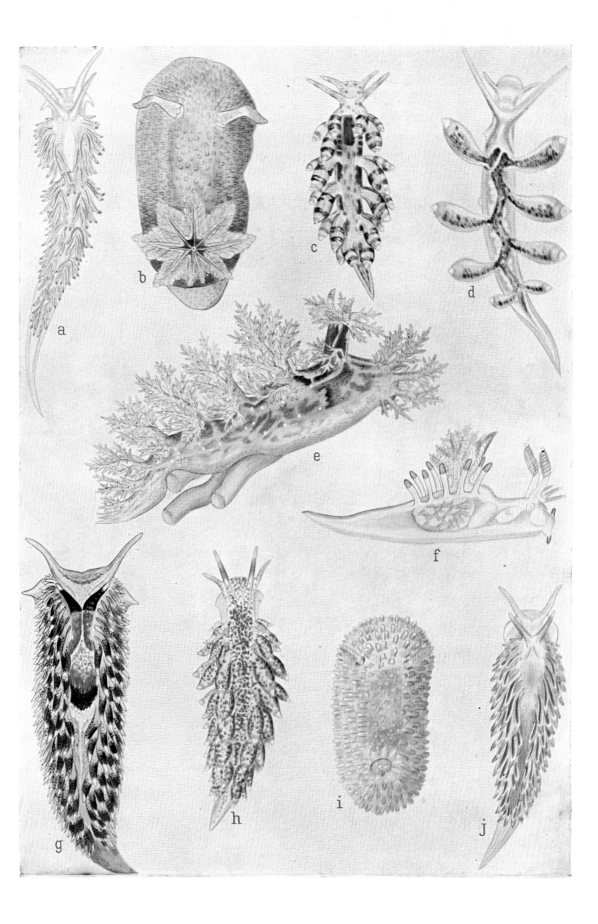

a

b

c

d

e

f

g

h

i

j

PLATE 16

PACIFIC COAST NUDIBRANCHS

a. HOPKINS DORIS, *Hopkinsia rosacea* MacFarland, 1 inch (Monterey to San Pedro, California), p. 307.

b. MACFARLAND'S GRAND DORIS, *Triopha grandis* MacFarland, 3 inches (California), p. 304.

c. NOBLE PACIFIC DORIS, *Archidoris nobilis* MacF., 4 inches (California), p. 300.

d. SAN DIEGO DORIS, *Diaulula sandiegensis* Cooper, 2½ inches (Alaska to California), p. 301.

e. ORANGE-SPIKED DORIS, *Polycera atra* MacF., ¾ inch (California), p. 305.

f. MACULATED DORIS, *Triopha maculata* MacF., 1 inch (California), p. 304.

g. MACFARLAND'S PRETTY DORIS, *Rostanga pulchra* MacF., ¾ inch (California), p. 300.

h. MONTEREY DORIS, *Archidoris montereyensis* Cooper, 1½ inches (California), p. 299.

i. HEATH'S DORIS, *Discodoris heathi* MacF., 1 inch (California), p. 300.

j. LAILA DORIS, *Laila cockerelli* MacF., ¾ inch (California), p. 304.

k. CARPENTER'S DORIS, *Triopha carpenteri* Stearns, 1 inch (California), p. 304.

l. PORTER'S BLUE DORIS, *Glossodoris porterae* Cockerell, ½ inch (California), p. 303.

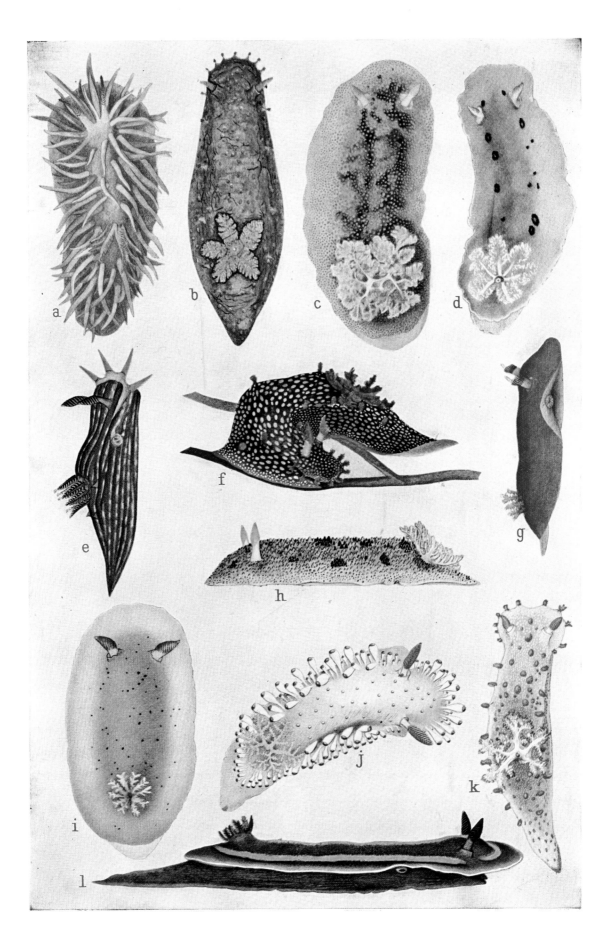

¾ to 1 inch in length, heavy for its size; whorls angular, with the upper third slightly concave to flat and the somewhat angular periphery flattish. Early whorls with minutely beaded threads, later whorls with fine, smoothish cords of various sizes. Nucleus tan to whitish. Color yellowish brown with a few spiral lines of alternating brown and white bars. Sometimes axially variegated. Dredged just offshore from 8 to 35 fathoms. Moderately common.

Calliostoma gemmulatum Carpenter   Gem Top-shell

Figure 33d

Cayucos, California, to the Gulf of California.

¾ inch in length, not as wide; characterized by its dark gray-green color and two extra-strong, beaded spiral cords. There are also 3 or 4 minor cords that are not so heavily beaded. Nucleus dark-tan. Moderately common in the littoral zone on rocks and wharf pilings.

Calliostoma supragranosum Carpenter   Granulose Top-shell

Plate 18s

Monterey, California, to Lower California.

½ inch in length, solid, glossy; characterized by numerous, fine, spiral cords which are sometimes weakly beaded, and by a wide, rather flattish periphery. Nucleus tan. Color light yellowish brown, commonly with a spiral row of subdued white spots at the lower periphery. Interior brightly nacreous. Moderately common on rocks at low tide.

Calliostoma annulatum Solander   Ringed Top-shell

Figure 33f

Alaska to San Diego, California.

1 to 1¼ inch in length, not quite so wide; characterized by its light weight, golden-yellow color with a mauve band at the periphery, and by the numerous, spiral rows of tiny, distinct beads (5 to 9 rows in the spire whorls). Nucleus pink. Dredged offshore and occasionally washed ashore. Formerly *C. annulatum* Martyn.

Calliostoma canaliculatum Solander 1786   Channeled Top-shell

Plate 3q; figure 33h

Alaska to San Diego, California.

1 to 1½ inches in length, not heavy, sides of whorl flat. Periphery of the last whorl sharp. Base of shell almost flat. Characterized by sharp, prominant, slightly beaded, spiral cords. Color yellowish tan. Nuclear whorls white. Moderately common offshore. Found on floating kelp weed. Formerly known as *C. canaliculatum* Martyn, and *doliarium* Holten 1802.

Calliostoma variegatum Carpenter   Variable Top-shell
Figure 33b

Alaska to southern California.

1 inch in length, similar to *doliarium*, but with smaller cords which are strongly beaded; nucleus pink; the sides of the spire slightly concave, and the periphery of the last whorl rounded. Uncommonly dredged in 15 to 400 fathoms.

Calliostoma gloriosum Dall   Glorious Top-shell
Figure 33a

San Francisco to San Diego, California.

1 inch in length, not quite so wide, rather light, with about 10 fine, spiral threads between sutures. The upper 5 are inclined to be minutely beaded. Periphery of last whorl moderately sharp. Columella white, fairly thick and with a swelling at the lower ⅔. Nuclear whorls white. Color of shell yellowish brown with darker purplish brown, slanting and rather elongate spots arranged in 2 spiral series. Moderately common in shallow water.

Calliostoma splendens Carpenter   Splendid Top-shell
Figure 33c

Monterey to Lower California.

¼ to ⅓ inch in length, equally wide, with about 5 to 6 whorls which bear between sutures 5 strong spiral cords. The upper 2 or 3 are finely beaded, the lower 2 or 3 are smooth and cord-like. Between the cords, the shell is brilliant orange-iridescent. General color a yellowish orange with large white maculations on the upper half of the whorls. Moderately common offshore, uncommonly washed ashore.

Calliostoma ligatum Gould   Ribbed Top-shell
Figure 33g

Alaska to San Diego, California.

¾ to 1 inch in length, equally wide, rather heavy; whorls quite well rounded; characterized by smooth, spiral, light-tan cords (6 to 8 on the spire whorls) on a background of chocolate. Sometimes flushed with mauve. No umbilicus. Aperture usually pearly-white. A very common littoral species from northern California north. Formerly *C. costatum* Martyn.

Subfamily GIBBULINAE
GENUS *Livona* Gray 1847

There is only one species in this genus, namely *L. pica* from the West

Indies. Although fairly good specimens are found without their soft parts in southern Florida and Bermuda, this species has been extinct in those areas for several hundred years. Living individuals may be found abundantly in the West Indies where they are used in chowders by some people. *Cittarium* Philippi 1847 is this genus.

Livona pica Linné   West Indian Top-shell

Figure 34

Southeast Florida (dead) and the West Indies (alive).

2 to 4 inches in length, heavy, rather rough, and with splotches of purplish black on dirty-white. Umbilicus round, narrof and very deep. Inner edge of lip with rich cobalt-blue mottlings. Operculum horny, large, round, multispiral and opalescent blue-green in life.

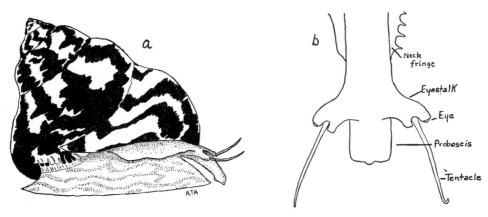

FIGURE 34. West Indian Top-shell, *Livona pica* Linné. a, shell with animal extended (3 inches); b, outline of head X2. (From Clench and Abbott 1943 in Johnsonia.)

GENUS *Norrisia* Bayle 1880

Norrisia norrisi Sowerby   Norris Shell

Plate 18m

Monterey, California, to Lower California.

1½ inches in length, slightly wider, heavy, smoothish with a glossy finish, especially on the underside. Lip sharp. Aperture thickened within and pearly. Umbilicus ovate, very deep, colored a greenish blue on the columellar side, bordered on the other side by glossy black-brown which fades into rich chestnut over the remainder of the shell. Operculum, multispiral, externally ornamented with spiral rows of dense bristles. Animal tinged with red. Moderately common among the kelp weed beds.

GENUS *Gaza* Watson 1879

Gaza superba Dall   Superb Gaza
                                                       Plate 3b

Northern Gulf of Mexico to the West Indies.

1 to 1½ inches in width. Spire somewhat elevated. Color old ivory with
a golden sheen. Early whorls faintly wine-colored. Although formerly
thought to be one of our rarest shells, it is now known to be relatively common in the Gulf of Mexico in 50 or more fathoms. It is indeed a beautiful
species.

The rare *Gaza* (*Callogaza* Dall 1881) *watsoni* Dall from deep water in
the West Indies is illustrated in figure 32a.

Subfamily MONODONTINAE
GENUS *Tegula* Lesson 1832

Tegula fasciata Born   Smooth Atlantic Tegula
                                                       Plate 17p

Southeast Florida and the West Indies.

½ to ¾ inch in width. Surface smooth; color yellowish to brown, with
a fine mottling of reds, browns and blacks; often with a narrow, pale, spiral
band of color on the periphery. Under the lens, spiral rows of alternating
red and white, short lines or dots may be seen. Some specimens may have
zigzag white bands. Interior of deep, round, smooth umbilicus and the callus
are white. Two teeth at the base of the columella. Thick adults may have
small teeth just inside the lower margin of the aperture. Whorl may be
slightly concave just below the suture. In the young only, the umbilicus
has two deep spiral grooves. Moderately common under rocks at low tide.

Tegula lividomaculata C. B. Adams   West Indian Tegula

Key West and the West Indies.

¾ inch in width and about ½ inch in length. Top of whorls sculptured
with about a dozen fairly regular, small, spiral cords. The angular periphery
of the whorls bears the largest cord. Umbilicus round, deep, and furrowed
on its sides by two spiral cords, the upper one ending at the columella in a
fairly sizeable bead. Columella set back quite far at its upper half; the lower
section bears the bead, and below that there are several, smaller, indistinct
beads. Color of shell grayish to brownish white with small mottlings of
reddish or blackish brown. Operculum, as in all *Tegula*, horny and multispiral. Formerly *scalaris* Anton (not Brocchi) and *indusi* "Gmelin." Common under rocks in the West Indies, but uncommon on the Lower Florida
Keys.

T. hotessierana Orbigny from the West Indies is similar, but rarely over ⅓ inch, with a more rounded periphery, with smaller, neater, equal-sized, smooth spiral threads, and dark bluish black in color, except for a whitish area around the narrow umbilicus. Uncommon.

Tegula excavata Lamarck   Green-base Tegula

Florida Keys? Caribbean area.

½ inch in length and width. Characterized by its bluish-gray color, corrugated sculpture (weak spiral cords and oblique lines of growth), its concave base, thin outer lip, and especially by the blue-green to iridescent-green circle of color around the very deep, round, narrow umbilicus. A variant exists in some areas which lacks the green, umbilical color and in which the spiral cords are stronger and the shell with axial, slanting bars of black-brown. Very common in the West Indies, along the rocky shores.

Tegula funebralis A. Adams   Black Tegula

Vancouver, B. C., to Lower California.

1 to 1½ inches in length, heavy, dark purple-black in color; smoothish, but with a narrow, puckered band just below the suture. Weak spiral cords rarely evident; coarse growth lines present in large, more elongate specimens. Base rounded. Umbilicus closed or merely a slight dimple. Columella pearly, with two small nodules at the base. A very common littoral, rock-loving species. Do not confuse with *T. gallina.*

Tegula gallina Forbes   Speckled **Tegula**
                                             Plate 18v

San Francisco to the Gulf of California.

1 to 1½ inches in length, very similar to *funebralis,* but a lighter, grayish green color with dense, zigzag, axial stripes of purplish. The shell surface is also coarser. A common, southern species found among littoral rocks.

Tegula brunnea Philippi   Brown Tegula

Crescent City to Santa Barbara Islands, California.

1 to 1½ inches in length, similar to *funebralis,* but light chestnut-brown in color with the base often glossy, brownish white. The umbilicus is closed, but usually with a dimple-like impression. Columella usually with only one small tooth near the base. Common at dead low tide on rocks. Usually heavily encrusted with algal growths.

Tegula aureotincta Forbes   Gilded Tegula
                                                              Plate 18k

Southern third of California to Mexico.

¾ to 1 inch in length, heavy; dark grayish to gray-green; characterized by a golden-yellow stain within the deep, round, narrow umbilicus, by the sky-blue band around the umbilicus, and by the 4 or 5 strong, smoothish, spiral cords on the periphery and the base. Top of whorls with weak, crude, slanting, axial wrinkles. A moderately common, littoral, rock-loving species.

Tegula ligulata Menke   Western Banded Tegula
                                                              Plate 18h

Monterey, California, to Acapulco, Mexico.

¾ inch in length, heavy; whorls and spire convex. Umbilicus very deep, round and fairly narrow. Whorls with numerous, beaded, spiral cords. Outer lip sharp, but thickened and pearly within. Lower part of lip with about 8 small nodules opposite the spiral threads which run back into the aperture. Color rusty-brown with black flecks. Compare with *aureotincta* whose umbilical area is stained with greenish blue and golden-yellow. A moderately common littoral, rock-dweller.

Subgenus *Chlorostoma* Swainson 1840

Tegula regina Stearns   Queen Tegula
                                                              Plate 3f

Catalina Island to the Gulf of California.

1½ inches in length, slightly wider; 6 to 7 whorls, spire flat-sided; base slightly concave. With numerous slanting, small, axial cords. The crenulated periphery slightly overhangs the suture of the whorls below. Base with strong, arched lamellae. Color dark purplish gray. Umbilical region stained with bright golden-yellow. A rather rare and choice collector's item secured by diving. It has also been washed ashore on Catalina Island.

Subgenus *Promartynia* Dall 1909

Tegula pulligo Gmelin 1791 (*marcida* Gould)   Dusky Tegula
                                                              Plate 18y

Alaska to Santa Barbara, California.

1 to 1½ inches in length, slightly wider. Resembles *brunnea*, but has a deep, round umbilicus and a thin, rather sharp columella. It is also very similar to *montereyi*, but its whorls are more rounded and its umbilicus is more smoothly rounded and without the white color and faint spiral ridges found in *montereyi*. This species is doubtfully placed here and perhaps

should be considered a typical *Tegula*. Moderately common, especially in the north.

Tegula montereyi Kiener   Monterey Tegula
Plate 18x

  Bolinas Bay, California, to Santa Barbara Island.

  1 to 1½ inches in length, about as wide. Conical in shape, with very flat-sided whorls and spire. Base almost flat. Surface smoothish, except for almost obsolete spiral threads. Umbilicus very deep, lined with 1 or 2 weak spiral cords. Columella arched, and with 1 prominent, pointed tooth. This rather rare species resembles a large *Calliostoma*. It is found on kelp in moderately deep waters.

Family TURBINIDAE
Subfamily LIOTIINAE

  The operculum in members of this subfamily is round, multispiral, and with a horny base on top of which are numerous rows of tiny calcareous beads.

GENUS *Cyclostrema* Marryat 1818

Cyclostrema cancellatum Marryat   Cancellate Cyclostreme

  Southeast Florida, the Bahamas to Jamaica.

  ½ inch in diameter, flat-topped, 4 whorls, opaque-white. Widely and deeply umbilicate. Axial sculpture of 15 to 17 rounded, low ribs which encircle the entire whorl and are made nodulose in crossing the 12 smaller spiral cords. Periphery squarish, with a cord above, below and at the center. Rare from 1 to 17 fathoms. *Cyclostrema* is a neuter, not feminine, word. *C. amabile* Dall from Cuba to Barbados is much rarer and differs in being smaller, in having a thicker, more rounded lip, and in lacking axial cords on top of the whorls. 25 to 80 fathoms.

GENUS *Liotia* Gray 1847

Liotia bairdi Dall   Baird's Liotia
Plate 17u

  North Carolina to Florida and Yucatan.

  ¼ inch in length, not quite so wide; thick, rose in color. Whorls globose, the last with about 10 spiral cords of tiny, prickly beads. Suture deeply channeled. Umbilicus very narrow and deep. Moderately common from 18 to 85 fathoms.

Liotia fenestrata Carpenter   Californian Liotia

Plate 18u

Monterey, California, to San Martin Island, Mexico.

⅛ inch in diameter; spire low, shell solid; deeply and narrowly umbilicate. Aperture circular, pearly within. Ash-white in color. Characterized by heavy cancellate sculpturing which makes the shell appear pitted by rows of deep, squarish holes. Uncommonly dredged from 10 to 25 fathoms. *L. cookeana* Dall is not this species, as is commonly thought, but is a *Cyclostrema*.

Genus *Arene* H. and A. Adams 1854

Arene cruentata Mühlfeld   Star Arene

Southeast Florida and the West Indies.

¼ inch in length, one third again as wide. 4 to 5 whorls angular with the periphery bearing a series of strong, triangular spines which are hollow on their anterior edges. Color white to cream with small, bright-red patches on top of the whorl. Below the main row of spines there is a minor spiral row of smaller spines. Suture channeled. Aperture circular, pearly within. Umbilicus round, deep, and bordered by 3 spiral, beaded cords. Uncommon under rocks.

The form *vanhyningi* Rehder from Sand Key, Key West, is pale gray-white with most red patches absent. It lacks fine, axial ridges on top of the whorl which are usually present in the typical form. Uncommon.

Arene venustula Aguayo and Rehder (Miami to Puerto Rico) is similar to *cruentata*, but smaller, much more squat, chalky-white, and with two peripheral rows of blunt spines. The rows are very close to each other. Rare, 20 fathoms.

Arene gemma Tuomey and Holmes   Gem Arene

Plate 17q

North Carolina to south half of Florida to Brazil.

⅛ inch or less, turbinate in shape; 3 spiral rows of neat, tiny beads on the squarish periphery. Suture minutely channeled and bounded below by a spiral row of whitish beads. Top slope of whorls and base of shell flattish. Axial threads on entire shell microscopic and crowded. Umbilicus round, deep, bordered by 7 to 9 distinct beads. Color of shell white to tan with minute specklings of red and/or brown. Commonly dredged from 3 to 100 fathoms.

Arene variabilis Dall Variable Arene
Plate 17s

North Carolina to southeast Florida and the West Indies.

$\frac{3}{16}$ inch in length, turbinate, similar to *A. gemma*, but pure white in color, with scale-like beads, suture more deeply channeled, and with a more rounded periphery. 12 very weak beads bordering the more open umbilicus. The 3 spiral rows of beads on the whorl may be almost smooth in some specimens. Very commonly dredged from 20 to 270 fathoms.

Subfamily *TURBININAE*
GENUS *Turbo* Linné 1758

Turbo castaneus Gmelin Chestnut Turban
Plate 3g

North Carolina to Florida, Texas and the West Indies.

1 to 1½ inches in length. Color orangish, greenish, brown or grayish, commonly banded with flame-like white spots. Aperture white. Callus on columella heavy. Lower lip projects downward. Operculum calcareous. The form named *crenulatus* Gmelin is merely less tuberculate.

Section *Taenioturbo* Woodring 1928

Turbo canaliculatus Hermann Channeled Turban
Plate 3a

Lower Florida Keys and the West Indies.

2 to 3 inches in length. A deep smooth channel runs just below the suture. Surface glossy. 16 to 18 strong, spiral, smooth cords on body whorl. Aperture white. Umbilicus narrow. Operculum pale-brown inside with 3 to 4 whorls, and white, smoothish and convex on the outside. This is the handsomest *Turbo* in the Western Atlantic, and considered a great rarity in American waters. Formerly *T. spenglerianus* Gmelin.

GENUS *Astraea* Röding 1798
Subgenus *Astralium* Link 1807

Astraea longispina Lamarck Long-spined Star-shell
Plate 3k, m

Southeast Florida and the West Indies.

2 to 2½ inches in width; shell low, almost flat on its underside. Periphery of whorls with strong, flattened, triangular spines. Either with or without an umbilicus. Aperture silvery inside. A form which has an elevated spire and is more spinose (pl. 3m) was known as *A. spinulosa* Lamarck. Short-spined specimens of this species are often erroneously called *A. brevi-*

spina Lamarck. The latter, however, is a distinct species from the West Indies which is characterized by a splotch of bright orange-red around the umbilical region (see pl. 3l).

Astraea americana Gmelin   American Star-shell
Plate 3i
Southeast Florida.

1 to 1½ inches in length, ¾ as wide. Characterized by its sharp-angled spire, flat sides, white to cream color, and by the numerous, long, wavy, weak, axial ribs. Base of shell with 5 to 8 small, finely fimbriated, spiral cords, and a small ridge at the base of the columella which has about a dozen small axial ridges. Commonly found under rocks at low tide on the Lower Florida Keys. Operculum variable, but usually thick, convex and with a small or large dimple.

The subspecies, *imbricata* Gmelin, from the West Indies has stronger, longer and fewer axial ribs which extend to the flat base of the shell and are hollow at their ends. The subspecies *guadeloupensis* Crosse from the Greater Antilles is intermediate between these two. Both moderately common at low water.

Subgenus *Lithopoma* Gray 1850

Astraea caelata Gmelin   Carved Star-shell
Plate 3h
Southeast Florida and the West Indies

2 to 3 inches in length and width. Similar to *A. tuber*, but with 9 to 10 spiral rows of numerous, hollow, scale-like spines on the lower ⅔ of the last whorl, 5 of which are on the base of the shell. Operculum thick, convex, and finely pustulose. Moderately common in the West Indies.

Astraea tuber Linné   Green Star-shell
Plate 3j
Southeast Florida and the West Indies.

1 to 2 inches in length, equally wide. Characterized by the peculiar green-and-white, cross-hatched color scheme, by the low, blunt, smooth axial ridges, and by the smoothish base of the shell. Sometimes mottled in soft browns. Common below low water in the West Indies, rare in Florida. Operculum with a thick, arched, tapering ridge on the exterior (like a large comma).

Subgenus *Pomaulax* Gray 1850

Astraea undosa Wood   Wavy Turban
Plate 18p

Ventura, California, to Lower California.

2 to 3 inches in length; characterized by a strong, wavy, overhanging periphery, and by the dark-brown, fuzzy periostracum. Base concave, with 3 small, indistinct spiral cords. Outside of operculum with 3 strong, prickly ridges. Common in shallow water, especially around Todos Santos Bay, Lower California.

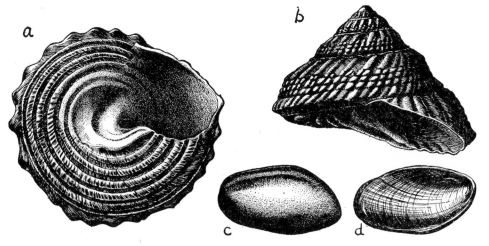

FIGURE 35. *Astraea gibberosa* Dillwyn, 2 inches. **c**, outer side of calcareous operculum; **d**, muscle attachment side.

Subgenus *Pachypoma* Gray 1850

Astraea gibberosa Dillwyn   Red Turban
Figure 35

Vancouver, B. C., to San Diego, California.

1½ to 3 inches in length, heavy, brick-red to reddish brown in color. Characterized by 5 to 6 strong, spiral cords on the flattish base. Operculum chitinous green on inner side; outer side swollen, smooth, enamel-white. Formerly *A. inaequalis* Martyn. Moderately common just offshore down to 40 fathoms.

GENUS *Homalopoma* Carpenter 1864

Shells small, turbinate in shape. Operculum calcareous, oval, thick; its exterior with a thick, paucispiral whorl. Underside of operculum convex with multispiral, chitinous whorling. *Leptothyra* Pease 1869 is this genus.

Homalopoma albida Dall   White Dwarf Turban

Southeast Florida, Cuba to Yucatan.

¼ inch in length, equally wide, very thick-shelled, resembling in shape a *Margarites*. Pure white in color. Whorls rounded, 5 to 6 in number, each bearing 5 to 6 strong, rounded, spiral cords, the lower 2 being below the periphery of the whorl. Aperture and parietal wall glossy, slightly opalescent. Columella arched, with a small tooth in the middle and a smaller one usually at the base. No umbilicus. Commonly dredged from 35 to 450 fathoms.

H. linnei Dall from southeast Florida to Barbados has 8 smaller, beaded spiral cords on the upper part of the whorls and 10 on the base, otherwise it is very similar to *albida*. It is quite rare.

Homalopoma carpenteri Pilsbry   Carpenter's Dwarf Turban

Plate 18i

Alaska to Lower California.

¼ to ⅜ inch (5 to 9 mm.) in length, solid, globose. Pinkish red to brownish red in color. Last whorl and base with 15 to 20, evenly sized, smooth, spiral cords separated from each other by a space about half as wide as the cords. Base of pearly columella with 2 or 3 exceedingly weak nodules. A very common species frequently washed ashore and inhabited by small hermit crabs from Monterey to Mexico. Do not confuse with *lurida*.

Homalopoma lurida Dall   Dark Dwarf Turban

Puget Sound to Lower California.

¼ inch (5 to 7 mm.) in length, similar to *carpenteri*, but half as large, black-brown in color, although occasionally whitish with red axial streaks. The spiral cords are usually fewer in number and more rounded. Moderately common in shallow water under rocks.

Homalopoma bacula Carpenter   Berry Dwarf Turban

Puget Sound to Lower California.

¼ inch or less in length, similar to *carpenteri* but with a flatter spire, and smoothish, except for numerous incised, spiral lines producing very weak threads. Color dark, rosy-brown. A moderately common shallow-water species, sometimes found with *carpenteri*. A thorough anatomical and life history study of this genus is needed to ascertain the validity of these species.

Family PHASIANELLIDAE
GENUS *Tricolia* Risso 1826

Tricolia affinis C. B. Adams   Checkered Pheasant

Lower Florida Keys and the West Indies.

⅛ to ¼ inch in length, moderately elongate, smoothish except for microscopic spiral grooves in some specimens. Color rose to brownish, sometimes whitish. Always with numerous small dots of pink, orange or a brownish color. Frequently with zigzag, axial bars of rose or brownish yellow. Often with irregular small spots or blotches of opaque-white. Umbilicus slit-like. Moderately common. *T. concinna* C. B. Adams is probably the same.

Tricolia tessellata Pot. and Mich. from the West Indies is somewhat the same, but it is characterized by distinct, revolving lines of orange or red that descend obliquely over the whorls. Common.

Tricolia pulchella C. B. Adams   Shouldered Pheasant
Plate 17r

Southeast Florida and the West Indies.

⅛ inch in length, spiral sculpture of numerous very small spiral cords, the largest being at the periphery of the whorl, thus giving the shell a slightly carinate shape. This carina is more pronounced in the early whorls and commonly bears a spiral row of tiny, white dots. Color variable, usually whitish gray with pink or brown axial mottlings and irregularly placed tiny dots of rose, yellow-brown or purplish brown. Umbilicus a mere chink. Operculum calcareous, convex, half smooth, the other half with fine, arched riblets. Common in shallow water among dead corals.

Tricolia compta Gould   Californian Banded Pheasant

Crescent City, California, to the Gulf of California.

¼ to ⅓ inch in length, resembling a moderately high-spired *Littorina*, but distinguished from that genus by its calcareous operculum. Shell smooth, in life covered by a thin gray-green, translucent periostracum. Characterized by the numerous, spiral lines of blackish green, red, brown or purplish which slant slightly downward, so that they are not parallel to the suture. Axial zigzag, wider bands are also present. Very abundant on eel-grass in shallow bays. Frequently washed ashore.

GENUS *Eulithidium* Pilsbry 1898

Eulithidium rubrilineatum Strong   Miniature Pheasant

Monterey to Lower California.

1/16 inch in length, depressed turbinate in shape, with 4 to 5 whorls. Characterized by its very small size, and by about a dozen obliquely set, spiral bright-red lines. The top of the whorls may be solid red and with large, opaque-white spots. Umbilicus a mere chink-like depression. Operculum calcareous and white. *Tricolia variegata* Carpenter is the same (not Lamarck). Common among weeds in tidepools and among kelps offshore.

Family NERITIDAE
Genus *Nerita* Linné 1758

Nerita peloronta Linné   Bleeding Tooth
Plate 4a

Southeast Florida, Bermuda and the West Indies.

¾ to 1½ inches in length; grayish yellow with zigzags of black and red. Characterized by the blood-red parietal area which bears 1 or 2 whitish teeth. Operculum: underside coral-pink; one half of outer side smooth and dark-orange, other half smoothish or papillose and brownish green. Very abundant along the rocky shores facing the open ocean. It is a popular souvenir.

Nerita versicolor Gmelin   Four-toothed Nerite
Plate 4b

South ¾ of Florida and the West Indies. Bermuda.

¾ to 1 inch in length; dirty-white with irregular spots of black and red arranged in spiral rows; spirally grooved; outer lip spotted with red, white and black on margin. Parietal area slightly convex, white to yellowish and with 4 (rarely 5) strong teeth. Operculum: exterior brownish gray, finely papillose and slightly concave. Commonly associated with *N. peloronta*. *Nerita variegata* Karsten (1789) is invalid, since it appears in a non-binomial work.

Nerita tessellata Gmelin   Tessellate Nerite
Plate 4f

Florida to Texas, the West Indies and Bermuda.

¾ inch in length, irregularly spotted with black and white, sometimes heavily mottled; coarsely sculptured with spiral cords of varying sizes. Parietal area concave, bluish-white and bearing 2 weak teeth in the middle. Operculum: exterior slightly convex, black in color. Commonly congregate in large numbers under rocks at low tide. Rare in northern Florida. Do not confuse with *N. fulgurans* Gmelin whose operculum is bluish white to yellowish gray, not black.

Nerita fulgurans Gmelin

Antillean Nerite
Plate 4c

Southeast Florida, the West Indies and Bermuda.

¾ to 1 inch in length, very similar to *N. tessellata*, but with a lighter-colored, yellowish gray operculum. The spiral ridges on the shell are more numerous, the color patterns blurred, the aperture relatively wider, and the teeth more prominent. This is a salt to brackish-water inhabitant of protected shores, and is abundant only in certain restricted localities. It is seldom represented or properly labeled in private collections.

Genus *Puperita* Gray 1857

Puperita pupa Linné

Zebra Nerite
Plate 4e

Southeast Florida and the West Indies.

⅓ to ½ inch in length, thin, smooth, chalky-white with black, axial, zebra-like stripes. Aperture and smooth operculum light-yellow. Lives in small, placid pools above the high-water mark. Common in the West Indies, rare in Florida.

Genus *Neritina* Lamarck 1816
Subgenus *Vitta* Mörch 1852

Neritina virginea Linné

Virgin Nerite
Plate 4i

Florida to Texas, the West Indies and Bermuda.

½ inch in length, smooth, glossy, very variable in color pattern and shades—blacks, browns, purples, reds, whites, olive—crooked lines, dots, mottlings, zebra-like stripes and sometimes spirally banded. Parietal area smooth, convex, white to yellow, and with a variable number of small, irregular teeth. Operculum usually black. A very common, widespread inhabitant of intertidal, brackish-water flats.

Neritina reclivata Say

Olive Nerite
Plate 4g

Florida to Texas and the West Indies.

½ inch in length, glossy, often with the spire eroded away. Ground color brownish green, olive or brownish yellow with numerous axial lines of black-brown or lavender. Operculum black to slightly brownish. Common in brackish water and also found in fresh-water springs near the seashore in Florida.

A globose form or subspecies (?) with a short spire and more convex whorls replaces the higher-spired, typical form from Texas to Panama, but may also appear in eastern Florida. It has been named *floridana* Reeve 1855, *rotundata* von Martens 1865 and *sphaera* Pilsbry 1931.

GENUS *Smaragdia* Issel 1869

Smaragdia viridis Linné   Emerald Nerite
                                                                Plate 4h
Southeast Florida, the West Indies and Bermuda.

¼ to ⅓ inch in length, glossy, smooth, pea-green, often with tiny chalk-white bars and rarely with purplish brown, narrow, zigzag bars. True *viridis* comes from the Mediterranean. Some workers separate our form as the subspecies *viridemaris* Maury 1917. *N. weyssei* Russell 1940 is a synonym.

Order *MESOGASTROPODA*
Superfamily *LITTORINACEA*
Family *LACUNIDAE*
GENUS *Lacuna* Turton 1827

Rather fragile, smooth periwinkles characterized by a shelf-like columella and a chink-like umbilicus. Periostracum smooth, fairly thin and light-brown. Operculum paucispiral and corneous. Cold-water inhabitants, usually dredged in areas of kelp weed.

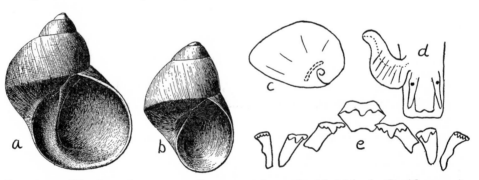

FIGURE 36. Northern Lacunas. **a**, *Lacuna carinata* Gould, ½ inch (Pacific coast); **b**, *Lacuna unifasciata* Cpr., ¼ inch (California); **c** to **e**, *Lacuna vincta* Turton, ⅜ inch (both coasts); **c**, operculum showing reinforcing bar (dotted); **d**, animal showing the penis on the right side; **e**, radula (a single row greatly enlarged).

Lacuna vincta Turton   Common Northern Lacuna
                                                    Plate 22p; figure 36c-e
Arctic Ocean to Rhode Island. Alaska to California.

¼ to ⅜ inch in length, 4 to 5 whorls, resembling a *Littorina*, but characterized by its fairly thin, but strong, translucent shell, its shelf-like columella along side of which is a long, narrow, deep umbilical chink. Outer lip fragile.

Shell smooth except for microscopic, spiral scratches. Color light-tan to brown with the spire tinted with purplish rose. Often confused with *Litiopa* which has a blade-like ridge on the columella just inside the aperture. Common from low water to 25 fathoms. Alias *L. divaricata* Fab. and *solidula* Loven.

Lacuna unifasciata Carpenter One-banded Lacuna
Figure 36b

Monterey, California, to Lower California.

¼ inch in length, moderately fragile, similar to the other Lacunas, but characterized by its very narrow, long, chink-like umbilicus and by the carinate periphery of the whorl which bears a fine, dark-brown spiral line. Early whorls usually pinkish, remainder yellowish tan. Umbilicus and columella white. The peripheral carina may be weak or obsolete, and the color line may consist of a series of faint, slanting streaks of light reddish brown. Very common in littoral seaweed and kelp in southern California.

Lacuna carinata Gould Carinate Lacuna
Figure 36a

Alaska to Monterey, California.

⅜ to ½ inch in length, 3 to 4 whorls, moderately fragile. Aperture semi-lunar, large. Outer lip thin. Columellar chink large, long and white. Shell smooth, chalky-white, but always covered by a thin, yellowish brown, smooth periostracum. Common on kelp weed. *L. porrecta* Cpr. and *striata* Gabb are the same. Do not confuse with *vincta* which has a higher spire and much narrower, brownish-tan umbilical chink.

Lacuna variegata Carpenter Variegated Lacuna

Puget Sound, Washington, to Santa Monica, California.

¼ inch in length, similar to *unifasciata*, but having a very deep umbilical chink which is bordered by a sharp ridge. The spiral carina at the level of the suture is very small, but quite sharp. The yellowish tan shell has mottlings or oblique bands of darker color. Moderately common in eel-grass along the shore.

Family LITTORINIDAE
GENUS *Haloconcha* Dall 1886

Haloconcha reflexa Dall Reflexed Haloconch

Alaska and the Bering Sea.

½ inch in length, fragile, 3 whorls; body whorl large. Resembles a *Vel-*

utina. Early whorls purplish brown, last whorl translucent, light chestnut-brown. Shell covered by a thin varnish of periostracum of the same color. At the edge of the thin, smooth outer lip, the periostracum is neatly curled back to form a minute ridge. Umbilicus slit-like, with the periostracum puckered along its length. Operculum littorinid, and is withdrawn well within the glossy-brown aperture. Not uncommon in shallow water.

Genus Littorina Ferussac 1821

Littorina littorea Linné   Common Periwinkle
Plate 19b

    Labrador to New Jersey. Western Europe.

    ¾ to 1 inch in length, thick, smoothish. Gray to brownish gray in color. Inside of aperture chocolate-brown. Columella and inner edge of aperture whitish. In young or perfect specimens there are fine, irregularly spaced, spiral threads with microscopic, wavy wrinkles in between. Introduced from Europe some time before 1840. A favorite food in Europe. Very common along the rocky shores of New England.

Littorina irrorata Say   Marsh Periwinkle
Plate 19c

    New York to north Florida to Texas.

    About 1 inch in length, thick-shelled, with numerous, regularly formed spiral grooves. Outer lip strong, sharp, slightly flaring, and with tiny grooves on the inside. Color usually grayish white with tiny, short streaks of reddish brown on the spiral ridges. Aperture yellowish white. Callus of inner lip and the columella pale reddish brown. Commonly found in large numbers among the sedges of brackish water marshes. Not recorded alive south of Indian River (east Florida) or Charlotte Harbor (west Florida).

Littorina ziczac Gmelin   Zebra Periwinkle
Plate 19e

    South half of Florida to Texas, the West Indies and Bermuda.

    Females about 1 inch, males about ½ inch in length. Shell fairly thick and strong. Base angulate; aperture purplish brown. Columella various shades of dark-brown. Outer shell white to bluish white with many narrow, zigzag, oblique lines of chestnut-brown or purplish brown. Early whorls uniformly pale reddish brown. Female shells: higher than wide, smoothish. Male shells: as high as wide, with strong spiral grooves. Operculum dark-brown. Abundant in crevices between tides in rocky areas. Introduced to the Pacific side of the Panama Canal. Do not confuse with the larger and thinner-shelled *L. angulifera* whose operculum is light-brown, not dark-brown, in color.

Littorina angulifera Lamarck

Angulate Periwinkle
Plate 19a

South half of Florida, the West Indies and Bermuda.

About 1 inch in length; thin-shelled but strong. First two or three whorls smooth, remainder with many fine, spiral grooves. Last whorl sometimes carinate. Color variable—whitish, yellowish or orange- to red-brown with darker, wavy, vertical, oblique stripes. Columella pale purplish with whitish edges. Operculum pale-brown. Common in mangrove areas where the waters are calm and brackish. It is found high above the high-tide mark clinging to wharf pilings, and is often seen on the trunks and branches of mangrove trees. Introduced to the Pacific side of the Panama Canal. *L. scabra* Linné is from the Indo-Pacific.

Littorina obtusata Linné

Northern Yellow Periwinkle
Plate 19f

Labrador to Cape May, New Jersey. Northwest Europe.

⅓ to ½ inch in length, equally wide, with a low spire; smoothish. Color variable but usually a uniform, bright, brownish yellow or orange-yellow. Sometimes with a white or brown spiral band. Columella whitish. Operculum bright yellow to orange-brown. This is *L. palliata* Say. A common coastal species associated with rockweeds.

Littorina mespillum Mühlfeld

Dwarf Brown Periwinkle
Plate 19k

Florida Keys and the Caribbean Area.

¼ inch in length, somewhat shaped like *obtusata*. Characterized by its dark-brown periostracum, glossy-brown columella and aperture, by its tiny, chink-like umbilicus, and by the presence, in some specimens, of rows of small, round blackish spots. Common in "splash-pools" from high-tide line to 6 or 7 feet above.

Littorina saxatilis Olivi

Northern Rough Periwinkle
Plate 19d

Arctic Seas to Cape May, New Jersey. Arctic Seas to Puget Sound.

¼ to ½ inch in length, resembling a "distorted, small *L. littorea.*" Adults characterized by poorly developed, smoothish, fine spiral cords. Color drab gray to dark-brown. Interior of aperture chocolate-brown. Females give birth to live, shelled young. Often found with *L. obtusata*, but not so common. This is *L. rudis* Maton and *L. groenlandica* Menke.

Littorina scutulata Gould   Checkered Periwinkle
                                                                        Plate 20c

Alaska to Lower California.

½ inch in length, moderately slender, semigloss finish and smooth. Color
light to dark reddish brown with small, irregular spots of bluish white. Columella white; interior of aperture whitish brown. A common littoral species.
Compare with *L. planaxis*.

Littorina planaxis Philippi   Eroded Periwinkle
                                                                        Plate 20a

Puget Sound to Lower California.

½ to ¾ inch in length, usually badly eroded; grayish brown with bluish
white spots and flecks. Characterized by the eroded, flattened area on the
body whorl just beside the columella. Interior of aperture chocolate-brown
with a white spiral band at the bottom. A common littoral, rock-loving species. Do not confuse with the smoother, higher-spired *L. scutulata*.

Littorina sitkana Philippi   Sitka Periwinkle
                                                                        Plate 20b

Bering Sea to Puget Sound, Washington.

¾ inch in length, solid, sharp lip, characterized by about a dozen strong
spiral threads on the body whorl. Columella whitish. Shell dark grayish to
rusty-brown; some with 2 or 3 wide spiral bands of whitish. A common
littoral species of the north.

GENUS *Nodilittorina* Martens 1897

Nodilittorina tuberculata Menke   Common Prickly-winkle
                                                                        Plate 19i

South Florida, the West Indies and Bermuda.

½ to ¾ inch in length. Shell rounded at the base. Several spiral rows
of small, fairly sharp nodules on the whorls. Columella flattened, forming a
slightly dished-out shelf. Color brownish gray. Operculum paucispiral. A
common rock-dwelling species found near the high-tide line. Do not confuse with the extremely similar *Echininus nodulosus* Pfr. which has a multispiral operculum, and whose columella is not shelved. Erroneously listed in
Johnsonia and other books as *Tectarius tuberculatus* Wood.

GENUS *Tectarius* Valenciennes 1833
Subgenus *Cenchritis* Martens 1900

Tectarius muricatus Linné   Beaded Periwinkle
                                                                        Plate 19g

Lower Florida Keys, the West Indies and Bermuda.

½ to 1 inch in length. Shell thick, with 11 rows of neat, rounded, whitish, evenly spaced beads on the last whorl. Columella grooved; umbilicus a narrow, oblique slit. Color of outer shell ash-gray. Interior dark-tan. Operculum paucispiral. One of the commonest West Indian littoral species, usually found well out of water on the rock cliffs.

GENUS *Echininus* Clench and Abbott 1942
Subgenus *Tectininus* Clench and Abbott 1942

Echininus nodulosus Pfeiffer   False Prickly-winkle
Plate 19h

Southeast Florida and the West Indies.

½ to 1 inch in length. Base of shell squarish. Whorls with 2 spiral, carinate rows of sharp nodules in addition to 2 or 3 rows of smaller, blunt nodules. Columella not shelved. Color grayish brown. Operculum multispiral. Lives well above high-tide mark on rocky shores. Be sure not to confuse with *Nodilittorina tuberculata* whose beads are lined up axially one under the other.

Superfamily RISSOACEA
Family RISSOIDAE
GENUS *Cingula* Fleming 1828

Extremely small shells, conic-ovate; aperture round, peristome complete; whorls moderately rounded. Nuclear whorls smooth. Umbilicus slit-like. There are about 15 confusing species on the west coast of America, most of which are found in Alaskan waters.

Cingula montereyensis Bartsch   Monterey Cingula

Moss Beach to Monterey, California.

4 mm. in length, light-brown, smooth. Suture slightly indented. Uncommon from shore to 15 fathoms.

Subgenus *Nodulus* Monterosato 1878

Cingula kelseyi Bartsch   Kelsey's Cingula

San Diego to Lower California.

2 mm. in length, translucent-white, with microscopic spiral striations and fine lines of growth. There are 4 other species in this subgenus which are found in Alaska (*C. asser* Bartsch, *C. kyskensis* Bartsch, *C. palmeri* Dall and *C. cerinella* Dall).

Subgenus *Onoba* H. and A. Adams 1854

Cingula aculeus Gould   Pointed Cingula

Nova Scotia to Maryland.

Extremely small, 2.5 mm. in length, elongate, about 5 whorls, no umbilicus. Whorls rounded. Suture well-impressed. Aperture ovate with a slightly flaring lip. Color light- to rusty-brown. Spiral sculpture of numerous, microscopic incised lines. Below the suture there are numerous, short, axial riblets. Common in shallow water.

GENUS *Amphithalamus* Carpenter 1865

Extremely small shells, less than 2 mm. in length, smooth, except for a faint cord or spiral thread on the periphery. Nucleus large, of 1½ whorls which are finely pitted like a thimble. The most striking character is a thin bridge separating the inner lip from the open umbilicus. There are 3 species in southern California:

Periphery without spiral line . . . .
. . . . *lacunatus* Carpenter (San Pedro south).
Periphery with thread or cord:
Periphery angulate . *inclusus* Carpenter (San Pedro south).
Periphery rounded . . . .
. . . . . . *tenuis* Bartsch (Monterey south).

GENUS *Rissoina* Orbigny 1840

Shells small, usually less than ⅛ inch in length, generally white in color, with strong or weak axial ribs, occasionally with fine spiral, incised lines. Aperture semilunar and somewhat flaring. Operculum corneous, thick, paucispiral, with a claviform process on the inner surface. We have presented nearly all of the species known to both sides of the United States in the form of a key (see pl. 22u).

Key to the Pacific Coast *Rissoina*

A. Color pure-white or bluish white . . . . B
Color yellow to light-red; 6 mm., Redondo Beach south . . . . .
. . . . *kelseyi* Dall and Bartsch

B. Axial ribs strong, less than 20 on the last whorl C
Axial ribs weak, numerous . . . . D

C. Interspaces with silky, wavy crinkles; 3 mm.; Coronado Islands . . .
. . . . *cleo* Bartsch
    Interspaces smooth, 3 mm.; Catalina Islands, south . *californica* Bartsch

D. Whorls decidedly inflated; 3 mm.; Monterey south . . *bakeri* Bartsch
    Whorls not inflated . . . . E

E. With very fine, numerous axial threads (48 to 55 on last whorl) . . F
    With coarse riblets (36 on last whorl; 14 on next to last); 3 mm. Alaska
    to Monterey . . . . *newcombei* Dall

F. Shell slender; 2 mm.; San Pedro south . . . . . *dalli* Bartsch
    Shell not as slender; 3.5 mm.; Redondo Beach south . . . .
. . . . *coronadoensis* Bartsch

Key to the Atlantic Coast *Rissoina*

A. Shell sculptured with riblets or spiral lines . . . . . . B
    Shell smooth, glossy-white; 4 mm.; Carolinas, Florida, and West Indies;
    syn.: *laevigata* C. B. Ads. . . . . . *browniana* Orbigny

B. With axial ribs more prominent than spiral threads C
    With axial ribs not more prominent than spiral threads . . . F

C. Axial ribs only . . . . D
    Axial ribs and spiral threads both present . . . . . . E

D. 4.5 to 6.0 mm.; white or stained yellow; 16 to 22 ribs; South Florida and
    the West Indies . . . . *bryerea* Montagu
    3.0 to 5.0 mm.; white; 11 to 14 ribs; suture sometimes deep; North Caro-
    lina to Florida and the West Indies . *chesneli* Michaud (Pl. 22u)

E. 4 to 5 mm.; ribs strong but disappearing on base; spiral threads strongest
    on base; white to rusty; southeast Florida and the West Indies . .
. . . . *multicostata* C. B. Adams
    6 to 7 mm.; 25 to 28 low, weak ribs, spirally striated or pitted between;
    glossy; white to yellowish; North Carolina to the West Indies . .
. . . . *decussata* Montagu

F. Sculpture strongly cancellate . . . . G
    Not strongly cancellate; low, spiral threads dominant; axial ribs faint;
    weakly cancellate; 5 to 10 mm.; southeast Florida and the West
    Indies . . . . *striosa* C. B. Adams

G. 5 to 7 mm.; white; strongly cancellate; depressed interspaces large and
    square; southeast Florida and the West Indies . *cancellata* Philippi

4 to 4.5 mm.; glassy-white; depressed interspaces small, rounded; Texas
to the West Indies and Bermuda   .   .   .   *sagraiana* Orbigny

Family *VITRINELLIDAE*
Genus *Vitrinella* C. B. Adams 1850

Shell minute, thin, depressed, umbilicate, and with 3 to 4 subtubular
whorls. The umbilicus has rather flattened walls and is usually bounded by a
spiral cord. The rounded aperture is oblique, with a thin lip, its upper margin
arching forward. Columella only moderately thickened. Operculum corneous, thin, multispiral. There are many species in American waters with
quite a number of genera and subgenera. The family is undergoing considerable change under the current research by H. A. Pilsbry. We are including only three examples of this interesting group. Consult recent numbers
of *The Nautilus*.

Subgenus *Vitrinella s. str.*

Vitrinella helicoidea C. B. Adams   Helix Vitrinella

North Carolina to Florida and the West Indies.

2 mm. in diameter, planorboid, 4 whorls, spire moderately raised. Translucent-white, glossy, smooth. Umbilicus round, very deep, moderately wide,
bounded by a small, spiral, smoothish thread. Wall of umbilicus flattish.
Columella strong, braced on the whorl above by a small, spreading callus.
Outer lip thin, sharp. Not uncommon in shallow water. This is the type of
the genus.

Subgenus *Circulus* Jeffreys 1865

Vitrinella multistriata Verrill   Threaded Vitrinella
                                                     Plate 17v
North Carolina to Florida.

5 mm. in diameter, planorboid, well-compressed, 4 whorls, opaque-white,
with a glossy sheen. Outer surface covered with numerous, crowded, spiral,
incised lines. Umbilicus with rounded sides, deep, rather narrow. 50 to 100
fathoms. Locally common.

Subgenus *Solariorbis* Conrad 1865

Vitrinella beaui Fischer   Beau's Vitrinella

North Carolina to Florida and the West Indies.

⅓ inch in diameter, strong, opaque-white, depressed, 4 whorls. Top of whorls rounded, slightly concave just below the fine suture; bearing 5 or 6 major, smooth, spiral threads on top with numerous, much finer threads between. Periphery bordered above and below by a major cord. Umbilicus widely funnel-shaped, deep. Outer lip crenulate above. Not uncommon in shallow water. One of our largest American Vitrinellid species. Provisionally placed in this subgenus.

Genus *Pseudomalaxis* P. Fischer 1885

Pseudomalaxis nobilis Verrill   Noble False Dial

Virginia to southeast Florida and the West Indies.

⅜ inch in diameter, dull-white, planorboid, with a very flat spire and a wide concave, non-umbilicate base. Periphery of shell flat, bordered above and below by one or two spiral cords of small beads. Aperture squarish. Operculum round, multispiral with a chitinous pimple on the inside. A rare and choice collector's item. Deep water. 70 fathoms.

Pseudomalaxis balesi Pilsbry and McGinty   Bales' False Dial

Palm Beach and along the Lower Keys, Florida.

1.8 mm. in diameter, 3 to 4 whorls, semitranslucent-white to burnt sienna. Sculpture of fine, spiral striae and strong, widely spaced, radial ribs. Peripheral zone flattened or concave between 2 projecting nodulose keels. Under rocks. Moderately common to rare. This genus was formerly placed in the family *Architectonicidae*.

Genus *Teinostoma* H. and A. Adams 1854

Shells usually about 2 to 3 mm. in diameter, depressed, glossy, white, usually smooth, and with an umbilical callus. They are very distinctive little shells, but require a high-powered lens for their inspection. We have figured only one species, but have included a key from the work of Pilsbry and McGinty (1945) (see pl. 17y).

Key to the Florida *Teinostoma*

1. Umbilical callus encircled by a keel. 1.7 mm.; Palm Beach to Cape Florida.
    12 to 50 faths. .   .   *T. (Annulicallus) lituspalmarum* Pils. and McG.
2. Umbilical callus and columellar lobe not closing the umbilicus completely; 3 mm.; southeast Forida. 80 fms. .   .   .   .
    .   .   .   .   *T. (Ellipetylus) cocolitoris* Pils. and McG.

3. Umbilicus closed by the callus, which passes smoothly into the base.
       (subgenus *Idioraphe*)

A. Periphery strongly carinate; 2 mm.; Destin, Florida. 20 fms.
       goniogyrus Pils. and McG.
AA. Periphery rounded or indistinctly rounded.

B. Surface spirally striate:

C. Umbilical callus extremely convex and thick; 2 mm. Palm
       Beach to Cape Florida   *pilsbryi* McGinty
CC. Umbilical callus strong, slightly convex:

D. Strongly spirally striate throughout; 2.3 mm.; Key
       Largo   *clavium* Pils. and McG.
DD. Weakly striate above only; 1.5 mm.; southeast Florida
       nesaeum Pils. and McG.

BB. No Spiral striations.

C. Diameter 1.8 to 2.2 mm.:

D. Rather globose, h/d ratio 75; shore to 50 fms.; south-
       east Florida   *parvicallum* Pils. and McG.
DD. Depressed, h/d ratio about 50:

E. Callus large. Lake Worth
       obtectum Pils. and McG.
EE. Callus small. Biscayne Bay, shore
       biscaynense Pils. and McG.

CC. Diameter 0.7 to 1.0 mm.; callus thick; Lower Keys, shore
       leremum Pils. and McG.

Superfamily CERITHIACEA
Family TURRITELLIDAE
GENUS *Tachyrhynchus* Mörch 1868

Tachyrhynchus erosum Couthouy   Eroded Turret-shell
                                               Plate 21l

Nova Scotia to Cape Cod, Massachusetts. Alaska to British Columbia.

¾ to 1 inch in length, elongate, ¼ as wide; 8 to 10 rounded whorls. No
umbilicus. Aperture round; columella smooth, slightly arched. Whorls with
5 to 6 smooth, flat-topped, spiral cords between sutures. Color cream to
chalky-white, with a thin, polished, gray-brown periostracum. Operculum
round, multispiral, chitinous, dark-brown. Common from 10 to 75 fathoms.
   T. reticulatum Mighels (Arctic Ocean to Maine; and Alaska) is similar,

but usually has spiral cords only on the base and one just above the suture, and has about 18 to 10 axial, rounded ribs per whorl. Some specimens show fine, spiral, incised lines. Common from 16 to 60 fathoms.

Tachyrhynchus lacteolum Carpenter   Milky Turret-shell

Alaska to Lower California.

½ inch in length, similar to *erosum*, but ⅓ as wide as long, and the cords between sutures are finely beaded. The beads are arranged more or less in axial rows. The last third of the body whorl bears weak, non-beaded spiral cords. This species differs from *reticulatum* in its smaller size, less slender shape, less convex whorls, and much finer sculpturing.

GENUS *Turritella* Lamarck 1799

Turritella acropora Dall   Boring Turret-shell
                                                             Plate 21j

North Carolina to Florida, Texas and the West Indies.

1 inch in length, resembling *exoleta*, but with convex whorls, and with numerous, fine, spiral threads a few of which, at the periphery, are slightly larger than the others. There is a very weak series of riblets just below the suture. Color yellowish to brownish orange. Common just offshore.

T. variegata Linné, the Variegated Turret-shell (pl. 21i) (West Indies), is similar, but up to 4 inches in length, with flat-sided whorls, and is mottled with mauve, white and dark-brown. Common.

Turritella exoleta Linné   Eastern Turret-shell
                                                             Plate 21h

South half of Florida and the West Indies.

2 inches in length, long, slender and with a sharp apex. Each whorl with a large, coarse cord above and below, with the part between the cords concave and occasionally crossed by microscopic, arched, brown, scale-like lamellae. Base of shell concave. Color glossy-white to cream with sparse, axial flammules of light yellow-brown. Moderately common from 1 to 100 fathoms. This species is placed in the subgenus *Torcula* Gray 1847.

Turritella cooperi Carpenter   Cooper's Turret-shell
                                                             Plate 20g

Monterey, California, to Lower California.

1 to 2 inches in length, 17 to 20 slightly convex whorls. Base concave. Columella and outer lip fairly fragile. Whorls with 2 or 3 small, spiral cords

and usually with a number of much smaller, variously sized threads. Color orangish to yellowish white with darker, axial flammules. Moderately common just offshore.

Turritella mariana Dall   Maria's Turret-shell
                                                                    Plate 20h

Catalina Island to Panama Bay, Panama.

1½ to 2½ inches in length, similar to *cooperi*, but with the whorls slightly concave due to the more prominent, irregularly beaded spiral cords. The aperture is not circular as in *cooperi*. Its color is usually much lighter. Uncommon 20 to 40 fathoms.

Family ARCHITECTONICIDAE
GENUS *Torinia* Gray 1842

Torinia bisulcata Orbigny   Orbigny's Sun-dial
                                                                    Plate 21x

North Carolina to Florida and the Gulf of Mexico.

¼ to ½ inch in diameter, spire flattened, each whorl with 5 crowded rows of neat, tiny, squarish beads. Periphery with a major, and below it a minor, beaded cord. Base rounded and with about 7 wide cords bearing beads. Umbilicus quite wide and very deep. Nuclear whorl glassy-white. Color of shell dull gray to dull cream. Operculum solid-conic, chitinous. Uncommon from 15 to 200 fathoms on mud bottom.

Torinia cylindrica Gmelin   Cylinder Sun-dial

Lower Florida Keys and the West Indies.

⅜ inch in length, equally wide; spire high; umbilicus narrow, round, very deep, bordered inside with 3 spiral, beaded cords. Columella with 4 small, depressed, spiral lines. Top of whorls with 4 spiral cords of closely packed, small beads. Color dark-gray to reddish brown with a cream base and with white spots on the periphery. Uncommon at low tide.

GENUS *Architectonica* Röding 1798

Architectonica nobilis Röding   Common Sun-dial
                                                                    Plate 4m

North Carolina to Florida, Texas and the West Indies.

1 to 2 inches in diameter, heavy, cream with reddish brown spots which are especially prominent just below the suture. Sculpture of 4 or 5 spiral cords which are usually beaded. Umbilicus round, deep and bordered by a

heavily beaded, spiral cord. Operculum, corneous, thin, paucispiral, brown and with lamellate growth lines. Moderately common in sand below low-water line. Known for years as *A. granulata* Lamarck which, however, is a later name.

Architectonica krebsi Mörch

Krebs' Sun-dial
Plates 4-o; 21y

North Carolina to southeast Florida and the West Indies.

½ inch in diameter, similar to *A. nobilis*, but glossy-smooth on top except for two microscopic spiral threads above the suture, with a smooth rounded base and with its deep umbilicus bordered by 2 beaded spiral rows, the innermost having about 30 bar-like beads (in contrast to 12 in *nobilis*). Operculum chitinous, brown and multispiral. Uncommon from 16 to 63 fathoms. Provisionally placed in this genus.

Architectonica peracuta Dall

Keeled Sun-dial
Plate 4n; figure 22g

Southeast Florida and the West Indies.

¾ inch in diameter, similar to *nobilis*, but smaller, with the spire much flatter, whorls almost smooth, periphery very sharp, and without color spots. Rare in 45 to 73 fathoms.

Family VERMETIDAE
GENUS *Petaloconchus* H. C. Lea 1843

Petaloconchus nigricans Dall

Black Worm-shell
Plate 21e

West Coast of Florida.

Lives in closely packed colonies of long, worm-like shells about 2 to 4 inches in length; each tube is about ⅛ inch in diameter. Shells rarely coiled, except at the beginning. Moderately fragile, gray to rusty-brown in color, and weakly sculptured spirally and longitudinally. Colonies of this species frequently form large reefs or banks.

Petaloconchus irregularis Orbigny

Irregular Worm-shell
Plate 21d

South half of Florida and the West Indies.

Similar to *nigricans*, but greatly coiled, and with heavier, larger shells which are strongly rugose. Occurs in compact masses attached to rocks and other shells. Color dark brownish. *P. erectus* Dall is smoother, pure white in color, and with the last part of the tube sticking straight up from the coiled mass (Erect Worm-shell); the latter occurs in deep water off southeast Florida.

Genus *Aletes* Carpenter 1857

Aletes squamigerus Carpenter   Scaled Worm-shell
                                                  Plate 20e

Forrester Island, Alaska, to Peru.

Grows in large, twisted masses. The shelly tubes are circular, ¼ to ½ inch in diameter. Sculpture of numerous, minutely scaled or rough, longitudinal cords. Color gray to pinkish gray. The last part of the shell which usually stands erect for ½ inch is smoothish. A very common, colonial species found in masses on wharf pilings or attached to rocks below the low-water line.

Genus *Spiroglyphus* Daudin 1800

Spiroglyphus lituellus Mörch   Flat Worm-shell
                                                  Plate 20d

Forrester Island, Alaska, to San Diego, California.

A small worm-tube mollusk found adhering to rocks and the shells of abalones in a tightly wound, flat spiral. The last whorl may grow up on top of the previous whorls and be erect for ¼ of an inch. Aperture circular, about ⅛ inch in diameter. Shell solid, with 2 large, scaled cords which give a somewhat squarish cross-section to the whole shell. Hollow scales and fimbriations present elsewhere. Color cream to purplish gray. Operculum horny, multispiral and brown. Moderately common.

There is a very similar species reported from the West Indies (*S. annulatus* Daudin).

Family *SILIQUARIIDAE*
Genus *Vermicularia* Lamarck 1799

Vermicularia spirata Philippi   West Indian Worm-shell
                                                  Plate 21c; figure 22i

Southeast Florida and the West Indies.

Evenly and closely spiraled for about ¼ inch, then becoming random and drawn out in its worm-like coiling. Shell rather thin, colored a translucent to opaque amber, orange-brown or yellowish. Early whorls dark, smooth, except for 1 (rarely 2) smooth, spiral cord on the middle of the whorl. Subsequent whorls with 2 major cords which soon lose their prominence. Smaller threads present, especially on the base of the shell. This is not the common West Florida species usually called *"spirata"* in other books. See *knorri* and also *fargoi*.

For anatomy and relationships in the worm-shells, see the excellent works by J. E. Morton (1951) in the Transactions of the Royal Society of New Zealand.

Vermicularia knorri Deshayes

Florida Worm-shell
Plate 21a

North Carolina to Florida and the Gulf of Mexico.

Differing from *spirata* in having the early, evenly coiled part pure white in color. The later whorls are very similar to *spirata*. Common in sponge masses, and frequently washed ashore.

Vermicularia fargoi Olsson

Fargo's Worm-shell
Plate 21b

West Coast of Florida to Texas.

Similar to *spirata* and *knorri*, but the "turritella" or wound stage is ¾ to 1 inch in length; the shell is thicker and sturdy, its color a drab grayish to yellowish brown. Early whorls tan to brown, with 2 (sometimes 3) spiral cords. Subsequent whorls with 3 major, brown-spotted, thick cords. Aperture with a squarish columella corner. Minute minor threads are between the main cords. Commonly found crawling on mud flats. A race occurs in Texas in which the "turritella" stage is much more slender.

GENUS *Tenagodus* Guettard 1770
(*Siliquaria* Bruguière 1789)

Tenagodus squamatus Blainville

Slit Worm-shell
Plate 21g

Southeast Florida and the West Indies.

A small worm-like shell with detached whorls throughout. Characterized by the long row of small holes or elongate slits on the middle of the whorl. Early whorls smooth, white; later whorls becoming very spinose and stained with brown. The coiling is very irregular and loose. Grows to about 5 or 6 inches in length. *T. modestus* Dall may be the young of this species.

Family CAECIDAE

These tiny, cucumber-shaped mollusks are occasionally found by screening the beach sand in warm water areas or by shaking out dead sponges. The Caecums begin life in a normal snail-like manner with a tiny, spiral shell, but within a few weeks they grow only in one direction to form a simple, slightly curved tube. The spiral apex is usually knocked off and the hole plugged with a septum. As additional growth takes place, the animal retreats gradually from the apical or rear portion and forms new, internal septa. The operculum is thin, circular, horny, and multispiral.

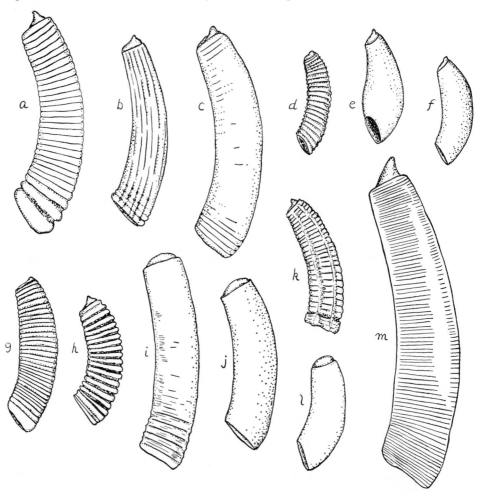

FIGURE 37. American Caecums. ATLANTIC: **a**, *Caecum floridanum* Stimpson; **b**, *C. cooperi* S. Smith; **c**, *C. carolinianum* Dall; **d**, *C. pulchellum* Stimpson; **e**, *C. nitidum* Stimpson; **f**, *C. nebulosum* Rehder (type). PACIFIC: **g**, *C. californicum* Dall; **h**, *C. dalli* Bartsch (type); **i**, *C. carpenteri* Bartsch (type); **j**, *C. occidentale* Bartsch (type); **k**, *C. heptagonum* Cpr.; **l**, *C. orcutti* Dall (type); **m**, *C. crebricinctum* Cpr. All X10.

GENUS *Caecum* Fleming 1817
Subgenus *Caecum s. str.*

Caecum floridanum Stimpson   Florida Caecum
                                                Figure 37a

North Carolina to Southern Florida.

3.0 mm. in length by 1.0 mm. in diameter, opaque-white, with 20 to 30 strong, axial rings, the last 3 or 4 being quite large.

Caecum cayosense Rehder | Key Caecum

Bonefish Key, Lower Florida Keys.

Shell similar to *C. floridanum*, but with about 14 very large, sharp axial rings.

Caecum californicum Dall | California Caecum
Figure 37g

Monterey to Lower California.

2.0 to 3.0 mm. in length. With 30 to 40 moderately developed, evenly spaced, rounded or squarish axial rings. Lip of aperture slightly thickened. Color a glossy, olive-brown.

Caecum dalli Bartsch | Dall's Caecum
Figure 37h

San Diego to Lower California.

About 3 mm. in length, usually with 18 to 24 moderately developed, evenly spaced, rounded or squarish, axial rings. Lip of aperture usually heavily developed in adults. Color tan. The number of raised rings varies from specimen to specimen, often in the same locality, and diligent search will usually bring to light any number desired. Extremes have been unwisely named (15 rings—*C. grippi* Bartsch; 17 to 19 rings—*C. licalum* Bartsch; and 19 to 22 rings—*C. diegense* Bartsch). Figure 37h is the holotype.

Subgenus *Micranellum* Bartsch 1920

Shells 3 to 7 mm. in length, opaque, with numerous, fine, closely packed, axial rings. About 8 Eastern Pacific and perhaps half a dozen Western Atlantic species in this subgenus.

Caecum pulchellum Stimpson | Beautiful Little Caecum
Figure 37d

Cape Cod south to North Carolina.

About 2 mm. in length, translucent-tan and glistening when alive; chalky-white when dead; with about 25 to 30 fine, closely set axial rings. Apex with a dome-shaped plug.

Caecum crebricinctum Carpenter | Many-named Caecum
Figure 37m

Monterey to Lower California.

6.0 mm. in length. Color pinkish brown to chalky-white with occasional darker brown mottlings. With about 100 fine, squarish, closely set axial rings. Plug with a rather long, oblique spur. Spur sometimes eroded down to a small sharp pimple (form *oregonense* Bartsch). Irregularities occur in the expansion of the tube; sometimes there is a more rapid expansion toward the anterior end (forms named as species: *C. pedroense* Bartsch and *C. barkleyense* Bartsch). *C. catalinense* Bartsch is probably this species, since many of the paratypes do not have the anterior end supposedly "bulbously expanded," and many specimens have about 100 axial rings, and not 75 as claimed. *C. rosanum* Bartsch appears to be a very long specimen (7 mm.) with sharply defined rings. Common.

Subgenus *Elephantanellum* Bartsch

3 to 5 mm. in length, white, with axial rings and longitudinal ribs. One species in southern California and 3 or 4 in the Western Atlantic. They resemble minute scaphopods, but are distinguished from them by the apical plug and small size of the shell.

Caecum cooperi S. Smith   Cooper's Atlantic Caecum

Figure 37b

South of Cape Cod to northern Florida.

4 to 5 mm. in length, slender, glossy, opaque-white; with about 15 strong, longitudinal ribs. Axial, raised rings are prominent near the aperture and sometimes give the shell a cancellate appearance at the anterior end. Apical plug with a fairly long, pointed prong. Common.

Caecum carpenteri Bartsch   Carpenter's Caecum

Figure 37i

San Pedro to Lower California.

3.5 to 4.8 mm. in length. First half to first ¾ of shell smooth, but at the apertural end developing about a dozen small, sharply defined axial rings. Longitudinal sculpture microscopic or absent. Color translucent-white to gray. This species is doubtfully placed in this subgenus.

Caecum heptagonum Carpenter   Heptagonal Caecum

Figure 37k

West Coast of Central America.

2.0 to 2.5 mm. in length. Opaque-white. 7-sided in cross-section. The 7 longitudinal ribs are strong, raised, and the spaces between them are flat or slightly concave. There are about 30 deeply cut circular lines around the shell, cutting across the ribs. Lip of aperture with one or two swollen axial rings.

Subgenus *Levia* de Folin 1875

Rather thick, glossy, slightly curved shells; aperture minutely constricted; sculpture absent except for microscopic growth lines. The shells are larger, heavier and not as bulbous as those in the subgenus *Fartulum*.

Caecum carolinianum Dall   Carolina Caecum
Figure 37c

North Carolina to southern Florida.

About 4.5 mm. in length, glossy, cream-white. Smooth except for microscopic growth lines. Apical plug sunk in at the posterior end of the shell and with a sharp, horn-like projection. Aperture minutely constricted.

Subgenus *Fartulum* Carpenter 1857

Shells very small, about 2 mm. in length, fragile, smooth, except for microscopic growth lines; not swollen in the middle; and with a nonconstricted aperture facing to one side (oblique).

Caecum nebulosum Rehder   Mottled Caecum
Figure 37f

Missouri Key, Florida Lower Keys.

1.5 to 2.0 mm. in length, fragile, translucent-tan with opaque-white mottlings. Not swollen in the center. Aperture oblique. Apex with a lopsided plug which has a single, weak spur. Found under flat rocks imbedded in tough, sticky marl.

Caecum orcutti Dall   Orcutt's Caecum
Figure 37l

San Pedro to Lower California.

2.0 to 2.5 mm. in length. Smooth, except for fine, circular scratches. Shell stubby, slightly compressed laterally; aperture oblique; apical plug dome-shaped. Color translucent-tan to yellow-brown. Moderately common.

Caecum occidentale Bartsch   Western Caecum
Figure 37j

Alaska to Lower California.

2.2 to 3.5 mm. in length. Smooth, except for fine, circular scratches. Shell elongate, round in cross-section. Aperture moderately oblique; apical plug dome-shaped with a tiny pimple on one side. Color translucent-tan to light-brown. Old specimens are whitish, often with a purplish stain. The shell has a white band behind the aperture. *C. hemphilli* Bartsch and *C. bakeri* Bartsch are probably diminutive forms of this species. The development of the small pimple on top of the dome-shaped plug is variable.

Subgenus *Meioceras* Carpenter 1858

Shells 2 to 4 mm. in length, very bulbous in the middle, smooth, and with an oblique, constricted aperture. Resembles a miniature *Cadulus* (Scaphopoda).

Caecum nitidum Stimpson   Little Horn Caecum
Figure 37e

Southern half of Florida and the West Indies.

2.0 mm. in length, glassy translucent-white with irregular specks or mottlings of chalk-white; bulbous in the center; apex with a lopsided, rounded plug which has a tiny projection on the highest side.

Caecum lermondi Dall from the west coast of Florida differs in having a single, moderately well-raised, circular hump around the middle of the shell. Uncommonly dredged just offshore.

Family *PLANAXIDAE*
Genus *Planaxis* Lamarck 1822

Planaxis lineatus da Costa   Dwarf Atlantic Planaxis

Lower Florida Keys and the West Indies.

¼ inch in length, thick and strong; glossy-smooth when the thin, smoothish, translucent periostracum is worn away. Color whitish cream with neat, spiral bands of brown (10 in last whorl, 5 showing in whorls above). Whorls in top of spire with 4 or 5 small spiral cords, later becoming obsolete. Aperture slightly flaring, enamel-white with 10 brown dots on the edge of the outer lip. Nuclear whorls very small, glossy, translucent-brown and sharply pointed.

Subgenus *Supplanaxis* Thiele 1929

Planaxis nucleus Bruguière   Black Atlantic Planaxis

Southeast Florida and the West Indies.

½ inch in length, resembling a thick, polished, dark-brown *Littorina* periwinkle. Characterized by 5 strong spiral cords which are developed on the outside of the body whorl only in the region behind the slightly flaring lip. 3 other cords are present just below the suture. Columella area dished; reinforced by the round, pillar-like columella. A small pimple is present near the posterior canal in the aperture. Outer lip with strong crenulations on the inside. Periostracum a soft gray-black felt. A common littoral species in the West Indies which bears its young in a brood pouch. Rare in Florida.

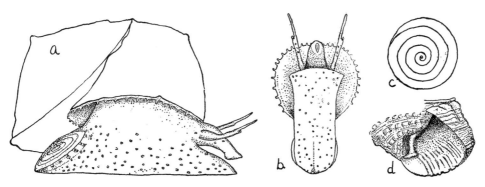

FIGURE 38. *Modulus modulus* Linné (southeast United States and the West Indies). **a,** side view of living animal; **b,** ventral view showing foot, head and under edge of mantle; **c,** operculum; **d,** apertural view of shell. X2. (From Abbott 1944 in *Johnsonia.*)

Family MODULIDAE
GENUS *Modulus* Gray 1842

Modulus modulus Linné   Atlantic Modulus
                                                          Plate 21f; figure 38
Florida to Texas and the West Indies.

About ½ inch in length. Characterized by the small, projecting, tooth-like, frequently brownish spine located on the lower end of the columella. Base of shell with about 5 strong, spiral cords. Top of whorls with low, slanting, axial ribs. Color grayish white with beach-worn specimens often exhibiting flecks of purple-brown. Found abundantly among weeds in shallow, warm waters.

Modulus carchedonius Lamarck, the Angled Modulus (Caribbean area) lives in deeper water, and differs in having the periphery of the shell well-angulated, the spiral cords smaller and neater, in lacking the strong, axial ribs, and in never having the columella tooth colored. Not too common.

Family POTAMIDIDAE
Subfamily POTAMIDINAE
(Horn Shells)
GENUS *Cerithidea* Swainson 1840

The horn shells are intertidal mud-lovers. The shells are elongate and with 10 to 15 convex whorls. Axial ribs are more prominent on the early whorls. Outer lip flares. Operculum horny, thin, paucispiral and with its nucleus at the center.

Subgenus *Cerithideopsis* Thiele 1929

Cerithidea costata da Costa   Costate Horn Shell
Plate 19u

West coast of Florida and the West Indies.

½ inch in length, translucent, pale yellowish brown. With 9 to 12 very convex whorls. Axial, curved ribs are round and distinct on the early whorls, fading out on the last two whorls. No old varices present. A common shallow-water, mud-loving species.

The subspecies *C. costata turrita* Stearns, the Turret Horn Shell from the Tampa-Sanibel region, has 15 to 20 (instead of 25 to 30) axial ribs on the next to the last whorl.

Cerithidea pliculosa Menke   Plicate Horn Shell
Plate 19t

Texas, Louisiana and the West Indies. Not Florida.

1 inch in length, brownish black in color. 11 to 13 slightly convex whorls. Several yellowish, former varices are present. Numerous spiral threads make the axial ribs slightly nodulose. Locally common. It may yet turn up in northwest Florida.

Cerithidea scalariformis Say   Ladder Horn Shell
Plate 19x

South Carolina to south half of Florida and the West Indies.

¾ to 1¼ inches in length. Pale russet-brown to slightly violaceous, usually with many conspicuous, dirty-white, spiral bands. 10 to 13 moderately convex whorls. Many coarse, axial ribs present which stop abruptly below the periphery of the whorl at a sharply marked, rounded spiral ridge. Base of shell with 6 to 8 spiral ridges. No former varices. Common on mud flats.

Cerithidea hegewischi californica Haldeman   California Horn Shell

Bolinas Bay, California, to Lower California.

1 to 1¼ inches in length, resembling our photo of *C. pliculosa* from the Atlantic (pl. 19t). Whorls 11, spirally and weakly threaded, and axially strongly ribbed (12 to 18 ribs per whorl). Dark-brown in color with 1 or

2 yellowish white, swollen varices on the spire. A very common species found in large colonies on mud flats.

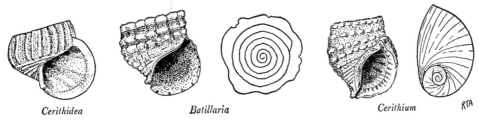

Cerithidea · · · · Batillaria · · · · Cerithium · · · · RTA

FIGURE 39. Last whorl and opercula in the Horn Shells. (From J. Bequaert in *Johnsonia*.)

Subfamily BATILLARIINAE
GENUS *Batillaria* Benson 1842

Cerithium-like in appearance. Siphonal canal very short and twisted to the left. Outer lip smooth inside. Operculum round, multispiral and horny, while in *Cerithidea* and *Cerithium* it is paucispiral.

Batillaria minima Gmelin · · · · · · · · · · · · · · · · · **False Cerith**
Plate 19s

South half of Florida and the West Indies.

½ to ¾ inch in length, resembling the Dwarf Cerith, *C. variabile* (see below). Color varies from black, gray to whitish, and often has black or white spiral lines. Finely nodulose with coarse axial swellings and uneven spiral threads. The siphonal canal is very short and twisted slightly to the left. Operculum multispiral. A very common intertidal species. Percy Morris (1951, pl. 31, fig. 15) labels this species as *Cerithidea turrita*.

Family CERITHIIDAE
GENUS *Cerithium* Bruguière 1789

Thericium Monterosato is this genus. The operculum is horny, thin, brown and paucispiral. Most species in the genus are shallow-water dwellers.

Cerithium floridanum Mörch · · · · · · · · · · · · · · · · **Florida Cerith**
Plate 19n

North Carolina to the south half of Florida.

1 to 1½ inches in length, elongate. Spire pointed, with 2 or 3 white, former varices on each whorl. Siphonal canal well-developed. With several spiral rows of 18 to 20 neat beads per whorl between which are fine, granulated spiral threads. Color whitish with mottlings and specklings of reddish brown. Distinguished from *C. literatum* by its more elongate shape and neater, smaller, more numerous beads. Common in shallow water.

Cerithium muscarum Say   Fly-specked Cerith
                                                             Plate 19m

South half of Florida and the West Indies.

1 inch in length, moderately elongate. Siphonal canal rather long and twisted to the left. 9 to 11 nodulated axial ribs on each whorl. Base of shell with a very strong spiral cord, often nodulated. Former varices rarely present. Apertural side of body whorl convex. Color slate- to brown-gray, usually with brown to reddish specks in spiral rows. Common in shallow, warm waters on the west coast of Florida.

Cerithium literatum Born   Stocky Cerith
                                                             Plate 19l

Southeast Florida, Bermuda and the West Indies.

1 inch in length, half as wide; siphonal canal short. Aperture side of body whorl slightly flattened. Usually 1 weak, former varix present. With numerous coarse spiral threads, and with a spiral row of 9 to 12 sharp, prominent nodules just below the suture. Sometimes a second, smaller row of spines is on the periphery. Color whitish with spiral rows of many black or reddish squares. Common in shallow water on the Lower Florida Keys.

Cerithium eburneum Bruguière   Ivory Cerith
                                                             Plate 19q

Southeast Florida, the Bahamas and Greater Antilles.

¾ to 1 (rarely 1½) inches in length, variable in shape, but usually moderately elongate. Each whorl has 4 to 6 spiral rows of from 18 to 22 small rounded beads. The beads are slightly larger in the middle row. There are usually a number of fairly large, former varices. Color variable: all white or cream, or with reddish brown blotches. Very common in shallow water. *C. versicolor* C. B. Adams is this species. Compare with *algicola* which may ultimately prove to be a genetic form of this species.

Cerithium algicola C. B. Adams   Middle-spined Cerith
                                                             Plate 19p

Southern third of Florida and the West Indies.

1 inch in length, similar to *eburneum*, but characterized by each whorl having the middle spiral row of 9 to 12 beads fairly large and pointed. These large beads may be axially drawn out to form low ribs. Former varices are not often present. Color as in *eburneum*. Common in the West Indies. *C. literatum* has its strongest row of spine-like beads just below the suture.

Cerithium variabile C. B. Adams   Dwarf Cerith
                                                             Plate 19-o

South half of Florida to Texas and the West Indies.

⅓ to ½ inch in length, not elongate. Apertural side of body whorl sometimes flat. 1 to 2 former varices on last whorl. 3 or 4 spiral rows of even-sized fine beads on the whorls of the spire. Color dark brown-black, but sometimes whitish with heavy specklings and mottlings and bands of reddish brown. Very common under rocks in warm water. Do not confuse with *Batillaria minima* (see above).

GENUS *Bittium* Gray 1847

Shell small, very slender, spire high and body whorl small. Whorls varicose. Nucleus of about 3 glassy, smooth whorls. Aperture ovate, the anterior canal broad and stout.

Subgenus *Bittium s. str.*

Bittium alternatum Say   Alternate Bittium

Gulf of St. Lawrence to Virginia.

Adults very small, ⅛ to ¼ inch in length, light- to dark-brown in color, sometimes translucent or with specklings. Suture impressed, whorls rounded. Sculpture on top whorls either cancellate or with 4 to 5 spiral rows of beads, or occasionally with axial, nodulated ribs. Base with small spiral cords. Outer lip flaring, thin and sharp. Columella short, twisted at the base and stained brown. Very abundant from tidal flats to 20 fathoms.

B. virginicum Henderson and Bartsch from Chincoteague, Virginia, is similar, but very elongate, more whorls, with a much more flaring and basally projecting lip, and with a large, whitish, former varix on the body whorl.

Bittium varium Pfeiffer   Variable Bittium
Plate 19r

Maryland to Florida, Texas and Mexico.

Adults similar to *alternatum*, but smaller (⅛ inch), nearly always with a former, thickened varix. The aperture is proportionately smaller and the base of the apertural lip is squarish instead of rounded. The last third of the body whorl is generally destitute of sculpturing. Common in eel-grass just below low tide.

Subgenus *Stylidium* Dall 1907

Bittium eschrichti Middendorf   Giant Pacific Coast Bittium

Alaska to Crescent City, California.

½ to ¾ inch in length, dirty whitish gray in color with an undertone of reddish brown. About a dozen whorls. With wide, flat-topped, raised spiral cords between which are depressed, squarish, spiral furrows half as wide as the cords. 4 to 5 cords between sutures. Common below low water. The subspecies *montereyense* Bartsch (Crescent City south to Lower California) is glossy, whitish with brown maculations and is proportionately shorter.

Bittium quadrifilatum Carpenter   Four-threaded Bittium

Monterey, California, to Lower California.

⅜ inch in length, similar to *attenuatum*, but earliest whorls with about a dozen smooth axial ribs which, however, in subsequent whorls become beaded as 4 to 5 small spiral threads cross them. The sculpturing may become faint at the very last third of the last whorl. Color reddish brown to gray. A very common littoral species.

Bittium attenuatum Carpenter   Slender Bittium

Forrester Island, Alaska, to Lower California.

⅓ inch in length, slender, yellowish brown to dark-brown. Sculpture variable. Nuclear whorls with two smooth spiral cords. Early whorls have 4 to 5 spiral rows of small beads, sometimes arranged axially. In the last whorl, the cords gradually become smooth and flat-topped and resemble those of *eschrichti*. Common just offshore to 35 fathoms.

Subgenus *Lirobittium* Bartsch 1911

Bittium interfossum Carpenter   White Cancellate Bittium

Monterey, California, to Lower California.

¼ inch in length, pure-white; whorls in spire with 2 rows of sharp beads connected by small axial and spiral threads or small cords. Base of shell with 3 very strong, rounded, smooth spiral cords. Moderately common under rocks at low tide.

Subfamily LITIOPINAE
Genus *Litiopa* Rang 1829

Litiopa melanostoma Rang   Brown Sargassum Snail
                                                    Plate 21k

Pelagic in floating sargassum weed.

¾₁₆ to ¼ inch in length, fragile, light-brown; moderately elongate, with 7 whorls, the last being quite large. Nuclear whorls extremely small. Surface glossy, smooth, except for numerous, microscopic, incised spiral lines. Characterized by the strong ridge just inside the aperture on the columella. Often washed ashore with floating sargassum weed, and frequently dredged in a dead condition at any depth. This is *L. bombix* Kiener and *L. bombyx* "Rang."

Subfamily CERITHIOPSINAE
GENUS *Cerithiopsis* Forbes and Hanley 1849

Cerithiopsis greeni C. B. Adams   Green's Miniature Cerith
Plate 19v

Cape Cod to both sides of Florida.

⅛ inch in length, elongate, slightly fusiform in shape, glossy-brown in color. 9 whorls, the first 3 embryonic, translucent-brown and smooth, the remainder with 2 to 3 spiral rows of large, glassy beads connected by weak spiral and axial threads. Columella arched in young specimens, but straight and continuous with the short siphonal canal in adults. Lip in adults smoothish, slightly flaring. *C. virginica* Henderson and Bartsch and *C. vanhyningi* Bartsch are possibly variations of this species. Common in shallow water.

Subgenus *Laskeya* Iredale 1918

Cerithiopsis subulata Montagu   Awl Miniature Cerith
Plate 19w

Massachusetts to the West Indies.

½ to ¾ inch in length, rather strong, slender and with about 14 whorls. Sides of whorls flattish, with 3 rows of distinct, raised, roundish beads (about 28 per row on the last whorl). There may be faint axial riblets connecting the beads. The middle row of beads may be reduced to a mere thread in specimens from southern localities. Base concave and with fine axial growth lines. Color chocolate-brown, with the beads a lighter shade. Some shells become eroded and colored an ash-gray or chalky-brown. *C. emersoni* C. B. Adams is probably a synonym. Common from 1 to 33 fathoms.

Cerithiopsis carpenteri Bartsch   Carpenter's Miniature Cerith

Crescent City, California, to Ensenada, Mexico.

¼ to ⅓ inch in length, dark chocolate-brown with whitish beads. Whorls in spire with 3 spiral rows of evenly sized, glassy, rounded beads.

Base of shell with 2 large, smoothish, spiral cords. *C. grippi* Bartsch and *C. pedroana* Bartsch are possibly dwarf forms of this species whose beaded sculpture is more variable than is generally suspected.

Genus *Seila* A. Adams 1861

Shell small, very slender, whorls flat-sided, nucleus glassy-smooth and of about 3 whorls. Small, short siphonal canal. Sculpture of strong spiral cords between which lie microscopic axial threads.

Seila adamsi H. C. Lea   Adams' Miniature Cerith

Plate 22t

Massachusetts to Florida, Texas and the West Indies.

¼ to ½ inch in length, resembling a miniature *Terebra*, with about a dozen whorls. Long, slender, flat-sided, dark-brown to light orange-brown in color, and characterized by 3 strong, squarish, spiral cords on each whorl (4 on the last whorl). Occasionally with minute axial threads showing between the spiral cords. Base of shell smoothish, concave. Outer lip fragile, wavy and sharp. Suture indistinct. This is *S. terebralis* C. B. Adams. Common from shore to 40 fathoms.

Seila montereyensis Bartsch   Monterey Miniature Cerith

Monterey, California, to the Gulf of California.

⅜ to ½ inch in length, yellowish to reddish brown. Whorls and spire flat-sided. Whorls in spire with 3 raised, flat-topped, evenly spaced, smooth cords between which are numerous, microscopic, axial threads. Last whorl with 5 cords. Base smoothish, concave. Common from low tide to 35 fathoms.

Genus *Alabina* Dall 1902

Shell small, slender. Nucleus slender, of 3 to 4 glassy, smooth whorls. Aperture subcircular. Lower part of the outer lip extended and flaring. Umbilicus very narrow and very small. With obscure, narrow, curved axial ribs and prominent spiral threads. This genus is put in the family Diastomidae by some workers.

Alabina tenuisculpta Carpenter   Sculptured Alabine

San Pedro, California, to Lower California.

¼ inch in length, slender, 8 to 9 whorls, ashen gray with a light-brown undertone. Outer lip thin; umbilicus small. Spiral sculpture of 4 to 6 weak cords or threads. Axial sculpture of weak, obsolete or sometimes strong, very tiny, rounded riblets. *A. t. diegensis* Bartsch is a strongly sculptured form of this species.

Family *TRIPHORIDAE*
GENUS *Triphora* Blainville 1828

Shell left-handed (sinistral), very small, and slender. Aperture subcircular. Siphonal canal short, curved backward, slightly emarginate, upper part almost or completely closed. Posterior canal very slightly developed. Sculpture of spiral rows of neat beads, often joined by axial threads.

Triphora nigrocincta C. B. Adams   Black-lined Trifora
Plate 19y

Massachusetts to Florida, Texas and the West Indies.

⅛ to ¼ inch in length, left-handed, with 10 to 12 slightly convex whorls; dark chestnut-brown with 3 spiral rows of prominent, grayish, glossy beads. Darker band of black-brown is just below the suture. Aperture and columella brown. A common species found on seaweed at low tide. Sometimes considered a subspecies of *perversa* from Europe.

Triphora decorata C. B. Adams   Mottled Trifora
Plate 19zz

Southeast Florida, the West Indies and Bermuda.

½ inch in length, left-handed, with about 20 flat-sided whorls which bear 3 spiral rows of large beads (28 per row per whorl). Color of shell cream to gray with large, irregular maculations of reddish brown. Moderately common from 1 to 40 fathoms. *T. ornata* Deshayes from the same area is very similar, but half as large, the spire slightly concave instead of being flat.

Triphora pulchella C. B. Adams   Beautiful Trifora
Plate 19z

Southeast Florida and the West Indies.

³⁄₁₆ inch in length, left-handed, spire slightly convex; 15 whorls slightly convex, and with 3 spiral rows of beads which are joined axially and spirally by small, low, smooth threads. Suture well-indented. Upper third of whorl, including beads, colored light-brown, lower two thirds white. Uncommon in shallow water down to 56 fathoms.

Triphora pedroana Bartsch   San Pedro Trifora

Redondo Beach, California, to Lower California.

¼ inch or less in length, slightly fusiform with very slightly convex sides to the spire. Suture almost impossible to see. Color glossy yellow-brown with 2 rows of glassy, whitish, rounded beads. A third much weaker row of beads, or an additional spiral thread, may appear in the last 2 or 3 whorls. Axial threads connecting the beads are weak and form small pits. Fairly common under stones along the low-tide zone.

Family JANTHINIDAE
Genus *Janthina* Röding 1798
Subgenus *Janthina* s. str.

Janthina janthina Linné   Common Purple Sea-snail
Plate 4j

Pelagic in warm waters; both coasts of the United States.

1 to 1½ inches in diameter. Whorls slightly angular. Two-toned, with purplish white above and deep purplish violet below. Outer lip very slightly sinuate. Common after certain easterly blows along the south-eastern United States, especially from April to May. This is *J. fragilis* Lamarck.

Subgenus *Violetta* Iredale 1929

Janthina globosa Swainson   Globe Purple Sea-snail
Plate 4k

Cast ashore along both coasts of the United States.

½ to ¾ inch in diameter. Whorls globose, well-rounded. Color violet throughout. Outer lip very slightly sinuate. Not very common.

Subgenus *Jodina* Mörch 1860

Janthina exigua Lamarck   Dwarf Purple Sea-snail
Plate 4l

Cast ashore in most warm seas.

¼ inch in length. Whorls slightly flattened from above. Outer lip with a prominent notch. Light-violet, banded at the suture. Fairly common. *J. bifida* Nuttall is probably this species.

NOTE
Superfamily PYRAMIDELLACEA

The families *Pyramidellidae*, *Aclididae*, *Eulimidae*, *Styliferidae* and *Entoconchidae*, most of which are small parasitic gastropods, have in the past been placed here among the prosobranchs, but are now considered to be opisthobranchs and related to the bubble shells. Recent work on the embryology and anatomy appears to justify this radical change in classification. They are located in this book on page 288.

Superfamily *EPITONIACEA*
Family *EPITONIIDAE*
(Wentletraps)
GENUS *Sthenorytis* Conrad 1862

Sthenorytis pernobilis Fischer and Bernardi   Noble Wentletrap
Figure 40c

North Carolina to southeast Florida and to Barbados.

1 to 1½ inches in length, solid, pure-white to grayish; angle of spire about 50 degrees. The 10 whorls are globose and each bears about 14 very large, thin, blade-like ribs. Apertural rim round, solid. A very choice collector's item. It is the only member of the genus in Western Atlantic waters, *S. cubana* Bartsch, *S. hendersoni* Bartsch and *S. epae* Bartsch being minor forms of this rare species.

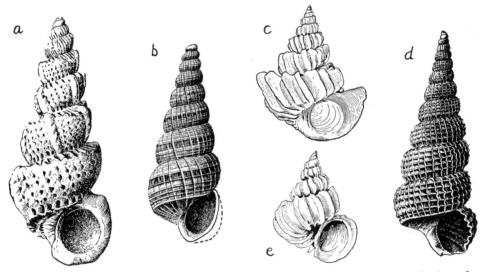

FIGURE 40. Atlantic Wentletraps. **a**, *Cirsotrema dalli* Rehder, 1½ inches; **b**, *Amaea mitchelli* Dall, 2 inches (Texas); **c**, *Sthenorytis pernobilis* Fischer and Bernardi, 1 inch; **d**, *Amaea retifera* Dall, 1 inch; **e**, *Epitonium krebsi* Mörch, ¾ inch.

GENUS *Cirsotrema* Mörch 1852

Cirsotrema dalli Rehder   Dall's Wentletrap
Plate 22c; figure 40a

North Carolina to southeast Florida and to Brazil.

1 to 1½ inches in length, rather slender, with a quite deep suture, thus giving the whorls a shouldered appearance. No umbilicus. Color a uniform, chalky grayish white. Outer lip with a thickened varix. Whorls with numerous groups of foliated costae. Surface pitted with small holes when the costae or ribs are closely crowded. Uncommon from 18 to 75 fathoms. *C. arcella* Rehder is believed to be the young of this species.

GENUS *Acirsa* Mörch 1857

Acirsa costulata Mighels and Adams   Costate Wentletrap

Arctic Ocean to Massachusetts.

¾ to 1¼ inches in length, rather turreted in shape and fairly thin in structure. 8 to 9 moderately convex whorls are devoid of sculpture except for weak, incised spiral lines and, in the early whorls, numerous but inconspicuous costae. Color straw to whitish, rarely with brown lines. Uncommon from low water to 50 fathoms. This is *Scalaria borealis* Beck.

GENUS *Opalia* H. and A. Adams 1853
Subgenus *Dentiscala* de Boury 1886

Opalia hotessieriana Orbigny   Hotessier's Wentletrap
Plate 22g

Southeast Florida and the Caribbean.

⅓ to ½ inch in length, moderately slender. Characterized by 10 to 14 large, square notches along the suture of each whorl. Ribs are rather weak. Surface, in fresh specimens, microscopically pitted. Color grayish white. Not uncommon from low water to 90 fathoms. *O. crenata* Linné (same range, but also the Eastern Atlantic) is larger, its whorls more strongly shouldered, and the notches at the suture are much weaker and more numerous.

Opalia wroblewskii Mörch   Wroblewski's Wentletrap
Plate 20j

Forrester Island, Alaska, to off San Diego.

1 to 1¼ inches in length, slender, heavy; looks beachworn; grayish white in color, often stained purple from the animal's dye gland. With 6 to 8 low, pronounced, axial, wide ribs. Base of shell bounded by a strong, smooth, low, spiral cord. Fairly common. *O. chacei* (Chace's Wentletrap) is probably a southern representative of this species.

Opalia insculpta Carpenter   Scallop-edged Wentletrap

Southern California to west Mexico.

½ to ¾ inch in length, dull whitish, 7 to 8 whorls, moderately slender. Characterized by the smoothish sides of the whorls and by the spiral ramp below the suture which bears 12 to 14 short, horizontal ribs per whorl. Early whorls may have weak axial ribs running from suture to suture. Spinal sculpture of microscopic, numerous scratches. *O. crenimarginata* Dall is this species. Very common among rocks at low tide.

GENUS Amaea H. and A. Adams 1854

Amaea mitchelli Dall   Mitchell's Wentletrap
Plate 22f; figure 40b

Texas coast to Yucatan.

1½ to 2½ inches in length, thin but strong; without an umbilicus. With about 15 rather strongly convex, pale-ivory whorls which have a dark brownish band at the periphery and a solid brown area below the basal ridge. About 22 low, irregular costae per whorl. Numerous spiral threads are fine, and produce a weak, reticulated pattern. Not very common, but occasionally washed up on Texas beaches.

Subgenus Scalina Conrad 1865
(Ferminoscala Dall 1908)

Amaea retifera Dall   Reticulate Wentletrap
Figure 40d

North Carolina to both sides of Florida to Barbados.

1 inch in length, elongate, thin but strong; with about 16 whorls which are beautifully reticulated by strong, sharp threads. Color straw to pale-brown with 2 light and narrow brownish bands, one above and one below the periphery. Commonly dredged off Florida from 13 to 120 fathoms.

GENUS Epitonium Röding 1798 (Scala)
Subgenus Epitonium s. str.

Epitonium krebsi Mörch   Krebs' Wentletrap
Figure 40e

South half of Florida to the Lesser Antilles.

½ to ¾ inch in length, stout. With umbilicus fairly narrow to wide, and very deep. 7 to 8 whorls attached by the costae (10 to 12 per whorl). China-white, rarely with a trace of brown to pinkish brown undertones. Moderately common from a few feet to 160 fathoms. *E. swifti* Mörch and *E. contorquata* Dall are this species.

Do not confuse with *E. occidentale* Nyst (Western Atlantic Wentletrap) from the same areas. It is not so stout, has 12 to 15 costae per whorl, a very small umbilicus or none, and the shoulder of the whorls is somewhat flattened. It is not common.

Epitonium tollini Bartsch   Tollin's Wentletrap

West Coast of Florida.

½ inch in length, slender, no umbilicus. 9 to 10 whorls strongly convex; suture deep. Each whorl has from 11 to 16 costae which are not shouldered, but are rounded, on top. They often line up one below the other. Outer lip thick and reflected. Inner lip much smaller. Color china-white, with the first few whorls a very faint amber-brown. Fairly common just off the outer beaches. Do not confuse with *E. humphreysi* whose costae are angular at the top.

Epitonium humphreysi Kiener   Humphrey's Wentletrap
                                         Plate 22d

Cape Cod, Massachusetts, to Florida and to Texas.

½ to ¾ inch in length, fairly slender, thick-shelled, and without an umbilicus. Color dull-white. Suture deep. The 9 to 10 convex whorls each have about 8 to 9 costae that are somewhat angled at the shoulder. Costae usually thick and strong. Outer and lower part of the apertural lip thickened and slightly flaring. Common from shore to 52 fathoms. Do not confuse with *E. angulatum* which is not so slender, is glossier, has thinner costae that are usually more angular at the shoulders.

Epitonium eburneum Potiez and Michaud listed by Percy Morris (1951, p. 122) does not occur in American waters. His illustrations are probably those of *E. commune* Linné from Europe.

Epitonium angulatum Say   Angulate Wentletrap
                                     Plate 22b

New York to Florida and to Texas.

¾ to 1 inch in length, moderately stout to somewhat slender, strong and without an umbilicus. 8 whorls with about 9 to 10 strong but thin costae which are very slightly reflected backwards and which are usually angulated at the shoulder, especially in the early whorls. The costae are usually formed in line with those on the whorl above and are fused at their points of contact. Outer lip thickened and reflected. Color china-white. One of the commonest Atlantic wentletraps found in shallow water to 25 fathoms. Do not confuse with *E. humphreysi*.

Epitonium foliaceicostum Orbigny   Wrinkled-ribbed Wentletrap

Southeast Florida to the Lesser Antilles.

½ to ¾ inch in length, moderately stout, without an umbilicus, and similar to *E. angulatum*, except that the 7 to 8 costae per whorl are thinner, more highly developed and usually quite angular. Moderately common from low water to 120 fathoms. Alias *muricata* Sby., *spina-rosae* Mörch and *pretiosula* Mörch.

Subgenus *Gyroscala* Boury 1887

Epitonium lamellosum Lamarck   Lamellose Wentletrap
                                                         Plate 22a

South half of Florida and the Caribbean. Also Europe.

¾ to 1¼ inches in length, without an umbilicus. 11 whorls whitish with irregular, brownish markings. Costae thin, high, always white. Characterized by a fairly strong, raised, spiral thread on the base of the shell. Moderately common from low water to 33 fathoms. Alias *E. clathrum* of authors, not Linné.

Epitonium rupicola Kurtz   Brown-banded Wentletrap
                                                         Plate 22e

Cape Cod, Massachusetts, to Florida and to Texas.

½ to 1 inch in length, moderately stout to slender, and without an umbilicus. Color whitish or yellowish with 2 brownish, spiral bands on each side of the suture. Color often diffused. About 11 globose whorls, each of which has from 12 to 18 weak or strong costae. Former, thickened varices sometimes present. Base of shell with a single, fine, spiral thread. Formerly known as *lineatum* Say and *reynoldsi* Sby. Common from low water to about 20 fathoms.

Epitonium indianorum Carpenter   Money Wentletrap

Forrester Island, Alaska, to Lower California.

1 inch in length, slender, pure white, of 11 whorls, each of which has 13 to 14 sharp costae which are slightly bent backwards. The tops of the costae are slightly pointed. Fairly common offshore.

Superfamily *HIPPONICACEA*
Family *HIPPONICIDAE*
GENUS *Cheilea* Modeer 1793

Cheilea equestris Linné   False Cup-and-saucer
                                                         Plate 21p

Southeast Florida and the West Indies.

½ to 1 inch in size, cap-shaped, dull-white, and with an internal, delicate, deep cup which has its anterior third neatly sliced away. The base of the cup is attached near the center of the inside of the shell but slightly off in the direction in which the apex of the shell points. Exterior has small, axial corrugations or tiny cords, rarely spinose. Nucleus minute, spiral and glassy-white. Uncommon except in the West Indies.

GENUS *Hipponix* Defrance 1819

Hipponix antiquatus Linné White Hoof-shell
Plate 21t

Southeast Florida and the West Indies. Crescent City, California, to Peru.

½ inch in size, white, heavy for its size, cap-shaped, and usually with a poorly developed spire which may be located either at one end of the shell or near the center. The nuclear whorls are spiral and glassy-white. There is a horseshoe-shaped muscle scar inside the shell. Axial sculpture of prominent, rugose ribs which are crossed by microscopic, incised lines. Periostracum absent or very thin and light-yellowish. Moderately common. Found clinging to rocks and other shells.

Some Pacific northwest specimens are limpet-like in shape, flattish, circular, gray-white, with the apex near the center of the shell, and with smoothish, strong, circular cords (form *cranoides* Carpenter). Another form bears foliaceous concentric lamellae which are finely striate axially (*serratus* Carpenter from Monterey to Panama).

Hipponix subrufus subrufus Lamarck Orange Hoof-shell

Southeast Florida and the West Indies.

½ inch in size, similar to *antiquatus*, but usually stained with light orange-brown, and with numerous, small spiral cords crossing concentric ridges of about the same size. This frequently gives a beaded surface. Periostracum fairly heavy, tufted and light brown. Moderately common.

Hipponix benthophilus Dall (Dall's Deepsea Hoof-shell) is well-spired in one plane and is entirely smooth. It is rare and comes from deep water off Florida and throughout the West Indies.

Hipponix subrufus tumens Carpenter Pacific Orange Hoof-shell

Crescent City, California, to Lower California.

Very close in characters to the Atlantic *subrufus subrufus,* but the shell is white in color (although the periostracum is yellow-brown), with more prominent spiral threads, and with coarser spiral threads in the young. Found offshore.

Hipponix barbatus Sowerby (Bearded Hoof-shell) from the same region is limpet-shaped, with coarse, nodulated ribs which are largest on the anterior slope of the shell. The edge of the shell is strongly serrated with cut lines. The periostracum is very shaggy especially on the middle of the anterior slope. Common.

Family *VANIKOROIDAE*
GENUS *Vanikoro* Quoy and Gaimard 1832

Vanikoro oxychone Mörch   West Indian Vanikoro

Southeast Florida and the West Indies.

⅓ inch in length, solid, strong and pure white. With 3 whorls. Characterized by its large aperture, by its deep, narrow, arched umbilicus and straight, rounded, pillar-like columella and by the 10 or 12 beaded, spiral cords on the last whorl. Small axial threads tend to give a slightly cancellate sculpture. Apex glassy and smooth. Suture well-indented. Uncommon in shallow water.

Superfamily *CALYPTRAEACEA*
Family *TRICHOTROPIDAE*
GENUS *Trichotropis* Broderip and Sby. 1829

Trichotropis borealis Broderip and Sowerby   Boreal Hairy-shell
Plate 24d

Arctic Seas to Maine. Arctic Seas to British Columbia.

½ to ¾ inch in length, with 4 to 5 carinate whorls. Shell not very strong, chalky-white and covered with a thick, brownish periostracum which has hairy spicules on the region of the shell's 3 major spiral cords. Umbilicus chink-like, bordered by a large, spiral cord. Spire usually eroded badly. Numerous, crowded axial threads present. A common cold-water species found from below low water to 90 fathoms.

Trichotropis cancellata Hinds   Cancellate Hairy-shell
Plate 24b

Bering Sea to Oregon.

¾ to 1 inch in length, 5 to 6 rounded whorls bearing between sutures 4 to 5 strong spiral cords, between which there may be small axial ribs which produce a cancellate sculpturing. Spire high, rather pointed. Aperture a little more than ⅓ the length of the shell. Periostracum thick, brown and with long spicules over the region of the cords. Commonly dredged in cold, shallow water.

Trichotropis bicarinata Sowerby   Two-keeled Hairy-shell
Plate 24a
Arctic Ocean to Alaska. Arctic Ocean to Newfoundland.

1½ inches in length, equally wide, with about 4 whorls. Characterized by 2 strong, spiral carinae at the periphery, by the wide, flattened columella and by the flaky, brown periostracum which is grossly spinose on the carinae. Uncommon just offshore in cold water.

Trichotropis insignis Middendorff   Gray Hairy-shell
Plate 24c
Alaska to northern Japan.

1 inch in length, similar to *T. bicarinata* but smaller, with a much heavier shell, weakly carinate with other numerous, uneven, spiral threads, and with a thin, grayish periostracum. Both this species and *bicarinata* are easily distinguished from the more common *cancellata* by their much shorter spires and large flaring apertures. Uncommon just offshore.

Family CAPULIDAE
GENUS *Capulus* Montfort 1810
Subgenus *Krebsia* Mörch 1877

Capulus incurvatus Gmelin   Incurved Cap-shell
North Carolina to southeast Florida and the West Indies.

½ inch in size, cap-shaped, white to cream, and with a very large, circular to slightly oval aperture. 1½ to 2 whorls. Spire small, usually tightly coiled, but sometimes partially free. Early whorls usually with small spiral cords, but these are frequently worn smooth. Sculpture of small, irregular, rounded growth lines which are crossed by numerous spiral cords which may be rounded or sharp. Periostracum thick, light-brown, with spirally arranged rows of small tufts. Muscle scar within the aperture is horseshoe-shaped with the swollen end just inside the columella. Uncommon on rocks just below low water. I believe that *C. intortus* Lamarck is merely a variant of this species. Compare with *Hipponix antiquata* which is much heavier, lacks the spiral cords and is more coarsely sculptured.

Capulus californicus Dall   Californian Cap-shell
Redondo Beach to Lower California.

1½ inches in diameter, ⅓ as high, obliquely ovate, fairly thin, and with a small, hooked-over apex. Shell white, covered by a soft, fuzzy, light-brown periostracum. Interior glossy-white. A rather rare species found in 20 to 30 fathoms attached to *Pecten diegensis*.

Family CALYPTRAEIDAE
GENUS *Calyptraea* Lamarck 1799

Calyptraea centralis Conrad   Circular Cup-and-saucer
Plate 21-o

North Carolina to Texas and the West Indies.

¼ to ½ inch in diameter, cap-shaped, with a circular base, and pure white in color. Apex central, small, minutely coiled and glossy-white. The shelly cup is attached to the inside of the shell and is flattish, arises near the center of the shell and flares out to the edge. Its free side is thickened into a columella-like, rounded edge. Commonly dredged in shallow water, especially off southeast Florida. Formerly known as *C. candeana* Orbigny.

Calyptraea fastigiata Gould   Pacific Chinese Hat
Plate 20l

Alaska to southern California.

⅓ to 1 inch in diameter, about ½ to ⅓ as high; the outline of the base of the shell is perfectly circular, and the apex is at the center of the shell. Interior glossy-white with the sinuate edge of the internal cup arising at the apex of the shell as a thickened, twisted columella and ending in fragile attachment near the edge of the shell. Young forms (*C. contorta* Cpr.) are relatively higher-spired. Exterior chalky-white with a thin, brownish periostracum. Dredged moderately commonly from 10 to 75 fathoms.

GENUS *Crucibulum* Schumacher 1817

Crucibulum auricula Gmelin   West Indian Cup-and-saucer
Plate 21s

West Florida to the Lower Keys and West Indies.

1 inch in diameter, similar to *C. striatum*, but the edges of the inner cup are entirely free. The edges of the main shell are crenulated, the external ribs are coarser, and the interior is sometimes pinkish. The outer surface may show coarse diagonal ribs if the specimen has lived attached to a scallop or other ribbed mollusk. Uncommonly dredged in shallow water and occasionally washed ashore.

Crucibulum spinosum Sowerby   Spiny Cup-and-saucer

Figure 7i

Southern California to Chili.

¾ to 1 inch in diameter, variable in height (⅓ to ¾ as high), and usually with an almost circular base. Exterior with a smoothish apical area, the remainder of the shell with radial rows of small prickles or sometimes erect, tubular spines. Interior glossy, chestnut-brown, sometimes with light radial rays, and with a delicate white cup attached by one side. A very common species from low water to 15 fathoms. Albino shells are sometimes found.

Subgenus *Dispotaea* Say 1826

Crucibulum striatum Say   Striate Cup-and-saucer

Plate 21r

Nova Scotia to South Carolina (and Florida?).

1 inch in diameter, cap-shaped, base round, edge smoothish and the slightly twisted apex near the center of the shell. Interior of shell with a small, shelly cup, of which only ⅔ is free from attachment to the main shell. Apex wax color and smooth; remainder of exterior with small, wavy, radial cords. Interior glossy, yellow-white or tinted with light orange-brown. Commonly dredged in shallow water.

Genus *Crepipatella* Lesson 1830

Crepipatella lingulata Gould   Pacific Half-slipper Shell

Plate 20k

Bering Sea to Panama.

½ to ¾ inch in diameter, thin, almost circular, low and with the apex near the edge of the shell. Characterized by its tannish to mauve-white, glossy interior which has a shallow deck which is attached to the main part of the shell only along one side. The middle of the deck often has a weakly raised ridge. Exterior wrinkled and brownish. A very common species found on rocks and on the shells of living gastropods.

Genus *Crepidula* Lamarck 1799

Crepidula fornicata Linné   Common Atlantic Slipper-shell

Plate 21m

Canada to Florida and to Texas.

¾ to 2 inches in size. Shelly deck extending over the posterior half on the inside of the shell. The deck is usually concave and white to buff. Its edge is strongly sinuate or waved in two places. Exterior dirty-white to tan, sometimes with brownish blotches and rarely with long color lines.

Variable in shape, rarely quite flat, sometimes high and arched. They may be corrugated if the individual has lived attached to a scallop or ribbed mussel. A common littoral species. When collecting on the west coast of Florida, do not confuse with *C. maculosa*. *C. fornicata* has been introduced to the West Coast of the United States.

Crepidula maculosa Conrad   Spotted Slipper-shell

West Coast of Florida to Vera Cruz, Mexico.

Resembling *C. fornicata*, but often spotted with small, mauve-brown blotches and sometimes streaked. The edge of the deck is straight or only very slightly convex. There is an oval muscle scar on the inside of the shell just below and in front of the right anterior edge of the deck and the main shell. The young are very much like southern forms of *C. convexa* Say.

Crepidula convexa Say   Convex Slipper-shell
                                                       Plate 21n

Massachusetts to Florida, Texas and the West Indies.

¼ to ½ inch in size, usually highly arched and colored a dark reddish to purplish brown. Interior, including the deck, chestnut to bluish brown. Some specimens may be spotted. The edge of the deck is almost straight. There is a small muscle scar inside the main shell on the right side just under the outer corner of the deck (see also *maculosa*). Some specimens are thick and heavy, others quite fragile, the latter type found attached to other shells. Common just offshore down to 116 fathoms.

The form *glauca* Say is ⅓ inch long, thin-shelled, usually dark-brown or translucent-tan, and with a white deck. It is found in over-crowded colonies on eel-grass where specimens become long and narrow. *C. acuta* Lea is this form also.

Crepidula aculeata Gmelin   Spiny Slipper-shell
                                                       Plate 21q

North Carolina to Florida, Texas and the West Indies.

½ to 1 inch in size, similar to *fornicata*, but characterized by its rough, spinose exterior, thinner and flatter shell and by its irregular edges. Color whitish, although often heavily mottled with reddish brown. The exterior is sometimes stained green by algal growths. A common species found attached to stones, mangroves and other shells. Occasionally dredged.

Crepidula onyx Sowerby   Onyx Slipper-shell
                                                       Plate 20f

Monterey, California, to Peru.

1 to 2 inches in length, fairly thick-shelled, characterized by its glossy, dark-chocolate to whitish brown interior, and by the large, slightly concave, pure-white deck inside which has a sinuate free edge. Very common from shallow estuaries to 50 fathoms on rocks, on other shells, or stacked up on top of each other.

Crepidula excavata Broderip   Excavated Slipper-shell

Monterey, California, to Peru.

1 inch in size, rather thin; back strongly arched with the apex distinct and hooked under itself near the posterior margin of the shell. Characterized by its light brownish white color, by the straight or slightly curved edge of the interior deck, and by a weak muscle scar on each side just under the deck. Found commonly attached to rocks and other shells.

Subgenus *Janacus* Mörch 1852

Crepidula plana Say   Eastern White Slipper-shell

Canada to Florida and the Gulf States. Rare in the West Indies.

½ to 1½ inches in size, very flat, either convex or concave, and always a pure milky white. The apex is very rarely turned to one side. It commonly attaches itself to the inside of large, dead shells, and rarely, if ever, "piles up" like *fornicata*. A common shallow-water species.

Crepidula nummaria Gould   Western White Slipper-shell

Alaska to Panama.

¾ to 1½ inches in length, characterized by its glossy-white underside, flattened shell, large deck which usually has a weak, raised ridge (or sometimes a hint of an indentation) running from the apical end forward to the leading edge. Exterior with or without a yellowish periostracum. Found in rock crevices and apertures of dead shells.

Superfamily STROMBACEA
Family XENOPHORIDAE
Genus *Xenophora* Fischer von W. 1807

This group of gastropods is noted for its peculiar habit of cementing to its own shell fragments of other shells, stones, bits of coral and coal. The animals resemble those of the Strombus conchs, but the operculum is much wider and not sickle-shaped. B. R. Bales once humorously observed:

"It is generally admitted that the camouflage of *Xenophora* is for protection rather than ornamentation, for it would be inconceivable that a female *Xenophora* would call over the back fence to her girl friend with, 'Come and see the perfect dream of a shell I picked up today and tell me if I have it on straight.'"

There are 3 species in the Atlantic, one a shallow water species, the other two (*longleyi* Bartsch, pl. 23d, and *caribaea* Petit, pl. 23e) deep water inhabitants.

Xenophora conchyliophora Born   Atlantic Carrier-shell
Plate 5b

North Carolina to Key West and the West Indies.

2 inches in diameter, not including foreign attachments. No umbilicus. From above, the shell with its attached rubble and shells looks like a small heap of marine trash. It will attach any kind of shell to itself, but in some areas has access to only one kind, say *Chione cancellata*. Animal bright-red. Seasonally not uncommon. Johnsonia is in error in calling this *trochiformis* Born 1778 (not 1780), which is the Peruvian shell known formerly as *Trochita radians* Lam.

Family APORRHAIDAE
Genus *Aporrhais* da Costa 1778

Aporrhais occidentalis Beck   American Pelican's Foot
Plate 23c

Labrador to off North Carolina.

2 to 2½ inches in length, spire high, whorls well-rounded and with about 15 to 25 curved axial ribs per whorl. Many minute spiral threads present. Outer lip greatly expanded and its edge heavily thickened. Color ashen-gray to yellowish white. Operculum small, corneous, brown, claw-like, but with smooth edges. Commonly dredged off New England from a few to 200 fathoms.

The form *mainensis* C. W. Johnson (Nova Scotia to Mt. Desert) differs in having 14 axial ribs, instead of about 22 to 25 as in the typical form, but specimens intergrade. The form *labradorensis* C. W. Johnson is smaller, more slender, and with up to 29 ribs per whorl.

Family STROMBIDAE
Genus *Strombus* Linné 1758

Strombus pugilis Linné   West Indian Fighting Conch
Plate 5g

Southeast Florida and the West Indies.

3 to 4 inches in length. Always with spines on the last whorl, but those on the next to the last whorl are nearly always the largest. Shoulder of outer lip nearly always turns slightly upwards. Color a rich cream-orange to salmon-pink throughout, except for a cobalt-blue splotch of color on the end of the canal. Periostracum very thin and velvety. This is primarily a West Indian species, and apparently will not interbreed with the mainland species, *S. alatus*. An aberrant form which has club-like spines was unnecessarily named *sloani* Leach 1814 and *peculiaris* M. Smith 1940. Percy Morris' colored figure (1951, pl. 19, fig. 9) is not *pugilis*, but *alatus*.

Strombus alatus Gmelin　　　　　　　　　　　　Florida Fighting Conch

Plate 5h

South Carolina to both sides of Florida and to Texas.

3 to 4 inches in length. With or without short spines on the shoulder of the last whorl. Shoulder of outer lip slopes slightly downward. Color a dark reddish brown, often mottled with orange-brown or having zigzag bars of color on the shiny parietal wall. Periostracum very thin and velvety. A very common shallow water species, especially on the west coast of Florida. Not found in the West Indies. Do not confuse with *S. pugilis*.

Strombus gigas Linné　　　　　　　　　　　　　　Queen Conch

Plate 23a

Southeast Florida and the West Indies. Bermuda.

6 to 12 inches in length. Characterized by its large size, large and flaring outer lip, and the rich pinks, yellows and orange shades in the aperture. Periostracum fairly thick and horny. It flakes off in dried specimens. A malform with flattened spines was named *horridus* M. Smith. A form with a deep channel at the suture occasionally turns up in the Bahamas. It was named *canaliculatus* L. Burry. *S. gigas verrilli* McGinty is a form of questionable value described from Lake Worth, Florida. Very common in the West Indies, becoming uncommon in the Florida Keys from over-fishing. Also called the Pink Conch.

Strombus costatus Gmelin　　　　　　　　　　　　　Milk Conch

Plate 23b

Southeast Florida and the West Indies.

4 to 7 inches in length. Shell very heavy usually, and with low, blunt spines. Parietal wall and thick outer lip highly glazed with cream-white enamel. Outer shell a yellowish white. The periostracum in dried specimens flakes off. Common in the West Indies. *S. spectabilis* A. H. Verrill is this species.

Strombus raninus Gmelin

Hawk-wing Conch
Plate 5c

Southeast Florida and the West Indies.

2 to 4 inches in length. Shell bluntly spinose with the last two spines on the last whorl by far the largest. Outer lip points upward at the top. Color of outer shell a brownish gray with chocolate-brown mottlings. Aperture cream-colored with a salmon-pink interior. Common in the West Indies. *S. bituberculatus* Lamarck is the same species.

Strombus gallus Linné

Rooster-tail Conch
Plate 5e

Southeast Florida (rare) and the West Indies.

4 to 6 inches in length, characterized by the long extension of the posterior end of the outer lip and the rather high spire. This species is not at all common, although it may be obtained in fair numbers along the north coast of Jamaica.

Superfamily CYPRAEACEA
Family LAMELLARIIDAE
GENUS *Lamellaria* Montagu 1815

Lamellaria diegoensis Dall

San Diego Lamellaria
Figure 43d

Southern California.

½ inch in length, equally wide, quite fragile and transparent-white in color. 3 whorls moderately globose, the last large. Aperture very large. Columella very thin. Surface smoothish, except for fine, irregular growth lines. Periostracum thin, clear and glossy. Uncommon offshore.

Lamellaria rhombica Dall (Washington to Lower California) is the same size, much flatter and thicker-shelled, and is opaque-white in color. Its columella is thicker and ridge-like. This species is more common than the preceding and is commonly washed ashore.

Subfamily VELUTININAE
GENUS *Velutina* Fleming 1821

Velutina laevigata Linné

Smooth Velutina
Plate 22n

Labrador to Cape Cod, Massachusetts. Alaska to Monterey, California.

½ to ¾ inch in length, very thin and fragile, translucent amber, and covered with a thick, brownish periostracum which is spirally ridged. Columella arched and narrow. Common offshore from 3 to 50 fathoms. *V. undata*

Brown (*zonata* Gould is the same) from the Gulf of St. Lawrence to Cape Cod is similar, but rarely over ⅓ inch in length, with a flattened shelf-like columella, and often colored with narrow, spiral bands of brown.

Family FOSSARIDAE
GENUS *Fossarus* Philippi 1841

Fossarus elegans Verrill and Smith
Elegant Fossarus
Plate 25c

Massachusetts to North Carolina.

2 to 3 mm. in length, turbinate in shape, with 4 whorls, chalky-white to gray in color and characterized by its delicate sculpturing which consists of 2 strong carinae on the periphery and 3 smaller ones below and a large one bordering the chink-like umbilicus. Between the cords are numerous, distinct, arched riblets. Outer lip thickened by a large varix. 2 or 3 smaller, former varices commonly present on the last whorl. The base of the arched columella is projecting. Uncommon below 70 fathoms.

Family ERATOIDAE
Subfamily ERATOINAE
GENUS *Erato* Risso 1826
Subgenus *Hespererato* Schilder 1932

Erato maugeriae Gray
Mauger's Erato
Plate 22w

North Carolina to Florida and the West Indies.

¹⁄₁₆ inch in length, resembling a small *Marginella*, but the curled-in, thickened outer lip has a row of about 15 small, even-sized teeth. Upper end of the outer lip is well-shouldered. Shell glossy, tan with a pinkish or yellowish undertone. Apex bulbous. Commonly dredged on either side of Florida from 2 to 63 fathoms.

Erato columbella Menke
Columbelle Erato

Monterey, California, to Panama.

¼ inch in length, glossy-smooth, slate-gray in color with a whitish, thickened outer lip. Spire elevated, nucleus brown. Outer lip markedly shouldered above, and bearing about a dozen extremely small teeth. Siphonal canal stained inside with purple-brown. Not uncommon from shore to 50 fathoms. Occasionally washed ashore with kelp weed.

Erato vitellina Hinds
Apple Seed
Plate 20-o

Bodega Bay, California, to Lower California.

½ inch in length, resembling a "beach-worn Columbella," and glossy-smooth. Body whorl with a large purple area bounded by a faint whitish line; remainder of shell, including the spire which is often glazed over, is dark brownish cream. Columella arched, bearing 5 to 8 small, whitish teeth. Lower ¾ of slightly incurled outer lip is with 7 to 10 small, whitish teeth. Moderately common in fairly shallow water. Occasionally washed ashore with kelp weed.

Subfamily *TRIVIINAE*
GENUS *Trivia* Broderip 1837

Resembling miniature cowries (*Cypraea*), but characterized by strong wrinkles or riblets running around the shell from the slit-like aperture to the center of the back of the shell. We have carefully reviewed and included all of the Western Atlantic species, but have not followed the Schilderian use of numerous genera, such as *Pusula* Jousseaume.

Trivia pediculus Linné   Coffee Bean Trivia
                                                               Plate 21bb

South half of Florida and the West Indies.

½ inch in length, characterized by its tan to brownish pink color with 3 pairs of large, irregular, dark-brown spots on the back, and in having 16 to 19 (usually 17) ribs crossing the outer lip. The center pair of spots on the back are the largest. Some specimens may be quite pink.

In some areas, a dwarf form of this species occurs (named *pullata* Sby.) which is ¼ inch in length, with a pink base, 13 to 17 riblets on the outer lip, and often with the brown mottlings spread over most of the back. Do not confuse this form with the species *suffusa* which is light-pink, with a white (not pink) outer lip crossed by 19 to 24 riblets, and with a pink blotch on each side of the anterior canal.

T. pediculus is a common species found from low water to 25 fathoms.

Trivia suffusa Gray   Suffuse Trivia
                                                               Plate 21aa

Southeast Florida and the West Indies.

¼ to ⅓ inch in length, elongate-globular, bright-pink with suffused brownish splotches and fine specklings. Anterior canal with a weak pinkish blotch on each side. Riblets on back somewhat beaded. Dorsal groove fairly well-impressed. Outer lip white and crossed by 18 to 23 (usually 20) riblets. Quite common in the Bahamas and Lesser Antilles. *T. armandina* Kiener is the same.

Trivia maltbiana Schwengel and McGinty   Maltbie's Trivia

Plate 21z

North Carolina to Florida and the Caribbean.

¼ to ½ inch in length, globose, slightly flattened above, and characterized by its pale tannish pink, translucent color, by its fine riblets, and by
having 24 to 28 ribs crossing the outer lip. Areas between the ribs are microscopically granular. Nuclear whorls visible through the last whorl. The dorsal groove is slight and the riblets nearly cross it. Moderately common just
offshore to 50 fathoms.

Trivia quadripunctata Gray   Four-spotted Trivia

Southeast Florida, Yucatan and the West Indies.

⅛ to ¼ inch in length, very similar to *suffusa*, but smaller, brighter pink,
and with 2 to 4 very small, dark red-brown dots on the center line of the
back. Riblets very fine, 19 to 24 crossing the outer lip. A very common
species frequently found on beaches with the color dots worn away and the
pink background rather faded. The riblets on the back are never pustulose
as they tend to be in *suffusa*, nor is there any fine color speckling.

Trivia antillarum Schilder   Antillean Trivia

Southeast Florida and the Antilles.

⅛ to ¼ inch in length, characterized by its deep reddish or brownish
purple color. Elongate-globular in shape. Riblets smooth. With or without
a faint dorsal groove over which the riblets usually cross. Outer lip with 18
to 22 teeth. Dredged from 30 to 100 fathoms and rarely cast upon the beach.
Formerly *T. subrostrata* Gray.

Trivia candidula Gaskoin   Little White Trivia

Plate 21cc

North Carolina to southeast Florida to Barbados.

⅛ to ¼ inch in length, characterized by its fairly globular shape, pure-
white color, somewhat rostrate ends and by the smooth riblets that pass over
the back. There is no dorsal furrow. Many specimens have only a few
rather strong riblets of which 17 cross the inside of the outer lip. Another
common form has more riblets (20 to 24 over the outer lip). It has been
named *leucosphaera* Schilder (*globosa* of authors, not Gray). The forms
intergrade. *Trivia nix* is also white, but is larger, more globose and with a
strong dorsal groove interrupting the ribs.

Trivia nix Schilder   White Globe Trivia

Southeast Florida and the West Indies.

⅜ inch in length, globular, pure-white in color. Characterized by about 22 to 26 riblets. Back with a strong groove interrupting the riblets. Alias *T. nivea* Gray. This is the largest and most globular of the white species found in the Western Atlantic. It is moderately uncommon.

Trivia ritteri Raymond   Ritter's Trivia

Monterey, California, to Lower California.

⅜ inch in length, globular, pure-white in color. Characterized by about 15 fine riblets that run over the bottom, sides and back of the shell without being interrupted by a dorsal groove. Uncommonly dredged on gravel bottom from 25 to 60 fathoms.

Trivia californiana Gray   Californian Trivia
Plate 20v

California to Lower California.

⅓ to just less than ½ inch in length, rotund, and characterized by its mauve color, white, slightly depressed crease on the midline of the back, and by the fairly coarse riblets crossing over the entire shell (outer lip with about 15). A common littoral species, often washed ashore with seaweed. Also lives as deep as 40 fathoms. *Trivia sanguinea* Sowerby, a more southerly species, is larger, deeper purple, without the prominent white streak on the back and with finer, more numerous riblets (outer lip with about 20).

Trivia solandri Sowerby   Solander's Trivia
Plate 20u

Catalina Island to Panama.

⅝ to ¾ inch in length, rotund, and characterized by the strong, raised, smooth riblets running over the lip and up onto the back. Dorsal groove deep, cream-colored and flanked by 8 to 10 cream nodules on each side. Ground color of shell dark purplish brown. Moderately common in the littoral zone.

Trivia radians Lamarck (Lower California to Ecuador) is larger, flatter, and with a brownish spot on the back which discolors the central groove. It is fairly common.

Family CYPRAEIDAE
Genus *Cypraea* Linné 1758
Subgenus *Trona* Jousseaume 1884

Cypraea zebra Linné   Measled Cowrie
                                         Plate 6d

Southeast Florida and the West Indies.

2 to 3½ inches in length, oblong, light-faun to light-brown, with large, round, white dots over the back. Toward the base of the shell these white dots have a brown center. The shell is darker brown, narrower and less inflated than *cervus*. Moderately common in intertidal waters. Formerly called *C. exanthema* Linné. A light orangish form, probably due to being buried in sand for some time, was described from Cuba (form *vallei* Jaume and Borro 1946).

Cypraea cervus Linné   Atlantic Deer Cowrie
                                         Plate 6f

Southern half of Florida and Cuba.

3 to 5 inches in length, similar to *zebra*, but usually with smaller and more numerous white spots, with a more inflated and larger shell, and never has ocellated spots on the base of the shell. Moderately common from low tide to several fathoms.

Subgenus *Luria* Jousseaume 1884

Cypraea cinerea Gmelin   Atlantic Gray Cowrie
                                         Plate 6c

Southeast Florida and the West Indies.

¾ to 1½ inches in length, rotund, with its back brownish mauve to light orange-brown which may be flecked with tiny, black-brown specks. Base cream to old ivory with light mauve-brown between some of the teeth, or sometimes with tiny flyspecks of brown. A moderately common species found under rocks on reefs.

Subgenus *Erosaria* Troschel 1863

Cypraea spurca acicularis Gmelin   Atlantic Yellow Cowrie
                                         Plate 6a

South half of Florida, Yucatan and the West Indies.

½ to 1¼ inches in length; back irregularly flecked and spotted with orange-brown and whitish. Base and teeth ivory-white. Lateral extremities often with small pie-crust indentations. Distinguished from *cinerea* in being flatter and without color on the base. A moderately common species found under rocks at low tide. True *spurca* L. is from the Mediterranean.

Cypraea mus Linné (pl. 6e) is often found in American collections although it is limited to the southern part of the Caribbean. It is 2 inches in length, mouse-gray (Mouse Cowrie), and has a pair of irregular black-brown stripes on the back. It is frequently deformed with one or two small bumps on the back.

Subgenus *Zonaria* Jousseaume 1884

Cypraea spadicea Swainson   Chestnut Cowrie

Plate 6b

Monterey, California, to Cerros Island, Lower California.

1 to 2 inches in length, half as high, with a hard, glossy enamel finish. Base white, with about 20 to 23 teeth on each side of the long, narrow aperture. Sides bluish to mauve-gray, above which there is dark-chocolate fading on top to light chestnut-brown with a bluish undertone. Moderately common at certain seasons at low tide among seaweed, and also down to 25 fathoms.

Family *OVULIDAE*
GENUS *Primovula* Thiele 1925
Subgenus *Pseudosimnia* Schilder 1927

Primovula carnea Poiret   Dwarf Red Ovula

Plate 22q

Southeast Florida, the West Indies and the Mediterranean.

1/3 to 1/2 inch in length. This species resembles a miniature cowrie. The body whorl is rotund, pink to yellow in color and with numerous, fine spiral, incised lines. Aperture narrow, arched, and with a canal at each end. Outer lip curled in like that of a cowrie, and with about 20 small, rounded, whitish teeth. Upper parietal wall with a large, rounded, short ridge or tooth. Apex not showing. Rare from 25 to 100 fathoms.

GENUS *Pedicularia* Swainson 1840
Subgenus *Pediculariella* Thiele 1925

Pedicularia decussata Gould   Decussate Pedicularia

Plate 7d

Georgia to southeast Florida and the West Indies.

1/4 to 1/2 inch in length, moderately thick-shelled, with a long and flaring aperture, and pure-white in color. Sculpture of fine reticulations with the spiral threads the strongest. Columella a straight ridge with the parietal wall concavely dished. The entire shell has a distorted, "squeezed" appearance. Nuclear whorls obese, translucent-brown, reticulated, and with a sinuate lip

when in its free-swimming, larval stage. An uncommon species found clinging to coral stems in moderately deep water. This is the only Eastern American species in this genus.

Pedicularia californica Newcomb　　　　　　　　　Californian Pedicularia
　　　　　　　　　　　　　　　　　　　　　　　　　　　　Plate 7b, c
　　Farallon Islands to San Diego, California.

⅜ to ½ inch in length, solid, aperture greatly enlarged and flaring. Apex
hidden by the expanded lip. Early whorls showing minute decussations, the
rest of the shell with small spiral threads. Interior uneven and glossy. Color
rose with the outer lip whitish. Uncommon. Found attached to red hydrocoralline, *Allopora californica* Verrill. We have also illustrated the form
ovuliformis Berry (pl. 7c).

GENUS *Neosimnia* Fischer 1884

Neosimnia acicularis Lamarck　　　　　　　　Common West Indian Simnia
　　　　　　　　　　　　　　　　　　　　　　　　　　　　Plate 7a
　　North Carolina to southeast Florida and the West Indies.

½ inch in length, narrow, glossy, thin-shelled but strong, and with a
long, toothless aperture. Color deep lavender or yellowish. Columella area
flattened or sometimes slightly dished and, in adults, always bordered by a
long, whitish ridge, one inside the aperture, the other on the body whorl.
Posterior end of columella sometimes slightly swollen. A common species
which attaches itself and its tiny egg-capsules to purple or yellow seafans.

Neosimnia uniplicata Sowerby　　　　　　　　　Single-toothed Simnia
　　　　　　　　　　　　　　　　　　　　　　　　　　　　Plate 7e
　　Virginia to both sides of Florida and the West Indies.

½ to ¾ inch in length, similar to *acicularis*, but with only the innermost,
longitudinal ridge on the columella, and with a twisted, spiral plication at the
posterior end of the columella. Moderately common on seafans.

Neosimnia piragua Dall　　　　　　　　　　　Dall's Treasured Simnia
　　　　　　　　　　　　　　　　　　　　　　　　　　　　Plate 7f
　　Between Jamaica and Haiti.

1 inch in length, extremely narrow, with the ends greatly produced.
Columella area bordered by two longitudinal ridges, the inner one tinted with
rose. Remainder of shell yellowish white. One of the rarest of the Western
Atlantic mollusks. 23 fathoms.

Neosimnia avena Sowerby · Western Chubby Simnia
Plate 7g

Monterey, California, to Panama.

½ inch in length, oblong, with short extremities. Lower end of the aperture wide, upper end narrow where the columella has a spiral swelling. Inner and lower part of the columella with a long, light-colored ridge. Exterior of whorls with numerous, microscopic, wavy, incised scratches. Color mauve to deep-rose with the varix and extremities a lighter pink. Rare in California, uncommon southward. *S. similis* Sowerby is probably this species.

Neosimnia loebbeckeana Weinkauff · Loebbeck's Simnia
Plate 7i

Monterey, California, to the Gulf of California.

¾ inch in length, translucent yellowish, rather fragile, and fusiform in shape with the extremities narrow and the middle gently swollen. Columella rounded, usually smoothish, but sometimes with a hint of flattening and subsequent thickening of the lower part of the columella. Upper end of the columella with a weak, spiral fold. Two subspecies have been described (*barbarensis* Dall and *catalinensis* Berry) but their distinctiveness has not been clearly demonstrated as yet. Not uncommonly dredged in association with seafans from 20 to 50 fathoms.

Neosimnia inflexa Sowerby · Inflexed Simnia
Plate 7h

Monterey, California, to Panama.

½ inch in length, a very vivid and dark lavender-rose or reddish purple. Shell elongate; columella flattened, bordered within and also somewhat on the body whorl by a long, axial, lighter-colored ridge. *N. variabilis* Carpenter is this species, and detailed studies of the animals may also show that *N. aequalis* Sowerby is a synonym.

GENUS *Cyphoma* Röding 1798

Cyphoma gibbosum Linné · Flamingo Tongue
Plates 8; 4r

North Carolina to southeast Florida and the West Indies.

¾ to 1 inch in length, glossy-smooth, chubby, and colored a rich cream-orange to apricot-buff except for a small whitish rectangle on the back. Callus on sides of shell indistinct and extending high up on the back with poorly defined edges. Mantle of animal pale-flesh with numerous squarish, black rings. Fairly common on gorgonians below low water.

Cyphoma mcgintyi Pilsbry   McGinty's Cyphoma
                                              Plates 8; 4s

Lower Florida Keys and the Bahamas.

Similar to *gibbosa*, but more elongate, whitish with tints of lilac or pink on the back. The side callus on the right is thick and narrow. Aperture cameo-pink. Mantle with numerous solid spots which are roughly round or in the shape of short bars. Not uncommon.

Cyphoma signatum Pilsbry and McGinty   Fingerprint Cyphoma
                                              Plate 4t

Lower Florida Keys.

Similar to *mcgintyi*, but with the transverse ridge on the back much weaker, and the anterior end of the aperture more dilated than in the two preceding species. Color light-buff with a cream-buff tint deep inside the aperture. Mantle pale-yellow with numerous, crowded, long, black transverse lines. The rarest of the Florida Cyphomas.

Superfamily *HETEROPODA*
Family *ATLANTIDAE*
GENUS *Atlanta* Lesueur 1817

Atlanta peroni Lesueur   Peron's Atlanta
                                              Figure 41

Atlantic and Pacific warm waters; pelagic.

½ inch in diameter, planorboid, compressed from above, fragile, transparent and glassy. Later whorls openly coiled but connected by a sharp peripheral keel. Outer lip notched in the region of the shell. Often washed ashore after storms, and frequently brought up in dredge hauls. Five other species have been reported from American waters.

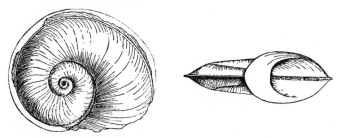

FIGURE 41. Shell of the heteropod, *Atlanta peroni* Lesueur, ½ inch.

GENUS *Oxygyrus* Benson 1835

Oxygyrus keraudreni Lesueur   Keraudren's Atlanta

Atlantic warm waters; pelagic.

½ inch in diameter, planorboid, nuclear whorls not visible; narrowly umbilicate on both sides. Whorls keeled only near the aperture. Body whorl near the aperture and the keel are corneous. No apertural slit. Operculum small, trigonal and lamellar. A common pelagic species, and the only one reported from our waters.

Family *CARINARIIDAE*
Genus *Carinaria* Lamarck 1801

Carinaria lamarcki Peron and Lesueur   Lamarck's Carinaria

Figure 42

Atlantic warm waters; pelagic.

Body up to 10 inches in length, tissues transparent; proboscis large and purple. Shell ⅕ the size of the animal, cap-shaped, very thin, fragile and transparent. Its apex is hooked. The shell is borne on top of the animal. This is a valuable collector's item, and in former years it brought fancy prices. Formerly known as *C. mediterranea* Lamarck and erroneously attributed under that name to Peron and Lesueur.

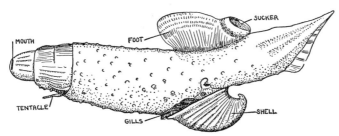

FIGURE 42. The heteropod, *Carinaria lamarcki* Peron and Lesueur, lives a pelagic life in warm seas. The animal may reach a length of 10 inches. It lives in an upside down position at the surface.

Superfamily *NATICACEA*
Family *NATICIDAE*
Subfamily *POLINICINAE*
Genus *Polinices* Montfort 1810

Polinices lacteus Guilding   Milk Moon-shell

Plate 22i

North Carolina to southeast Florida and the West Indies.

¾ to 1½ inches in length, glossy, milk-white, umbilicus deep with its upper portion covered over by the heavy callus of the parietal wall. Periostracum thin, smooth, yellowish. Operculum, corneous, thin, transparent, either wine-red or amber. Common in sandy, intertidal areas. *P. uberinus*

Orbigny is ¼ to ½ inch in length, with its umbilicus not so much covered, and it may be only a form of this species.

Polinices immaculatus Totten   Immaculate Moon-shell

Gulf of St. Lawrence to North Carolina.

⅜ inch in length, subovate, smooth, milk-white and glossy when deprived of its thin greenish-yellow periostracum. The ivory-white, thickened callus does not encroach upon the small, round, deep umbilicus. Operculum corneous, thin, light-brown. Commonly dredged off New England, and often found in fish stomachs.

Polinices brunneus Link   Brown Moon-shell
                                                   Plate 5j
Southeast Florida (rare), West Indies (Texas?).

1 to 2 inches in length, heavy, glossy-smooth, with a deep, white umbilicus and small, low spiral callus. Exterior tan to orange-brown. Operculum corneous, thin, amber-brown.

Polinices uberinus Orbigny   Dwarf White Moon-shell

North Carolina to southeast Florida and the Caribbean.

½ inch in length, very similar to *lacteus*, but the umbilical opening is larger, the callus is button-shaped and located against the columella near the center, and there is a large, rounded ridge running back from the callus into the umbilicus. Commonly dredged from 15 to 100 fathoms. Rarely in beach drift.

Subgenus *Neverita* Risso 1826

Polinices duplicatus Say   Shark Eye
                                                  Plates 5k; 22h
Cape Cod to Florida and the Gulf States.

1 to 2½ inches in length, glossy-smooth; umbilicus deep but almost covered over by a large, button-like, brown callus. Color slate-gray to tan; base of shell often whitish. Columella white. The shell is generally flattened and much wider than high, but some specimens (pl. 22h) are as wide as high and globose in shape. Operculum corneous, brown, and thin. This is a very common sand-lover found along our eastern coasts. Compare young specimens with *Natica livida*.

Subgenus *Glossaulax* Pilsbry 1929

Polinices reclusianus Deshayes
Recluz's Moon-shell
Plate 20i

Crescent City, California, to Lower California.

1½ to 2½ inches in length, very heavy for its size. Spire moderately to quite well elevated. Exterior semi-glossy, grayish with rusty-brown or greenish stains. Characterized by a large, tongue-like callus, brownish or white in color, which may or may not cover the entire umbilicus. There is a strong white, reinforcing callus at the top of the inside of the aperture. Operculum translucent, reddish brown. The shape of shell and degree of development of the umbilical callus is variable, and has received various names—*altus* Arnold and *imperforatus* Dall. A common shallow water species also found as deep as 25 fathoms.

Polinices draconis Dall
Drake's Moon-shell
Figure 43a

Alaska to Lower California.

2 to 2½ inches in length, very similar to *Lunatia lewisi*, but with a wider more elongate umbilicus, and with a very small, almost obsolete callus above the umbilicus. Uncommon in waters from 10 to 25 fathoms.

GENUS *Sigatica* Meyer and Aldrich 1886

Sigatica carolinensis Dall
Carolina Moon-shell
Plate 22l

North Carolina to southeast Florida and the West Indies.

¼ inch in length, white, glossy, ovate, fairly thin-shelled; umbilicus deep, round, without a callus. Characterized by 2 smooth nuclear whorls, followed by 3 whorls which are finely grooved by about 20 spiral lines. Suture well-channeled. Operculum paucispiral, corneous, its early whorls thickened and raised somewhat. *S. holograpta* McGinty is so similar that it may well be this species. Dredged 20 to 95 fathoms; not uncommon.

S. semisulcata Gray from West Florida and the West Indies reaches ½ inch in size, has 5 to 6 spiral lines cut into the top third of the whorl, and a few within the umbilicus. Often confused with *Polinices lacteus*. Rare.

GENUS *Amauropsis* Mörch 1857

Amauropsis islandica Gmelin
Iceland Moon-shell
Plate 22r

Arctic Seas to off Virginia.

1 to 1½ inches in length, ¾ as wide, rather thin, but strong. Suture

smooth and narrowly channeled. Shell smooth, yellowish white and covered with a thin, yellowish brown periostracum which flakes off when dry. Umbilicus absent or a very slight slit. Operculum paucispiral, horny, translucent-brown and with microscopic, spiral lines. A moderately common, cold-water species found just offshore down to 70 fathoms.

Its counterpart, *A. purpurea* Dall, common in Alaska, is very similar, but ¾ inch in length and with a greenish and darker periostracum.

GENUS *Eunaticina* Fischer 1885

Eunaticina oldroydi Dall   Oldroyd's Fragile Moon-shell
                                                              Figure 43e

Oregon to San Diego, California.

1½ to 2½ inches in length, resembling *Lunatia lewisi*, but much lighter in weight, with a more pointed spire, without the heavy, brownish callus and having, instead, the upper part of the columella expanded into a white, thin area which partially obscures the umbilicus. Micro-sculpturing on shell exterior is prominent. Moderately common; dredged offshore 30 to 70 fathoms.

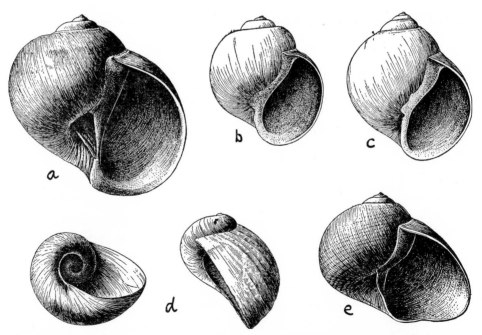

FIGURE 43. **a**, *Polinices draconis* Dall, 2 inches (Pacific Coast); **b**, *Natica clausa* Brod. and Sby., 1 inch (Arctic waters); **c**, *Lunatia pallida* Brod. and Sby., 1½ inches (Arctic waters); **d**, *Lamellaria diegoensis* Dall, ½ inch (California); **e**, *Eunaticina oldroydi* Dall, 2 inches (Pacific Coast).

Genus *Lunatia* Gray 1847

Lunatia heros Say   Common Northern Moon-shell

Figure 22a

Gulf of St. Lawrence to off North Carolina.

2 to 4½ inches in length, not so wide; globular in shape; umbilicus deep, round, not very large, and only slightly covered over by a thickening of the columellar wall. Color dirty-white to brownish gray. Aperture glossy, whitish or with tan or purplish brown stains. Periostracum thin, light yellow-brown. Operculum corneous, light-brown and thin. A very common intertidal species in the New England area. The egg case is a wide, circular ribbon of sand, about the thickness of an orange peel and easily bent when damp. The tiny eggs are embedded in the ribbon.

Lunatia triseriata Say   Spotted Northern Moon-shell

Plate 22m

Gulf of St. Lawrence to North Carolina.

½ inch in length, similar to young *heros* but the last whorl usually has three spiral rows of 12 to 14 bluish or reddish brown, squarish spots. The borders of the egg case are crenulated in contrast to the smooth borders of that in *heros*. This is a moderately deep-water species. Not uncommon from 1 to 63 fathoms.

Lunatia groenlandica Möller   Greenland Moon-shell

Plate 22k

Arctic Seas to off New Jersey.

¾ to 1 inch in length, 4 to 5 well-rounded whorls. Spire about ¼ the total length of the shell. Umbilicus very small, mostly covered over by the callus-like swelling of the top of the columella. Suture fine, deeply indented, bordered below by a weakly raised spiral swelling. Shell white, covered by a thin, smooth, greenish-yellow periostracum. Operculum chitinous, translucent, light-tan, paucispiral. Moderately common offshore.

Lunatia lewisi Gould   Lewis' Moon-shell

Plate 24n

British Columbia to Lower California.

3 to 5 inches in length, moderately heavy. Whorls globose, slightly shouldered a little distance below the suture. Umbilicus deep, round and narrow. Characterized by the brown-stained, rather small, button-like callus partially obscuring the top edge of the umbilicus. A very common species found in shallow water to 25 fathoms. They are more commonly found in the summer months. Do not confuse with *P. draconis*.

Lunatia pallida Broderip and Sowerby Pale Northern Moon-shell

Figure 43c

Arctic Seas to off North Carolina. Arctic Seas to California.

1½ to 1¾ inches in length, not quite so wide, smooth, pure-white in color, and covered with a thin, yellowish white periostracum. Parietal wall moderately thickened with a white glaze. Umbilicus almost closed to slightly open. Commonly dredged offshore in cold northern waters. In the Atlantic, this species rarely exceeds 1 inch in length. *P. groenlandica* Möller may be this species.

Subfamily SININAE
Genus *Sinum* Röding 1798

Sinum perspectivum Say Common Baby's Ear

Plate 22s

Virginia to Florida and the Gulf States. The West Indies.

1 to 2 inches in maximum diameter, but very flat, with very large white aperture and strongly curved columella. Numerous fine spiral lines on top of whorls. Color dull-white with a light-brown, thin periostracum. Animal envelops the shell. Commonly found in shallow, sandy areas, especially in the Carolinas and the west coast of Florida.

Sinum maculatum Say Maculated Baby's Ear

Carolinas and west coast of Florida.

Similar to *perspectivum*, but shell not so flat, with weaker spiral sculpture, and colored dull-brown or with yellowish brown maculations.

Sinum scopulosum Conrad Western Baby's Ear

Monterey to Todos Santos Bay, Lower California.

1 to 1¼ inches in length, 4 whorls, the early ones being very smooth, the last whorl very large. Numerous spiral grooves can be seen with the naked eye. Shell chalky-white, but usually covered with a thin, yellowish, translucent periostracum. The spire is more elevated and the whorls more inflated than those in *S. debile* Gould, from Catalina Island to the Gulf of California. *S. scopulosum* is moderately common, and is the same as *S. californicum* Oldroyd.

Subfamily *NATICINAE*
Genus *Natica* Scopoli 1777
Subgenus *Naticarius* Dumeril 1806

Natica canrena Linné   Colorful Atlantic Natica
Plate 51

North Carolina to Key West and the West Indies.

1 to 2 inches in length, glossy-smooth, except for weak wrinkles near the suture. Color pattern variable; sometimes with axial, wavy, brown lines and with 4 spiral rows of arrow-shaped or squarish brown spots. Umbilicus and its large round, internal callus white. Exterior of hard operculum with about 10 spiral grooves. Uncommon in eastern Florida; common in the West Indies.

Natica livida Pfeiffer   Livid Natica
Plate 22-o

Southeast Florida, Caribbean and Bermuda.

½ inch in length, glossy-smooth, exterior lead-gray with vague, spiral, darker-gray bands. Aperture and columella brown; callus which almost fills the umbilicus characteristically dark to light chocolate-brown. Moderately common on intertidal sand flats. Do not confuse with *Polinices duplicatus*, which is much flatter and has a corneous operculum, but which also has a brown to purplish brown callus.

Subgenus *Cryptonatica* Dall 1892

Natica clausa Broderip and Sowerby   Arctic Natica
Figure 43b

Arctic Ocean to North Carolina. Arctic Ocean to off southern California.

1 to 1¼ inches in length, fairly thin, smooth, yellow-white, with a smooth, gray to yellowish-brown periostracum. Umbilicus sealed over by a small, flat callus. Operculum, calcareous, thin, slightly concave, smooth, white and paucispiral. Commonly dredged in moderately deep water, and occasionally found intertidal north of Massachusetts. The sand-collar egg-case has smooth edges, and has a pimpled surface caused by the small compartments of young.

Natica pusilla Say   Southern Miniature Natica
Plate 22j

Cape Cod to Florida, the Gulf States, and the West Indies.

¼ to ⅓ inch in length, glossy-smooth, similar to *clausa*, but more ovate, often with a small, open chink next to the umbilical callus, and is a much smaller shell. Nucleus of operculum often stained with brown. Color white, but often with weak, light-brown color markings. Commonly dredged in shallow water to 18 fathoms.

Superfamily *TONNACEA*
Family *CASSIDIDAE*
GENUS *Sconsia* Gray 1847

Sconsia striata Lamarck   Royal Bonnet
Plate 9h

Southeast Florida to off Texas and the West Indies.

1½ to 2½ inches in height. Shell hard, polished, often with numerous fine, spiral incised lines. Usually two old varices are present. Rare, but recently being brought in by shrimp trawlers in the Gulf of Mexico. A choice deep-water species.

GENUS *Morum* Röding 1798

Morum oniscus Linné   Atlantic Wood-louse
Plate 25s

Southeast Florida and the West Indies.

¾ to 1 inch in height. Whorls with 3 spiral rows of rather prominent bulbous low tubercles. Parietal wall glazed over and ingrained with numerous white dots which are developed into minutely raised pustules. Color (with thin, velvety, gray periostracum removed) whitish with specklings or mottlings of brown or black-brown. Nucleus papilliform, white or pink. Operculum very small, corneous, and with its nucleus on the side. Nocturnal. Just below low-tide mark under coral slabs.

GENUS *Phalium* Link 1807

These are miniature helmet shells which rarely exceed a length of 5 inches. The Scotch Bonnet of Florida (*Phalium granulatum*) is well-known to most collectors. This genus differs from *Cassis* in having much smaller shells which do not have an extended, upturned siphonal canal and do not develop a massive parietal shield. Typical *Phalium* which has 4 or 5 tiny spines on the base of the outer lip (as for example in the Indo-Pacific genotype, *P. glaucum* Linné) is not represented in American waters. Our two species belong to the subgenus *Semicassis* which lacks these tiny spines. Operculum as in *Cassis*.

Subgenus *Semicassis* Mörch 1852

Phalium granulatum Born   Scotch Bonnet
Plate 9e

North Carolina to the Gulf of Mexico and the West Indies.

1½ to 3 inches in length, with about 20 spiral grooves on the body whorl. Weak axial ribs sometimes present which make the shell coarsely beaded. Lower parietal area pustulose. Outer lip may be greatly thickened occasionally. Not uncommonly washed ashore. It is also present on the west coast of Central America as the subspecies *centiquadrata* Valenciennes. Formerly known as *Semicassis abbreviata* Lamarck and *S. inflatum* Shaw.

Phalium cicatricosum Gmelin   Smooth Scotch Bonnet
Plate 9f

Southeast Florida, Bermuda and the Caribbean.

Shell 1½ to 2 inches in length, similar to *P. granulatum* but without the spiral grooves; sometimes smaller specimens have nodules on the shoulder of the whorl. Rare in Florida. Meuschen named this shell first, but his works are now ruled out as invalid. The nodulated, smaller variety was named *peristephes* Pilsbry and McGinty.

Genus *Cassis* Scopoli 1777

The helmet shells are large, handsome mollusks which have been used by man for centuries. Large numbers of cameos are still cut from them, the meat is often used in chowders, and the uncut shells serve as attractive doorstops or mantel-pieces. In the Pacific, they are sliced in half and the body whorl used either as a cooking container or boat-bailer. The half dozen known species are found only in the West Indies and Indo-Pacific area. They live in moderately deep water and although sometimes are obtained in knee-deep waters, they usually must be dived for in 10 to 20 feet of water. The helmet shells are carnivorous and include the spiny sea urchins in their diet. Operculum semicircular, corneous and concentric.

Cassis tuberosa Linné   King Helmet

Southeast Florida and the West Indies.

Adults 4 to 9 inches in length, massive, with a finely reticulated sculpture. Color brownish cream with black-brown patches on the lip and a large patch of brown at the center of the parietal shield. This species may be easily confused with the Flame Helmet (*Cassis flammea* Linné) which occurs in the Bahamas and Antilles. The latter lacks the reticulated sculpture, lacks brown color between the teeth on the outer lip, has a rounded (not triangular) parietal shield and is from 3 to 5 inches in length. Rare in Florida (10 fathoms), common to the south in shallow water.

Cassis madagascariensis Lamarck   Emperor Helmet
Plate 23v

Southeast Florida, the Bahamas and the Greater Antilles.

Adults 4 to 9 inches in length, massive. Three spiral rows of large blunt spines; the topmost spine of the first row generally the largest. Color pale-cream on the outer surface. Parietal shield and outer lip pale- to deep-salmon. Teeth white, brown sometimes between them. Moderately common from 5 to 10 fathoms in the Bahamas. Very rare in Florida where it is replaced by the Clench's Helmet, the subspecies *spinella* Clench.

Cassis madagascariensis spinella Clench   Clench's Helmet

Off Beaufort, North Carolina (fossil?), and the Florida Keys.

Similar to the typical *madagascariensis*, but with numerous small, evenly sized spines, more noticeable on the top row. Frank Lyman has collected this novel form or subspecies by the dozen in 20 feet of water off the Florida Keys. It is not a rarity and has been in old collections for many years. We have seen specimens labelled as coming from the Bimini Islands, Bahamas, but the record needs confirmation.

Genus *Cypraecassis* Stutchbury 1837

Cypraecassis testiculus Linné   Reticulated Cowrie-helmet
                                                   Plate 9c
Southeast Florida, Bermuda and the West Indies.

1 to 3 inches in length. Body whorl closely sculptured by small, distinct, longitudinal ridges which are crossed by a dozen or so spiral grooves, thus producing a reticulated surface. The shoulder of the body whorl in a very few specimens may have pinched-up, low tubercles or ribs. It is only a form. Entire animal light brownish orange, with underside of foot smeared with darker shades of orange. No periostracum. No operculum. Eggs laid under small rocks in greenish-brown clusters of 100 or so, teardrop-shaped, translucent capsules. Reef inhabitant, below low-water level.

Family CYMATIIDAE
Genus *Argobuccinum* Bruguière 1792
Subgenus *Fusitriton* Cossmann 1903

Argobuccinum oregonense Redfield   Oregon Triton
                                                      Plate 24g
Bering Sea to San Diego, California.

4 to 5 inches in length, about 6 whorls. Characterized by its fusiform shape, convex whorls, which each bear 16 to 18 axial ribs nodulated by the crossing of smaller spiral pairs of threads. The epidermis is heavy, spiculose, bristle-like and gray-brown. Aperture and siphonal canal interiors are

Amauropis - Genus

Be good to your Hanes® pantyhose and they'll be good to you.

You can make your beautiful Hanes pantyhose last longer if you put them on right.

Which means you should put them on while you're sitting down.

Then using both hands, gather one leg of the panty stocking down to the toe and slip over the foot. Smooth the toe and heel in place, and guide stocking over instep and up the ankle. Then do the same for the other leg.

Alternating from side to side, draw pantyhose gently up each leg. Stand and make sure entire leg of pantyhose is stretched to its full length. Then ease panty over hips and up to waist.

If more adjustment is needed, inch garment up from ankle to waist. Never pull or yank your pantyhose.

To take them off, put thumb inside waistband, slide over hips and roll down to ankles. Then take one more second to return pantyhose to original shape before setting aside for laundering.

And here's something else to remember. There are many beautiful Hanes styles to choose from:

HANES ULTRA SHEER. The sheerest Hanes of all. Made for dressing up and going out and looking very beautiful.

HANES SOFT SHEER. Cantrece® II adds a rich matte look and a soft supple feeling.

HANES TOP CONTROL. With even more spandex added to the panty, you get a total feeling of confidence.

HANES SHEER TOE TO WAIST. Made without a heel or a toe or a pantyline in sight.

ALIVE® SHEER SUPPORT. A beautifully sheer pantyhose with graduated support. The kind that puts its greatest pressure at the ankle then relaxes it all the way up your legs. So they feel marvelous. In 3 styles: reinforced heel and toe, nude heel, and all-sheer.

HANES EVERYDAY. This durable Hanes pantyhose has a great stretchy fit and a short short panty that won't show no matter what you wear. At a price just right for everyday. In 2 styles: reinforced toe or sandalfoot.

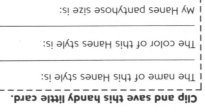

Clip and save this handy little card.

The name of this Hanes style is:

_____

The color of this Hanes style is:

_____

My Hanes pantyhose size is:

_____

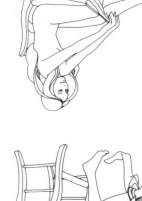

40–122

Printed in U.S.A.

enamel-white. Enamel, single tooth on parietal wall near the top of the aperture. Operculum chitinous, thick, brown. A common offshore species in its northern range.

GENUS *Cymatium* RÖDING 1798
Subgenus *Cymatium* s. str.

Cymatium femorale Linné   Angular **Triton**
Plate 5d

Southeast Florida and the West Indies.

3 to 7 inches in length, with 2 or 3 former varices; outer lip flaring, thickened into a noduled varix which is drawn up to a point posteriorly. Columella with 1 small fold and above it sometimes several much smaller ones. Color varies from brownish to reddish orange. Not uncommon in the West Indies in shallow water among eel-grass.

Subgenus *Lampusia* Schumacher 1817

Cymatium martinianum Orbigny   Atlantic Hairy **Triton**
Plate 9l

North Carolina to southeast Florida and the West Indies.

1½ to 3 inches in length; old varices strong, beaded, and spaced ⅔ of a whorl apart. Spiral sculpture of a dozen or so squarish, irregularly sized, weakly beaded cords. Aperture orange-brown with the parietal area dark-brown between the white teeth. Periostracum very thick, matted, light-brown. The embryonic shell is about 4 mm. in length, glossy-brown, with a flaring lip which has a small stromboid notch. *Dissentoma prima* Pilsbry 1945 is this species. *C. velei* Calkins is also a synonym. *C. aquatile* Reeve and *C. pileare* Linné are names applied to Indo-Pacific forms, and not this species. Common in shallow water.

Do not confuse with *C. chlorostomum* which has just inside its outer lip a series of single, rather large, whitish teeth, instead of smaller, paired, yellowish brown teeth.

Cymatium gracile Reeve   Dwarf Hairy **Triton**
Plate 25n

North Carolina to Key West and the West Indies.

1 inch in length, with only one or no former varix. Whorls squarish at the shoulder where there are 2 spiral rows of prominent beads. The last whorl has only 1 row of about 6 to 8 rather large tubercles in addition to spiral and axial threads. Siphonal canal moderately long, slender. Color whitish with 1 or 2 orange-brown bars on the varix. Periostracum rather thick, gray-brown. Uncommon below low-water line.

Cymatium chlorostomum Lamarck   Gold-mouthed Triton
Plate 25q

Southeast Florida and the West Indies. Bermuda.

¾ to 2½ inches in length; coarsely corrugated by spiral, noduled cords; varices spaced ⅔ of a whorl apart. Shell ash-gray with brown flecks and characterized by an orange mouth with white teeth. A common West Indian species which is also abundant in the Indo-Pacific. See differentiating remarks under *C. martinianum*.

Subgenus *Tritoniscus* Dall 1904

Cymatium labiosum Wood   Lip Triton
Plate 25m

Florida Keys and the West Indies.

Shell ¾ inch in length, much like *Cymatium gracile,* but heavier, with a much shorter siphonal canal, slightly umbilicated, and with strong spiral cords on the base of the shell. Uncommon in intertidal reef areas.

Subgenus *Gutturnium* Mörch 1852

Cymatium muricinum Röding   Knobbed Triton
Plate 25r

Southeast Florida and the West Indies. Bermuda.

1 to 2 inches in length; characterized by a thickened cream parietal shield and a long, bent back siphonal canal. Color ash-gray, sometimes dark-brown with a narrow cream, spiral band. Not uncommon in intertidal reef areas. Interior of aperture brownish red to yellowish white. *C. tuberosum* Lamarck is a later name for this species.

Subgenus *Ranularia* Schumacher 1817

Cymatium cynocephalum Lamarck   Dog-head Triton
Plate 9j

Southeast Florida and the West Indies.

1½ to 2½ inches in length; with globular whorls which are squarish at the shoulder. Siphonal canal long and slender. Usually with only one former varix. Apical whorls cancellate; last whorl with slightly noduled, spiral cords. Parietal wall with an oval splotch of dark-brown, over which run light-orange spiral cords. Uncommon in Florida. The subgenus *Tritonocauda* Dall 1904 is the same as *Ranularia*.

Genus *Distorsio* Röding 1798

Distorsio clathrata Lamarck   Atlantic Distorsio
Plate 25aa

North Carolina to Florida, the Gulf States and the Caribbean.

¾ to 2½ inches in length; whorls distorted, aperture with grotesque arrangement of the teeth; siphonal canal twisted. Whorls with coarse reticulate pattern. Parietal shield glossy, reticulated with raised threads, colored white to brownish white. Differs from *constricta mcgintyi* in having a less distorted body whorl which is more evenly rounded and more evenly knobbed or reticulated. The parietal wall is generally reticulated instead of pustuled. Dredged from 5 to 65 fathoms. Frequently brought in by shrimp fisherman.

Distorsio constricta mcgintyi Emerson and Puffer 1953   McGinty's Distorsio
Plate 25z

North Carolina to south half of Florida.

1 to 2 inches in length, very close to *clathrata*, but the body whorl is very distorted, bulging and with cruder nodules. The upper and inner corner of the aperture usually has only one small, short, white tooth, while in *clathrata* there are usually 2 fairly large, longer, obliquely set teeth. The lower parietal wall has a deep, smooth, wide groove separating the two axial rows of teeth. Commonly dredged from 25 to 125 fathoms. Formerly called *D. floridana* Olsson and McGinty 1951 (not *floridana* Gardner 1947). Typical *constricta* Broderip is from the Eastern Pacific.

GENUS *Charonia* Gistel 1848

Charonia tritonis nobilis Conrad   Trumpet Triton
Plate 5f

Southeast Florida and the West Indies.

Adults 1 to 1½ feet in length. The early whorls are purplish pink. In old specimens these are usually lost. Adults usually have a swollen, angular shoulder on the last whorl, a feature which distinguishes our Atlantic subspecies from the typical *tritonis* Linné of the Indo-Pacific area. *C. atlantica* Bowdich is a synonym of the Pacific subspecies, despite the name. Rare in Florida; moderately common in the West Indies below low water.

Family *BURSIDAE*
GENUS *Bursa* Röding 1798 (= *Ranella*)
Subgenus *Bursa s. str.*

Bursa thomae Orbigny   St. Thomas Frog-shell

Southeast Florida and the West Indies.

½ to 1 inch in length. Characterized by the varices being placed axially one below the other and by the delicate lavender aperture. Rare in moderately shallow water. The posterior siphonal canal is prominent and not attached to the body whorl.

Subgenus *Tutufa* Jousseaume 1881

Bursa tenuisculpta Dautz. and Fischer   Fine-sculptured Frog-shell

Southeast Florida and the West Indies.

2 to 3 inches in length; with 5 to 7 spiral rows of numerous, evenly sized beads. Old varices spaced ⅔ of a whorl apart so that the varices do not line up under each other. Color dull ash-gray. Dredged on rare occasions.

Subgenus *Colubrellina* Fischer 1884

Bursa corrugata Perry   Gaudy Frog-shell
Plate 9k

Southeast Florida and the Caribbean. Lower California to Ecuador.

2 to 3 inches in length; flattened laterally; with 2 prominent, knobbed varices on each whorl. Just in front of each varix there is a sharp frill. There are generally 1 or 2 rows of blunt nodules on the whorls. This is a rare species in the Atlantic, but more frequently encountered on the west coast of Central America. Alias *caelata* Broderip, *ponderosa* Reeve and *louisa* M. Smith.

Bursa granularis Röding   Granular Frog-shell
Plate 25-o
Southeast Florida and the West Indies.

¾ to 2 inches in length, flattened laterally. Varices axially placed one below the other. Color orange-brown with 3 narrow, white bands which appear as prominent white squares on the varices. Spiral sculpture of several rows of small beads, those on the periphery of the whorl having the largest beads. Teeth in aperture white. Uncommon. Alias *cubaniana* Orbigny and *affinis* Broderip.

Subgenus *Bufonaria* Schumacher 1817

Bursa spadicea Montfort   Chestnut Frog-shell
Plate 25p
Southeast Florida and the Caribbean.

1 to 2 inches in length, flattened laterally; with strong, rounded varices, 2 on each whorl and lined up axially one under the other. Surface covered with spiral rows of numerous, small beads. Posterior siphon has one wall next to the body whorl. Color yellowish with diffused markings of orange-brown. Rare. Dredged off Florida in moderately deep water. Alias *B. crassa* Dillwyn.

Subgenus *Crossata* Jousseaume 1881

Bursa californica Hinds   Californian Frog-shell
Plate 20r

Monterey, California, to the Gulf of California.

3 to 5 inches in length, moderately heavy, tan-cream in color and with about 6 whorls, each of which has 2 varices, one opposite the other. The last varix has 4 to 5 large nodules; in the spire only one nodule shows. Between the varices there are 2 stout spines. White aperture with a posterior canal almost the size of the anterior (siphonal) canal. Lip crenulate. Common offshore, occasionally washed ashore. A scavenger.

Family *TONNIDAE*
Genus *Tonna* Brünnich 1772

Tonna maculosa Dillwyn   Atlantic Partridge Tun
Plate 9d

Southeast Florida and the West Indies.

2 to 5 inches in length, thin but strong. Nuclear whorls golden-brown and glassy-smooth. Periostracum thin and usually flakes off in dried specimens. *Dolium album* Conrad is only an albino form. *Tonna perdix* Linné is not this species, but an Indo-Pacific shell which has a more pointed spire, clearer squares of color and fewer spiral ribs. Our species is fairly common, especially in the West Indies. Adults do not have an operculum in this genus.

Tonna galea Linné   Giant Tun
Plate 23f

North Carolina to Florida, the Gulf States and the West Indies.

5 to 7 inches in length, thin but rather strong, although the lip is easily broken. Ground color whitish to light coffee-brown, sometimes slightly mottled. With 19 to 21 broad, flattish ribs. This species also occurs in the Mediterranean and the Indo-Pacific. The subspecies, *brasiliana* Mörch, known only from Brazil, has a pushed-down, flattish spire. Not uncommon below low water.

Genus *Eudolium* Dall 1889

Eudolium crosseanum Monterosato   Crosse's Tun
Plate 23g

Off New Jersey to the Lesser Antilles.

2 to 3½ inches in length, moderately thin-shelled, but strong. Each of the 6 whorls bears numerous, spiral ridges and fine threads. Nuclear whorls

smooth, dark-brown. Outer lip turned back, slightly thickened and with its inner edge crenulated. Color white to light-cream, with the ridges straw-yellow. Periostracum thin and light yellowish brown. No operculum in adults. Uncommon from 96 to 300 fathoms. Very rare in private collections.

Family FICIDAE
Genus *Ficus* Röding 1798

Ficus communis Röding   Common Fig Shell
Plate 9i

North Carolina to the Gulf of Mexico. The Bahamas.

3 to 4 inches in length, thin, rather fragile, and with spiral threads which are sometimes made reticulate by axial threads. Uncommon, except on the west coast of Florida where it is washed ashore in great numbers. No operculum present. Formerly known as *Pyrula* and *Ficus papyratia* Say, but the latter name is preceded by two earlier names, *communis* Röding 1798 and *reticulata* Lamarck 1816 (as well as 1822).

Carol's Fig Shell (named after Mrs. Richard W. Foster), *Ficus carolae* Clench, is very rare, and is irregularly spotted with reddish brown on the inside of the shell. It was first discovered by Mr. Leo L. Burry of Sarasota off Key Largo, Florida, in 100 fathoms.

Order NEOGASTROPODA
Superfamily MURICACEA
Family MURICIDAE
Subfamily RAPANINAE
Genus *Forreria* Jousseaume 1880

Forreria belcheri Hinds   Giant Forreria
Plate 24i

Morro Bay, California, to Lower California.

3 to 6 inches in length, solid, smoothish, cream-brown; surface with 10 prominent, pointed, scale-like spines on the shoulder of each whorl. These are the tops of the varices which flatten out and are welded closely to the lower part of the whorl. Former siphonal canals prominent to the left of a narrow, not deep umbilicus. Interior enamel-white. Common in intertidal areas near oyster bars. Also down to 15 fathoms.

Subgenus *Austrotrophon* Dall 1902

Forreria cerrosensis cerrosensis Dall   Cerros Forreria
Figure 44a, b

Off southern Lower California.

3 inches in length, with a rather long siphonal canal, and bearing 8 long axial ribs (or former varices) which are blade-like and are curled into long upswept, large spines. With or without small, low, spiral threads which may be more pronounced on the last whorl. Color yellowish to brownish white. Sometimes with subdued, wide brown lines. The form *pinnata* Dall is a smaller, 8 to 9 ribbed, spirally threaded variant from the same region. Uncommon from 21 to 74 fathoms.

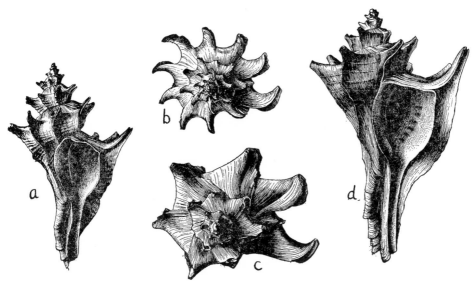

FIGURE 44. a and b, *Forreria cerrosensis* Dall and, c and d, its subspecies, *catalinensis* Oldroyd, from California. Reduced ½.

Forreria cerrosensis catalinensis Oldroyd   Catalina Forreria

Figure 44c, d

  Southern third of California.

  2 to 3 inches in length, similar to the typical *cerrosensis*, but with 7 ribs, sturdier shell, with less development of the blade-like ribs, and often with more brownish coloration. Formerly thought to be *Boreotrophon triangulatus* Cpr. Moderately common offshore, sometimes cast ashore.

Subfamily MURICINAE
GENUS *Murex* Linné 1758
Subgenus *Murex s. str.*

Murex cabriti Bernardi   Cabrit's Murex

Plate 10h

  South ¾ of Florida and the Lesser Antilles.

1 to 3 inches in length. The long, slender siphonal canal bears 3 rows of long and slender spines. Color drab-white, sometimes a pale yellowish pink between the varices. Each varix has 3 to 4 sharp, long spines. Uncommonly dredged in moderately deep water. Considered a collector's item. Do not confuse with the less rare *Murex tryoni* Hidalgo of southeast Florida and the West Indies, which is much the same except that it lacks spines all down the siphonal canal and is almost smooth between the varices. Compare also with the next species which is common.

Murex recurvirostris rubidus F. C. Baker　　　　　　　Rose Murex

South half of Florida and the Bahamas.

1 to 2 inches in length. Siphonal canal rather long and slender, and with 3 pairs of short prickly spines near the top. The knobby varices may have a short spine at the top. Between the varices are 3 low knobby ribs, 2 of which are larger than the third. Color variable: cream, pink, orange or red. A spiral band of darker color is found in some specimens. It lives in shallow, sandy areas and is commonly washed ashore. Alias *messorius* Sby., *anniae* M. Smith, *delicatus* M. Smith and *citrinus* M. Smith.

True *recurvirostris* Broderip is found in the Eastern Pacific. *M. recurvirostris sallasi* Rehder and Abbott from the Yucatan, Mexico, area has 3 equal-sized, finely beaded ribs between each varix, is brightly colored with shell-pink and occasionally has fine spiral lines of brown.

Murex beaui Fischer and Bernardi　　　　　　　Beau's Murex
　　　　　　　　　　　　　　　　　　　　　　　Plate 10d

South Florida, the Gulf of Mexico, and the West Indies.

3 to 5 inches in length. The spiny varices usually have prominent, thin, wavy webs. Between the varices there are 5 or 6 rows of low, evenly sized and evenly spaced knobs. Color cream to pale brownish. Uncommon offshore.

Subgenus *Hexaplex* Perry 1811
Section *Phyllonotus* Swainson 1833

Murex pomum Gmelin　　　　　　　　　　　　Apple Murex
　　　　　　　　　　　　　　　　　　　　　　　Plate 10l

North Carolina to Florida and the West Indies.

2 to 4½ inches in length. Sturdy with a rough surface. No long spines. Colored dark-brown to yellowish tan. Aperture glossy, ivory, buff, yellow or orangish with a dark-brown spot on the upper end of the parietal wall.

Outer lip crenulate and with 3 or 4 daubs of dark-brown. A very common shallow water species.

Section *Muricanthus* Swainson 1840

Murex fulvescens Sowerby   Giant Eastern Murex
Plate 10b

North Carolina to Florida and to Texas.

5 to 7 inches in length. Characterized by the large shell, and the strong, straight, rather short spines. Exterior milky-white to dirty-gray. Aperture enamel white. Thin spiral color lines are usually prominent on the whorls. Fairly common along the shallow areas of northeastern Florida where they are found abundantly during the breeding season. Well-known to the shrimp fishermen whose nets often ensnare them. *Murex burryi* Clench and Farfante is probably the young of this species.

Section *Murexiella* Clench and Farfante 1945

Murex hidalgoi Crosse   Hidalgo's Murex
Figure 45a

North Carolina to the Lesser Antilles.

1 to 1½ inches in length. Spines frondose and long, with webbing in between which is exquisitely sculptured with scale-like lamellations. Color grayish white to cream. This is probably the rarest of our eastern Murex species. Recently, one specimen was dredged off northeast Florida in a few fathoms of water.

Subgenus *Chicoreus* Montfort 1810

Murex brevifrons Lamarck   West Indian Murex
Plate 10a

Lower Florida Keys and the West Indies.

3 to 6 inches in length. Numerous, stout, fairly long spines on the varices which arch backwards and bear sharp fronds. Raised, spiral lines prominent between the varices. Color variable from cream to dark-brown. Uncommon in the Lower Keys, but fairly common to abundant in the West Indies. Percy Morris (1951, pl. 14, fig. 1) labels this species as *Murex florifer*.

Murex florifer Reeve   Lace Murex
Plate 10e

South half of Florida and the West Indies.

1 to 3 inches in length. Aperture small, nearly round. 8 to 10 crowded,

frondose, scaly spines bordering the outer lip and siphonal canal. Top spine sometimes twice as long as the others. Color dark-brown, light-brown, or whitish and, in the latter case, the nuclear whorls at the spire are pinkish. Usually 1 axial low ridge between each varix, although occasionally with more and smaller axial ribs.

The Lace Murex is one of Florida's most common species in this genus. It lives in a wide variety of habitats from mangrove, muddy areas to protected rocks and frequently in clear, sandy areas. The ecological variety, which is whitish and with reduced spines, was named *arenarius* by Clench and Farfante. This species differs from the 4 to 5 inch-long *M. brevifrons* in being smaller, in having closely crowded scaly spines, and in having a round instead of elongate operculum. For many years this species was called *rufus* Lamarck 1822 (not *rufus* Montagu 1803).

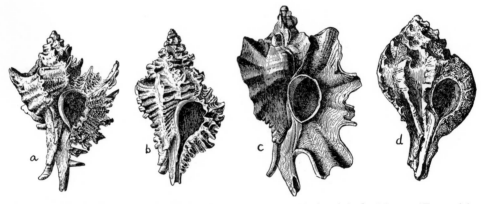

FIGURE 45. **a,** *Murex (Murexiella) hidalgoi* Crosse (Atlantic); **b,** *Murex (Favartia) cellulosus* Conrad (Atlantic); **c,** *Murex (Pterynotus) trialatus* Sby., form *carpenteri* Dall (Pacific); **d,** *Murex (Maxwellia) santarosana* Dall (Pacific). All reduced ½.

Subgenus *Favartia* Jousseaume 1880

Murex cellulosus Conrad   Pitted Murex
                                                              Figure 45b

North Carolina to Florida, the Gulf of Mexico and the West Indies.

1 inch in length. Shell rough, with 5 to 7 poorly developed fluted varices. It rarely develops spines, but when present they are short and stubby with a thin webbing connecting each spine in the varix. The siphonal canal strongly upturned. Aperture small, almost round. Color a dull grayish white.

This is one of the smallest and most compact species of Murex on the Atlantic Coast and is often found in shallow, intertidal waters, especially near oyster beds where it probably does moderate damage to young oysters. Its identification is made difficult when the siphonal canal has been broken off completely. A deep water form, *M. cellulosus leviculus* Dall (pl. 25j), of more delicate sculpturing and with brown markings, is found off the coast of North Carolina and eastern Florida. An inch-long, chubby subspecies, *nuceus* Mörch (pl. 25i), with a shorter and wider siphonal canal and heavily scaled varices, occurs in the West Indies but has been collected by Dr. J. S. Schwengel on Tea Table Key, Lower Florida Keys, and off Fort Walton by Mr. L. A. Burry,

Subgenus *Pterynotus* Swainson 1833
Section *Pteropurpura* Jousseaume 1880

Murex bequaerti Clench and Farfante   Bequaert's Murex

North Carolina south to Key West.

1 to 2½ inches in length. Spire high. No spines. Each varix is a high, rounded, thin plate or web. Between these varical webs there is a single, low, rounded nob. Color a uniform cream-white. A bizarre species which is the least spinose of our American forms. It is being collected in dredging operations along the west coast of Florida in increasing numbers, although it remains a rarity. It was named after one of our foremost malacologists at Harvard University, Dr. Joseph C. Bequaert. Dall identified this species as *Murex macropterus* Desh.

Murex trialatus Sowerby   Western Three-winged Murex
Figure 45c
Northern California to Lower California.

2 to 3 inches in length, with 3 large, wavy, wing-like varices per whorl. Siphonal canal closed along its length. The body whorl between each varix is smoothish, with or without one low, rounded tubercle, and sometimes with 2 to 5 weak, spiral cords or threads. Anterior face of each varix with fine, crowded, axial fimbriations. Color grayish, dark- or light-brown, or with white spiral bands.

Typical *trialatus* from southern California and Lower California reaches a length of 3 inches, is generally dark chestnut to blackish brown with 4 to 6 narrow white bands, and has very fine spiral threads which are sometimes scaled, beaded or smooth.

The subspecies *carpenteri* Dall (fig. 45c)—*tremperi* Dall and *petri* Dall are ecological forms or color varieties, as is the all-white *alba* Berry—has larger wings which are smooth on the posterior face. The color is generally light yellowish brown, all-white or with 2 wide white bands. Common off-shore.

M. erinaceoides rhyssus Dall is similar to *trialatus*, but with scaly fimbriations all over, sometimes with numerous, rather strong spiral cords, and grayish white in color. Southern California, south. Uncommon offshore.

Subgenus *Maxwellia* Baily 1950

Murex gemma Sowerby · Gem Murex

Plate 24e

Santa Barbara, California, to Lower California.

1 to 1¼ inches in length, moderately high-spired, with 6 varices per whorl. The varices are swollen, roundish and smooth and connect with each other in the middle area of the whorl, but in the area of the suture, and again near the base of the shell, the varix is thin, elevated and curled back and may bear one or several small spines. There are several spiral low cords colored blackish blue which are more obvious on the middle or smoother part of the whorl. The spire appears to have squarish pits crudely dug out. Very common along rocky areas under protective rubble and masses of worm tubes.

Murex santarosana Dall · Santa Rosa Murex

Figure 45d

Santa Barbara Islands to Lower California.

1½ inches in length, spire low, and with 6 curled-back, spined varices per whorl. Anterior surface of varices strongly fimbriated. Narrow intervarical space smooth. Color brownish white. Uncommon to rare on gravel bottom just offshore to 30 fathoms. Do not confuse with the common *gemma*.

Murex festivus Hinds · Festive Murex

Plate 24l

Morro Bay, California, to Lower California.

1½ to 2 inches in length, spire high, 3 varices per whorl. Color brownish cream with numerous fine, dark spiral lines. Varix, with its thin, fimbriated surface, curled backwards. One very large, rounded nodule between varices. Very common on rocks or mud flats and down to 75 fathoms.

Genus *Boreotrophon* Fischer 1884

Boreotrophon clathratus Linné · Clathrate Trophon

Arctic Seas to Maine.

1 to 2 inches in length, with rounded whorls, slightly flaring lip, and numerous axial, foliated ribs. Chalk-white in color. There are several forms

described, and we have figured the subspecies *scalariformis* Gould, commonly dredged in the Grand Banks.

Boreotrophon scitulus Dall

Handsome Trophon
Figure 46d

Alaska to San Diego, California.

1 to 1½ inches in length, rather fragile, pure-white in color, with a rather long siphonal canal. Characterized by 5 or 6 spiral rows of long, delicate, anteriorly hollowed spines. In the spire only two rows show. Operculum thin, light-brown, chitinous, ungulate. Rare, 50 to 250 fathoms.

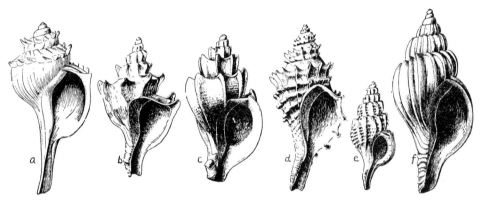

FIGURE 46. Pacific Coast Trophons. **a,** *Boreotrophon dalli* Kobelt; **b,** *B. triangulatus* Cpr.; **c,** *B. multicostatus* Esch.; **d,** *B. scitulus* Dall; **e,** *B. orpheus* Gould; **f,** *B. pacificus* Dall. All about natural size.

Boreotrophon multicostatus Eschscholtz

Many-ribbed Trophon
Figure 46c

Bering Sea to San Pedro, California. Northern Japan.

1 to 1¼ inches in length, short canal, deep suture, 5 or 6 whorls are shouldered above. Characterized by the flaring thin lip, brownish aperture, 8 to 10 lamella-like ribs per whorl, and weak microscopic spiral threads. Moderately common; littoral in Alaska, 10 to 30 fathoms in Puget Sound.

Boreotrophon stuarti E. A. Smith

Stuart's Trophon
Plate 24j

Alaska to San Diego, California.

2 inches in length, waxy texture, pure-white to yellow-cream, with 9 to 11 strong, lamella-like, high-shoulder ribs per each of the 7 whorls. Whorls in spire cancellated by the 2- or 3-spiral raised cords. Body whorl with 5 very weak spiral rounded threads. *B. smithi* Dall is the same. Uncommon from low tide (in Alaska) to 25 fathoms.

Boreotrophon pacificus Dall   Northwest Pacific Trophon
Figure 46f

Alaska to off Lower California.

¾ to 1 inch in length, similar to *multicostatus*, with the spire ⅖ the length of the shell, canal twice as long, with 12 to 14 ribs which are shouldered further below the suture. Suture slightly indented. Fairly common at low tide in Alaska. Occurs in very deep water farther south.

Boreotrophon dalli Kobelt   Dall's Trophon
Figure 46a

Arctic Ocean to Fuca Strait, Washington

2 inches in length, spire ¼ and canal nearly ½ the length of the shell. 5 whorls globose, crowned at the shoulder with 12 to 20 short spines per whorl. The ribs over the whorl are moderately to obsoletely developed. Rare in about 30 to 50 fathoms.

Boreotrophon triangulatus Carpenter   Triangular Trophon
Figure 46b

Monterey to San Pedro, California.

Nearly 1 inch in length, 7 whorls, with 8 delicate axial ribs which bear rather short erect protuberances on the shoulder. Nuclear whorl smooth, followed by squarish whorls which bear ribs from suture to suture, and which are smooth in between. Siphonal canal rather short. Grayish white with enamel-white aperture. *B. peregrinus* Dall is the same. Formerly confused with *Forreria* (*Austrotrophon*) *cerrosensis cerrosensis* and its form, *pinnatus* Dall, and its subspecies *catalinensis* Oldroyd. The latter group has longer spines, longer siphonal canal and the ribs are only on the periphery of the early whorls.

Boreotrophon orpheus Gould   Orpheus Trophon
Figure 46e

Alaska to Redondo Beach, California.

¾ to 1 inch in length, resembling *pacificus* but with a spire ³⁄₇ the length of the shell, and with 3 strong, but small, spiral cords showing in the spire, but with about the same number and type of axial ribs. It resembles an immature *stuarti*, but has more and smaller ribs and its whorls are not so sharply shouldered on top nor so strongly cancellate in the spire. Moderately common from 12 to 80 fathoms.

GENUS *Trophonopsis* Bucquoy, Dautz. and Dollfuss 1882

Trophonopsis lasius Dall   Sandpaper Trophon

Bering Sea to Lower California.

1½ to 2 inches in length; spire half the length of the grayish white shell; 6 whorls with numerous, indistinct to moderately well-developed, rounded ribs and more numerous, small, frequently scaled, spiral cords—all of which gives the shell a rough, sandpaper feel. Whorls in spire shouldered slightly. Aperture enamel-white. References in 1937 and earlier to *Trophon tenuisculptus* Carpenter are this species. The true *tenuisculptus* is an *Ocenebra*.

GENUS *Muricopsis* Bucquoy, Dautz. and Dollfuss 1882

Muricopsis hexagona Lamarck   Hexagonal Murex
Plate 25h

Florida Keys and the West Indies.

1 to 1½ inches in length, elongate, heavy, with a high spire, and sharply spinose on each of the 7 axial ribs on each whorl. Exterior chalk-white or tinted with orange-brown. Aperture white. A moderately common reef species. The genus *Muricidea* Mörch 1852, not Swainson 1840, is the same as this genus.

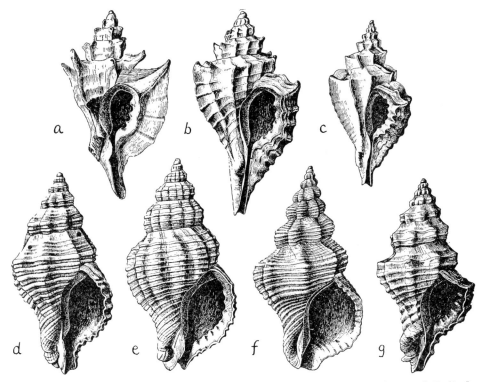

FIGURE 47. Oyster Drills of the Atlantic Coast. **a**, *Eupleura stimpsoni* Dall; **b**, *Eupleura caudata* Say; **c**, *Eupleura sulcidentata* Dall; **d**, *Urosalpinx perrugata* Conrad; **e**, *Urosalpinx cinerea* Say; **f**, *Pseudoneptunea multangula* Philippi; **g**, *Muricopsis ostrearum* Conrad. All about 1 inch in length.

PLATE 17

ATLANTIC COAST UNIVALVES

a. ANTILLEAN LIMPET, *Acmaea antillarum* Sby., 1 inch (Florida and West Indies), p. 106.

b. DWARF SUCK-ON LIMPET, *Acmaea leucopleura* Gmelin, ½ inch (Florida and West Indies), p. 106.

c. JAMAICA LIMPET, *Acmaea jamaicensis* Gmelin, ½ inch (Florida and West Indies), p. 106.

d. KNOBBY KEYHOLE LIMPET, *Fissurella nodosa* Born, 1 inch (Florida and West Indies), p. 100.

e. ROSY KEYHOLE LIMPET, *Fissurella rosea* Gmelin, 1 inch (east Florida to Brazil), p. 100.

f. BARBADOS KEYHOLE LIMPET, *Fissurella barbadensis* Gmelin, 1 inch (Florida and West Indies), p. 100.

g. WOBBLY KEYHOLE LIMPET, *Fissurella fascicularis* Lam., 1 inch (Florida and West Indies), p. 101.

h. SOWERBY'S FLESHY LIMPET, *Lucapina sowerbii* Sby., ¾ inch (Florida to Brazil), p. 98.

i. FILE FLESHY LIMPET, *Lucapinella limatula* Reeve, ½ inch (North Carolina to West Indies), p. 98.

j. NORTHERN BLIND LIMPET, *Lepeta caeca* Müller, ⅓ inch (Arctic to Massachusetts), p. 107.

k. CANCELLATE FLESHY LIMPET, *Lucapina suffusa* Reeve, 1 inch (Florida and West Indies), p. 98.

l. LISTER'S KEYHOLE LIMPET, *Diodora listeri* Orb., 1½ inches (Florida and West Indies), p. 96.

m. CAYENNE KEYHOLE LIMPET, *Diodora cayenensis* Lam., 1 inch (Virginia to Brazil), p. 96.

n. DYSON'S KEYHOLE LIMPET, *Diodora dysoni* Reeve, ½ inch (Florida and West Indies), p. 97.

o. RUFFLED RIMULA, *Emarginula phrixodes* Dall, ⅓ inch (North Carolina to West Indies), p. 94.

p. SMOOTH ATLANTIC TEGULA, *Tegula fasciata* Born, ½ inch (southeast Florida and West Indies), p. 118.

q. GEM ARENE, *Arene gemma* Tuom. and Holm., ⅛ inch (North Carolina to Brazil), p. 122.

r. SHOULDERED PHEASANT, *Tricolia pulchella* C. B. Ads., ⅛ inch (Florida and West Indies), p. 127.

s. VARIABLE ARENE, *Arene variabilis* Dall, ¼ inch (North Carolina to West Indies), p. 123.

t. NORTHERN ROSY MARGARITE, *Margarites costalis* Gould, ⅜ inch (Arctic to Massachusetts; also Alaska), p. 107.

u. BAIRD'S LIOTIA, *Liotia bairdi* Dall, ¼ inch (North Carolina to Mexico), p. 121.

v. THREADED VITRINELLA, *Vitrinella multistriata* Verrill, ⅛ inch (North Carolina to Florida), p. 138.

w. SCULPTURED TOP-SHELL, *Calliostoma euglyptum* A. Ads., ¾ inch (North Carolina to Texas), p. 112.

x. LAMELLOSE SOLARELLE, *Solariella lamellosa* Verr. and Smith, ⅛ inch (Massachusetts to North Carolina), p. 110.

y. SHORT-SPIRED TEINOSTOMA, *Teinostoma cryptospira* Verr., 2 mm (North Carolina to Florida), p. 139.

PLATE 18

PACIFIC COAST UNIVALVES

a. KEYHOLE LIMPET, *Megathura crenulata* Sby., 4 inches (Monterey, California, south), p. 99.

b. ROUGH KEYHOLE LIMPET, *Diodora aspera* Esch., 2 inches (Alaska to Lower California), p. 97.

c. VOLCANO SHELL, *Fissurella volcano* Reeve, 1 inch (Crescent City, California. south), p. 100.

d. TWO-SPOTTED KEYHOLE LIMPET, *Megatebennus bimaculatus* Dall, ½ inch (entire coast), p. 99.

e. HARD-EDGED KEYHOLE LIMPET, *Lucapinella callomarginata* Dall, 1 inch (California, south), p. 99.

f. FINGERED LIMPET, *Acmaea digitalis* Esch., 1 inch (Alaska to Mexico), p. 103.

g. TEST'S LIMPET, *Acmaea conus* Test, ¾ inch (California to Lower California), p. 102.

h. WESTERN BANDED TEGULA, *Tegula ligulata* Menke, ¾ inch (Monterey, California, south), p. 120.

i. CARPENTER'S DWARF TURBAN, *Homalopoma carpenteri* Pils., ¹/₃ inch (entire coast), p. 126.

j. GIANT OWL SHELL, *Lottia gigantea* Gray, 3 inches (Crescent City, California, south), p. 101.

k. GILDED TEGULA, *Tegula aureotincta* Forbes, 1 inch (Southern California, south), p. 120.

l. ROUGH LIMPET, *Acmaea scabra* Gould, 1 inch (Washington to Lower California), p. 103.

m. NORRIS SHELL, *Norrisia norrisi* Sby., 1½ inches (Monterey, California, south), p. 117.

n. SHIELD LIMPET, *Acmaea pelta* Esch., 1½ inches (Alaska to Lower California), p. 102.

o. FILE LIMPET, *Acmaea limatula* Cpr., 1½ inches (Northern California, south), p. 102.

p. WAVY TURBAN, *Astraea undosa* Wood, 3 inches (Ventura, California, south), p. 124.

q. MASK LIMPET, *Acmaea persona* Esch., 1½ inches (Alaska to Monterey, California), p. 103.

r. WHITE CAP LIMPET, *Acmaea mitra* Esch., 1 inch (Alaska to off Lower California), p. 101.

s. GRANULOSE TOP-SHELL, *Calliostoma supragranosum* Cpr., ½ inch (Monterey California, south), p. 115.

t. FENESTRATE LIMPET, *Acmaea fenestrata* Reeve, 1 inch (California, south), p. 102.

u. CALIFORNIAN LIOTIA, *Liotia fenestrata* Cpr., ⅛ inch (Monterey, California, south), p. 122.

v. SPECKLED TEGULA, *Tegula gallina* Forbes, 1 inch (San Francisco, California, south), p. 119.

w. CRIBRARIA LIMPET, *Acmaea fenestrata cribraria* Gould, 1 inch (Alaska to California), p. 102.

x. MONTEREY TEGULA, *Tegula montereyi* Kiener, 1½ inches (California), p. 121.

y. DUSKY TEGULA, *Tegula pulligo* Gmelin, 1 inch (Alaska to California), p. 120.

z. SEAWEED LIMPET, *Acmaea insessa* Hinds, ¾ inch (Alaska to Lower California), p. 105.

z.z. UNSTABLE LIMPET, *Acmaea instabilis* Gould, 1 inch (Alaska to San Diego, California), p. 105.

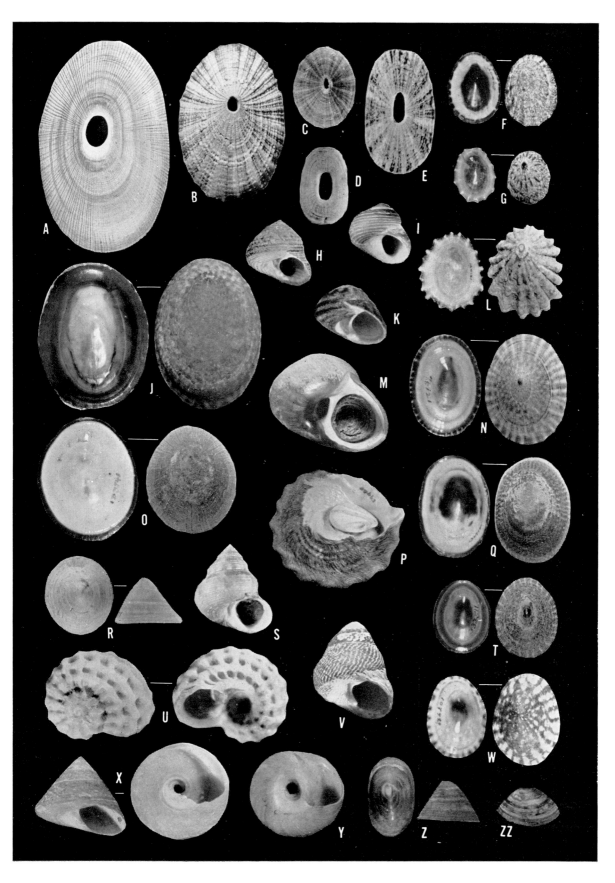

PLATE 19

ATLANTIC COAST UNIVALVES

a. ANGULATE PERIWINKLE, *Littorina angulifera* Lam., 1 inch (Florida and West Indies), p. 133.

b. COMMON PERIWINKLE, *Littorina littorea* L., 1 inch (Labrador to New Jersey; Europe), p. 132.

c. MARSH PERIWINKLE, *Littorina irrorata* Say, 1 inch (New York to Texas), p. 132.

d. NORTHERN ROUGH PERIWINKLE, *Littorina saxatilis* Olivi, ½ inch (Arctic New Jersey), p. 133.

e. ZEBRA PERIWINKLE, *Littorina ziczac* Gmelin, 1 inch (Florida to Texas and south), p. 132.

f. NORTHERN YELLOW PERIWINKLE, *Littorina obtusata* Linné, ½ inch (Arctic to New Jersey), p. 133.

g. BEADED PERIWINKLE, *Tectarius muricatus* Linné, 1 inch (Florida Keys, south), p. 134.

h. FALSE PRICKLY-WINKLE, *Echininus nodulosus* Pfr., ¾ inch (Florida Keys, south), p. 135.

i. COMMON PRICKLY-WINKLE, *Nodilittorina tuberculata* Mke., ½ inch (Florida Keys, south), p. 134.

j. SPOTTED PERIWINKLE, *Littorina meleagris* Pot. and Mich., ⅓ inch (Florida Keys, south).

k. DWARF BROWN PERIWINKLE, *Littorina mespillum* Mühlfeld, ¼ inch (Florida Keys, south), p. 133.

l. STOCKY CERITH, *Cerithium literatum* Born, 1 inch (Southeastern Florida and West Indies), p. 154.

m. FLY-SPECKED CERITH, *Cerithium muscarum* Say, 1 inch (Florida and West Indies), p. 154.

n. FLORIDA CERITH, *Cerithium floridanum* Mörch, 1½ inches (North Carolina to Florida), p. 153.

o. DWARF CERITH, *Cerithium variabile* C. B. Adams, ½ inch (Florida to Texas, south), p. 154.

p. MIDDLE-SPINED CERITH, *Cerithium algicola*, C. B. Adams, 1 inch (Florida and West Indies), p. 154.

q. IVORY CERITH, *Cerithium eburneum* Brug., 1 inch (Southeastern Florida and West Indies), p. 154.

r. VARIABLE BITTIUM, *Bittium varium* Pfr., ⅛ inch (Maryland to Texas and Mexico), p. 155.

s. FALSE CERITH, *Batillaria minima* Gmelin, ½ inch (Florida and West Indies), p. 153.

t. PLICATE HORN SHELL, *Cerithidea pliculosa* Menke, 1 inch (Florida to Texas, south-, p. 152.

u. COSTATE HORN SHELL, *Cerithidea costata* Da Costa, ½ inch (West Florida, south , p. 152.

v. GREEN'S MINIATURE CERITH, *Cerithiopsis greeni* C. B. Adams, ⅛ inch (Massachusetts to West Indies , p. 157.

w. AWL MINIATURE CERITH, *Cerithiopsis subulata* Montagu, ½ inch (Massachusetts to West Indies , p. 157.

x. LADDER HORN SHELL, *Cerithidea scalariformis* Say, 1 inch (South Carolina to Florida, Cuba), p. 152.

y. BLACK-LINED TRIFORA, *Triphora nigrocincta* C. B. Adams, ¼ inch (Massachusetts, south), p. 159.

z. BEAUTIFUL TRIFORA, *Triphora pulchella* C. B. Adams, ¼ inch (Florida, south), p. 159.

zz. MOTTLED TRIFORA, *Triphora decorata* C. B. Adams, ½ inch (Florida and West Indies), p. 159.

PLATE 20

PACIFIC COAST UNIVALVES

a. ERODED PERIWINKLE, *Littorina planaxis* Phil., ¾ inch (Washington to Lower California), p. 134.

b. SITKA PERIWINKLE, *Littorina sitkana* Phil., ¾ inch (Alaska to Washington), p. 134.

c. CHECKERED PERIWINKLE, *Littorina scutulata* Gld., ½ inch (Alaska to Mexico), p. 134.

d. FLAT WORM-SHELL, *Spiroglyphus lituellus* Mörch (Alaska to California), p. 144.

e. SCALED WORM-SHELL, *Aletes squamigerus* Cpr. (Alaska to Peru), p. 144.

f. ONYX SLIPPER-SHELL, *Crepidula onyx* Sby., 2 inches. Left shell lived on scallop (California to Chili), p. 171.

g. COOPER'S TURRET-SHELL, *Turritella cooperi* Cpr., 2 inches (California to Mexico), p. 141.

h. MARIA'S TURRET-SHELL, *Turritella mariana* Dall, 2½ inches (California to Panama), p. 142.

i. RECLUZ'S MOON-SHELL, *Polinices reclusianus* Desh., 2 inches (California to Lower California), p. 187.

j. WROBLEWSKI'S WENTLETRAP, *Opalia wroblewskii* Mörch, 1 inch (Alaska to San Diego, California), p. 162.

k. PACIFIC HALF-SLIPPER SHELL, *Crepipatella lingulata* Gld., ¾ inch (Alaska to Panama), p. 170.

l. PACIFIC CHINESE HAT, *Calyptraea fastigata* Gld., 1 inch (Alaska to California), p. 169.

m. GOULD'S DOVE-SHELL, *Nitidella gouldi* Cpr., ½ inch (Alaska to California), p. 222.

n. PACIFIC MUD NASSA, *Nassarius tegula* Reeve, ¾ inch (California to Lower California), p. 238.

o. APPLE SEED, *Erato vitellina* Hinds, ½ inch (California to Lower California), p. 176.

p. IDA'S MITER, *Mitra idae* Melville, 2 inches (California), p. 249.

q. BEATIC DWARF OLIVE, *Olivella beatica* Cpr., ½ inch (Alaska to California). p. 247.

r. CALIFORNIAN FROG-SHELL, *Bursa californica* Hinds, 4 inches, with cluster of eggs laid on clam shell (California to Lower California), p. 199.

s. GIANT WESTERN NASSA, *Nassarius fossatus* Gould, 2 inches (Vancouver, B. C., to Lower California), p. 240.

t. Top view of living animal of *Oliva*, p. 245.

u. SOLANDER'S TRIVIA, *Trivia solandri* Sby., ¾ inch (Catalina Island, California to Panama), p. 179.

v. CALIFORNIAN TRIVIA, *Trivia californiana* Gray, ½ inch (California to Mexico), p. 179.

w. X-ray photo of SPINDLE SHELL, *Fusinus*, p. 243.

PLATE 21

ATLANTIC COAST UNIVALVES

a. FLORIDA WORM-SHELL, *Vermicularia knorri* Desh., 5 inches (North Carolina to Gulf of Mexico), p. 145.

b. FARGO'S WORM-SHELL, *Vermicularia fargoi* Olsson, 3 inches (West Florida to Texas), p. 145.

c. WEST INDIAN WORM-SHELL, *Vermicularia spirata* Phil., 4 inches (Florida Keys and West Indies), p. 144.

d. IRREGULAR WORM-SHELL, *Petaloconchus irregularis* Orb., 3 inches (Florida and West Indies), p. 143.

e. BLACK WORM-SHELL, *Petalconchus nigricans* Dall, 3 inches West Florida), p. 143.

f. ATLANTIC MODULUS, *Modulus modulus* L., ½ inch (Florida to Brazil), p. 151.

g. SLIT WORM-SHELL, *Tenagodus squamatus* Blainville, 4 inches (Florida and West Indies), p. 145.

h. EASTERN TURRET-SHELL, *Turritella exoleta* L., 2 inches (Florida and West Indies), p. 141.

i. VARIEGATED TURRET-SHELL, *Turritella variegata* L., 4 inches (West Indies), p. 141.

j. BORING TURRET-SHELL, *Turritella acropora* Dall, 1 inch (North Carolina to West Indies), p. 141.

k. BROWN SARGASSUM SNAIL, *Litiopa melanostoma* Rang, ³/₁₆ inch (Pelagic, warm seas), p. 156.

l. ERODED TURRET-SHELL, *Tachyrhynchus erosum* Couthouy, 1 inch (Nova Scotia to Massachusetts, p. 140.

m. COMMON ATLANTIC SLIPPER-SHELL, *Crepidula fornicata* L., 2 inches (Canada to Texas), p. 170.

n. CONVEX SLIPPER-SHELL, *Crepidula convexa* Say, ½ inch (Massachusetts to West Indies), p. 171.

o. CIRCULAR CUP-AND-SAUCER, *Calyptraea centralis* Conrad, ½ inch (North Carolina to Texas, south), p. 169.

p. FALSE CUP-AND-SAUCER, *Cheilea equestris* L., ¾ inch (Florida Keys and West Indies), p. 165.

q. SPINY SLIPPER-SHELL, *Crepidula aculeata* Gmelin, ¾ inch (North Carolina to Texas, south), p. 171.

r. STRIATE CUP-AND-SAUCER, *Crucibulum striatum* Say, 1 inch (Nova Scotia to South Carolina), p. 170.

s. WEST INDIAN CUP-AND-SAUCER, *Crucibulum auricula* Gmelin, 1 inch (West Florida, south), p. 169.

t. WHITE HOOF-SHELL, *Hipponix antiquatus* L., ½ inch (Florida, south; California, south), p. 166.

u. GRACEFUL MELANELLA, *Melanella gracilis* C. B. Ads., ¼ inch (North Carolina to Gulf of Mexico), not in text.

v. FAT MELANELLA, *Melanella gibba* De Folin, ¼ inch (North Carolina to Gulf of Mexico), not in text.

w. TWO-LINED MELANELLA, *Melanella bilineata* Alder, ½ inch (North Carolina to West Indies), not in text.

x. ORBIGNY'S SUN-DIAL, *Torinia bisulcata* Orb., ¹/₃ inch (North Carolina to Gulf of Mexico), p. 142.

y. KREBS' SUN-DIAL, *Architectonica krebsi* Mörch, ½ inch (North Carolina to West Indies), p. 143.

z. MALTBIE'S TRIVIA, *Trivia maltbiana* Schw. and McG., ½ inch (North Carolina to Florida, south), p. 178.

aa. SUFFUSE TRIVIA, *Trivia suffusa* Gary, ¹/₃ inch (Southeastern Florida and West Indies), p. 177.

bb. COFFEE BEAN TRIVIA, *Trivia pediculus* L., ½ inch (Florida to Brazil), p. 177.

cc. LITTLE WHITE TRIVIA, *Trivia candidula* Gaskoin, ¼ inch (North Carolina to West Indies), p. 178.

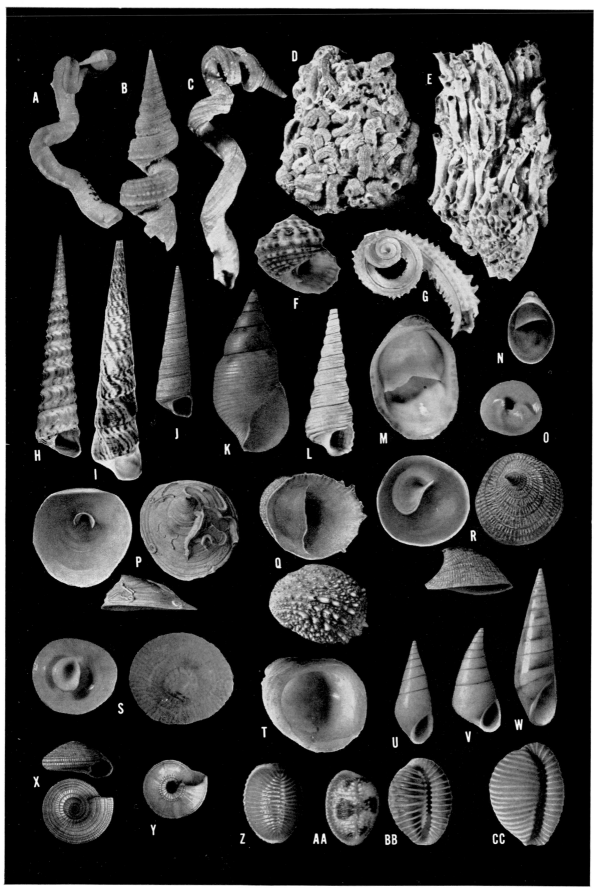

PLATE 22

ATLANTIC COAST UNIVALVES

a. LAMELLOSE WENTLETRAP, *Epitonium lamellosum* Lam., 1 inch (Florida and Caribbean), p. 165.

b. ANGULATE WENTLETRAP, *Epitonium angulatum* Say, 1 inch (New York to Texas), p. 164.

c. DALL'S WENTLETRAP, *Circostrema dalli* Rehder, 1 inch (North Carolina to Brazil), p. 161.

d. HUMPHREY'S WENTLETRAP, *Epitonium humphreysi* Kiener, ¾ inch (Massachusetts to Texas), p. 164.

e. BROWN-BANDED WENTLETRAP, *Epitonium rupicola* Kurtz, 1 inch (Massachusetts to Texas), p. 165.

f. MITCHELL'S WENTLETRAP, *Amaea mitchelli* Dall, 2 inches (Texas), p. 163.

g. HOTESSIER'S WENTLETRAP, *Opalia hotessieriana* Orb., ½ inch (Southeastern Florida, south), p. 162.

h. SHARK EYE, *Polinices duplicatus* Say, 2 inches (high-spired form) (Massachusetts to Texas), p. 186.

i. MILK MOON-SHELL, *Polinices lacteus* Guilding, 1 inch (North Carolina to West Indies), p. 185.

j. SOUTHERN MINIATURE NATICA, *Natica pusilla* Say, ¹/₃ inch (Massachusetts to Gulf and West Indies), p. 191.

k. GREENLAND MOON-SHELL, *Lunatia grönlandica* Moller, 1 inch (Arctic to off New Jersey), p. 189.

l. CAROLINA MOON-SHELL, *Sigatica carolinensis* Dall, ¼ inch (North Carolina to West Indies), p. 187.

m. SPOTTED NORTHERN MOON-SHELL, *Lunatia triseriata* Say, ½ inch (Canada to off North Carolina), p. 189.

n. SMOOTH VELUTINA, *Velutina laevigata* L., ¾ inch (Arctic to Massachusetts; also to California), p. 175.

o. LIVID NATICA, *Natica livida* Pfr., ½ inch (Southeastern Florida and West Indies), p. 191.

p. COMMON NORTHERN LACUNA, *Lacuna vincta* Turton, ⅜ inch (Arctic to Rhode Island; also to California), p. 130.

q. DWARF RED OVULA, *Primovula carnea* Poiret, ½ inch (Southeastern Florida and West Indies), p. 181.

r. ICELAND MOON-SHELL, *Amauropsis islandica* Gmelin, 1 inch (Arctic to Massachusetts), p. 187.

s. COMMON BABY'S EAR, *Sinum perspectivum* Say, 1½ inches (Virginia to Texas, south), p. 190.

t. ADAM'S MINIATURE CERITH, *Seila adamsi* H. C. Lea, ⅛ inch (Massachusetts to Texas, south), p. 158.

u. CHESNEL'S RISSOINA, *Rissoina chesneli* Mich., ¹/₅ inch (North Carolina to Texas, south), p. 137.

v. VARIABLE DWARF OLIVE, *Olivella mutica* Say, ¹/₃ inch (North Carolina to Texas, south), p. 246.

w. MAUGERI'S ERATO, *Erato maugeriae* Gray, ¹/₁₆ inch (North Carolina to West Indies), p. 176.

x. JASPER CONE, *Conus jaspideus* Gmelin, ¾ inch (Florida and West Indies), p. 262.

y. STEARNS' CONE, *Conus stearnsi* Conrad, ½ inch (Florida to Mexico), p. 262.

z. WARTY CONE, *Conus verrucosus* Hwass, ¾ inch (Southeastern Florida and West Indies), p. 263.

PLATE 23

ATLANTIC COAST UNIVALVES

a. QUEEN CONCH, *Strombus gigas* L., 10 inches (also young) (Florida Keys and West Indies), p. 174.

b. MILK CONCH, *Strombus costatus* Gmelin, 6 inches (Southern Florida and West Indies), p. 174.

c. AMERICAN PELICAN'S FOOT, *Aporrhais occidentalis* Beck, 2 inches (also young) (off New England), p. 173.

d. LONGLEY'S CARRIER-SHELL, *Xenophora longleyi* Bartsch, 7 inches (Gulf of Mexico, south), p. 173.

e. CARIBBEAN CARRIER-SHELL, *Xenophora caribaea* Petit, 4 inches (off Florida and Cuba), p. 173.

f. GIANT TUN, *Tonna galea* L., 6 inches (North Carolina to Texas and West Indies; Pacific), p. 199.

g. CROSSE'S TUN, *Eudolium crosseanum* Montero., 3 inches (off New Jersey to West Indies), p. 199.

h. WEST INDIAN CROWN CONCH, *Melongena melongena* L., 4 inches (Florida Keys, south), p. 235.

i. KNOBBED WHELK, *Busycon carica* Gmelin, 8 inches (Cape Cod, Massachusetts to Northeastern Florida), p. 235.

j. NEW ENGLAND NASSA, *Nassarius trivittatus* Say, ¾ inch (Nova Scotia to South Carolina), p. 239.

k. PERVERSE WHELK, *Busycon perversum* L., form *eliceans* Mont., 6 inches Northeastern Florida), p. 236.

l. CARIBBEAN VASE, *Vasum muricatum* Born, 4 inches (South Florida and West Indies), p. 245.

m. PYGMY COLUS, *Colus pygmaea* Gould, 1 inch (Quebec to off North Carolina), p. 229.

n. CHANNELED WHELK, *Busycon canaliculatum* L., 6 inches (Cape Cod to Northeastern Florida), p. 236.

o. LIGHTNING WHELK, *Busycon contrarium* Conrad, 8 inches (South Carolina to Texas), p. 236.

p. EASTERN MUD NASSA, *Nassarius obsoletus* Say, ¾ inch (Quebec to Northeastern Florida), p. 240.

q. COMMON EASTERN NASSA, *Nassarius vibex* Say, ½ inch (North Carolina to Texas, south), p. 237.

r. VARIABLE NASSA, *Nassarius ambiguus* Pult., ½ inch (North Carolina to West Indies), p. 239.

s. BROWN-CORDED NEPTUNE, *Neptunea decemcostata* Say, 3 inches (Nova Scotia to Massachusetts), p. 229.

t. HAIRY COLUS, *Colus pubescens* Verrill, 2 inches (Quebec to off North Carolina), p. 229.

u. FAT COLUS, *Colus ventricosus* Gray, 2½ inches (Nova Scotia to New York), not in text.

v. EMPEROR HELMET, *Cassis madagascariensis* Lam., 9 inches (Southeastern Florida and West Indies), p. 193.

w. LONGHORNED SMOKE SHELL, *Typhis longicornis* Dall, ½ inch (Florida and West Indies), not in text.

x. STIMPSON'S COLUS, *Colus stimpsoni* Mörch, 4 inches (Labrador to off North Carolina), p. 227.

y. FLORIDA HORSE CONCH, *Pleuroploca gigantea* Kiener, 20 inches (North Carolina to Florida), p. 242.

PLATE 24

PACIFIC COAST UNIVALVES

a. TWO-KEELED HAIRY-SHELL, *Trichotropis bicarinata* Sby., 1½ inches (North Pacific and North Atlantic), p. 168.

b. CANCELLATE HAIRY-SHELL, *Trichotropis cancellata* Hinds, 1 inch (Alaska to Oregon), p. 167.

c. GRAY HAIRY-SHELL, *Trichotropis insignis* Midd., 1 inch (Alaska to Japan), p. 168.

d. BOREAL HAIRY-SHELL, *Trichotropis borealis* B. and Sby., ¾ inch (Arctic to Washington; also to Maine), p. 167.

e. GEM MUREX, *Murex gemma* Sby., 1 inch (Santa Barbara to Lower California), p. 206.

f. NUTTALL'S THORN PURPURA, *Pterorytis nuttalli* Conrad, 1½ inches (California to Lower California), p .219.

g. OREGON TRITON, *Argobuccinum oregonensis* Redfield, 4 inches (Alaska to San Diego, California), p. 194.

h. FOLIATED THORN PURPURA, *Pterorytis foliata* Gmelin, 3 inches (Alaska to San Pedro, California), p. 218.

i. GIANT FORRERIA, *Forreria belcheri* Hinds, 5 inches (California to Lower California), p. 200.

j. STUART'S TROPHON, *Boreotrophon stuarti* E. A. Smith, 2 inches (Alaska to San Diego, California), p. 207.

k. POULSON'S DWARF TRITON, *Ocenebra poulsoni* Cpr., 2 inches (Santa Barbara, California, to Lower California), p. 218.

l. FESTIVE MUREX, *Murex festivus* Hinds, 2 inches (Morro Bay, California, to Lower California), p. 206.

m. GRACEFUL DWARF TRITON, *Ocenebra gracillima* Stearns, ⅓ inch (Monterey, California, to Gulf of California), p. 217.

n. LEWIS MOON-SHELL, *Lunatia lewisi* Gould, 4 inches (Canada to Lower California), p. 189.

o. SPOTTED THORN DRUPE, *Acanthina spirata* Blainville, 1 inch (Washington to California), p. 211.

p. LEFT-HANDED BUCCINUM, *Volutopsius harpa* Mörch, 4 inches (Alaska), p. 226.

q. COMMON NORTHWEST NEPTUNE, *Neptunea lyrata* Gmelin, 5 inches (Arctic to Washington), p. 230.

r. PRIBILOFF NEPTUNE, *Neptunea pribiloffensis* Dall, 5 inches (Alaska to British Columbia), p. 230.

s. FAT NEPTUNE, *Neptunea ventricosa* Gmelin, 3 inches (Arctic and Alaska), p. 230.

t. GLACIAL BUCCINUM, *Buccinum glaciale* L., 3 inches (Arctic to Washington; also to Nova Scotia), p. 226.

u. SILKY BUCCINUM, *Buccinum tenue* Gray, 2 inches (Arctic to Washington; also to Maine), p. 225.

v. BAER'S BUCCINUM, *Buccinum baeri* Midd., 1½ inches (Bering Sea, Alaska), p. 226.

w. KELLET'S WHELK, *Kelletia kelleti* Forbes, 4 inches (State Barbara, California, to Mexico), p. 231.

x. LIVID MACRON, *Macron lividus* A. Ads., 1 inch (Monterey, California, to Lower California), p. 234.

y. COOPER'S NUTMEG, *Narona cooperi* Gabb, 2 inches (Monterey, California, to Lower California), p. 253.

z. SANTA BARBARA SPINDLE, *Fusinus barbarensis* Trask, 4 inches (Oregon to California), p. 244.

zz. GIANT PANAMA SPINDLE, *Fusinus dupetit-thouarsi* Kiener, 6 inches (off Lower California), not in text.

PLATE 25

ATLANTIC COAST UNIVALVES

a. FLORIDA ROCK-SHELL, *Thais haemastoma floridana* Conrad, 2 inches (North Carolina to West Indies), p. 213.

b. DELTOID ROCK-SHELL, *Thais deltoidea* Lam., 1 inch. (Southeastern Florida to Brazil), p. 214.

c. ELEGANT FOSSARUS, *Fossarus elegans* Verr. and Smith, 3 mm (Massachusetts to North Carolina), p. 176.

d. RED-MOUTHED ROCK-SHELL, *Thais h. haemastoma* L., 3 inches (Europe; South America), p. 213.

e. SCALY DOGWINKLE, *Thais lapillus* L., form *imbricata* Lam. (North Atlantic to New York), p. 215.

f. RUSTIC ROCK-SHELL, *Thais rustica* Lam., 1½ inches (Southeastern Florida to Brazil), p. 214.

g. ATLANTIC DOGWINKLE, *Thais lapillus* L., 1 inch (North Atlantic to New York), p. 214.

h. HEXAGONAL MUREX, *Muriscopsis hexagona* Lam., 1 inch (Florida Keys and West Indies), p. 209.

i. MÖRCH'S PITTED MUREX, *Murex cellulosus nuceus* Mörch, 1 inch (Florida to West Indies), p. 205.

j. DALL'S PITTED MUREX, *Murex cellulosus leviculus* Dall, ½ inch (North Carolina to West Florida), p. 205.

k. DELTOID ROCK-SHELL, *Thais deltoidea* Lam., 1 inch (Southeastern Florida to Brazil), p. 214.

l. WIDE-MOUTHED PURPURA, *Purpura patula* L., 3 inches (Southeastern Florida and West Indies), p. 213.

m. LIP TRITON, *Cymatium labiosum* Wood, ¾ inch (North Carolina to West Indies), p. 196.

n. DWARF HAIRY TRITON, *Cymatium gracile* Reeve, 1 inch (North Carolina to West Indies), p. 195.

o. GRANULAR FROG-SHELL, *Bursa granularis* Röding, 2 inches (Southeastern Florida, south), p. 198.

p. CHESTNUT FROG-SHELL, *Bursa spadicea* Montfort, 2 inches (Southeastern Florida and Caribbean), p. 198.

q. GOLD-MOUTHED TRITON, *Cymatium chlorostomum* Lam., 2 inches (Southeastern Florida, south), p. 196.

r. KNOBBED TRITON, *Cymatium muricinum* Röding, 2 inches (Southeastern Florida and West Indies), p. 196.

s. ATLANTIC WOOD-LOUSE, *Morum oniscus* L., ¾ inch (and young) (Southeastern Florida, south), p. 192.

t. INTRICATE BAILY-SHELL, *Bailya intricata* Dall, ½ inch (south half of Florida), p. 231.

u. CANDé'S PHOS, *Antillophos candei* Orb., 1 inch (North Carolina to Cuba), p. 231.

v. BLACKBERRY DRUPE, *Drupa nodulosa* C. B. Adams, ¾ inch (Florida and West Indies), p. 211.

w. WHITE-SPOTTED ENGINA, *Engina turbinella* Kiener, ½ inch (Florida Keys, south), p. 232.

x. ARROW DWARF TRITON, *Colubraria lanceolata* Menke, 1 inch (North Carolina to West Indies), p. 232.

y. TINTED CANTHARUS, *Cantharus tinctus* Conrad, 1 inch (North Carolina to West Indies), p. 233.

z. FLORIDA DISTORSIO, *Distorsio constricta mcgintyi* Em. and Puff., 2 inches (North Carolina to Florida), p. 197.

aa. ATLANTIC DISTORSIO, *Distorsio clathrata* Lam., 2 inches (North Carolina to Texas, south), p. 196.

bb. COMMON DOVE-SHELL, *Columbella mercatoria* L., ½ inch (Southeastern Florida and West Indies), p. 220.

cc. RAVENEL'S DOVE-SHELL, *Mitrella raveneli* Dall, ⅜ inch (North Carolina to Florida), p. 223.

dd. GLOSSY DOVE-SHELL, *Nitidella nitidula* Sby., ½ inch (Southeastern Florida and West Indies), p. 222.

ee. GREEDY DOVE-SHELL, *Anachis avara* Say, ½ inch (New Jersey to Texas), p. 221.

ff. WELL-RIBBED DOVE-SHELL, *Anachis translirata* Rav., ⅓ inch (Massachusetts to Texas, south), p. 221

gg. LUNAR DOVE-SHELL, *Mitrella lunata* Say, ³/₁₆ inch (Massachusetts to Texas, south), p. 223.

hh. WHITE-SPOTTED DOVE-SHELL, *Nitidella ocellata* Gmelin, ½ inch (Florida Keys and West Indies), p. 222.

PLATE 26

ATLANTIC COAST UNIVALVES

a. SULCATE MITER, *Mitra sulcata* Gmelin, ½ inch (Southeastern Florida to West Indies), p. 249.

b. BEADED MITER, *Mitra nodulosa* Gmelin, 1 inch (North Carolina to West Indies), p. 248.

c. HENDERSON'S MITER, *Mitra hendersoni* Rehder, ¾ inch. Holotype (Southeastern Florida, south), p. 249.

d. BARBADOS MITER, *Mitra barbadensis* Gmelin, 1 inch (Southeastern Florida to West Indies), p. 249.

e. WHITE-RIBBED, *Mitra albicostata* C. B. Adams, ½ inch (Florida Keys and West Indies), not in text.

f. WHITE-SPOTTED DRILLIA, *Monilispira albomaculata* C. B. Adams, ½ inch (West Indies), p. 271.

g. GRAY ATLANTIC AUGER, *Terebra cinerea* Born, 1½ inches (Southeastern Florida to West Indies), p. 266.

h. SHINY ATLANTIC AUGER, *Terebra hastata* Gmelin, 1¼ inches (Southeastern Florida to West Indies), p. 266.

i. COMMON ATLANTIC AUGER, *Terebra dislocata* Say, 1½ inches (Virginia to Texas, south), p. 265.

j. CONCAVE AUGER, *Terebra concava* Say, ¾ inch (North Carolina to West Florida), p. 266.

k. FINE-RIBBED AUGER, *Terebra protexta* Conrad. 1 inch (North Carolina to Texas), p. 266.

l. SPINED SPINDLE-BUBBLE, *Rhizorus acutus* Orb., 3 mm (North Carolina to West Indies), p. 281.

m. WATSON'S CANOE-BUBBLE, *Scaphander watsoni* Dall, ¾ inch (off North Carolina to Florida), p. 281.

n. OYSTER TURRET, *Crassispira ostrearum* Stearns, ½ inch (North Carolina to Cuba), p. 270.

o. GIANT CANOE-BUBBLE, *Scaphander punctostriatus* Mighels, 1½ inches (off entire coast), p. 281.

p. COMMON WEST INDIAN BUBBLE, *Bulla occidentalis* A. Adams, ¾ inch (North Carolina to West Indies), p. 277.

q. ORBIGNY'S BARREL-BUBBLE, *Cylichna bidentata* Orb., 3 mm (North Carolina to Texas, south), p. 282.

r. GOULD'S BARREL-BUBBLE, *Cylichna gouldi* Couthouy, ⅜ inch (Arctic to Massachusetts), p. 282.

s. EASTERN PAPER-BUBBLE, *Haminoea solitaria* Say, ½ inch (Massachusetts to North Carolina), p. 279.

t. ADAM'S BABY-BUBBLE, *Acteon punctostriatus* C. B. Adams, 5 mm (Massachusetts to West Indies), p. 275.

u. MINIATURE MELO, *Micromelo undata* Brug., ½ inch (Florida Keys and West Indies), p. 276.

v. ORBIGNY'S HELMET-BUBBLE, *Ringicula semistriata* Orb., 3 mm (North Carolina to West Indies), p. 276.

w. BUSH'S BARREL-BUBBLE, *Pyrunculus caelatus* Bush, 3 mm (North Carolina to Southeastern Florida), p. 280.

x. CHANNELED BARREL-BUBBLE, *Retusa canaliculata* Say, 5 mm (Nova Scotia to Texas, south), p. 280.

y. COMMON PAPER NAUTILUS, *Argonauta argo* L., 5 inches. Shell of female filled with eggs (Tropical Seas), p. 485.

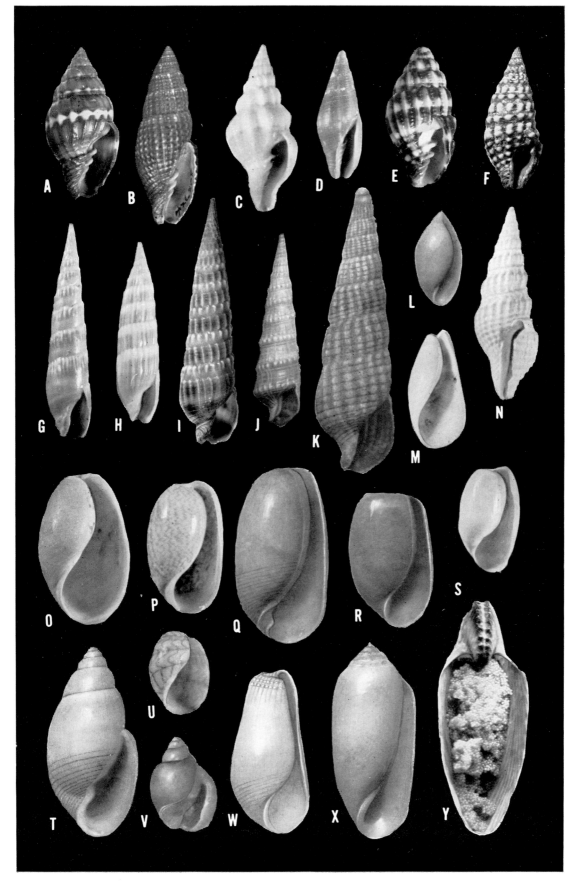

PLATE 27

ATLANTIC COAST BIVALVES

a. COMMON AWNING CLAM, *Solemya velum* Say, ¾ inch (Nova Scotia to Florida), p. 333.

b. SHORT YOLDIA, *Yoldia sapotilla* Gould, 1 inch (Arctic to North Carolina), p. 340.

c. POURTALES' GLASS SCALLOP, *Propeamussium pourtalesianum* Dall, ½ inch (off Florida), p. 369.

d. MYALIS YOLDIA, *Yoldia myalis* Couthouy, 1 inch (Arctic to Cape Cod), p. 340.

e. BROAD YOLDIA, *Yoldia thraciaeformis* Storer, 2 inches (Arctic to North Carolina; also Alaska), p. 340.

f. SULCATE LIMOPSIS, *Limopsis sulcata* V. and B., ½ inch (Massachusetts to West Indies), p. 347.

g. UNDULATE BITTERSWEET, *Glycymeris undata* L., 2 inches (North Carolina to West Indies), p. 348.

h. DECUSSATE BITTERSWEET, *Glycymeris decussata* L., 2 inches Florida and West Indies), p. 348.

i. COMB BITTERSWEET, *Glycymeris pectinata* Gmelin, ¾ inch (North Carolina to West Indies), p. 348.

j. MOSSY ARK, *Arca umbonata* Lamarck, 2 inches (North Carolina to West Indies), p. 342.

k. DOC BALES' ARK, *Barbatia tenera* C. B. Adams, 1 inch (Florida and West Indies), p. 343.

l. ICELAND SCALLOP, *Chlamys islandica* Müller, 3 inches (Arctic to Massachusetts; also to Washington), p. 365.

m. ATLANTIC DEEPSEA SCALLOP, *Placopecten magellanicus* Gmelin. 9 inches (Labrador to North Carolina), p. 366.

n. TURKEY WING, *Arca zebra* Swainson, 3 inches (North Carolina to West Indies), p. 342.

o. CUT-RIBBED ARK, *Anadara lienosa floridana* Conrad, 4 inches (North Carolina to Texas), p. 344.

p. EARED ARK, *Anadara notabilis* Röding. 2 inches (Florida to Brazil), p. 344.

q. RED-BROWN ARK, *Barbatia cancellaria* Lam., 1 inch (Florida and West Indies), p. 343.

r. WHITE BEARDED ARK, *Barbatia candida* Helbling, 2 inches (North Carolina to Brazil), p. 342.

s. TRANSVERSE ARK, *Anadara transversa* Say, 1 inch (Cape Cod to Texas), p. 345.

t. BLOOD ARK, *Anadara ovalis* Brug., 2 inches (Cape Cod to Texas and West Indies), p. 345.

u. WHITE MINIATURE ARK, *Barbatia domingensis* Lam., ½ inch (North Carolina to West Indies), p. 343.

v. SAW-TOOTHED PEN SHELL, *Atrina serrata* Sby., 6 inches (North Carolina to Florida), p. 360.

w. AMBER PEN SHELL. *Pinna carnea* Gmelin, 5 inches (Florida and West Indies), p. 360.

x. STIFF PEN SHELL, *Atrina rigida* Solander, 7 inches (North Carolina to West Indies), p. 360.

y. INCONGRUOUS ARK, *Anadara brasiliana* Lam., 2 inches (North Carolina to Texas to Brazil), p. 346.

z. PONDEROUS ARK, *Noetia ponderosa* Say, 2 inches (Virginia to Texas), p. 346.

PLATE 28

ATLANTIC COAST BIVALVES

a. COMMON EASTERN OYSTER, *Crassostrea virginica* Gmelin, 3 inches (entire Atlantic Coast), p. 375.

b. SPONGE OYSTER, *Ostrea permollis* Sby., 2 inches (North Carolina to West Indies), p. 374.

c. CRESTED OYSTER. *Ostrea equestris* Say, 2 inches (North Carolina to West Indies), p. 373.

d. COON OYSTER, *Ostrea frons* L., 2 inches. Two forms. (Florida and West Indies), p. 373.

e. DISCORD MUSCULUS, *Musculus discors* L., 1 inch (North Atlantic and North Pacific), p. 355.

f. SMOOTH MUSCULUS, *Musculus laevigatus* Gray, 1 inch (North Atlantic and North Pacific), p. 355.

g. BLACK MUSCULUS, *Musculus niger* Gray, 2 inches (Arctic to North Carolina; also to Washington), p. 355.

h. ATLANTIC RIBBED MUSSEL, *Modiolus demissus* Dillwyn, 3 inches (Canada to South Carolina), p. 351.

i. ATLANTIC PAPER MUSSEL, *Amygdalum papyria* Conrad, 1 inch (Maryland to Texas), p. 353.

j. GLANDULAR CRENELLA, *Crenella glandula* Totten, 1/3 inch (Labrador to North Carolina), p. 350.

k. GIANT DATE MUSSEL, *Lithophaga antillarum* Orb., 4 inches (Gulf of Mexico, south), p. 357.

l. FLATTENED CARDITA, *Venericardia perplana* Conrad, 1/4 inch (North Carolina to Florida), p. 380.

m. BLACK DATE MUSSEL, *Lithophaga nigra* Orb., 2 inches (Florida and West Indies), p. 355.

n. MAHOGANY DATE MUSSEL, *Lithophaga bisulcata* Orb., 1 inch (Florida to Argentina), p. 357.

o. LENTIL ASTARTE, *Astarte subequilatera* Sby., 1 inch (Arctic Seas to off Georgia), p. 376.

p. CORAL-BORING CLAM, *Coralliophaga coralliophaga* Gmelin, 1½ inches (Gulf and West Indies), p. 382.

q. BOREAL ASTARTE, *Astarte borealis* Schumacher, 1½ inches (Arctic to Massachusetts; to Alaska), p. 375.

r. WAVED ASTARTE, *Astarte undata* Gould, 1 inch (Gulf of Maine), p. 376.

s. SMOOTH ASTARTE, *Astarte castanea* Say, 1 inch (Nova Scotia to Cape Cod, Massachusetts), p. 376.

t. NORTHERN CARDITA, *Venericardia borealis* Conrad, 1 inch (Labrador to North Carolina), p. 379.

u. GLASSY LYONSIA, *Lyonsia hyalina* Conrad, ½ inch (Nova Scotia to South Carolina), p. 468.

v. LEA'S SPOON-CLAM, *Periploma leanum* Conrad, 1 inch (Nova Scotia to North Carolina), p. 474.

w. PAPER SPOON CLAM, *Periploma papyratium* Say, 1 inch (Labrador to Rhode Island), p. 472.

x. UNEQUAL SPOON CLAM, *Periploma inequale* C. B. Adams, 1 inch (South Carolina to Texas), p. 473.

y. CONRAD'S THRACIA, *Thracia conradi* Couthouy, 3 inches (Nova Scotia to New York), p. 471.

PLATE 29

ATLANTIC COAST BIVALVES

a. GIANT ROCK SCALLOP, *Hinnites multirugosus* Gale, 7 inches (Alaska to Lower California), p. 369.

b. GIANT PACIFIC SCALLOP, *Pecten caurinus* Gould, 7 inches (Alaska to California), p. 361.

c. HEMPHILL'S LIMA, *Lima hemphilli* Hert. and Strong, 1 inch (California to Mexico), p. 371.

d. FALSE JINGLE SHELL, *Pododesmus macroschismus* Desh., 3 inches (Alaska to Lower California), p. 372.

e. PERUVIAN JINGLE SHELL, *Anomia peruviana* Orb., 2 inches (California to Peru), p. 372.

f. NATIVE PACIFIC OYSTER, *Ostrea lurida* Cpr., 3 inches (Alaska to Lower California), p. 374.

g. GIANT PACIFIC OYSTER, *Crassostrea gigas* Thunberg, 8 inches (Canada to California), p. 375.

h. CALIFORNIAN DATE MUSSEL, *Botula californiensis* Phil., 1 inch (California south), p. 356.

i. KELSEY'S DATE MUSSEL, *Lithophaga plumula kelseyi* Hert. and Strong, 1 inch (California), p. 357.

j. SCISSOR DATE MUSSEL, *Lithophaga aristata* Dill., 1 inch (Florida and California, south), p. 357.

k. FALCATE DATE MUSSEL, *Botula falcata* Gould, 3 inches (Oregon to Lower California), p. 356.

l. STOUT CARDITA, *Venericardia ventricosa* Gould, ¾ inch (Washington to California), p. 379.

m. NORTHERN UGLY CLAM, *Entodesma saxicola* Baird, 4 inches (Alaska to California), p. 469.

n. CALIFORNIAN SUNSET CLAM, *Gari californica* Conrad, 3 inches (Alaska to California), p. 441.

o. CALIFORNIAN TULIP MUSSEL, *Modiolus fornicatus* Cpr., 1 inch (California), p. 352.

p. CALIFORNIAN MUSSEL, *Mytilus californianus* Conrad, 8 inches (Alaska to Mexico), p. 354.

q. NUTTALL'S BLADDER CLAM, *Mytilimeria nuttalli* Conrad, 1 inch (Alaska to Lower California), p. 469.

r. CARPENTER'S CARDITA, *Cardita carpenteri* Lamy, ½ inch (Canada to Lower California), p. 378.

s. PURPLISH PACIFIC TAGELUS, *Tagelus subteres* Cpr., 2 inches (California to Lower California), p. 441.

t. ROCK-DWELLING SEMELE, *Semele rupicola* Dall, 1 inch (California to Mexico), p. 435.

u. CALIFORNIAN TAGELUS, *Tagelus californianus* Conrad, 3 inches (California to Panama), p. 440.

v. BLUNT JACKKNIFE CLAM, *Solen sicarius* Gould, 3 inches (Canada to Lower California), p. 444.

w. ROSE PETAL SEMELE, *Semele rubropicta* Dall, 2 inches (Alaska to Mexico), p. 435.

x. NUTTALL'S MAHOGANY CLAM, *Sanguinolaria nuttalli* Conrad, 3 inches, (California to Lower California), p. 439.

y. PACIFIC RAZOR CLAM, *Siliqua patula* Dixon, 5 inches (Alaska to California), p. 442.

z. BARK SEMELE, *Semele decisa* Gould, 3 inches (San Pedro, California to Lower California), p. 435.

PLATE 30

ATLANTIC COAST BIVALVES

a. BROAD-RIBBED CARDITA, *Cardita floridana* Conrad, 1 inch (Florida to Mexico), p. 378.

b. LINDSLEY'S CRASSINELLA, *Crassinella mactracea* Lindsley, $^1/_3$ inch (Massachusetts to New York), p. 377.

c. NORTHERN DWARF COCKLE, *Cerastoderma pinnulatum* Conrad, $^1/_2$ inch (Labrador to North Carolina), p. 404.

d. STOUT TAGELUS, *Tagelus plebeius* Solander, 3 inches (Cape Cod to Texas), p. 440.

e. ATLANTIC RUPELLARIA, *Rupellaria typica* Jonas, 1 inch (North Carolina to the West Indies), p. 420.

f. ATLANTIC RAZOR CLAM, *Siliqua costata* Say, 2 inches (Canada to New Jersey), p. 442.

g. PURPLISH TAGELUS, *Tagelus divisus* Spengler, 1 inch (Cape Cod to Mexico), p. 440.

h. SMALL FALSE DONAX, *Heterodonax bimaculatus* Linné, ¾ inch (Florida and West Indies), p. 441.

i. ST. MARTHA'S RAZOR CLAM, *Solecurtus sanctaemarthae* Orb., 1½ inches (North Carolina to Brazil), p. 445.

j. CANCELLATE SEMELE, *Semele bellastriata* Conrad, ½ inch (North Carolina to West Indies), p. 435.

k. ATLANTIC JACKKNIFE CLAM, *Ensis directus* Conrad, 8 inches (Labrador to South Carolina), p. 443.

l. DWARF TIGER LUCINA, *Codakia orbiculata* Montagu, ¾ inch (North Carolina to West Indies), p. 391.

m. CROSS-HATCHED LUCINA, *Divaricella quadrisulcata* Orb., ¾ inch (Massachusetts to West Indies), p. 391.

n. GREEN JACKKNIFE CLAM, *Solen viridis* Say, 2 inches (Rhode Island to Texas), p. 444.

o. CRESTED TELLIN, *Tellidora cristata* Recluz, 1 inch (North Carolina to Texas), p. 430.

p. DENTICULATE WEDGE SHELL, *Donax denticulata* Linné, 1 inch (West Indies), p. 438.

q. TEXAS WEDGE SHELL, *Donax variabilis roemeri* Philippi, ¾ inch (Texas coast), p. 437.

r. COQUINA SHELL, *Donax variabilis* Say, ¾ inch (Virginia to Texas), p. 437.

s. GRANULAR POROMYA, *Poromya granulata* Nyst, $^1/_3$ inch (Massachusetts to Cuba), p. 475.

t. GRANT AND GALE MACOMA, *Macoma planiuscula* Grant and Gale, 1 inch (Arctic to Oregon), p. 433.

u. MEROPSIS TELLIN, *Tellina meropsis* Dall, ½ inch (San Diego, California, south), p. 426.

v. COMMON ATLANTIC ABRA, *Abra aequalis* Say, ¼ inch (North Carolina to Texas, and south), p. 437.

w. DALL'S LITTLE ABRA, *Abra lioica* Dall, ¼ inch (off Massachusetts to West Indies), p. 437.

x. NORTHERN DWARF TELLIN, *Tellina agilis* Stimpson, $^1/_3$ inch (Canada to Maryland), p. 422.

y. FLORIDA MARSH CLAM, *Pseudocyrena floridana* Conrad, 1 inch (West Florida to Texas), p. 381.

z. GIBB'S CLAM, *Eucrassatella speciosa* A. Ads., 2 inches (North Carolina to West Indies), p. 377.

aa. FLORIDA LUCINA, *Lucina floridana* Conrad, 1½ inches (West Florida to Texas), p. 387.

bb. CAROLINA MARSH CLAM, *Polymesoda caroliniana* Bosc, 1 inch (Virginia to Texas), p. 381.

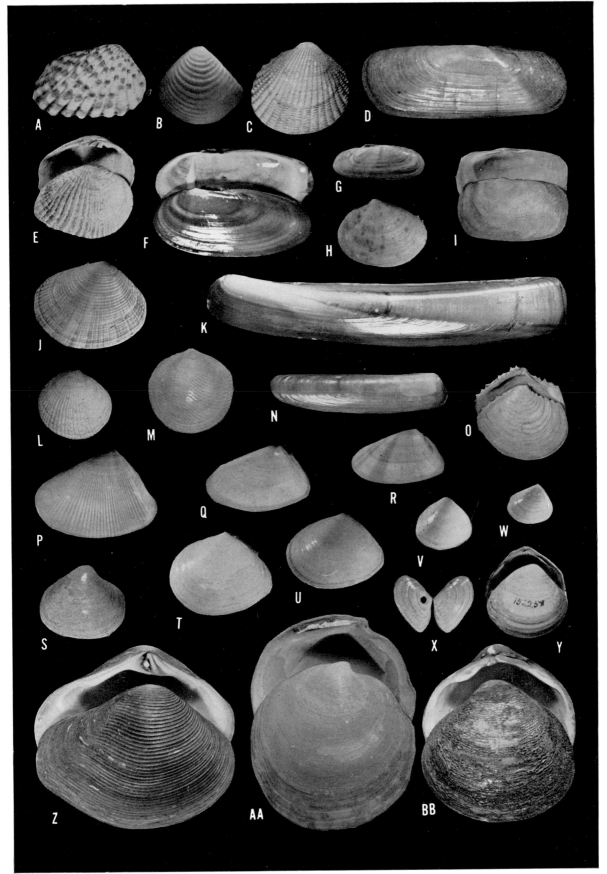

PLATE 31

PACIFIC COAST BIVALVES

a. GIANT PACIFIC COCKLE, *Trachycardium quadragenarium* Con., 5 inches (California), p. 398.

b. NUTTALL'S COCKLE, *Clinocardium nuttalli* Conrad, 4 inches (Alaska to California), p. 403.

c. CALIFORNIAN LUCINA, *Codakia californica* Conrad, 1½ inches (California), p. 390.

d. FRILLED VENUS, *Chione gnidia* Brod. and Sby., 3 inches (Lower California to Peru), not in text.

e. PACIFIC COAST BITTERSWEET, *Glycymeris subobsoleta* Cpr., 1 inch (Alaska to Lower California), p. 349.

f. SMOOTH PACIFIC VENUS, *Compsomyax subdiaphana* Cpr., 2 inches (Alaska to Lower California), p. 411.

g. NUTTALL'S LUCINA, *Phacoides nuttalli* Con., 1 inch (Santa Barbara, California, south), p. 388.

h. PISMO CLAM, *Tivela stultorum* Mawe, 5 inches (San Mateo, California south), p. 412.

i. FRILLED CALIFORNIAN VENUS, *Chione calif. undatella* Sby., 2 inches (California, south), p. 408.

j. COMMON CALIFORNIAN VENUS, *Chione californiensis* Brod., 2 inches (San Pedro, California, south), p. 407.

k. SMOOTH PACIFIC VENUS, *Chione fluctifraga* Sby., 2½ inches (San Pedro, California, south), p. 408.

l. COMMON WASHINGTON CLAM, *Saxidomus nuttalli* Con., 4 inches (Alaska to Lower California), p. 417.

m. and n. COMMON PACIFIC LITTLE-NECK, *Protothaca staminea* Con., 2 inches (entire Pacific Coast), p. 410.

o. WHITE ROCK VENUS, *P. staminea* form *ruderata* Desh (most common in Alaska), p. 410.

p. CALIFORNIAN WEDGE SHELL, *Donax californicus* Con., 1 inch (Santa Barbara, California, south), p. 438.

q. COMMON PACIFIC WEDGE SHELL, *Donax gouldi* Dall, ¾ inch (California to Mexico), p. 438.

r. CALIFORNIAN IRUS VENUS, *Irus lamellifera* Conrad, 1 inch (California), p. 142.

s. WAVY PACIFIC THRACIA, *Cyathodonta undulata* Con., 1 inch (California to Mexico), p. 472.

t. WEST COAST RUPELLARIA, *Rupellaria tellimyalis* Cpr., 1 inch (California to Mexico), p. 420.

u. MODEST TELLIN, *Tellina modesta* Cpr., ¾ inch (Alaska to Lower California), p. 425.

v. CALIFORNIA CUMINGIA, *Cumingia californica* Conrad, 1 inch (California to Chili), p. 436.

w. STIMPSON'S SURF CLAM, *Spisula polynyma* form *alaskana* Dall, 4 inches (Alaska), p. 446.

x. COMMON PACIFIC SPOON CLAM, *Periploma planiusculum* Sby., 1 inch (California to Peru), p. 473.

y. SALMON TELLIN, *Tellina salmonea* Cpr., ½ inch (Alaska to California), p. 426.

z. PACIFIC GAPER, *Schizothaerus nuttalli* Conrad, 7 inches (Washington to Lower California), p. 450.

PLATE 32

ATLANTIC COAST BIVALVES

a. GIANT ATLANTIC COCKLE, *Dinocardium robustum* Solander, 4 inches (Virginia to Texas), p. 401.

b. VANHYNING'S GIANT COCKLE, *D. robustum vanhyningi* Cl. and Smith, 4 inches (West Florida), p. 401.

c. FRILLED PAPER COCKLE, *Papyridea semisulcata* Sby., ½ inch (Florida and West Indies), p. 398.

d. GREENLAND COCKLE, *Serripes groenlandicus* Brug., 3 inches (Quebec to Massachusetts; Alaska), p. 401.

e. ICELAND COCKLE, *Clinocardium ciliatum* Fabr., 3 inches (Arctic to Massachusetts; also to Washington), p. 403.

f. OCEAN QUAHOG, *Arctica islandica* L., 3 inches (Arctic to off North Carolina), p. 381.

g. SOUTHERN QUAHOG, *Mercenaria campechiensis* Gmelin, 5 inches (and young) (Virginia to Florida), p. 406.

h. NORTHERN QUAHOG, *Mercenaria mercenaria* L., 4 inches (Canada to east Florida), p. 406.

i. GRAY PYGMY VENUS, *Chione grus* Holmes, ⅜ inch (North Carolina to Louisiana), p. 408.

j. ROSTRATE CUSPIDARIA, *Cuspidaria rostrata* Spengler, ¾ inch (Massachusetts to West Indies), p. 476.

k. TEXAS VENUS, *Callocardia texasiana* Dall, 3 inches (Gulf of Mexico), p. 416.

l. MORRHUA VENUS, *Pitar morrhuana* Linsley, 1½ inches (Canada to North Carolina), p. 414.

m. PRINCESS VENUS, *Antigona listeri* Gray, 3 inches (Southeastern Florida and West Indies), p. 404.

n. QUEEN VENUS, *Antigona rugatina* Heilprin, 1½ inches (North Carolina to West Indies), p. 405.

o. DWARF SURF CLAM, *Mulinia lateralis* Say, ⅓ inch (Maine to Texas), p. 449.

p. ATLANTIC SURF CLAM, *Spisula solidissima* Dill., 6 inches (Nova Scotia to South Carolina), p. 446.

q. CHANNELED DUCK CLAM, *Labiosa plicatella* Lam., 3 inches (North Carolina to West Indies), p. 449.

r. ARCTIC WEDGE CLAM, *Mesodesma arctata* Con., 1½ inches (Greenland to Virginia), p. 451.

s. FRAGILE ATLANTIC MACTRA, *Mactra fragilis* Gmelin, 2 inches (North Carolina to West Indies), p. 445.

t. CAMPECHE ANGEL WING, *Pholas campechiensis* Gmelin, 3 inches (North Carolina to Mexico), p. 462.

u. GIANT FALSE DONAX, *Iphigenia brasiliensis* Lam., 2½ inches (Florida and West Indies), p. 439.

v. TRUNCATE GAPER CLAM, *Mya truncata* L., 2 inches (Arctic to Massachusetts; also to Washington), p. 455.

w. STRIATE MARTESIA, *Martesia striata* L., 1 inch (Florida and West Indies), p. 464.

x. COMMON SOFT-SHELL CLAM, *Mya arenaria* L., 4 inches (Labrador to North Carolina), p. 455.

y. ATLANTIC GROOVED MACOMA, *Apolymetis intastriata* Say, 2 inches (Florida and West Indies), p. 434.

z. FALSE ANGEL WING, *Petricola pholadiformis* Lam., 2 inches (Canada to West Indies), p. 420.

Muricopsis ostrearum Conrad Mauve-mouth Drill
Figure 47g

West coast of Florida to the Florida Keys.

1 inch in length, extraordinarily like *Urosalpinx perrugata*, but more elongate, with a longer siphonal canal which is bent slightly back, and with a light-mauve aperture. Moderately common from low tide area to 35 fathoms. *M. floridana* Conrad is this species.

GENUS *Pseudoneptunea* Kobelt 1882

Pseudoneptunea multangula Philippi False Drill
Figure 47f

North Carolina to both sides of Florida and the West Indies.

1 to 1¼ inches in length, rather broad, with a short, fairly open siphonal canal. Outer lip sharp, finely crenulated. At the base of the columella there is a single, small fold. 8 to 9 short, rounded axial ribs on the periphery of the whorl. Spiral cords weak. Color gray with red-brown specklings; sometimes solid yellow-orange. Moderately common.

Subfamily PURPURINAE
GENUS *Drupa* Röding 1798
Subgenus *Morula* Schumacher 1817

Drupa nodulosa C. B. Adams Blackberry Drupe
Plate 25v

South half of Florida and the West Indies.

½ to 1 inch in length, elongate, grossly studded with round, black beads. Aperture purplish black. Outer lip thick, and with 4 to 5 relatively large, white beads. A common shallow-water species found under rocks.

GENUS *Acanthina* Fischer von Waldheim 1807
Subgenus *Acanthinucella* Cooke 1918

Acanthina spirata Blainville Spotted Thorn Drupe
Plate 24-o

Puget Sound to San Diego, California.

1 to 1½ inches in length, rather low-spired, solid, smoothish, except for numerous, poorly developed, spiral threads. Spine on lower, outer lip is strong, behind which on the base of the outside of the body whorl is a weak, spiral groove. Whorls slightly shouldered. Color bluish gray with numerous rows of small, red-brown dots. Aperture within is bluish white. A round-shouldered, smaller form (*punctulata* Sby. or *lapilloides* Conrad) is found

from Monterey to Lower California. A common southern species found above high-tide mark on rocks; also on mussel beds.

Acanthina paucilirata Stearns   Checkered Thorn Drupe

San Pedro, California, to Lower California.

⅓ to ½ inch in length, characterized by about 6 spiral rows of small squares of black-brown on a cream-white background. Early whorls cancellate, later whorls smoothish except for 4 or 5 very small, smooth, raised, spiral threads. Top of whorl slightly concave. Spine at base of outer lip small and needle-like. Aperture dentate, brownish with black squares on the outer lip. Siphonal canal short. Common above high-tide mark in southern California.

GENUS *Urosalpinx* Stimpson 1865

Urosalpinx cinerea Say   Atlantic Oyster Drill
Figure 47e

Nova Scotia to southern Florida. Introduced to San Francisco and to England.

½ to 1 inch in length; without varices; outer lip slightly thickened on the inside and sometimes with 2 to 6 small, whitish teeth. Siphonal canal moderately short and straight. With about 9 to 12 rounded, axial ribs per whorl and with numerous, strong, spiral cords. Color grayish or yellowish white, often with irregular, brown, spiral bands. Aperture tan to dark-brown. This common species is very destructive to oysters. It occurs from intertidal areas down to about 25 feet or more. Females grow faster and hence are larger than the males. They may reach an age of 7 years. The drills move inshore to spawn. Each female spawns once a year (May to September in Virginia; June to September in Canada and England). The female deposits 25 to 28 leathery, vase-shaped capsules, each containing 8 to 12 eggs. *U. follyensis* B. Baker is an ecologic form.

Urosalpinx perrugata Conrad   Gulf Oyster Drill
Figure 47d

West coast of Florida (to Louisiana?).

Similar to *cinerea*, but with 6 to 9 axial ribs which are quite large at the periphery of the whorl. The spiral cords are fewer and stronger. Aperture rosy-brown or yellow-brown. Outer lip more thickened on the inside and usually with 6 small, whitish teeth. This may be a subspecies of *cinerea*. Common on mudflats. Always compare with *Muricopsis ostrearum* Conrad which resembles this species very closely.

Urosalpinx tampaensis Conrad   Tampa Drill

Tampa Bay area, west Florida.

½ to 1 inch in length; the light-brown aperture is thickened and the outer lip has 6 small, white teeth. With about 9 to 11 sharp, axial ribs per whorl, crossed by about 9 to 10 equally strong spiral cords on the last whorl, thus giving the shell a cancellate sculpture. The whorls in the spire show only 2 spiral, nodulated cords. Exterior dark-gray. Common on mudflats.

Genus *Purpura* Bruguière 1789

Purpura patula Linné   Wide-mouthed Purpura
Plate 25l

Southeast Florida and the West Indies.

2 to 3½ inches in length; without an umbilicus. Exterior dull, rusty-gray. Columella salmon-pink. Inner borders of aperture with splotches of blackish brown. Common in the West Indies, uncommon in Florida. The animal exudes a harmless liquid which stains the hands and collecting bag a permanent violet.

The subspecies *pansa* Gould (west coast of Mexico south to Columbia) is similar in most respects, but the columella is colored a whitish cream.

Genus *Thais* Röding 1798
Subgenus *Stramonita* Schumacher 1817

Thais haemastoma floridana Conrad   Florida Rock-shell
Plate 25a

North Carolina to Florida, and the Caribbean.

2 to 3 inches in length, solid, smooth to finely nodulose. Color light-gray to yellowish with small flecks and irregular bars of brownish. Interior of aperture salmon-pink, often with brown between the denticulations of the outer lip. Running inside the aperture high up on the body whorl above the parietal area, there is a strong spiral ridge. Some specimens have a faint fold or plica on the base of the columella. This is a very common species, but quite variable in shape and color pattern. Typical *haemastoma* Lamarck occurs in the Mediterranean and West Africa. See additional remarks under *Thais rustica.*

Thais haemastoma haysae Clench   Hays' Rock-shell

Northwest Florida to Texas.

This subspecies is characterized by its large size (up to 4½ inches in length), strongly indented suture, and rugose sculpture with a row of double, strong nodules on the shoulder of the whorls. M. D. Burkenroad (1931) has given a long account of the biology of this oyster pest.

Thais rustica Lamarck

Rustic Rock-shell
Plate 25f

Southeast Florida and the West Indies. Bermuda.

1½ inches in length, irregularly sculptured with 2 spiral rows of blunt spines, one on the shoulder, the other at the center of the body whorl. Color dirty-gray to dull mottled brown. Interior of aperture whitish but generally margined with spots of dark-brown along the outer lip. Parietal wall glossy-white. This species is smaller and more nodulose than *floridana* and always has a white aperture. Erroneously called *Thais undata* Lamarck which is an Indo-Pacific species.

Three confusing species are often found together in southern Florida. The young of *Cantharus tinctus* (see p. 233) has the lower third of its white columella turned away (to the left) 20 degrees from the axis of the shell, and its early whorls in the spire are not shouldered. *Thais haemastoma floridana* is characterized by its almost straight, cream to orange columella, and by the numerous raised cream or white spiral ridges on the inside of the outer lip. *Thais rustica* has a stouter, white columella which is slightly twisted and purple-stained at the lower, inner corner. Its outer lip teeth occur in groups of 2 or 3 and near the edge of the lip are stained by heavy blotches of purple-brown.

Subgenus *Mancinella* Link 1807

Thais deltoidea Lamarck

Deltoid Rock-shell
Plate 25b, k

Jupiter Inlet, Florida, to the West Indies. Bermuda.

1 to 2 inches in length, heavy, and coarsely sculptured with two spiral rows of large, blunt spines. Parietal wall tinted with lavender, mauve or rose. Interior of aperture glossy-white. Exterior grayish white with mottlings of black or dull brown. Columella with a small but distinct ridge at the base which forms the margin of the siphonal canal. This is an abundant species where intertidal rocks are exposed to the ocean surf.

Subgenus *Polytropa* Swainson 1840

Thais lapillus Linné

Atlantic Dogwinkle
Plate 25g

Southern Labrador to New York. Norway to Portugal.

1 to 2 inches in length, roughly sculptured or smoothish. Commonly

with rounded, spiral ridges. Sometimes imbricated with small scales (form named *imbricata* Lamarck, pl. 25e). Color usually dull-white, but sometimes yellowish, orange or brownish. Rarely with dark-brown or blackish spiral bands. This species has also been known as *Nucella lapillus*. For biology see H. B. Moore (1936 and 1938).

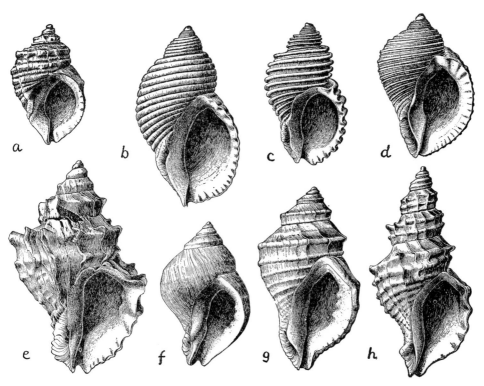

FIGURE 48. Pacific coast Dogwinkles. **a,** *Thais emarginata* Deshayes; **b** and **c,** *T. canaliculata* Duclos; **d,** *T. lima* Gmelin; **e** to **h,** forms of *T. lamellosa* Gmelin. All about natural size.

Thais lamellosa Gmelin   Frilled Dogwinkle
                                            Figures 48e-h

Bering Straits to Santa Cruz, California.

1½ to 5 inches in length, solid, usually with a fairly high, pointed spire. Columella almost vertical and straight, not flattened. Size, details of shape, sculpturing and color very variable. White, grayish, cream or orange, sometimes spirally banded. Smoothish or with variously developed foliated, axial ribs. Sometimes spinose. A very common rock-loving species. Formerly *Purpura crispata* Martyn.

Thais canaliculata Duclos   Channeled Dogwinkle
Figure 48b, c

Aleutian chain to Monterey, California.

1 inch in length, moderately globose, its spire higher than that of *emarginata*, but lower than that of *lamellosa*. Columella arched, flattened below. Characterized by about 14 to 16 low, flat-topped, closely spaced, spiral cords on the body whorl. Suture slightly channeled. Color white or orange-brown, often spirally banded. Moderately common on rocks and mussel beds. Do not confuse with *lima* from Alaska.

Thais emarginata Deshayes   Emarginate Dogwinkle
Figure 48a

Bering Sea to Mexico.

1 inch in length, with a rather short spire and with globose whorls. Aperture large. Columella strongly arched, and flattened and slightly concave below. Sculpturing variable, but characteristically with coarse spiral cords, usually alternatingly small and large. Cords often scaled or coarsely noduled. Exterior yellow-gray to rusty-brown, often with darker, narrow spiral bands. Interior and columella light- to chestnut-brown. Exceedingly common in many places along the coast where there are rocks.

Thais lima Gmelin   File Dogwinkle
Figure 48d

Alaska and Japan to northern California.

1 to 2 inches in length, very similar to *canaliculata*, but with 17 to 20 round-topped spiral cords, often smooth, sometimes minutely fimbriated. Cords often alternate in size. Color whitish or orange-brown, rarely banded. Common intertidally. Compare with *canaliculata*.

Genus *Ocenebra* Gray 1847

Tritonalia Fleming 1828 may also be used as a name for this genus, although *Ocenebra* would seem to be the wiser choice and will probably be the final choice. *Ocinebra* is a misspelling.

Ocenebra interfossa Carpenter   Carpenter's Dwarf Triton
Figure 49a

Alaska to Lower California.

½ to ¾ inch in length, spire half the length of the shell; light-gray in color, delicately sculptured. 8 to 11 axial ribs on the body whorl crossed by about a dozen strong, microscopically scaled spiral cords. The surface is often fimbriated axially between the cords. Siphonal canal moderately long, usually sealed over. Littoral to several fathoms. Common. There are 2 named varieties of doubtful biological significance: *atropurpurea* Cpr. and *clathrata* Dall.

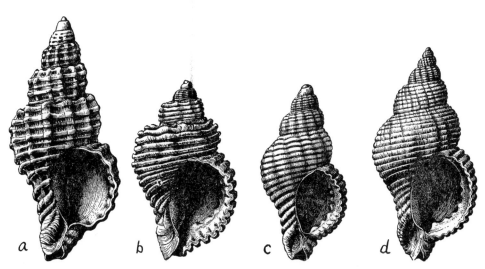

FIGURE 49. Dwarf Tritons of the Pacific coast. **a**, *Ocenebra interfossa* Cpr.; **b**, *O. circumtexta* Stearns; **c** and **d**, *O. lurida* Midd. All natural size.

Ocenebra lurida Middendorff   Lurid Dwarf Triton
                                                        Figure 49c, d

    Alaska to Catalina Island, California.

    1 to 1⅓ inches in length; 5 to 6 whorls; moderately elongate spire whose whorls show the axial ribs more prominently than the numerous fine spiral threads. Suture well-impressed. Body whorl with 8 to 10 rounded ribs which are strongest and shouldered just below the suture but which fade out below the periphery of the whorl. The smaller, smooth, spiral cords are elevated and prominent, often with numerous tiny axial lamellae between them. 6 to 8 small teeth on the inside of the outer lip. Color variable, whitish to rusty-brown, sometimes banded. Periostracum dark-brown and fuzzy. Siphonal canal usually sealed over. Very common from northern California north. Littoral to 30 fathoms.

Ocenebra gracillima Stearns   Graceful Dwarf Triton
                                                        Plate 24m

    Monterey, California, to the Gulf of California.

    ⅓ inch in length, similar to *lurida* in shape; with 5 whorls, those in the spire weakly cancellate with the axial ribs the strongest. Last whorl without axial ribs, except for a rather strong, rounded varix behind the outer lip. Last whorl with about a dozen or so light-brown, spotted, spiral threads over a background of light yellowish gray. 3 to 5 fairly large teeth on the inside of the outer lip. Siphonal canal short, sealed over in adults. Periostracum thin, fuzzy, grayish with mauve-brown undertones. Interior of aperture light mauve-brown, usually with a whitish, spiral band on the middle of the body whorl. *O. stearnsi* Hemphill is the same. Very common in rocky rubble and on wharf pilings.

Ocenebra circumtexta Stearns   Circled Dwarf Triton
                                                  Figure 49b

Moss Beach, California, to Lower California.

¾ to 1 inch in length, spire ⅓ the length of the shell. Characterized by very strong, rough spiral cords (15 on the body whorl, 6 on the whorls above). Under a lens the cords are seen to consist of arched, crowded, raised axial lamellae. The cords are often cream-white with the interspaces black-brown. Axial ribs wide, low, rounded and 7 to 9 per whorl. Some specimens are banded or have large, red-brown spots. A white form of this species was unnecessarily named *citrica* Dall. A very abundant species on rocks at low tide to 30 fathoms.

Ocenebra poulsoni Carpenter   Poulson's Dwarf Triton
                                                  Plate 24k

Santa Barbara, California, to Lower California.

1½ to 2 inches in length; a sturdy shell with a semi-gloss finish. 8 to 9 nodulated, rounded axial ribs per whorl crossed by numerous very fine, incised spiral lines and 4 or 5 larger, rounded, raised spiral cords. The latter make nodules on the ribs. Siphonal canal narrowly open. Periostracum thin, grayish or brownish and smoothish. When the periostracum is absent, the shell is glossy-white with numerous, fine spiral lines of dark- to yellow-brown. Aperture white. An exceedingly common species found in nearly every region on rocks and wharf pilings.

GENUS *Pterorytis* Conrad 1862

Pterorytis foliata Gmelin   Foliated Thorn Purpura
                                               Plate 24h

Alaska to San Pedro, California.

2 to 3 inches in length, with 3 large, thin, foliaceous varices per whorl which are finely fimbriated on the anterior side. Numerous spiral cords are rather prominent and of various sizes. Siphonal canal closed, its anterior tip turned up and to the right. Base of outer lip with a moderately strong spine.

Aperture white. Exterior white to light-brown. Common in the Puget Sound area and to the north on rocks near shore. Also down to 35 fathoms. Appears in some books as *Purpura* or *Ceratostoma foliatum* Martyn.

Pterorytis nuttalli Conrad

Nuttall's Thorn Purpura

Plate 24f

Point Conception, California, to Lower California.

1½ to 2 inches in length, similar to *Pter. foliata* (and somewhat resembling *Ocenebra poulsoni*), but with much more poorly developed varices, and with one prominent, noduled rib between each varix. Spine on outer lip usually long and sharp. Exterior yellowish brown, sometimes spirally banded. Siphonal canal closed along its length. A common littoral species in the southern part of its range.

GENUS *Eupleura* H. and A. Adams 1853

Eupleura caudata Say

Thick-lipped **Drill**

Figure 47b

South of Cape Cod to south half of Florida.

½ to 1 inch in length; apex pointed; siphonal canal moderately long, almost closed, coming to a sharp point below. Last varix large, rounded and with small nodules. Inside of outer lip with about 6 small, bead-like teeth. Whorls with spiral cords and strong axial ribs. There are 4 to 6 axial ribs between the last 2 varices. *E. etterae* B. B. Baker is a large, ecologic form of this species.

Eupleura sulcidentata Dall

Sharp-ribbed **Drill**

Figure 47c

West coast of Florida.

Similar to *E. caudata*, but with the spiral sculpture almost absent. There are only 2 or 3 axial ribs between the last 2 varices. The varices are thin and sharp, and the entire shell is slightly compressed laterally (has a more oval outline from an apical or top view than does *caudata*). The axial ribs are often sharp and may bear a small spine at the top. Color gray, chocolate-brown, tan or rarely pinkish, and sometimes with a narrow spiral brown band. Moderately common. *E. stimpsoni* Dall from deep water in the Gulf of Mexico is figured on p. 209.

Family MAGILIDAE
GENUS *Coralliophila* H. and A. Adams 1853

Coralliophila abbreviata Lamarck

Short Coral-shell

Southeast Florida and the West Indies.

¾ to 1½ inches in length, solid, grayish white, rather misshapen, with rounded or squared shoulders, and with or without weak, rounded axial ridges. Spiral sculpture of crowded, variously sized spiral cords which are made up of numerous microscopic scales. Aperture enamel-white, rounded above and constricted into a short siphonal canal below. Umbilicus small, shallow and funnel-shaped. A common species found living in the bases of seafans. *C. deburghiae* Reeve is a deep-water form or species with long, triangular ribs projecting straight out from the periphery of the whorl. It is uncommon.

Subgenus *Latiaxis* Swainson 1840

Coralliophila costata Blainville   California Latiaxis

Point Conception, California, to Panama.

1 to 1¾ inches in length, variable in shape and the development of frills and spines. Deep-water forms (called *hindsi* Carpenter) bear triangular, flattened, up-turned spines on the periphery of the whorl. Spiral cords are strongly scaled. Color light-gray with an enamel-white aperture. Moderately common offshore. A choice collector's item.

Superfamily *BUCCINACEA*
Family *COLUMBELLIDAE*
(*Pyrenidae*)
Genus *Columbella* Lamarck 1799

Columbella mercatoria Linné   Common Dove-shell

Plate 25bb

Southeast Florida and the West Indies.

½ to ¾ inch in length, solid, squat, highly colored with white and brown, interrupted spiral bars over yellow, pink or orange background. Sometimes only maculated with one color (orange, brown or yellow). Outer lip thick, bearing about a dozen white teeth. A common shallow-water species frequently cast up on beaches. Not found on the west coast of Florida.

Columbella rusticoides Heilprin   Rusty Dove-shell

Key West north along the west coast of Florida.

Similar to *C. mercatoria,* but much more slender, smooth on the center of the body whorl, and with mauve-brown marks between the apertural teeth. Also more faintly colored and lacking spiral bars or lines of brown. Common down to 20 feet.

GENUS *Anachis* H. and A. Adams 1853

Anachis avara Say — Greedy Dove-shell

Plate 25ee

New Jersey to both sides of Florida to Texas.

⅜ to ½ inch in length, moderately elongate, with about a dozen smooth axial plications on the upper half of each whorl. Spiral, incised lines very weak or absent, but strong at the base. Aperture narrow, a little less than half the length of the shell. Weak, smooth varix present. The 3 nuclear whorls are smooth and translucent-white. Next few whorls with numerous axial riblets. Color yellowish brown to dark gray-brown over which may be seen a faint pattern of irregular, white, large dots. Sometimes with dark-brown specklings. 4 weak teeth inside inner lip. A very common low-tide mark species.

Anachis obesa C. B. Adams — Fat Dove-shell

Virginia to Florida, the Gulf States and the West Indies.

³⁄₁₆ to ¼ inch in length, moderately wide, dull grayish with 1 or 2 subdued, spiral brown bands in some specimens. Small, sharp axial ribs are numerous; spiral incised lines numerous, not crossing ribs. There is a fairly strong, occasionally knobbed, spiral cord immediately below the suture. Varix large, smooth and rounded. Body whorl behind it usually smoothish. Parietal shield faintly developed, but with a sharp edge. Inner wall of outer lip with about 3 to 5 small teeth. The form *ostreicola* Melville from northwest Florida is dark-brown and with stronger spiral threads or wider incised lines. This is a common shallow-water species.

Anachis translirata Ravenel — Well-ribbed Dove-shell

Plate 25ff

Massachusetts to northeast Florida.

Similar to *avara*, but with about twice as many axial ribs which run the entire depth of each whorl, and with strong incised spiral lines throughout. Outer lip with about a dozen tiny teeth. Color drab yellowish to brown, but sometimes with wide, subdued, spiral bands of darker brown. Dredged commonly just offshore usually from 2 to 20 fathoms.

Anachis penicillata Carpenter — Penciled Dove-shell

San Pedro to Lower California.

⅕ inch in length, rather slender, with 6 whorls, of which the first 2 nuclear ones are smooth, the remainder with about 15 strong axial riblets per whorl, which are made slightly uneven by numerous very fine spiral threads. Color translucent-cream with sparse spottings of brown. Common under rocks between tide marks.

Genus *Nitidella* Swainson 1840

Nitidella ocellata Gmelin   White-spotted Dove-shell
                                                       Plate 25hh

Lower Florida Keys and the West Indies.

½ inch in length, smooth, characteristically dark black-brown with numerous small white dots which may be quite large just below the suture. Outer lip thick, with 5 or 6 small whitish teeth. Aperture short, narrow, purplish brown within. When beachworn, the color is reddish or yellowish brown. Common under rocks at low tide. Formerly known as *cribraria* Lamarck.

Nitidella nitidula Sowerby   Glossy Dove-shell
                                                       Plate 25dd

Southeast Florida and the West Indies.

½ inch in length, characterized by the long aperture (¾ that of the entire shell) and by the very glossy shell. Color whitish with heavy mottlings of light-yellow to mauve-brown. Outer lip with about a dozen small teeth. Common in the West Indies on rocks at low tide.

Nitidella gouldi Carpenter   Gould's Dove-shell
                                                       Plate 20m

Alaska to San Diego, California.

½ inch in length, 7 whorls are smoothish and slightly convex. Spire almost flat-sided. Base of shell on exterior of canal with about 9 fine, incised spiral lines. Bottom of white columella with a single, low plait. Outer lip simple, sharp and often reinforced within by 4 or 5 weak pustules. Shell whitish with faint brown maculations, covered with a yellowish-gray periostracum. Fairly common from just offshore to 300 fathoms.

Nitidella carinata Sowerby   Carinate Dove-shell

San Francisco to Lower California.

⅓ inch in length, glossy, brightly variegated with orange, yellow, white and brown. Shoulder of last whorl usually strongly swollen. Exterior of canal with about a dozen spiral, incised lines. Both ends of the aperture are stained dark-brown. Outer lip thickened, crooked, and with about a dozen small spiral threads or teeth inside. Fairly common in shallow water. *N.*

gausapata Gould (California to Alaska) is similar, but without the swollen shoulder.

GENUS *Mitrella* Risso 1826
Subgenus *Astyris* H. and A. Adams 1853

Mitrella lunata Say   Lunar Dove-shell
Plate 25gg

Massachusetts to Florida, Texas and the West Indies.

³⁄₁₆ to ¼ inch in length, glossy, smooth, translucent-gray and marked with fine, axial zigzag brown to yellow stripes. Base of shell with fine, incised spiral lines. Aperture constricted, slightly sinuate. Outer lip with 4 small teeth on the inside. No prominent varix. Nuclear whorls very small and translucent. Color sometimes milky-white or mottled in brown. A very common species found at low tide.

Mitrella raveneli Dall   Ravenel's Dove-shell
Plate 25cc

North Carolina to both sides of Florida.

⅜ inch in length, resembling *lunata*, but translucent-whitish, without the mottlings, normally a slightly larger shell, more elongate, with a longer siphonal canal, and with a rather thin outer lip which lacks the deep sinuation found in the upper portion of the outer lip as in *lunata*. Commonly dredged from 5 to 90 (rarely to 200) fathoms. Rarely washed ashore.

Mitrella tuberosa Carpenter   Variegated Dove-shell

Alaska to the Gulf of California.

¼ inch in length, slender, with a narrow, pointed, flat-sided spire. Shell smooth and usually glossy. Outer lip slightly thickened and with small teeth within. Color a translucent yellowish tan with opaque, light-brown flammules and maculations. Sometimes all brown with tiny white dots. Early whorls in worn specimens have a lilac tinge. Periostracum thin and translucent. Common in shallow water; 7 to 30 fathoms. *M. variegata* Stearns may be this species.

GENUS *Amphissa* H. and A. Adams 1853

This genus was formerly placed in the family *Buccinidae*.

Amphissa versicolor Dall   Joseph's Coat Amphissa
Figure 50a

Oregon to Lower California.

½ inch in length, rather thin, but quite strong; surface glossy. 7 whorls. Suture well-impressed. Whorls in spire and upper third of body whorl with about 15 obliquely slanting, strong, rounded, axial ribs. Numerous spiral, incised lines are strongest on the base of the body whorl. Lower columella area with a small shield. Outer lip thickened within by about a dozen small white teeth. Color pinkish gray with indistinct mottlings of orange-brown. A common littoral to shallow-water species.

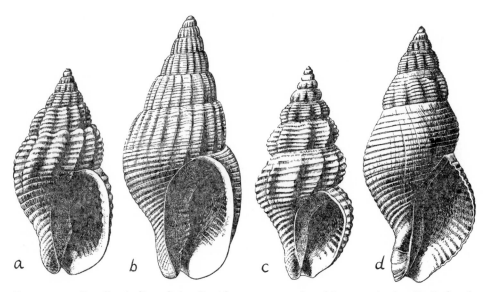

FIGURE 50. Small whelks of the Pacific coast. **a,** *Amphissa versicolor* Dall; **b,** *A. columbiana* Dall; **c,** *A. undata* Cpr.; **d,** *Searlesia dira* Reeve. All X3.

Amphissa columbiana Dall   Columbian Amphissa
                                                Figure 50b

Alaska to San Pedro, California.

1 inch in length, similar to *versicolor,* but characterized by its large size, numerous, weak, vertical, axial ribs (20 to 24 on the next to the last whorl, and missing on the last part of the last whorl), and by the low, rounded varix behind the outer lip. Color yellow-brown with indistinct mauve mottlings. Periostracum thin, yellowish brown. Moderately common in shallow water from Oregon to Alaska.

Amphissa undata Carpenter   Carpenter's Amphissa
                                                Figure 50c

Monterey, California, to Lower California.

⅓ to ½ inch in length, similar to *versicolor,* especially in color, but with a much higher spire, stronger axial ribs, and much stronger, more acute, spiral cords. Moderately common from 25 to 265 fathoms.

Amphissa bicolor Dall   Two-tinted Amphissa

Farralon Islands, to San Diego, California.

½ to ⅝ inch in length, similar to *versicolor*, but thinner-shelled, usually with fewer ribs, glossy-white in color, but covered with a pale-straw periostracum, and without the small teeth inside the outer lip. Dredged commonly from 40 to 330 fathoms.

Family BUCCINIDAE
GENUS *Buccinum* Linné 1758

Buccinum undatum Linné   Common Northern Buccinum

Arctic Seas.to New Jersey. Europe.

2 to 4 inches in length, solid, chalky gray to yellowish with a moderately thick, gray periostracum. Axial ribs 9 to 18 per whorl, low, extending ¼ to ½ way down the whorl. Spiral cords small, usually about 5 to 8 between sutures. Outer lip slightly or well sinuate and somewhat flaring. Aperture and parietal wall enamel-white. Anterior canal short. 1½ nuclear whorls fairly large, smooth and translucent-white. Operculum oval, concentric, chitinous, and light yellow-brown. A very variable shell which sometimes lacks the axial ribs but may have numerous spiral threads. Common just offshore to several fathoms in cold water.

Buccinum tenue Gray   Silky Buccinum
Plate 24u

Arctic Seas to Washington State. Arctic Seas to the Gulf of Maine.

1½ to 2½ inches in length; aperture ½ the length of the shell. Outer lip slightly sinuate, thin and slightly flaring. Axial ribs small, numerous, intertwining and extending from suture to suture. Spiral sculpture of microscopic, beaded threads giving a silky appearance. Color light-brown. Common offshore. Compare with *plectrum* Stimpson.

Buccinum plectrum Stimpson   Plectrum Buccinum
Figure 51a

Arctic Seas to Puget Sound. Arctic Seas to the Gulf of St. Lawrence.

2 to 3 inches in length; aperture a little more than ⅓ the entire length of the shell. Outer lip strongly sinuate, thickened and flaring. Axial ribs

small (but larger and fewer than those in *tenue*), and limited to the upper fourth of the whorl. Spiral sculpture of numerous rough, but microscopic, incised lines. Color grayish white. Common offshore in cold water. Do not confuse with *B. tenue*.

Buccinum glaciale Linné   Glacial Buccinum
                                                    Plate 24t

Arctic Seas to Washington State. Arctic Seas to the Gulf of St. Lawrence.

2 to 3 inches in length, fairly thick-shelled, but light in comparison to its size. Characterized by its thick, flaring, turned-back outer lip, and by the 2 wavy, strong spiral cords on the periphery of the whorl. Spiral incised lines numerous. Color mauve-brown. Aperture cream with a purplish flush within. Common from low tide to several fathoms in the Arctic region.

Buccinum baeri Middendorff   Baer's Buccinum
                                                    Plate 24v
Bering Sea.

1 to 2 inches in length, resembling a thin, beach-worn *Thais lima*. Aperture about ⅔ the length of the shell. Rather thin, smoothish, except for microscopic, incised spiral lines. Color drab grayish with purplish to reddish undertones. Periostracum, when present, is thin, translucent and light-brown. Operculum ⅕ the size of the aperture. Commonly found washed ashore on the beaches of Alaska and the Aleutians. A very unattractive shell.

GENUS *Volutopsius* Mörch 1857
Subgenus *Volutopsius s. str.*

Volutopsius castaneus Mörch   Chestnut Buccinum
                                                    Figure 51c
Alaska.

2½ to 3½ inches in length, rather solid; with 4 whorls; aperture large and slightly flaring. Interior brownish white enamel. Columella slightly arched, white within, brown on the parietal wall. Exterior surface brownish and smoothish, except for coarse axial wrinkles appearing more as deformities in growth. Moderately common on rocks below low-water mark.

Subgenus *Pyrulofusus* Mörch 1869

Volutopsius harpa Mörch   Left-handed Buccinum
                                                    Plate 24p
Alaska.

3 to 4 inches in length, characteristically sinistral (left-handed), 4 whorls, with the smoothish nucleus indented. Sculpture of 6 to 12 very oblique, low, rounded axial ribs and numerous (often paired) raised spiral threads. Color ash-gray with a light brownish yellow periostracum. Interior of aperture tinted with tan. Operculum much smaller than the aperture. Dextral (right-handed) specimens are rarities. Fairly common in deep water.

GENUS *Jumala* Friele 1882
(*Beringius* Dall)

Jumala crebricostata Dall

Thick-ribbed Buccinum

Figure 51b

Alaska.

4 to 5 inches in length, moderately heavy, with 5 to 6 whorls. Characterized by the very strong, rounded spiral cords (3 to 4 between sutures) which on the base of the shell tend to be flat-topped. Periostracum grayish brown, thin and semi-glossy. A very handsome, but not commonly acquired species which occurs from 80 to 100 fathoms.

Jumala kennicotti Dall

Kennicott's Buccinum

Figure 51g

Alaska.

5 to 6 inches in length, not very heavy. Characterized by about 9 strong, arched, somewhat rounded axial ribs extending from suture to suture, and, on the body whorl, extending ¾ the way down the whorl. Spiral sculpture of microscopic scratches, except on the base where there are a dozen or so weak threads. Periostracum light-brown, thin, and usually flakes off in dried specimens. Shell chalky and whitish gray in color. Not uncommon in several fathoms of water; rarely in very shallow water.

GENUS *Colus* Röding 1798

Colus stimpsoni Mörch

Stimpson's Colus

Plate 23x

Labrador to off North Carolina.

3 to 5 inches in length, fusiform, moderately strong, chalk-white in color, but covered with a semi-glossy, light- to dark-brown, moderately thin periostracum. Length of aperture about half the length of the entire shell. Sculpture of numerous incised spiral lines. Fairly common from 1 to 471 fathoms.

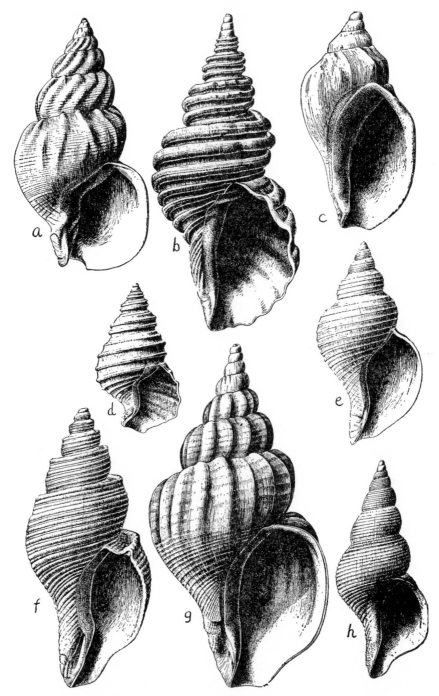

FIGURE 51. Pacific Buccinums and Neptunes. **a**, *Buccinum plectrum* Stimpson; **b**, *Jumala crebricostata* Dall; **c**, *Volutopsius castaneus* Mörch; **d**, *Neptunea eucosmia* Dall; **e**, *Neptunea phoenicea* Dall; **f**, *Neptunea tabulata* Baird; **g**, *Jumala kennicotti* Dall; **h**, *Colus spitzbergensis* Reeve. All reduced about ½.

Colus pubescens Verrill — Hairy Colus

Plate 23t

Gulf of St. Lawrence to off North Carolina.

2 to 2½ inches in length, very similar to *stimpsoni*, but the aperture is about ½ the entire length of the shell, the suture more abruptly impressed, the whorls slightly more convex and the siphonal canal usually, but not always, more twisted. Very commonly dredged from 18 to 640 fathoms.

Colus pygmaea Gould — Pygmy Colus

Plate 23m

Gulf of St. Lawrence to off North Carolina.

Less than 1 inch in length, with 6 to 7 fairly convex whorls, fairly fragile, chalk-white, with spiral incised lines, and covered with a light olive-gray, thin, velvety periostracum. Aperture slightly more than ½ the length of the entire shell. Commonly dredged from 1 to 640 fathoms.

Colus caelata Verrill and Smith (Massachusetts to North Carolina, deep water) is about the same size, but is characterized by about 12 strong axial ribs per whorl in addition to numerous fine spiral threads. It is chalky-white to gray.

Colus spitzbergensis Reeve — Spitzbergen Colus

Figure 51h

Bering Sea to Washington State. Arctic Seas to Gulf of St. Lawrence.

2½ to 3 inches in length, rather light-shelled, and with 6 fairly well-rounded whorls. Spire long and of about 30 to 35 degrees. Siphonal canal short; columella almost straight. Outer lip flaring, slightly thickened. Sculpture of numerous (12 to 14 between sutures) low, flat-topped, small, equally sized spiral cords. Chalk-gray with a reddish to yellowish brown, thin periostracum. Commonly dredged from 1 to 142 fathoms.

GENUS *Neptunea* Röding 1798

Neptunea decemcostata Say — Brown-corded Neptune

Plate 23s

Nova Scotia to Massachusetts.

3 to 4½ inches in length, rather heavy. Characterized by its grayish-white, rather smooth shell which bears 7 to 10 very strong, reddish-brown, spiral cords. The upper whorls show 2 to 3 cords. There is an additional band of brown just below the suture. A common cold-water species found offshore, but occasionally washed up on New England beaches.

Neptunea ventricosa Gmelin   Fat Neptune
Plate 24s

Arctic Ocean and Bering Sea.

3 to 4 inches in length, heavy, with a large, ventricose body whorl. Axial ribs or growth lines coarse and indistinct, rarely lamellate. Shoulders sometimes weakly nodulated. Spiral cords absent or very weak. Color a dirty-brownish white. Aperture white or flushed with brownish purple. Moderately common offshore. This is *Chrysodomus satura* Martyn and its several poor varieties.

Neptunea pribiloffensis Dall   Pribiloff Neptune
Plate 24r

Bering Sea to British Columbia.

4 to 5 inches in length, similar to *N. lyrata*, but with a lighter shell, with weaker and more numerous spiral cords, and with more numerous and stronger secondary spiral threads. Outer lip more flaring and the siphonal canal with more of a twist to the left. Fairly commonly dredged from 50 to 100 fathoms.

Neptunea lyrata Gmelin   Common Northwest Neptune
Plate 24q

Arctic Ocean to Puget Sound, Washington.

4 to 5 inches in length, ¾ as wide, solid, fairly heavy. With 5 to 6 strongly convex whorls, bearing about 8 strong to poorly developed, raised spiral cords (2 of which usually show in each whorl in the spire). Faint, quite small, spiral threads are also present. Exterior dull whitish brown. Aperture enamel-white with a tan tint. Fairly common in Alaska from shore to 50 fathoms. This is *Chrysodomus lirata* Martyn.

Subgenus *Ancistrolepis* Dall 1894

Neptunea eucosmia Dall   Channeled Neptune
Figure 51d

Alaska to Oregon.

1½ inches in length, solid, outer lip sharp, strong and crenulated. Siphonal canal short, wide and slightly twisted. Spiral cords strong. Suture channeled. Shell chalk-white, but covered with a rather thick, yellow-brown to gray periostracum which is axially lamellate and bears minute, erect hairs. Aperture glossy-white. Not uncommonly dredged from 62 to 780 fathoms. *N. californica* Dall and *bicincta* Dall appear to be this species.

Subgenus *Sulcosipho* Dall 1916

Neptunea tabulata Baird   Tabled Neptune

Figure 51f

British Columbia to San Diego, California.

3 to 4 inches in length, moderately solid, with 8 whorls, colored white with a thin brown periostracum. Characterized by the wide, flat channel next to the suture. It is bounded by a raised, scaly or fimbriated spiral cord. Remainder of whorl with numerous sandpapery spiral threads. A choice collector's item, not uncommonly dredged from 30 to 200 fathoms.

GENUS *Kelletia* Fischer 1884

Kelletia kelleti Forbes   Kellet's Whelk

Plate 24w

Santa Barbara, California, to San Quentin Bay, Mexico.

4 to 5 inches in length, characterized by its very heavy, white, fusiform shell, its fine, wavy suture, and by the sharp, crenulated outer lip. Whorls slightly concave between the suture and the shouldered periphery, which bears 10 strong, rounded knobs per whorl. Base with about 6 to 10 incised, spiral lines. Aperture glossy and white. Very commonly caught in traps from 10 to 35 fathoms. There are no other recent species in the genus. Often misspelled with two t's.

GENUS *Bailya* M. Smith 1944

Bailya intricata Dall   Intricate Baily-shell

Plate 25t

Southern half of Florida.

½ inch in length, fairly strong, pure white in color and with cancellate sculpturing. Last whorl with 12 to 14 low axial ribs which are crossed by about a dozen spiral cords (between which there may be a much smaller thread). At their intersection there are small beads. Outer lip with a frilled, rounded varix. Columella smooth. Weak spiral cord present on inside of aperture on the upper parietal wall. Whorls slightly shouldered. No notch in lower part of outer lip. Nuclear whorl smooth, glassy and rounded. Uncommon from 1 to 50 fathoms.

Bailya parva C. B. Adams from the West Indies differs in not having its whorls shouldered, having weaker spiral cords, and in occasionally having brown coloring.

GENUS *Antillophos* Woodring 1928

Antillophos candei Orbigny   Candé's Phos

Plate 25u

North Carolina to south Florida and Cuba.

1 to 1¼ inches in length, slightly less than half as wide; strong, heavy and pure white. Last whorl with about 13 small spiral cords and about 24 stronger axial ribs. Where they cross, there are small, rounded beads. Outer lip near the low part has a shallow notch. Inside the lip are about a dozen prominent, spiral ridges. Columella with 2 low spiral ridges near the base, sometimes weaker ones above. Upper parietal wall with a strong spiral cord running back into the aperture. Nuclear whorls smooth, glossy, white and slightly carinate. Very commonly dredged from 20 to 100 fathoms.

Genus *Engina* Gray 1839

Engina turbinella Kiener   White-spotted Engina
Plate 25w

Lower Florida Keys and the West Indies.

⅓ to ½ inch in length, dark purple-brown with about 10 low, white knobs per whorl on the periphery. Base with 2 to 4 spiral rows of much smaller white knobs. Microscopic spiral threads numerous. Aperture thickened and constricted by 4 to 5 whitish teeth on the outer lip and by a twist of the columella just above the narrow siphonal canal. Do not confuse with *Mitra sulcata* which has several columellar plications. Common under rocks at low tide.

Genus *Searlesia* Harmer 1916

Searlesia dira Reeve   Dire Whelk
Figure 50d

Alaska to Monterey, California.

1 to 1½ inches in length, half as wide, with the brown aperture half the length of the dark gray, fusiform shell. Outer lip thin but strong and with fine serrations which extend back into the shell as small spiral threads. Columella arched, chocolate-brown and glossy. Whorls in spire with 9 to 11 low, rounded axial ribs, and all of the exterior with numerous fine, unequal-sized spiral threads. Siphonal canal short and slightly twisted to the left. A common shallow-water species commonly from northern California to the north.

Genus *Colubraria* Schumacher 1817

Colubraria lanceolata Menke   Arrow Dwarf Triton
Plate 25x

North Carolina to both sides of Florida and the West Indies.

¾ to 1 inch in length, slender, with 7 whorls. Aperture long and narrow. Varix strong and curled back. Parietal shield elevated into a collar.

Former distinct varices present every ⅔ of a whorl. Sculpture very finely cancellate and beaded. Nucleus brown, smooth and bulbous. Shell ash-gray with occasional orange-brown smudges. Uncommonly dredged on rocky bottom just offshore.

Colubraria testacea Mörch from Tortugas and the West Indies is ½ to 1½ inches in length, fatter, with numerous, small beads and with wider varices. *C. swifti* Tryon (subgenus *Monostiolum* Dall), also from the West Indies, is ⅜ inch long, without former varices, axially ribbed and heavily maculated with brown. Both are uncommon.

Genus *Pisania* Bivona-Bernardi 1832

Pisania pusio Linné   Miniature Triton Trumpet

Plate 13-o

Southeast Florida and the West Indies.

1 to 1½ inches in length, sturdy, smooth, and usually with a glossy finish. The outer lip is weakly toothed within, and the upper parietal wall has a small, white, swollen tooth near the top of the aperture. Color variable, but usually purplish brown with narrow spiral bands of irregular dark and light spots commonly chevron-shaped. Moderately common below low-water line in the region of coral reefs.

Genus *Cantharus* Röding 1798
Subgenus *Pollia* Sowerby 1834

Cantharus tinctus Conrad   Tinted Cantharus

Plate 25y

North Carolina to both sides of Florida and the West Indies.

¾ to 1¼ inches in length, heavy, spire evenly conic; aperture with a small canal at the top. Axial ribs low and weak. Spiral cords numerous and weak, forming weak beads as they cross the ribs. Inside of outer lip with small teeth which are strongest near the top. Color of shell variegated with yellow-brown, blue-gray and milky-white. Fairly common in shallow water. The young are easily confused with *Thais*. See remarks under *T. rustica*, p. 214.

Cantharus auritula Link   Gaudy Cantharus

Southeast Florida and the West Indies.

Similar to *C. tinctus*, but broader, with shouldered whorls, with about 9 stronger axial ribs per whorl and with about 10 sharp spiral threads on the last whorl. Outer lip turned in as the varix is formed. Color brighter. Posterior canal longer. Common in the West Indies; intertidal.

Cantharus cancellarius Conrad   Cancellate Cantharus

West coast of Florida to Texas and Yucatan.

Similar to *C. tinctus*, but with a lighter shell, higher spire, and with sharp, spiral threads and narrow, axial ribs which cross to make a beaded and cancellate sculpturing. Posterior siphonal canal absent or weak. Varix very weak. Moderately common in shallow water.

GENUS *Macron* H. and A. Adams 1853

Macron lividus A. Adams   Livid Macron
                                                   Plate 24x

Monterey, California, to Lower California.

¾ to 1 inch in length, half as wide, strong, with 5 whorls which are covered with a thick, felt-like, dark-brown periostracum. Shell yellowish to bluish white. Outer lip sharp, strong, and near its base bearing a small, spiral thread. Columella strongly concave and white. Upper end of aperture narrow, with a small, short channel and with a white, tooth-like callus on the parietal wall. Siphonal canal short and slightly twisted. Base of shell with a half dozen incised spiral lines. Operculum chitinous, brown, thick, oval and with the nucleus at one end. Very common under stones at low tide.

Family *MELONGENIDAE*
GENUS *Melongena* Schumacher 1817
(*Galeodes*)

Melongena corona Gmelin   Common Crown Conch
                                                   Figure 52

Florida, the Gulf States and Mexico.

2 to 4 inches in length, very variable in size, color, shape and production of spines. Dirty-cream with wide, spiral bands of brown, purplish brown or dark bluish black. Pure white "albinos" are infrequent. Shoulder and base of shell with 1, 2, 3 or 4 rows of semi-tubular spines which may point upward or horizontally. Numerous varieties have been named which do not even warrant subspecific standing: *minor* Sowerby (dwarf); *estephomenos* Melville (dwarf and narrow); *altispira* Pilsbry and Vanatta (long and narrow); *bispinosa* Philippi (2 rows of spines); *inspinata* Richards spineless shoulder); and *martiniana* Philippi. A very common species in Florida. Used extensively in the shellcraft industry.

Melongena corona perspinosa Pilsbry and Vanatta appears to be a good subspecies. (Tampa south to Lossmans Key, Florida). Up to 4½ inches in length, heavier and wider than *corona*, with a wider aperture, and with shoulder spines standing out at right angles, and with 2 or 3 rows of smaller spines below the larger ones. A descendant possibly of the Pliocene subspecies *subcoronata* Heilprin. Soft parts and radula described and figured in Frank Lyman's excellent *Shell Notes*, vol. 2, no. 2-3, 1948 (published privately by Lyman, Lantana, Florida).

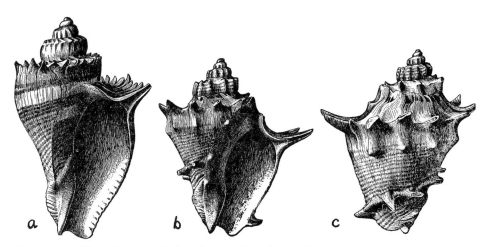

FIGURE 52. Two forms of the Crown Conch, *Melongena corona* Gmelin, from Florida. **a**, sandy area; **b** and **c**, from oyster bed. Reduced ½.

Melongena melongena Linné   West Indian Crown Conch
                                                              Plate 23h
Florida Keys (?) and the West Indies.

3 to 6 inches in length, similar to *corona*, but heavier, with rounded shoulders; smaller, more solid spines, and with a distinct channel at the suture. Common in the Greater Antilles.

GENUS *Busycon* Röding 1798
(*Fulgur* Montfort)

Busycon carica Gmelin   Knobbed Whelk
                                                              Plate 23i
South shore of Cape Cod to central east Florida.

Adults 5 to 9 inches in length; characterized by having low tubercles on the shoulder of the whorl and in being right-handed. Aperture light orange-yellow, but sometimes brick-red. The young show axial streaks of brownish purple. Common in shallow water.

Busycon contrarium Conrad

Lightning Whelk
Plate 23-o

South Carolina to Florida and the Gulf States.

4 to 16 inches in length, left-handed, with a row of moderately small, triangular knobs at the shoulder. Color grayish white with long, axial, wavy streaks of purplish brown which are blurred along their posterior edge. Albino shells are rare. Off Yucatan and rarely in Florida, right-handed specimens are found. Their siphonal canal is longer than that in *B. carica*, and the shell is lighter than that of *perversum*. A very common species in west Florida.

Busycon perversum Linné

Perverse Whelk
Plate 23k

Both sides of central Florida.

4 to 8 inches in length, very heavy and with a glossy finish. This species should not be confused with the common *contrarium*. This species can be either left-handed (formerly known as *kieneri* Philippi 1848) or right-handed (formerly known as *eliceans* Montfort 1810, pl. 23k). The name *B. perversum* or *Fulgur perversa* in most old popular books refers to *B. contrarium*. The perverse whelk is an uncommon species. It is characterized by the heavy, polished shell and the swollen, rounded ridge around the middle of the whorl. Dredged from 4 to 10 fathoms.

Subgenus *Busycotypus* Wenz 1943

Busycon canaliculatum Linné

Channeled Whelk
Plate 23n

Cape Cod to St. Augustine, Florida.

5 to 7½ inches in length, characterized by a deep, rather wide channel running along the suture and by the heavy, felt-like, gray periostracum. Common in shallow, sandy areas. Left-handed specimens are rare. The subgenera *Fulguropsis* E. S. Marks 1950 and *Sycofulgur* Marks 1950 are the same as Wenz's subgenus.

Busycon spiratum Lamarck

Pear Whelk
Plate 9g

North Carolina to Florida and the Gulf States.

3 to 4 inches in length; characterized by its smooth, rounded shoulders and the deep, but narrow channel at the suture. Periostracum thin and velvety. Do not confuse with *Ficus* which is a much more fragile shell. Common in shallow, sandy, clear water areas. The animal is cream-gray. Known in all previous popular books as *B. pyrum* Dillwyn. In the western

part of the Gulf of Mexico, specimens often have a weak keel on the shoulder (form or subspecies *plagosum* Conrad).

Busycon coarctatum Sowerby   Turnip Whelk
Plate 1a

Yucatan area, Mexico.

Until 1950 this was considered a very rare species, but dredging activities of shrimp trawlers have brought a large number of them to light. Characterized by its turnip-like shape, single row of numerous small, dark-brown spines, and by its golden-yellow aperture. 5 inches in length.

Family *NASSARIIDAE*
Genus *Nassarius* Dumeril 1806
Subgenus *Nassarius s. str.*

Nassarius vibex Say   Common Eastern Nassa
Plate 23q

Cape Cod to Florida, the Gulf States and the West Indies.

½ inch in length, heavy, with a well-developed parietal shield. Last whorl with about a dozen, poorly developed, axial ribs which are coarsely beaded. Color gray-brown to whitish with a few splotches or broken bands of subdued, darker brown. A common sand or mud-flat species. Some specimens have numerous weak spiral cords. Parietal shield sometimes yellowish.

Nassarius acutus Say   Sharp-knobbed Nassa
Figure 53c

West coast of Florida to Texas.

¼ inch in length, characterized by its glossy shell, its strong, pointed beads, and in occasionally having a narrow, brown, spiral thread connecting the beads. Moderately common. Fossil specimens are twice as large.

Nassarius insculptus Carpenter   Smooth Western Nassa
Figure 53f

Point Arena, California, to Lower California.

¾ inch in length, outer lip thickened, parietal wall thick, white but not very wide. Body whorl smoothish, except for weak, fine spiral threads. Axial ribs numerous only on early whorls. Color white. covered by a yellowish white periostracum. Moderately common; dredged from 20 to 200 fathoms.

Nassarius tegula Reeve

San Francisco to Lower California.

¾ inch in length, moderately heavy, with a heavy, whitish or brown-stained parietal callus. Body whorl smoothish around the middle, but with a spiral row of fairly large nodes below the suture. In the spire, the nodes are usually divided in two. Base of body whorl with a few weak, spiral threads. Outer lip thick. Color olive-gray to brownish, often with a narrow, whitish or purplish, spiral band. A common mud-flat species.

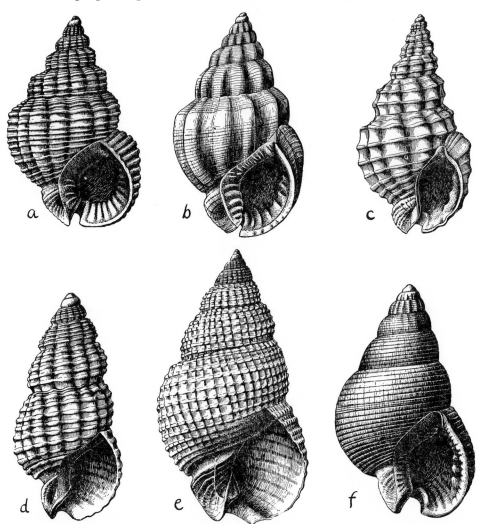

FIGURE 53. Nassa Mud Snails. ATLANTIC: **a**, *Nassarius ambiguus* Pulteney; **b**, *N. ambiguus* form *consensus* Rav.; **c**, *N. acutus* Say. PACIFIC: **d**, *N. mendicus* Gould; **e**, *N. perpinguis* Hinds; **f**, *N. insculptus* Cpr. All X3.

Subgenus *Hinia* Gray 1847

Nassarius ambiguus Pulteney 1799   Variable Nassa
Plate 23r; figure 53a

North Carolina to Florida and the West Indies.

½ inch in length, relatively light-shelled, usually pure white in color, but occasionally with 1 or 2 narrow, spiral bands of light yellowish brown. Number of strong, axial ribs per whorl varies from 8 to 12. Upper part of whorl sometimes shouldered. Numerous spiral, rounded cords are strong or weak. Parietal shield enamel-white, usually not well-developed. *N. consensus* Ravenel is possibly only a form of this unusually variable species (fig. 53b).

Nassarius trivittatus Say   New England Nassa
Plate 23j

Nova Scotia to South Carolina.

¾ inch in length, rather light-shelled, 8 to 9 whorls; nuclear whorls smooth. Whorls in spire with 4 to 5 rows of strong, distinct beads. Parietal wall thinly glazed with white enamel. Outer lip sharp and thin. Whorls slightly channeled just below the suture. Color light-ash to yellowish gray. Common from shallow water to 45 fathoms.

Nassarius perpinguis Hinds   Western Fat Nassa
Figure 53e

Puget Sound to Lower California.

¾ to 1 inch in length, fairly thin, with a rather fragile outer lip. Similar to *N. californianus*, but with much finer sculpture (usually finely cancellate or minutely beaded), and yellowish white in color with 2 or 3 narrow, spiral bands of orange-brown, one of which borders the suture. The sculpture is variable with spiral threads often predominant. Very abundant along most of the coast. Intertidal flats to 50 fathoms.

Nassarius californianus Conrad   Californian Nassa

Squaw Creek, Oregon, to Lower California.

1 inch in length, without a thick parietal shield and the outer lip not thickened. Shell with numerous, rather coarse beads arranged in 20 to 30 axial, slanting ribs. 11 to 12 spiral threads on the last whorl; 5 to 7 on the whorls above. Color white with an ashy or yellow-gray periostracum. Moderately common just offshore to 35 fathoms. Compare with *perpinguis*.

Nassarius mendicus Gould

Western Lean Nassa

Figure 53d

Alaska to Lower California.

½ to ¾ inch in length, with a moderately high spire. Outer lip not thickened. Sculpture consists of numerous, small beads which are formed by the crossing of about a dozen small axial ribs and smaller spiral threads. Color yellowish gray. Common in shallow water in the north.

The subspecies or form *cooperi* Forbes has weaker spiral threads and about 7 to 9 strong, whitish, smoother axial ribs which persist to the last of the body whorl. Color grayish yellow to whitish, often with fine, spiral, brown or mauve lines. Very common in the south.

Subgenus *Zaphon* H. and A. Adams 1853

Nassarius fossatus Gould

Giant Western Nassa

Plate 20s

Vancouver Island to Lower California.

1½ to 2 inches in length, orange-brown to brownish white in color. Early whorls coarsely beaded; last whorl with about a dozen coarse, variously sized, flat-topped spiral threads and with about a dozen short axial ribs on the top third of the last whorl. Outer lip with a jagged edge and constricted at the top. The largest and one of the common intertidal Nassa snails on the Pacific coast.

Subgenus *Ilyanassa* Stimpson 1865

Nassarius obsoletus Say

Eastern Mud Nassa

Plate 23p

Gulf of St. Lawrence to northeast Florida. Introduced to the Pacific coast.

¾ to 1 inch in length, usually covered with mud and algae, and has its spire eroded at the tip. Color dark black-brown. Sculpture of numerous rows of weak beads. Parietal wall thickly glazed with brown and gray. Columella with a single, strong spiral ridge near the base. Outer lip with half a dozen small grayish teeth which run back into the aperture. Very common on oozy, warm mud flats.

Family *FASCIOLARIIDAE*
Subfamily *FASCIOLARIINAE*
Genus *Leucozonia* Gray 1847

Leucozonia nassa Gmelin

Chestnut Latirus

Plate 11d

Florida to Texas and the West Indies.

1½ inches in length, heavy, squat, with its whorls shouldered by about 9 large nodules. Characterized by its semi-glossy, chestnut-brown color with a faint, narrow spiral band of whitish at the base of the shell which terminates into a small, distinct spine on the outer lip. Columella with 4 weak folds at the base. Aperture yellowish tan within. Common among rocks at low tide. Alias *L. cingulifera* Lamarck.

Leucozonia ocellata Gmelin White-spotted Latirus

Plate 11e

West coast of Florida and the West Indies.

1 inch in length, ⅔ as wide, squat and heavy. Color dark-brown to blackish with a row of about 8 large, white, rounded nodules at the periphery and about 3 or 4 spiral rows of smaller white squares on the base of the shell. Base of columella with 3 small folds. Apex usually worn white. A common intertidal species found under rocks.

GENUS *Latirus* Montfort 1810

Latirus mcgintyi Pilsbry McGinty's Latirus

Plate 11b

Southeast Florida.

½ to 2½ inches in length, elongate, heavy, and with about 10 whorls. Color cream with a solid yellow-brown periostracum. Aperture bright yellow. Umbilicus variable, but usually funnel-shaped. Each whorl with 8 low, rounded ribs which are noduled by 2 spiral cords in the upper whorls and 4 cords on the wide periphery of the last whorl. Numerous fine spiral threads present. Lower columella with 2 weak folds. Uncommon.

Latirus infundibulum Gmelin Brown-lined Latirus

Plate 11a

Florida Keys and the West Indies.

3 inches in length, heavy, resembling a *Fusinus* in shape, but characterized by the 3 weak folds on the columella, the light-tan to light-brown shell which bears small, darker brown, wavy, glossy, smooth spiral cords. 7 to 8 strong axial nodules per whorl. Umbilicus imperfect, sometimes funnel-shaped. Moderately common in the West Indies, rare in Florida.

Latirus brevicaudatus Reeve Short-tailed Latirus

Plate 11f

Lower Florida Keys and the West Indies.

1 to 2½ inches in length, rather broad, with a short siphonal canal, with 8 to 9 rounded, long axial ribs crossed by numerous spiral threads. Color

light-chestnut, reddish brown or dark-brown. Not so shouldered as, and less coarsely sculptured than, *mcgintyi*. It is much stouter and not so elongate as *infundibulum*, but like that species may have narrow, brown spiral lines or threads. Moderately common in the West Indies.

GENUS *Fasciolaria* Lamarck 1799

Fasciolaria tulipa Linné   True Tulip
Plate 13b

North Carolina to south half of Florida and West Indies.

3 to 5 inches in length, with 2 or 3 small spiral grooves just below the suture, between which the shell surface is often crinkled. Sometimes with broken spiral color lines. A beautiful orange-red color variety is not uncommon on the Lower Keys. Common. Giants reach a length of 10 inches.

Fasciolaria hunteria Perry   Banded Tulip
Plate 13c

North Carolina to Florida and the Gulf States.

2 to 4 inches in length, whorls entirely smooth near the suture. The widely spaced, rarely broken, distinct, spiral, purple-brown lines are characteristic. Albino shells are rare. A common western Florida species which lives in warm, shallow areas. Formerly *F. distans* Lamarck, a later name.

The subspecies *branhamae* Rehder and Abbott from Yucatan to off west Texas has a much longer siphonal canal and the spiral color lines are also on the siphonal canal. Intergrades exist in Louisiana and Alabama. Branham's Tulip is moderately common.

GENUS *Pleuroploca* P. Fischer 1884

Pleuroploca gigantea Kiener   Florida Horse Conch
Plate 13a

North Carolina to both sides of Florida.

Almost 2 feet in length, although usually about 1 foot. Outer surface dirty-white to chalky-salmon, and covered with a fairly thick, black-brown periostracum which flakes off in dried specimens. The young (up to about 3½ inches) have a thinner periostracum and the entire shell is a bright orange-red. A form which lacks the nodules on the last whorl was named *reevei* Philippi 1851. *P. papillosa* Sowerby 1825 is insufficiently described to apply with any certainty to this species.

A similar, large species, *P. princeps* Sowerby (the Panama Horse Conch), occurs from the Gulf of California to Ecuador. Its operculum has deep, rounded grooves. Both of these Horse Conchs were previously put in the genus *Fasciolaria*.

Subfamily *FUSININAE*
GENUS *Fusinus* Rafinesque 1815

Fusinus timessus Dall   Turnip Spindle
Plate 11g

Gulf of Mexico.

About 3 inches in length, solid, pure white, with a thin, gray periostracum. Aperture round with a flaring, raised parietal wall which, like the inside of the outer lip, is enamel-white and bears numerous spiral threads. Each whorl with 10 to 12 low, short axial ribs at the periphery. Upper whorls with 8 to 9 small, but sharp and slightly wavy, smooth spiral cords. Last whorl and the long siphonal canal with a total of about 30 to 40 small cords between which is often a very fine one. Dredged uncommonly from 20 to 50 fathoms.

Fusinus eucosmius Dall   Ornamented Spindle
Plate 11c; figure 22k

Gulf of Mexico.

3 inches in length, with about 12 rounded whorls and with a small, roundish aperture located at the middle of the shell. Siphonal canal long, its diameter about equal to that of the aperture. Whorls with 8 large, rounded axial ribs which in the upper whorls are crossed by about 6 strong, sharp, slightly wavy spiral threads. Apex often leaning to one side. Color all white with a rather heavy, grayish-white to yellowish periostracum. Rather commonly dredged offshore, but still a collector's item.

Subgenus *Barbarofusus* Grabau and Shimer 1909

Fusinus harfordi Stearns   Harford's Spindle
Figure 54a

Mendocino County, California.

2 inches in length, heavy, exterior dark, orange-brown, with 11 to 12 wide, rounded axial ribs crossed by small, sharply raised, finely scaled spiral cords. Rare in moderately deep water.

Fusinus kobelti Dall   Kobelt's Spindle
Figure 54b

Monterey to Catalina Island, California.

2½ inches in length, heavy, similar to *harfordi*, but with a longer siphonal canal, fewer and larger axial ribs (8 to 10 per whorl), colored white, except for several orange-brown spiral cords. Periostracum rather thick, opaque and light-brown. The spiral cords in *harfordi* are much larger and with squarish tops. Moderately common in shallow water to 35 fathoms.

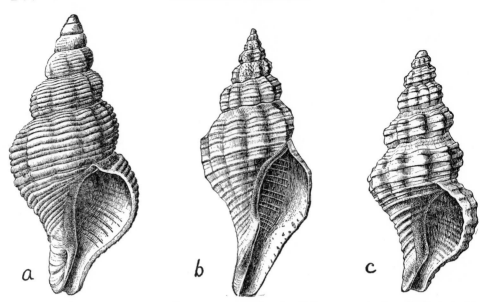

FIGURE 54. Californian Spindles. **a,** *Fusinus harfordi* Stearns, 2 inches; **b,** *F. kobelti* Dall, 2½ inches; **c,** *Aptyxis luteopicta* Dall, ¾ inch.

Fusinus barbarensis Trask — Santa Barbara Spindle

Plate 24z

Oregon to San Diego, California.

4 to 5 inches in length, almost ¼ as wide, 9 to 10 rounded whorls, the early ones with about 10 low, axial ribs which are very weak or absent in the last 2 whorls. Spiral threads prominent and numerous. Color dirty gray-white, sometimes with a pinkish or yellowish cast. Dredged from 50 to 200 fathoms, and occasionally brought up in fishermen's nets.

GENUS *Aptyxis* Troschel 1868

Aptyxis luteopicta Dall — Painted Spindle

Figure 54c

Monterey, California, to Lower California.

¾ inch in length, strong, with a thin outer lip. Color dark purplish brown with an indistinct, wide, spiral band of cream at the periphery. Common from low tide to 20 fathoms.

Family XANCIDAE
Subfamily XANCINAE
GENUS *Xancus* Röding 1798

Xancus angulatus Solander — West Indian Chank

The Bahamas, Key West, Cuba, Yucatan and Bermuda.

7 to 14 inches in length, very heavy. Color cream-white with a thick, light-brown periostracum. Interior often tinged with glossy, pinkish cream or deep, brownish orange. Columella bears 3 strong, widely spaced folds. Middle of whorl on inside of aperture often with a spiral, weak ridge. A left-handed specimen of this species would be worth its weight in silver. Once called *Turbinella scolyma* Gmelin. Common in the Bahamas and Cuba.

Subfamily *VASINAE*
Genus *Vasum* Röding 1798

Vasum muricatum Born

Caribbean Vase

Plate 231

South half of Florida and the West Indies.

2½ to 4 inches in length, heavy. Blunt spines are at the shoulder and near the base. Shell chalk-white, covered by thick, black-brown periostracum. Aperture glossy-white and with a purplish tinge. Columella with 5 strong folds, the first and third being the largest. Rather common, often in pairs, in shallow water. Preys on worms and clams.

The subspecies *coestus* Broderip 1833 (Panamanian Vase) occurs from the Gulf of California to Panama, and differs only in having 4 (rarely 5) columella folds and in having heavier spiral cords. It is common.

Superfamily *VOLUTACEA*
Family *OLIVIDAE*
Genus *Oliva* Bruguière 1789

Oliva sayana Ravenel

Lettered Olive

Plate 12a

North Carolina to Florida and the Gulf States.

2 to 2½ inches in length, moderately elongate, with a glossy finish and with rather flat sides. Color grayish tan with numerous purplish brown and chocolate-brown, tent-like markings. A common species found at night crawling in sand in shallow water. Formerly called *O. litterata* Lamarck. Do not confuse with *O. reticularis* which is generally smaller, which has a much more shallow canal at the suture, whose apical whorls are slightly convex instead of slightly concave, and whose sides of the whorls are more convex. Dead specimens buried for a long time in bay mud may take on an artificial black coloration.

Oliva reticularis Lamarck

Netted Olive

Plate 12c

Southeast Florida and the West Indies.

1½ to 1¾ inches in length, similar to *sayana*, but smaller, more globose, with an oily finish and generally more lightly colored. The Golden Olive or Golden Panama is merely a rare orange form of this species. Sometimes pure white or very dark-brown in color. A common West Indian species. See remarks under *sayana*.

Genus *Olivella* Swainson 1831

Distinguished from the genus *Oliva* by its much smaller shell and in possessing an operculum.

Olivella mutica Say — Variable Dwarf Olive
Plate 22v

North Carolina to Florida, Texas and the West Indies.

¼ to ½ inch in length, half as wide, with a sharp apex. Strong, glossy callus is present on the parietal wall at the upper end of the aperture. Variable in color: ashy grays and chocolate-browns to yellowish or whitish with wide bluish-gray spiral bands. Sometimes brightly banded with white and browns. A very common species found in warm, shallow waters.

Olivella nivea Gmelin — West Indian Dwarf Olive
Plate 11h, j

Southeast Florida, the West Indies and Bermuda.

½ to 1 inch in length, whorls about 7, apex sharply pointed; nucleus small, white, tan or purple. Suture channel is deep and fairly wide; with a strongly concave, etched, spiral indentation on the side of the preceding whorl. Color variable, usually cream-white with orange, tan or purple occurring in clumps in a spiral series just below the suture and just above the fasciole (that raised spiral ridge at the base of the shell). Fasciole lacks color. Common from shore to 25 fathoms. Compare with *jaspidea* which has a more bulbous apex.

Olivella jaspidea Gmelin — Jasper Dwarf Olive
Plate 11-i

Southeast Florida to Barbados.

½ to ¾ inch in length, whorls about 5, apex blunt, nuclear whorls large. Color variable, usually grayish white with small, dull maculations of purplish brown. Fasciole at base of columella with irregular, brown spots and bars. A common West Indian species found in shallow water in sand. Compare with *nivea*.

Olivella moorei Abbott — Moore's Dwarf Olive

Off Key Largo to Key West, Florida.

¼ inch in length, apex bulbous. Characterized by its translucent shell with numerous, long, wavy, axial flammules of reddish brown on the sides of the whorls. Dredged from 115 to 144 fathoms. Named for Hilary B. Moore of the University of Miami, Florida.

Olivella floralia Duclos   Common Rice Olive

North Carolina to both sides of Florida and the West Indies.

⅜ to ½ inch in length, slender, fusiform and with a sharp apex. Color all white, but often with a dull bluish undertone. Apex white, orange or dull purplish. Columella with numerous, very small folds. Common in shallow water.

Olivella baetica Carpenter   Beatic Dwarf Olive
                                                         Plate 20q
Kodiak Island, Alaska, to Lower California.

½ to ¾ inch in length, moderately elongate, rather light-shelled, glossy, and colored a drab-tan with weak purplish brown maculations often arranged in axial flammules which may be more pronounced near the suture. Columellar callus weakly developed, the lower end with a double-ridged spiral fold. Fasciole white, often stained with brown. Early whorls usually purplish blue. *O. porteri* Dall is the same.

Olivella pedroana Conrad   San Pedro Dwarf Olive

Oregon to Lower California.

⅜ to ½ inch in length, resembling *O. baetica*, but much heavier, much stouter, with a heavy callus, and colored light-buff to clouded, brownish gray with long, distinct, axial, zigzag stripes of darker brown. Fasciole and callus always white. The lowest columellar spiral ridge is single or rarely double. Moderately common from 1 to 15 fathoms. *O. pycna* Berry is the same, and matches the neotype designated by Woodring in 1946. *O. intorta* Carpenter is also this species.

Subgenus *Callianax* H. and A. Adams 1853

Olivella biplicata Sowerby   Purple Dwarf Olive
                                                         Plate 12i
Vancouver Island to Lower California.

1 to 1¼ inches in length, globular to elongate, quite heavy. Upper columella wall with a heavy, low, white callus. Lower end of columella with a

raised, spiral fold which is cut by 1, 2, or 3 spiral, incised lines. Color variable, but usually bluish gray or whitish brown with violet stains around the
fasciole and lower part of the aperture. Brown and pure-white specimens are
sometimes found. Abundant in summer months in sandy bays and beaches.
Sometimes dredged down to 25 fathoms on gravel bottom.

Family *MITRIDAE*
Genus *Mitra* Lamarck 1799

Mitra florida Gould   Royal Florida Miter

Plate 13i

South half of Florida and the West Indies.

1½ to 2 inches in length, with about 6 whorls. Characterized by its
smooth, white, glossy whorls which bear on the last one about 16 spiral rows
of evenly spaced, small, roundish dots of orange-brown. There are also odd
patches of light orange-brown. 9 columella folds, the lower 7 being very
weak. An uncommon species considered a choice collector's item. Formerly
known as *M. fergusoni* Sowerby.

Mitra swainsoni antillensis Dall   Antillean Miter

North Carolina to Florida and the West Indies.

3 inches in length, about ¼ as wide, with the aperture half as long as
the entire shell. 10 whorls smooth, except for 5 or 6 weak spiral threads on
the upper fourth of the whorl. Columella with 4 slanting, spiral folds, the
largest being the uppermost. Color grayish white with a light-brown to olive
periostracum. Short siphonal canal slightly recurved. Rare.

Mitra nodulosa Gmelin   Beaded Miter

Plate 26b

North Carolina to Florida and the West Indies.

¾ to 1 inch in length, solid, glossy, orange to brownish orange in color,
and with about 17 long, axial riblets which are rather neatly beaded. Suture
deep, with the whorls slightly shouldered. Columella folds 3, large and
white. A common species frequently washed ashore or found under rocks at
low tide.

Mitra styria Dall   Dwarf Deepsea Miter

Lower Florida Keys and the West Indies.

½ inch in length, fusiform in shape, moderately fragile and ashen-white in color. 10 whorls. Characterized by the numerous, very small, beaded, axial riblets and the thin, gray periostracum. Columella folds 5, the lower 2 being very weak. Nuclear whorls small, smooth and pointed. Commonly dredged from 30 to 333 fathoms.

Mitra barbadensis Gmelin

Barbados Miter
Plate 26d

Southeast Florida and the West Indies.

1 to 1½ inches in length, slender, with the aperture wide below and half the length of the entire shell. Characterized by its yellow-brown to fawn color which has an occasional fleck of grayish white. Aperture tan within. Columella with 5 slanting folds. The sides of the spire are almost flat. Weak spiral threads are often present especially in the earlier whorls. A common species under rocks at low tide.

Mitra hendersoni Rehder

Henderson's Miter
Plate 26c

Southeast Florida and the West Indies.

½ to ¾ inch in length, fusiform in shape, with 8 whorls, each bearing a dozen sharp axial ribs which extend halfway down the whorl. Numerous microscopic, spiral cords present. Columella with 4 folds. Color drab pinkish gray with the upper half of the whorl bearing a wide, lighter, spiral band. Moderately common offshore in several fathoms.

Mitra sulcata Gmelin

Sulcate Miter
Plate 26a

Southeast Florida and the West Indies.

½ inch in length, rather fusiform in shape, with axial ribs as in *hendersoni*, but without spiral threads. 4 columella folds large and dark-brown. Color of shell dark chocolate-brown with a narrow, white, spiral band on the upper half of the whorl. Moderately common below low-water line under rocks in sand. Do not confuse with *Engina turbinella* which has no columella folds. *Mitra albocincta* C. B. Adams is probably this species.

Mitra idae Melville

Ida's Miter
Plate 20p

Farallon Islands to San Diego, California.

2 to 3 inches in length, heavy, elongate. With 3 columella folds. Color mauve-brown, but usually covered with a thick, finely striate, black periostracum. Uncommon offshore.

Family VOLUTIDAE
Subfamily VOLUTINAE
GENUS *Voluta* Linné 1758

Voluta musica Linné　　　　　　　　　Common Music Volute

Plate 13g

Caribbean area.

2 to 2½ inches in length, heavy and with a polished finish. 3 nuclear whorls bulbous and yellowish. 3 postnuclear whorls plicate at the shoulder. Columella with about 9 evenly spaced folds. Characterized by the pinkish cream background and 2 to 3 spiral bands of fine lines which are dotted with darker brown (the musical notes). A moderately common West Indian species not found in the United States, but a favorite with collectors. A number of useless names have been applied to the numerous variations of this species. This is one of the few volutes to have an operculum.

Voluta virescens Solander　　　　　　　Green Music Volute

Lower Florida Keys (rare) and the Caribbean.

2 inches in length, moderately heavy with the aperture ⅘ the total length of the shell. Whorls flat-sided and with weak, axial nodules high on the shoulder. Numerous spiral, incised lines and fine threads present. Columella with about a dozen folds of variable sizes. Exterior dull greenish brown with weak, narrow, spiral bands of lighter color dotted with black-brown. Aperture pale cream to gray within. A rare species in southeast Florida, but not uncommon along the northern coast of South America.

Subfamily SCAPHELLINAE
GENUS *Scaphella* Swainson 1832

Scaphella junonia Shaw　　　　　　　　The Junonia

Plate 13f

North Carolina to both sides of Florida to Texas.

5 to 6 inches in length, rather solid and smooth. 4 folds on the columella. Characterized by the cream background and the spiral rows of small reddish brown dots. Moderately common from 1 to 30 fathoms, but rarely washed ashore. A golden form occurs off Alabama (subspecies *johnstoneae* Clench 1953) and specimens from Yucatan have a white background with smaller spots (subspecies *butleri* Clench 1953). About 50 specimens a year are found on west Florida beaches, and many more are brought in by fishermen.

Subgenus *Aurinia* H. and A. Adams 1853

Scaphella dohrni Sowerby   Dohrn's Volute
Plate 13j

Off the south half of Florida.

3 to 4 inches in length, similar to *junonia,* but much lighter in weight, much more slender, with a higher spire and with numerous exceedingly fine, incised (cut) spiral lines. Some specimens have the early whorls slightly angled and with short axial ribs. This is the form named *florida* Clench and Aguayo and is probably not a good species. A rare species which is appearing in private collections more and more.

Scaphella dubia Broderip   Dubious Volute

Off south half of Florida and the Gulf of Mexico.

4 inches in length, similar to *dohrni,* but more slender, with fewer spots, and with 6 to 7, instead of 9 to 10, rows of spots. A rare and exquisite species from moderately deep water.

Scaphella schmitti Bartsch   Schmitt's Volute
Plate 13e

Off Tortugas, Florida.

5 inches in length. Under the brownish periostracum the shell is chalky, pale salmon and with 4 or 5 spiral rows of weak, brown, square spots. A thick, yellowish-gray glaze overlays the periostracum on the parietal side of the body whorl. Columella straight, while in *R. georgiana* Clench from Georgia to east Florida it is arched. Both quite rare, the former in 80 fathoms of water. The genus *Rehderia* Clench 1946 was unfortunately erected upon an ecological or pathological character and should be considered a synonym of *Scaphella.*

GENUS *Arctomelon* Dall 1915
(*Boreomelon* Dall 1918)

Arctomelon stearnsi Dall   Stearns' Volute

Alaska.

4 to 5 inches in length, strong; exterior chalky-gray with mauve-brown undertones. Aperture semi-glossy, light-brown. Columella brownish with 2 moderately large folds and a weak one below. Nucleus bulbous, chalky-white. Uncommon offshore down to 100 fathoms.

Family CANCELLARIIDAE
GENUS *Cancellaria* Lamarck 1799

Cancellaria reticulata Linné   Common Nutmeg
Plate 13k

North Carolina to both sides of Florida.

1 to 1¾ inches in length, strong, with numerous spiral rows of small, poorly shaped beads which, with the weak axial and spiral threads, give a reticulate appearance. Columella with 2 folds, the uppermost being very strong and furrowed by 1 or 2 smaller ridges. Color cream to gray with heavy, broken bands and maculations of dark orange-brown. Rarely all white. Common in shallow water to several fathoms. *C. conradiana* Dall is probably only a form of this species.

The subspecies *adelae* Pilsbry from the Lower Florida Keys is smoothish, except for incised lines on the body whorl. The aperture is faintly flushed with pink. Uncommon. (Adele's Nutmeg).

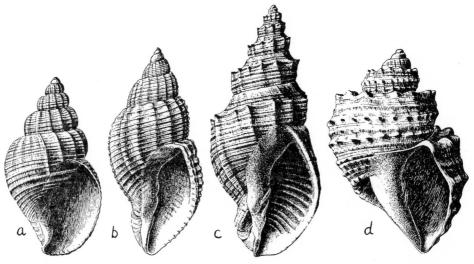

FIGURE 55. Cancellarid Shells. **a**, *Admete couthouyi* Jay (Atlantic and Pacific; ¾ inch); **b**, *Cancellaria crawfordiana* Dall (California; 1½ inches); **c**, *Narona cooperi* Gabb (California; 2 inches); **d**, *Trigonostoma tenerum* Philippi (Florida; ¾ inch).

Subgenus *Massyla* H. and A. Adams 1854

Cancellaria crawfordiana Dall   Crawford's Nutmeg
Figure 55b

Bodega to San Diego, California.

1 to 2 inches in length, heavy, white in color, but covered with a thick, rather fuzzy, gray-brown periostracum. Aperture enamel-white. Uncommon from 16 to 204 fathoms.

GENUS *Narona* H. and A. Adams 1854
Subgenus *Progabbia* Dall 1918

Narona cooperi Gabb   Cooper's Nutmeg
                                                       Plate 24y; figure 55c

Monterey, California, to Lower California.

2 to 2½ inches in length, moderately heavy; columella with 2 small spiral folds. Whorls slightly shouldered, with about a dozen to 15 narrow axial ribs which at the top bear a single, low, sharp knob. Color brownish cream with a dozen or so narrow, brown spiral bands. Aperture orange-cream. Outer lip sometimes with numerous white, glossy, spiral cords on the inside. An uncommon, deep-water species, occasionally brought up in fish nets. Said to grow to 7 inches in length.

GENUS *Trigonostoma* Blainville 1827

Trigonostoma tenerum Philippi   Philippi's Nutmeg
                                                       Figure 55d

Southern half of Florida.

¾ inch in length, fairly thin, but a quite strong shell. 4 whorls, strongly shouldered with the upper part of the whorl smooth and flat, and the sides with 3 to 5 spiral rows of strong nodules or blunt beads. Umbilicus very deep and funnel-shaped. Color light orangish brown. Uncommon just off-shore.

Trigonostoma rugosum Lamarck (in the subgenus *Bivetiella* Wenz 1943) is similar, but heavier, whitish with brownish maculations, without an umbilicus, and with about 8 strong axial ribs crossed by spiral threads. Known as the Rough Nutmeg. It is rare in most areas of the West Indies, and has not been reported from the United States.

GENUS *Admete* Kröyer 1842

Admete couthouyi Jay   Common Northern Admete
                                                       Figure 55a

Arctic Seas to Massachusetts. Arctic Seas to San Diego, California.

½ to ¾ inch in length, moderately thick, with 6 whorls. Suture wavy, well-impressed. Sculpture coarsely reticulate, often beaded or with the axial cords the strongest. Columella strongly arched and bearing 2 to 5 very weak, spiral folds near the middle. Shell dull white, covered with a fairly thick, gray-brown periostracum. Commonly dredged in cold waters. There are several other deep-water species on both of our coasts but they occur in very deep water.

Family MARGINELLIDAE
Genus *Marginella* Lamarck 1799
Subgenus *Eratoidea* Weinkauff 1879

Marginella haematita Kiener   Carmine Marginella
                                                  Figure 56a

Southeast Florida and the West Indies.

¼ inch in length, characterized by its glossy, bright and deep rose color, 4 strong columella teeth, pointed spire and thickened outer lip whose inner edge bears about 15 small, round teeth. Uncommon from 25 to 90 fathoms. *M. philtata* M. Smith and *M. jaspidea* Schwengel are probably this species.

Marginella denticulata Conrad   Tan Marginella
                                                  Figure 56c

North Carolina to both sides of Florida and the West Indies.

⅜ inch in length, similar to *haematita*, but with a longer spire, only 7 to 9 teeth on the outer lip, with a shallow U-shaped notch at the top of the aperture, and the entire shell is yellow-tan to whitish. Uncommon from low tide to 600 fathoms. *M. eburneola* Conrad is this species.

Marginella aureocincta Stearns   Golden-lined Marginella
                                                  Figure 56b

North Carolina to both sides of Florida and the West Indies.

¹⁄₁₆ inch (4.0 mm.) in length; aperture half the length of the entire shell; spire pointed. Outer lip thickened, with about 4 very small teeth just inside the aperture. Columella with 4 strong folds or teeth. Color translucent-white, with two distinct, narrow spiral bands of light tan-orange on the body whorl (1 showing in the whorls of the spire). A very common species from low-water line to 90 fathoms.

Genus *Prunum* Herrmannsen 1852

Prunum carneum Storer   Orange Marginella
                                                  Plate 11k

Southeast Florida and the West Indies.

¾ inch in length, very glossy; outer lip thickened, smooth and white. Apex half covered by a callus of enamel. Lower third of columella with 4 strong, slanting teeth. Shell bright orange with a faint, narrow, whitish, spiral band on the middle of the whorl and one just below the suture. Uncommon in Florida on reef flats to 6 fathoms.

Prunum roosevelti Bartsch and Rehder   Roosevelt's Marginella
                                                  Plate 11-o

The Bahamas.

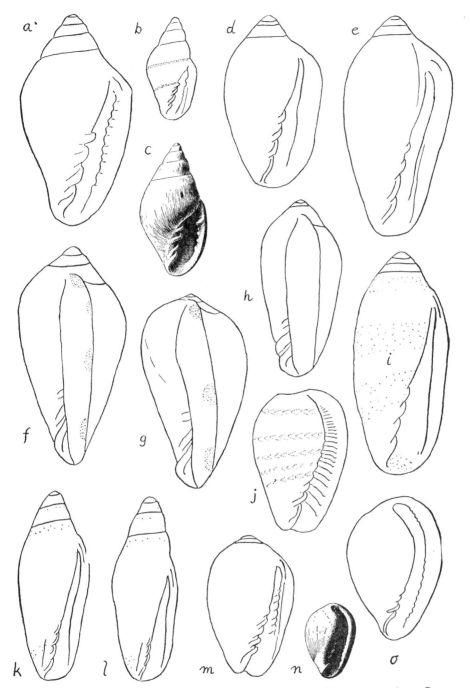

FIGURE 56. Marginellas. **a,** *Marginella haematita* Kiener; **b,** *M. aureocincta* Stearns, **c,** *M. denticulta* Conrad; **d,** *Prunum bellum* Conrad; **e,** *Prunum amabile* Redfield; **f,** *P. limatulum* Conrad; **g,** *P. apicinum* Menke; **h,** *P. virginianum* Conrad; **i,** *Hyalina avena* Val.; **j,** *Persicula catenata* Montagu; **k,** *H. avenacea* Deshayes; **l,** *H. torticula* Dall; **m,** *Persicula minuta* Pfeifer; **n,** *Gibberulina pyriformis* Cpr.; **o,** *G. ovuliformis* Orb. (**m** and **o** X10, the others X5).

1 inch in length, extremely close to *carneum*, differing only in being larger, and in having a brown spot on the apex and 2 large chocolate spots on the outer lip. There may be also 2 smaller spots at the anterior end of the shell. Apparently rare and possibly a color form of *carneum*. I have seen only 3 specimens.

Prunum labiatum Valenciennes   Royal Marginella
Plate 11l

Off Texas to Central America.

1 to 1¼ inches in length, similar to *carneum*, but stouter, lip orange-brown, body whorl whitish gray with 3 darker, subdued spiral bands. Outer lip with small teeth on its inner edge. Very uncommon, but has been found off Yucatan by shrimp fishermen. The Texas record is open to question.

Subgenus *Leptegouana* Woodring 1928

Prunum guttatum Dillwyn   White-spotted Marginella
Plate 11m

Southeast Florida and the West Indies.

½ to ¾ inch in length; outer lip smooth, white and with 2 or 3 brown spots on the lower half. 4 columella teeth. Color of body whorl pale whitish with 3 obscure bands of light pinkish brown, and irregularly spotted with weak, opaque-white, roundish dots. Not uncommon in shallow water.

Prunum bellum Conrad   La Belle Marginella
Figure 56d

Off North Carolina to Key West.

¼ inch in length, glossy, white, sometimes with a bluish-gray undertone. Sometimes with a rose tint on the body whorl. Spire moderately elevated. Outer lip thickened, without teeth. Lower half of columella with 4 strong, equally sized teeth. Commonly dredged from 1 to 200 fathoms.

Prunum amabile Redfield   Queen Marginella
Figure 56e

Off North Carolina to Key West.

⅜ inch in length, similar to *bellum*, but with a shorter spire, more slanting columellar teeth, colored a translucent-tan with a heavy suffusion of orange on the shoulder of the whorl which becomes lighter on the lower part of the whorl. There is a fairly well-developed, white callus on the parietal wall. Uncommonly dredged from 25 to 125 fathoms.

Prunum apicinum Menke   Common Atlantic Marginella

Plate 11n; figure 56g

North Carolina to Florida, the Gulf States and the West Indies.

½ inch in length, glossy, with a dark nuclear whorl. Outer lip thickened, smooth, white, with 2 small, red-brown dots near the middle and a smaller one at the very top and very bottom. Body whorl golden to brownish orange with 3 subdued, wide bands of darker color. A very common, shallow-water species. About 1 in every 5000 specimens is sinistral.

Prunum limatulum Conrad   Boreal Marginella

Figure 56f

Virginia to South Carolina.

½ inch in length, similar to *apicinum*, but with a higher spire, milky-cream color, with 3 faint, spiral bands of mauve or weak orange. Outer lip not sinuate, and is usually marked with 4 spots. Nucleus white, while in *apicinum* it is usually bright pink. Not uncommon from 18 to 132 fathoms. *Marginella borealis* Verrill is the same.

Prunum virginianum Conrad   Virgin Marginella

Figure 56h

North Carolina to west Florida and Yucatan.

¼ inch in length, similar to *apicinum*, but without spots on the thick varix; the third columella tooth is the largest; color of last whorl whitish to cream, often with a faint curdling of darker orange-cream, and with a deeper, suffused band just below the suture and at the base of the shell. Moderately common, 14 to 56 fathoms.

Genus Persicula Schumacher 1817

Persicula catenata Montagu   Princess Marginella

Figure 56j

Southeast Florida and the West Indies.

¼ inch in length, glossy; apex sheared off and sealed over by a weak callus. Columella teeth 7. Inside of outer lip with about 20 to 25 small teeth. Color translucent grayish with 7 spiral rows of teardrop-shaped, opaque-white spots and with 2 very subdued, wide spiral bands of light-brown. Uncommon in shallow water to 92 fathoms.

Subgenus *Gibberula* Swainson 1840

Persicula minuta Pfeiffer   Snowflake Marginella

Figure 56m

South half of Florida and the West Indies.

⅛ inch in length, resembling a miniature *apicinum*, but pure white in color. Like *Gibberulina ovuliformis*, but the aperture is not so long and has microscopic, spiral teeth inside the thin, curled-in outer lip. Columella with 4 oblique folds. Alias *M. lavalleana* Orb. Common in shallow water to 40 fathoms. The subgenus *Granula* Jousseaume 1874 is this subgenus.

Persicula jewetti Carpenter   Jewett's Marginella

Monterey, California, to Lower California.

¼ inch (5.0 mm.) in length, snow-white, glossy, rather stout. Apex smoothed over and obscured. Outer lip smooth, slightly curled inward. Columella with 3 or 4 rather distinct, slanting, spiral folds with several smaller ones higher on the columella. Common from low tide to several fathoms.

There are 3 similar, small and white species which are very difficult to separate; and according to some workers, size and proportionate dimensions are of significance:

> *P. regularis* Cpr. (Regular Marginella), Monterey to Lower California. Length 3.3 mm., ratio of diameter to length 1 to 1.5. Low tide to 30 fathoms. Common.
> *P. subtrigona* Cpr. (Triangular Marginella), Monterey to Lower California. Length 3.5 mm., ratio of diameter to length 1 to 1.25. Low tide to 50 fathoms. Uncommon.
> *P. politula* Dall (Polite Marginella), Santa Barbara to Lower California. Length 3.0 mm., ratio of diameter to length 1 to 2. Low tide to 20 fathoms. Uncommon.

GENUS *Hyalina* Schumacher 1817
Subgenus *Volvarina* Hinds 1844

Hyalina avena Valenciennes   Orange-banded Marginella

Plate 11p; figure 56i

North Carolina to Key West and the West Indies.

¼ to ½ inch in length, slender; spire pointed, but short. Outer lip curled in, white and smooth. Aperture narrow above, wide below. 3 to 4 slanting, columellar teeth. Color whitish, cream or yellowish with 4 to 6 spiral bands of subdued orange-tan. A moderately common, shallow-water species. The pink variety, especially common in Yucatan has been given the name *beyerleana* Bern.

Hyalina veliei Pilsbry   Velie's Marginella

West coast of Florida.

½ inch in length, somewhat like our figure 56e, but with a higher, more pointed spire. Shell quite thin for a Marginella; color yellowish to whitish and somewhat translucent. Outer lip thickened, pushed in at the middle and white in color. Columella with 4 very distinct folds. Common in shallow water inside dead *Pinna* shells on mangrove mud flats.

Hyalina avenacea Deshayes Little Oat Marginella

Figure 56k

North Carolina to both sides of Florida and the West Indies.

⅓ to ½ inch in length, slender, very similar to *H. avena*, but usually smaller, with a longer spire, more slender anterior end, and pure, opaque-white in color, except for a very faint hint of straw color below the suture, again at the middle of the body whorl and also near the base. Common from shallow water to 750 fathoms. This is *avenella* Dall and *succinea* Conrad.

Hyalina torticula Dall Knave Marginella

Figure 56l

Off eastern Florida.

⅓ inch in length, slender, fusiform, with a tall spire which is leaning to one side. Color opaque-white, glossy, and with a hint of straw-colored bands. Possibly a sport of *avenacea*. Uncommon in deep water.

Hyalina californica Tomlin Californian Marginella

Santa Monica, California, to Mexico.

⅓ inch in length, slender, aperture ⅚ the length of the entire shell, with 4 whorls, and colored a grayish to bright-orange with 3 distinct or obscure, rather wide, spiral bands of white. Lower third of columella white and with 4 distinct, spiral folds. Outer lip smooth, rounded, pushed in slightly, especially near the central portion. Moderately common in rocky rubble under stones at dead low tide.

GENUS *Gibberulina* Monterosato 1884
(*Cypraeolina* Cerulli-Irelli 1911)

Gibberulina ovuliformis Orbigny Teardrop Marginella

Figure 56-o

North Carolina to both sides of Florida and the West Indies.

⅛ inch (2.5 mm.) in length, globular, glossy, opaque-white. Aperture as long as the shell. Apex hidden under top of outer lip. Upper part of whorl slightly shouldered. Lower third of columella with 3 or 4 small, slanting teeth. Outer lip thickened. Do not confuse with *Persicula minuta* Pfr. Common in shallow water to several fathoms. Alias *lacrimula* Gould, *hadria* Dall and *amianta* Dall.

Gibberulina pyriformis Carpenter　　　　　　　Pear-shaped Marginella
　　　　　　　　　　　　　　　　　　　　　　　　　Figure 56n

　　Izhut Bay, Alaska, to Gulf of California.

⅛ inch (3 mm.) in length; aperture as long as the shell. Glossy, translucent milk-white. Lower columella with 4 fairly strong folds with several microscopic teeth farther above. Outer lip curled in and with about 30 microscopic teeth. Animal black. Very common all along the Pacific coast from low-tide line to 40 fathoms. On mud, gravel or backs of abalones.

Family CONIDAE
Genus *Conus* Linné 1758

Conus spurius atlanticus Clench　　　　　　　Alphabet Cone
　　　　　　　　　　　　　　　　　　　　　　　　　Plate 14p

　　Florida and the Gulf of Mexico.

2 to 3 inches in length; spire slightly elevated in the center. Top of whorls smooth, except for tiny growth lines. Color white with spiral rows of orange-yellow squares. Interior of aperture white. A rather common and attractive species found in shallow water. True *spurius spurius* Gmelin from the Bahamas and Antilles differs only in having the spots merging into occasional mottlings. Another race occurs off Yucatan in which the spots are sometimes smaller and a rather dark bluish purple.

Conus aureofasciatus Rehder and Abbott　　　Golden-banded Cone
　　　　　　　　　　　　　　　　　　　　　　　　　Plate 14g

　　Tortugas to off Yucatan, Mexico.

2 to 3 inches in length, similar in shape to *spurius*, although sometimes more slender. Characterized by several spiral bands of light-yellow. Dredged in several fathoms of water. Uncommon to rare. It is possible that this species may be only a freak color form of *spurius*.

Conus daucus Hwass　　　　　　　　　　　　　Carrot Cone
　　　　　　　　　　　　　　　　　　　　　　　　　Plate 14a

　　Both sides of Florida and the West Indies.

1 to 2 inches in length. Spire rather low, sometimes almost worn flat. Shoulder even and sharp. Spire with small, spiral threads. Color deep, solid orange to lemon-yellow, rarely with a lighter band. Spiral rows of minute

brown dots sometimes present on sides. Interior of aperture pinkish white. Color of spire is orange with large white splotches. Uncommon below 15 feet of water.

Conus juliae Clench
Julia's Cone
Plate 14b

Off northeast Florida to Tortugas.

1½ to 2 inches in length. Spire moderately high, flat-sided and with about 10 to 12 whorls. Shoulders of whorls slightly rounded; sides nearly flat. Color a pale pinkish brown to orangish with a moderate and indistinct band of cream or white at the mid area. This is overlaid with a series of fine spiral, broken lines or dots of brown. Spire whitish with axial, zigzag, reddish brown streaks. A choice collector's item. Named after Mrs. William J. Clench, a great contributor to the cause of malacology.

Conus floridanus Gabb
Florida Cone
Plate 14d

North Carolina to both sides of Florida.

1½ to 1¾ inches in length. Spire well-elevated and slightly concave. Sides of whorls flat. The top of each whorl in the spire is concave and also has faint lines of growth. Color variable: usually white with elongate, rather wide patches of light orange-yellow to yellow. Spire with splashes of color. There is usually a white, spiral band around the middle of the whorl which may have small dots of yellowish brown. Moderately common in shallow water to 7 fathoms.

Conus floridanus floridensis Sowerby (pl. 14e) is an extremely dark color form with spiral rows of reddish brown dots and heavier mottlings.

C. floridanus burryae Clench is another color form from off the Lower Florida Keys in which the spiral rows of brownish dots merge into solid lines. The lower end of the shell in very dark brown to deep brownish black. Uncommon.

Conus sennottorum Rehder and Abbott
Sennotts' Cone
Plate 14h

Gulf of Mexico, from Tortugas to Yucatan.

1 inch in length, with a glossy, smooth finish. Slightly turnip-shaped. Color variable: white to bluish white with spiral rows of very small brown dots. Yellowish-brown maculations may be present. Moderately common in 18 fathoms off Yucatan. Named after John and Gladys Sennott.

Conus sozoni Bartsch
Sozon's Cone
Plate 14c

South Carolina to Key West and the Gulf of Mexico.

2 to 4 inches in length. Spire elevated, slightly concave, with the top of each whorl also concave and with fine, arched lines of growth. There are 10 to 12 small spiral ridges at the lower end of the shell. Sides of whorls flat. Color as shown in the photograph, with the two whitish spiral bands being characteristic. Large and perfect specimens are collector's items, although individuals less than 2 inches in length are rather commonly dredged in 50 feet of water off both sides of Florida. Beach specimens have been collected on rare occasions. Named after the sponge diver, Sozon Vatikiotis.

Conus regius Gmelin　　　　　　　　　　　　　　　Crown Cone
Plate 14m

Southern Florida and the West Indies.

2 to 3 inches in length. Spire low; shoulders of whorls with low, irregular knobs or tubercles. Color very variable even in the same locality. A rare yellowish color form (*citrinus* Gmelin, not Clench 1942) occurs in the Lower Florida Keys, Cuba and the Antilles. The interior of the aperture of this species is white. Uncommon in Florida.

Conus mus Hwass　　　　　　　　　　　　　　　Mouse Cone
Plate 14-o

Southeast Florida and the West Indies.

1 to 1½ inches in length. Spire elevated somewhat. Shoulders of whorls with low, irregular, white knobs, between which are brown splotches. Color a dull bluish gray with olive-green or brown mottlings. Interior of aperture with 2 wide spiral bands of subdued brown. Periostracum thick, velvety and yellowish to greenish brown. The name *Conus citrinus* Gmelin (erroneously applied to this species in Johnsonia and other books) is actually the yellow form of *regius*. The Mouse Cone is very common in intertidal, reef areas.

Conus stearnsi Conrad　　　　　　　　　　　　　Stearns' Cone
Plate 22y

North Carolina to both sides of Florida to Yucatan.

½ to ¾ inch in length. A small, slender, graceful cone with a high spire. Top of whorls concave. Sides almost flat. Color usually dull grayish with rows of tiny, white squares and with dull, yellowish brown streaks or mottlings. Highly colored specimens may have rich reddish brown mottlings. Moderately common from shallow water to 30 feet in sand. Do not confuse with *jaspideus*.

Conus jaspideus Gmelin　　　　　　　　　　　　Jasper Cone
Plates 14n; 22x

South half of Florida and the West Indies.

½ to ¾ inch in length; very similar to *stearnsi*, but generally more brightly hued with larger, reddish brown mottlings. The shell is fatter, with more rounded sides, and has strong, spiral lines cut into the sides, usually right up to the shoulder. A very common shallow, sand-loving species. *C. peali* Green is the same species.

Conus verrucosus Hwass   Warty Cone
Plate 22z

Southeast Florida and the West Indies.

¾ to 1 inch in length. A heavy, small cone with a rather high spire and slightly rounded sides. It has small knobs on the shoulder of the last whorl and about 10 spiral rows of distinct warts on the sides. Color white to yellowish with large, brown or yellow mottlings. Uncommon just offshore along the Lower Keys. Common in the West Indies.

A color form, *vanhyningi* Rehder, is a deep, rich peach with the interior of the aperture also pink.

Conus stimpsoni Dall   Stimpson's Cone
Plate 14j

Southeast Florida and the Gulf of Mexico.

1½ to 2 inches in length. A simple cone with a sharp, slightly concave, rather high spire, and with flat sides. It is usually smooth, but may have 15 to 20 cut spiral lines on the sides. Color is an even wash of yellowish white, but sometimes with 2 or 3 slightly darker, wide, yellowish spiral bands. Periostracum gray and rather thick. We have figured the holotype (specimen which Dall used in describing the species). Uncommon in rather deep water down to 30 fathoms.

Conus villepini Fischer and Bernardi   Villepin's Cone
Plate 14f

Tortugas to Yucatan.

1½ inches in length. Spire rather well elevated, very slightly concave. Each whorl in the spire is concave, with 3 to 4 spiral threads, and with fine, arched growth lines. Sides of shell smooth and slightly convex. There are about 9 indistinct spiral threads at the bottom end of the shell. Color of the thin periostracum is light yellowish brown. Shell light grayish white with a faint pinkish undertone. There are 3 or 4 long, irregular, axial streaks of dark reddish brown on the sides of the last whorl. Interior of aperture blushed with rosy-white. We have illustrated the holotype of *amphiurgus* Dall in color which is a synonym. Rare in deep water.

Conus mazei Deshayes　　　　　　　　　　　　　　　Maze's Cone
Plate 14k

Southeast Florida and the West Indies.

1½ to 2 inches in length. A long, narrow, and very handsome species which has rows of delicate beads on the very high spire. This is probably the most valuable cone in Florida waters. A few fortunate collectors in Florida have dredged this unusual cone.

Conus granulatus Linné　　　　　　　　　　Glory-of-the-Atlantic Cone
Plate 14l

Southeast Florida and the West Indies.

1 to 1¾ inches in length. A fairly slender cone with rounded whorls in the spire which have spiral threads. Colored a brilliant orange-red to bright-red with flecks of brown and gold. Coarse spiral threads are usually present on the sides. Interior of aperture with a rosy-pink blush. A perfect specimen of this species is, indeed, a collector's item. It is very rare in Florida and not at all common in the West Indies. It lives in reefs just offshore.

Conus austini Rehder and Abbott　　　　　　　　　　Austin's Cone

Tortugas to Yucatan and West Indies.

2 to 2½ inches in length, pure white in color, although some may have a yellow-brown apex. Characterized by numerous odd-sized spiral threads on the sides. Sides of whorls flat to slightly rounded. Shoulders sharp to slightly rounded. Top of whorls slightly concave, with one smooth spiral carina and several much smaller threads. Shell sometimes with axial puckerings or rib-like wrinkles. Periostracum velvety and grayish brown. Rare off Florida but common in 20 fathoms off Yucatan.

Conus clarki Rehder and Abbott　　　　　　　　　　Clark's Cone
Plate 14i

Off Louisiana.

1 to 1½ inches in length, whitish in color and with small weak spots, rather turnip-shaped, similar to *austini*, but with 27 to 30 very strong, squarish spiral cords on the sides. The cords, and especially the one at the shoulder, are strongly beaded. Between the cords there are microscopic, axial threads. Periostracum gray. Apparently rare offshore in 29 fathoms. This and the preceding species were named after Austin H. Clark, scientist, author and gentleman. *C. frisbeyae* Clench and Pulley 1952 is unquestionably this species.

Conus californicus Hinds — Californian Cone

Farallon Islands, California, to Lower California.

¾ to 1 inch in length. Spire moderately elevated and slightly concave. The shoulders of the shell are rounded, the sides very slightly rounded. The chestnut to pale-brown, velvety periostracum is rather thick. Shell grayish white in color. Interior whitish with a light-brown tint. Rather common in shallow water along certain parts of southern California.

Family *TEREBRIDAE*
GENUS *Terebra* Bruguière 1789

Terebra dislocata Say — Common Atlantic Auger

Plate 26i

Virginia to Florida, Texas and the West Indies.

1½ to 2 inches in length, slender. Whorls with about 25 axial ribs per whorl which are divided ⅓ to ½ their length by a deep, impressed, spiral line. Many specimens show prominent, squarish, raised spiral cords between the ribs. Columella with 2 fused spiral folds near the base. Color a dirty, pinkish gray, but sometimes orangish. A common shallow-water species.

Terebra taurina Solander — Flame Auger

Plate 13h

Southeast Florida, the Gulf of Mexico and the West Indies.

4 to 6 inches in length, heavy, rather slender. Characterized by a cream color with 2 spiral rows of axial, red-brown bars, the upper series being twice as long as the lower one. Upper whorls faintly and axially ribbed. Upper half of each whorl swollen and with a single incised line. *T. flammea* Lamarck and *T. feldmanni* Röding are this species. Formerly considered quite rare, but now not infrequently dredged in the Gulf of Mexico.

Terebra floridana Dall — Florida Auger

Off South Carolina to south Florida.

2 to 3 inches in length, very long and slender. Color light-yellow to yellowish white. Each whorl has just below the suture a row of about 20 oblong, slightly slanting, smooth axial ribs. Below this, and separated from it by an impressed line, is a similar row of much shorter, axial ribs. The lower third of the whorl is marked by 3 or 4 raised, spiral threads only. Columella with a single, strong fold near the bottom. A fairly rare species.

Terebra concava Say

Concave Auger
Plate 26j

North Carolina to both sides of Florida.

¾ inch in length, slender, about 12 whorls, semi-glossy, and with slightly concave whorls. Whorls in spire with a large, heavily nodulated or beaded, swollen spiral cord just below the suture. Above the suture there is a spiral series of 20 very small beads per whorl. The concave middle of the whorl bears about 5 microscopic, incised spiral lines. Color yellowish gray. Common in shallow water. Do not confuse with the larger yellow *T. floridana* which has 2 spiral rows of elongate beads just below the suture.

Terebra protexta Conrad

Fine-ribbed Auger
Plate 26k

North Carolina to Florida and Texas.

¾ to 1 inch in length, about 13 whorls, dull-white in color and with a well-indented suture. Whorls in spire slightly concave with about 22 fine axial ribs running from suture to suture, but which are broken weakly by 7 to 9 incised spiral lines. The upper line is about ¼ the way down the whorl.

Several forms exist which have been given names: form *lutescens* Smith has about 30 to 32 finer axial riblets per whorl which are made slightly beaded by the spiral lines; in the form *limatula* Dall, the ribs and the spiral threads are about equal in size and give a reticulated pattern. All occur together in fairly deep water and are common.

Terebra hastata Gmelin

Shiny Atlantic Auger
Plate 26h

Southeast Florida and the West Indies.

1¼ to 1½ inches in length. Characterized by its smooth, highly glossy finish, its numerous axial ribs which extend from suture to suture, and by its bright yellowish color and white band below the suture. Columella smooth-ish and white. This is the "fattest" species in the western Atlantic, and is fairly common in the West Indies.

Terebra cinerea Born

Gray Atlantic Auger
Plate 26g

Southeast Florida and the West Indies.

1 to 2 inches in length, slender, with flat-sided whorls and a sharp apex. Numerous small riblets extend halfway down the whorls (about 45 to 50 per whorl). Color all cream or bluish brown; sometimes with darker spots below the suture. Surface with exceedingly fine, numerous rows of pin-pricks which give the shell a silky appearance under the lens. Moderately common in shallow water. Compare with *salleana* Deshayes.

Terebra salleana Deshayes — Sallé's Auger

North Florida to Texas and Colombia.

1 to 1⅓ inches in length, similar to *cinerea*, but always a dark bluish gray or brownish, with fewer, larger punctations, with about 30 ribs per whorl, and with a purple, not white, nucleus. Common in shallow water.

Terebra pedroana Dall — San Pedro Auger

Redondo Beach, California, to Lower California.

1 to 1¼ inches in length, strong, slender, with about 12 whorls and colored grayish to whitish yellow or brownish. Sculpture between sutures of first a fairly broad row of well to poorly developed nodules (about 15 to 18 per whorl), followed below by a flat area which is weakly and axially wrinkled or ribbed and with numerous, fine, spiral, incised lines. Siphonal canal bounded by a sharp spiral line on the outer shell. Fairly common in shallow water.

Family TURRIDAE

The family *Turridae* is a very large and diverse group of toxoglossate gastropods which are very difficult to classify. A book of this size cannot do justice to the many interesting species found in our waters. The family probably contains no less than 500 genera and subgenera and several thousand species. An interesting and valuable review of the family is given by A. W. Powell in the *Bulletin of the Auckland Institute and Museum*, no. 2, pp. 1 to 188, 1942. Those interested should consult the works of Grant and Gale, Bartsch, Dall, Rehder, and Woodring. We have included here only a very sketchy representation of our American Turrid fauna.

Subfamily TURRINAE

Shells rather large, usually with a long, slender canal. Sinus on or adjacent to peripheral keel; deep and V-shaped. Operculum leaf-shaped with an apical nucleus. Radula with only 2 marginals which are wish-bone in shape.

GENUS *Gemmula* Weinkauff 1875

Gemmula periscelida Dall — Atlantic Gem Turret

Figure 57c

North Carolina to Tortugas, Florida.

1½ to 2 inches in length, heavy and with the sinus or anal notch well below the suture. Color ash-gray. See illustration. Rare in 100 fathoms.

Genus *Polystira* Woodring 1928

Polystira albida Perry   White Giant Turret
Plate 13l

South Florida, the Gulf of Mexico and the West Indies.

3 to 4 inches in length, pure-white in color. *P. virgo* Lamarck, and "Wood" are this species. Not uncommonly dredged in the Gulf of Mexico.

Polystira tellea Dall   Delicate Giant Turret
Plate 13m

Off southeast Florida.

3 to 3½ inches in length. Grayish white. Sculpture not so distinct nor so smooth as in *albida*. Not uncommonly dredged off Key West. Do not confuse this and the preceding species with *Fusinus couei* (pl. 13d).

Subfamily *COCHLESPIRINAE*

Shell with a long canal. Sinus on the shoulder, rounded, broad and shallow to rather deep. Operculum variable. Radula with 2 strong marginals and a very large central. Shell thin with a sharply angled periphery.

Genus *Ancistrosyrinx* Dall 1881

Ancistrosyrinx radiata Dall   Common Star Turret
Figure 57e

South Florida, the Gulf of Mexico and the West Indies.

½ inch in length. A delicate, glossy, translucent and highly ornamented species. Anterior canal very long. Shoulders keeled, with numerous, small, sharp, triangular spines. Commonly dredged from 30 to 170 fathoms.
    A. elegans Dall (Elegant Star Turret) from about 200 fathoms off Key West is 2 inches in length, more elongate, with more numerous and duller spines on the sharp shoulder. Very rare.

Subfamily *CLAVINAE*

Shell between ¼ and ½ inch in length, spire tall and the anterior canal short. Sinus on the shoulder, moderately to deeply U-shaped, often rendered subtubular by a parietal tubercle. Operculum with an apical nucleus. Radula variable.

Genus *Crassispira* Swainson 1840

Crassispira ebenina Dall   Dall's Black Turret
Figure 57j

Southeast Florida and the West Indies.

¾ inch in length; a solid brown-black in color and with a slight sheen.

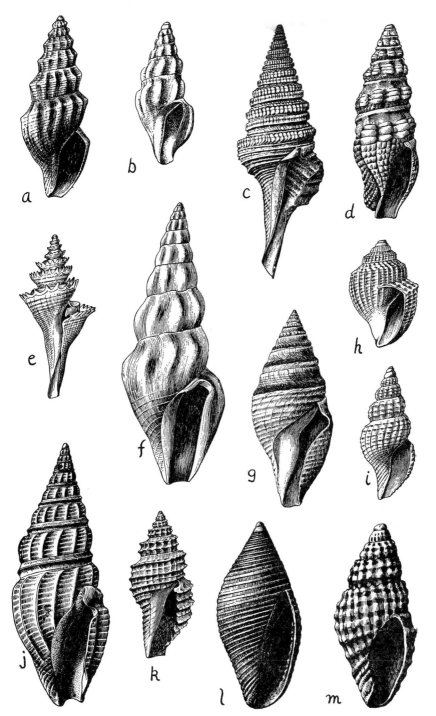

FIGURE 57. American Turret and Mangelia Shells. ATLANTIC: **a,** *Kurtziella limonitella* Dall, ⅜ inch; **b,** *Inodrillia aepynota* Dall, ½ inch; **c,** *Gemmula periscelida* Dall, 2 inches; **d,** *Monilispira leucocyma* Dall, ⅓ inch; **e,** *Ancistrosyrinx radiata* Dall, ½ inch; **f,** *Cerodrillia thea* Dall, ½ inch; **g,** *Genota viabrunnea* Dall, 2 inches; **h,** *Gymnobela blakeana* Dall, ¼ inch; **i,** *Mangelia morra* Dall, ¼ inch; **j,** *Crassispira ebenina* Dall, ¾ inch; **k,** *Mangelia corbicula* Dall, ⅓ inch. PACIFIC: **l,** *Mitromorpha filosa* Cpr., ⅜ inch; **m,** *Mitromorpha aspera* Cpr., ⅜ inch.

15 short axial ribs per whorl. Spiral threads numerous and fine. Sinus small, its posterior end round, its opening narrow. Not uncommon below low water under rocks.

C. sanibelensis Bartsch and Rehder is similar, but 1 inch in length, with 9 longer and wider axial ribs, with a large slit, and colored orange-chestnut with white between the ribs. Uncommon around Sanibel Island.

Subgenus *Crassispirella* Bartsch and Rehder 1939

Crassispira ostrearum Stearns   Oyster Turret
Plate 26n

North Carolina to south half of Florida. Cuba.

⅓ to ⅔ inch in length; light yellow-brown to chestnut. Sinus U-shaped. About 20 weakly beaded axial ribs per whorl. Just below the suture there is a single, smooth, strong spiral cord. Spiral threads moderately strong to weak (16 to 20 on the last whorl, 4 between sutures). Lower part of outer lip thin and strongly crenulate or wavy. Common from low water to 90 fathoms. *C. tampaensis* Bartsch and Rehder is very similar, and may be this species.

Genus *Cerodrillia* Bartsch and Rehder 1939

Cerodrillia perryae Bartsch and Rehder   Perry's Drillia

West coast of Florida.

½ inch in length, flesh-colored, with a broad, golden-brown band around the periphery. 8 to 9 axial ribs per whorl. Faint spiral lines present. Not uncommon. *C. thea* has shorter axial ribs and is uniform chocolate-brown.

Cerodrillia thea Dall   Thea Drillia
Figure 57f

West coast of Florida.

½ inch in length, thick-shelled, with a glossy-brown finish, and with the short, slanting ribs cream in color. Outer lip prominent. Sinus deep and U-shaped. Uncommon in shallow water.

Genus *Monilispira* Bartsch and Rehder 1939

Monilispira albinodata Reeve   White-banded Drillia

Southeast Florida and the West Indies.

½ inch in length, resembling a *Cerithium* in shape; color dark blackish brown with a white band bearing about 13 knobs per whorl. Last whorl with 2 or 3 spiral white bands. Fairly common in shallow water under rocks. *M. albomaculata* C. B. Adams is a similar species from the West Indies and is figured on plate 26f.

Monilispira leucocyma Dall

White-knobbed Drillia

Figure 57d

South half of Florida and the West Indies.

⅓ inch in length. One nuclear whorl smooth. Shell dark to light grayish brown with the nodules white. Aperture dark-brown. A common shallow-water species.

GENUS *Inodrillia* Bartsch 1943
Subgenus *Inodrillara* Bartsch 1943

Inodrillia aepynota Dall

Tall-spired Turret

Figure 57b

North Carolina to northeast Florida.

½ inch in length; chalk-white to pinkish white. Moderately common from 63 to 120 fathoms.

Subfamily *CONORBIINAE*

Shell conoidal; sinus broad and shallow, occupying the width of the shoulder. Operculum absent in *Genota*. Radula with 2 slender marginals only.

GENUS *Genota* H. and .A. Adams 1853

Genota viabrunnea Dall

Brown-banded Genota

Figure 57g

Southeast Florida and the West Indies.

1½ to 2 inches in length, heavy and thick-shelled. Sculpture of numerous spiral rows of very fine, glossy beads. Color yellowish to orangish white with a spiral, suffused band of light-brown well below the suture. Nucleus dark-brown and with tiny arched, smooth ribs. Anal sinus very wide. Rare from 100 to 350 fathoms.

Subfamily *MANGELIINAE*

Shell small, ovate or fusiform, with a short canal and without an operculum. Sinus on shoulder usually very shallow. Radula with 2 slender marginals.

Genus *Mangelia* Risso 1826

Mangelia morra Dall

Morro Mangelia

Figure 57i

Off north Carolina to Tortugas.

¼ inch in length, yellowish tan. Anal notch deep. 16 to 450 fathoms. Common. Provisionally placed in this genus. *Mangilia* is a misspelling.

Genus *Glyphostoma* Gabb 1872

Glyphostoma gabbi Dall

Gabb's Mangelia

Florida, the Gulf of Mexico and the West Indies.

⅓ inch in length. The 3 nuclear whorls are smooth and with a single, strong carina at the periphery. Shell white with 2 wide spiral bands of rose-brown on the whorl. The upper one is interrupted by about 15 short white ribs per whorl. Fine spiral threads numerous. Notch deep, with thickened, rounded sides. Varix strong. Moderately common from 30 to 150 fathoms.

Genus *Rubellatoma* Bartsch and Rehder 1939

Rubellatoma rubella Kurtz and Stimpson

Reddish Mangelia

North Carolina to southeast Florida.

¼ inch in length. Sinus shallow and U-shaped. Axial ribs long and rounded (about 9 per whorl). Spiral sculpture of numerous incised lines. Color grayish cream with light reddish between the ribs. Commonly dredged from 9 to 80 fathoms.

R. diomedea Bartsch and Rehder from Sanibel Island is extremely similar, but is more brightly colored with a wide spiral band of reddish brown. Uncommon to rare.

Genus *Kurtziella* Dall 1918

Kurtziella limonitella Dall

Punctate Mangelia

Figure 57a

North Carolina to both sides of Florida.

⅜ inch in length, semi-translucent and yellowish white. Sinus widely V-shaped. Between the strong, rounded, axial ribs there are numerous rows of microscopic opaque-white punctations. Uncommon from a few to 48 fathoms.

Subfamily BORSONIINAE

Shell biconic or fusiform in shape; sinus on the shoulder, poorly developed. Operculum present or absent. Radula with 2 slender marginals. Shell usually with columella plications.

GENUS *Gymnobela* Verrill 1884

Gymnobela blakeana Dall

Blake's Turret
Figure 57h

North Carolina to the Lower Florida Keys.

¼ inch in length. Sinus barely discernible. Shell thin but strong. Color translucent-white or chalky-white. Nuclear whorls distinct, without strong sculpturing and light-brown in color. Uncommon from 70 to 140 fathoms.

GENUS *Mitromorpha* P. P. Carpenter 1865

Mitromorpha filosa Carpenter

Filose Turret
Figure 57l

Monterey, California, to the Gulf of California.

⅜ inch in length, solid, light orange-brown in color. Spiral cords may be slightly beaded in some specimens. Uncommon offshore.

Mitromorpha aspera Carpenter

Beaded Turret
Figure 57m

Monterey, California, to the Gulf of California.

⅜ inch in length, strongly beaded and somewhat cancellate, with a glossy finish and light orange-brown in color. Moderately common offshore.

Subfamily DAPHNELLINAE

Shell fusiform or ovate, canal short. Operculum absent. Sinus adjoining the suture. The protoconch has diagonally cancellate sculpturing. Radula with 2 slender, curved marginals only.

GENUS *Daphnella* Hinds 1844

Daphnella lymneiformis Kiener

Volute Turret

Southeast Florida and the West Indies.

⅓ to ½ inch in length; resembles a miniature, elongate *Scaphella* volute-shell. With about 8 whorls, the nuclear ones smoothish, the next 4 with strong, axial ribs, but the last 2 whorls with only numerous fine spiral threads crossed by exceedingly fine growth lines. Aperture elongate, rather expanded and a little flaring below. Sinus moderately large and simple. Color cream with yellowish brown maculations. Uncommon from shallow water to 25 fathoms.

Subclass OPISTHOBRANCHIA
(Bubble-shells, Pteropods, Sea Slugs)
Order ONCHIDIATA
Family ONCHIDIIDAE
GENUS *Onchidella* Gray 1850

Without a shell, animal slug-like, low, oval, with two short tentacles or eyestalks at the end of which are the eyes. Mantle entirely covering the back; respiratory, anal and female genital pores at the posterior underside; male pore below the right tentacle and above the sensory lobe. Shallow water to intertidal. Formerly placed in the pulmonates, but now believed to be an early offshoot of the opisthobranchs. See Freter, 1943.

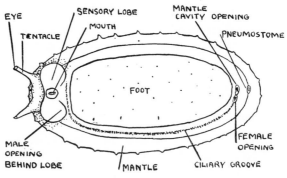

FIGURE 58. Underside of the marine slug, *Onchidella*, ½ inch.

Onchidella floridana Dall		Florida Onchidella

West coast of Florida, the Lower Keys and Bermuda.

½ inch in length, uniform slaty-blue to dark-gray; underside bluish white, with a greenish tinge to the veil. Dorsal surface velvety. Mantle margin with about 100 whitish, elongate tubercles. Common along the shore at low tide. Lives in rock crevices in nests, returning home after browsing at low tide.

Onchidella carpenteri Binney		Carpenter's Onchidella

Puget Sound to Lower California.

5 mm. in length; body oblong, with its ends circularly rounded; upper surface regularly arched; uniform smoke-gray in color. Fresh specimens are needed to make a better description. Littoral to shallow water. Habits not known.

Onchidella borealis Dall   Northwest Onchidella

Alaska to Coos Bay, Oregon.

8 to 12 mm. (½ inch) in length; back regularly arched but a little pointed in the middle, smooth or very finely granulose, tough and coriaceous. Color black or gray, with dots and streaks of yellowish white; foot light-colored, also the head and tentacles. On rocks near high-tide mark. Gregarious. Common.

Order *TECTIBRANCHIA*
(Bubble-shells, Sea-hares)
Family *ACTEONIDAE*
GENUS *Acteon* Montfort 1810

External shell with a prominent spire; cephalic disk divided; operculum thin, corneous. Erroneously spelled *Actaeon*.

Acteon punctostriatus C. B. Adams   Adams' Baby-bubble
Plate 26t

Cape Cod to the Gulf of Mexico and the West Indies.

3 to 6 mm. in length, solid, moderately globose, with a rather high spire. Columella with a single, twisted fold. Lower half of body whorl with numerous spiral rows of fine, punctate dots. Color white. Commonly found from low tide to 60 fathoms.

Acteon punctocaelatus Carpenter   Carpenter's Baby-bubble

British Columbia to Lower California.

10 to 20 mm. (¾ inch) in length, solid, oblong, 4 to 5 whorls, with two broad, ashy or brown spiral zones and about 26 spiral grooves on the body whorl. Columella obliquely truncated at base, and with one spiral fold. Base stained orange. Commonly found in shallow water in sand. *A. vancouverensis* Oldroyd is the same species.

Acteon candens Rehder   Rehder's Baby-bubble

North Carolina to southeast Florida and Cuba.

7 to 10 mm. in length, very similar to *punctostriatus*, but larger, very much thicker-shelled, glossy, opaque milk-white with light orange-brown suffusions on the body whorl. Commonly dredged in a few fathoms of water.

Family RINGICULIDAE
Genus *Ringicula* Deshayes 1838

Ringicula semistriata Orbigny   Orbigny's Helmet-bubble
                                                    Plate 26v
North Carolina to southeast Florida and the West Indies.

2 to 3 mm. in length, thick-shelled, resembling a miniature *Phalium* or Scotch Bonnet. 4 globose whorls, spire elevated. Aperture oblong; columella thickened by 3 folds, 1 above, 2 below. Outer lip very thick, swollen in the middle by a large tooth. Whorls, white, smooth, except for fine striations on the base. Not uncommonly dredged from 34 to 107 fathoms.

R. nitida Verrill (Verrill's Helmet-bubble from Maine to the Gulf of Mexico. 100 to 500 fathoms) is exteriorly smooth, with a simple, thickened outer lip, and with 2 smaller, spiral ridges on the columella.

Family HYDATINIDAE
Genus *Micromelo* Pilsbry 1894

Micromelo undata Bruguière   Miniature **Melo**
                                                    Plate 26u
Lower Florida Keys and the West Indies.

½ inch in length, oval, rather thin and moderately fragile. Characterized by its whitish to cream color overlaid by 3 widely spaced, fine spiral lines of red and by many or few axial, wavy, lighter red flammules or lines. Uncommon. Found at low tide.

Genus *Hydatina* Schumacher 1817

Hydatina vesicaria Solander   Brown-lined Paper-bubble
                                                    Plate 13q
South half of Florida and the West Indies.

1 to 1½ inches in length, very thin, fragile, globose. Periostracum thin, buff to greenish. Shell characterized by many close, wavy, brown spiral lines. Animal large and colorful. Foot very broad. Moderately common in certain shallow, warm-water areas where they burrow in silty sand. Formerly called *H. physis* Linné which, however, is believed to be limited to the Indo-Pacific.

Family DIAPHANIDAE
GENUS *Diaphana* Brown 1827

Diaphana minuta Brown — Arctic Paper-bubble
Figure 59b

Arctic Seas to Connecticut. Europe.

3 to 5 mm. in length, globose, thin, fragile, and transparent-tan in color. Last whorl globose below, constricted somewhat above. Apex large, globose, obliquely and mammillarly projecting. Suture deep. Columella long, straight, not thickened, the edge partly closing the narrow umbilicus. Moderately common from 6 to 16 fathoms. *Diaphana debilis* Gould, *D. hiemalis* Couthouy and *D. globosa* Loven are considered synonyms of this species by Lemche (1948) and other modern workers.

Family BULLIDAE
GENUS *Bulla* Linné 1758

The names *Vesica* Swainson 1840 and *Bullaria* Rafinesque 1815 have been ill-advisably used for this genus. Fortunately, the name *Bulla* has been conserved for this group of bubble-shells by the International Commission for Zoological Nomenclature.

Bulla striata Bruguière — Striate Bubble
Plate 13p

West coast of Florida to Texas and the West Indies.

¾ to 1 inch in length, similar to *occidentalis*, but larger, heavier, and with the spiral grooves well-marked toward the base of the shell and within the apical perforation. The whorls are compressed at the apical end. Columella usually with a brown-stained callus. Locally common. *B. amygdala* Brug. is probably a smooth form of this species.

Bulla occidentalis A. Adams — Common West Indian Bubble
Plate 26p

North Carolina to southeast Florida and the West Indies.

½ to 1 inch in length, smooth, varying from fragile to quite strong, and from cylindrical (young) to fairly swollen. Apex deeply and narrowly perforate. Color very variable, but usually whitish with mottlings, zebra stripes and obscure bands of brown. Surface with numerous, microscopic striations. This is a very common bubble-shell which is found most easily at night and at low tide on grassy, mud flats. The author is often misnamed as "C. B. Adams."

Bulla gouldiana Pilsbry   California Bubble

Santa Barbara to the Gulf of California.

1½ to 2 inches in length, rotund, fragile. Grayish brown with darker, streaked mottlings which are bordered posteriorly with cream. Periostracum dark-brown and microscopically crinkled. Collected abundantly at night. *Bulla punctulata* A. Adams from Lower California south is much heavier and constricted or narrowed at the top third of the shell.

Family ATYIDAE
Genus *Atys* Montfort 1810

Atys caribaea Orbigny   Caribbean Paper-bubble
                                                    Figure 59c

Southeast Florida and the West Indies.

⅓ to ½ inch in length, fragile, translucent milk-white, oval-oblong, smooth except for a dozen or so very fine, incised spiral lines at both ends.

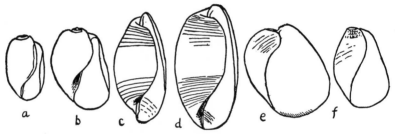

FIGURE 59. Paper-bubbles of the Atlantic Coast. **a**, *Retusa obtusa* Montagu; **b**, *Diaphana minuta* Brown; **c**, *Atys caribaea* Orbigny; **d**, *Atys sandersoni* Dall; **e**, *Philine quadrata* S. Wood; **f**, *Philine lima* Brown. All X5.

Spire concealed, marked by a twisted, spiral, funnel-like umbilicus. Columella acute, a little separated by a deep, narrow umbilicus. Common from shallow water to 90 fathoms.

Atys sandersoni Dall   Sanderson's Paper-bubble
                                                    Figure 59d

North Carolina to southeast Florida and the West Indies.

¼ to ⅓ inch in length, similar to *caribaea*, but thicker-shelled, with flatter sides, deeper and wider umbilicus, and with more numerous and finer spiral lines at each end. Fairly common from shallow water to over 100 fathoms.

Genus *Haminoea* Turton and Kingston 1830

Key to the Atlantic Species

(To determine on which side of the apical perforation the lip arises, hold the shell with the apex toward you and the apertural lip facing to the right.)

a. Apertural lip arising on the left side of the perforation, and angled near its insertion:

   b. Shell with numerous fine spiral grooves; ⅓ inch; yellowish to whitish; southeast Florida and the West Indies . *elegans* Gray

   bb. Shell smooth; ⅜ inch; West Indies .   . *glabra* A. Adams

aa. Apertural lip arising on right side; not angled:

   c. Well-grooved spirally:

      d. Sides of whorls globose; ½ inch; amber to whitish; Cape Cod to North Carolina; common .   . *solitaria* Say (pl. 26s)

      dd. Sides of whorls flattish; ⅓ to ½ inch, translucent-white; west Florida to Texas; common .   . . *succinea* Conrad

   cc. Spiral striae absent or excessively fine; ½ inch; translucent greenish yellow; globose; Gulf to West Indies . . *antillarum* Orbigny

Pacific Coast Species

Haminoea virescens Sowerby   Sowerby's Paper-bubble

Puget Sound to Mexico.

½ inch in length, very fragile, a translucent greenish yellow in color. Aperture very large and open. Upper part of outer lip high and narrowly winged. No apical hole. A common, littoral species on the open coast. *H. cymbiformis* Cpr. and *H. olgae* Dall are the same.

Haminoea vesicula Gould   Gould's Paper-bubble

Alaska to the Gulf of California.

¾ inch in length, very fragile, similar to *virescens*, but with a barrel-shaped whorl (from an apertural view), proportionately smaller aperture, with a tiny apical perforation, and with a lower, more rounded wing on the upper part of the outer lip. Shell color much the same, but the thin periostracum is often rusty-brown or yellowish orange. A common, littoral bay species.

Family RETUSIDAE
Genus *Retusa* Brown 1827

Retusa obtusa Montagu   Arctic Barrel-bubble

Figure 59a

Arctic Seas to off North Carolina.

3 mm. in length, fairly fragile, smooth, stubby and with the spire commonly slightly sunk or only a little elevated. Columella smooth. A chinklike umbilicus is present. Color translucent-white with yellowish brown staining. Common from shore to 90 fathoms. *R. pertenuis* Mighels and *R. turrita* Möller are this species.

Retusa sulcata Orbigny   Sulcate Barrel-bubble

North Carolina to southeast Florida and the West Indies.

2 mm. in length. Characterized by its small size, fine axial threads, white color, oblong shape, flat sides and deeply sunken spire. Moderately common from 3 to 95 fathoms.

Retusa canaliculata Say   Channeled Barrel-bubble

Plate 26x

Nova Scotia to Florida, Texas and the West Indies.

4 to 6 mm. in length, solid, oblong with its spire moderately elevated, but almost invariably eroded. Glossy smooth, except for microscopic growth lines. Outer lip thin, advanced above. Columella of a single, raised, strong spiral ridge. Suture slightly channeled. Nucleus (when present) very small and pimple-like. Color white to cream, commonly with dark, rust-brown staining. *Acteocina candei* Orbigny is probably only a southern representative of this species. A common shallow-water species which was formerly placed in the genus *Acteocina*.

Genus *Pyrunculus* Pilsbry 1894

Pyrunculus caelatus Bush   Bush's Barrel-bubble

Plate 26w

North Carolina to southeast Florida.

3 mm. in length, pyriform in shape, rather thick and opaque-white. Spire concealed within a very deep pit. Rather rare from 15 to 43 fathoms.

Genus *Rhizorus* Montfort 1810
(*Volvula* A. Adams, not Gistel)

Rhizorus oxytatus Bush   Southern Spindle-bubble

North Carolina to southeast Florida and Cuba.

3 to 4 mm. in length, fragile, translucent-white, spindle-shaped, with a sharp, long, spike-like apex. Glossy and with 4 or 5 very fine, indistinct, punctate, spiral lines at each end. Outer lip thin, following the curvature of the body whorl to just below the middle where it continues in a straight line. Umbilicus chink-like. Periostracum thin and pale-straw. Common from 5 to 100 fathoms. *R. bushi* Dall comes from deep water off North Carolina. It is larger, with a long apical process and with a long, chink-like umbilicus. The genus *Volvulella* Newton 1891 is *Rhizorus*.

Rhizorus acutus Orbigny

Spined Spindle-bubble
Plate 26l

North Carolina to southeast Florida and the West Indies.

2 to 3 mm. in length, spindle-shaped, rather oblong, fragile, except for minute spiral lines at each end. Upper end of aperture ends in a very sharp, rather prolonged spike. Umbilicus rarely, if ever, present. Commonly dredged down to 150 fathoms. *R. minutus* Bush is probably the young. *R. aspinosus* Dall from the same region has the process poorly developed, and may be a form of this species.

Family SCAPHANDRIDAE
GENUS *Scaphander* Montfort 1810

Scaphander punctostriatus Mighels

Giant Canoe-bubble
Plate 26-o

Arctic Seas to Florida and the West Indies.

1 to 1½ inches in length, very lightweight, but moderately strong. Ovate-oblong. Apex with a slightly sunken area. Aperture constricted above, roundly open below. Columella simple, rounded. Shell smoothish, except for numerous, spiral rows of microscopic, elongate punctations. Color chalk-white, with a straw periostracum. Fairly common from 20 to 1000 fathoms.

Scaphander nobilis Verrill (Noble Canoe-bubble) from off New England is the same size or smaller, has a proportionately much larger aperture, its outer lip is wing-like above, and the microscopic punctations are round. It is uncommon. A figure of *S. watsoni* Dall is on plate 26m.

Family ACTEOCINIDAE
GENUS *Acteocina* Gray 1847

Acteocina culitella Gould

Western Barrel-bubble

Kodiak Island, Alaska, to Lower California.

½ to ¾ inch in length, moderately solid, oblong but more constricted at the upper portions. Spire of 5 whorls, elevated, pointed and with a tiny, pimple-like nucleus (usually eroded in northern specimens). Suture narrowly and deeply channeled. Body whorl swollen at the lower half. With numerous microscopic, wavy, incised spiral lines. Color yellowish, sometimes with numerous golden-yellow, fine spiral lines. Columella is a single, raised spiral cord. Common in shallow water. *A. cerealis* Gould is probaby the same species.

GENUS *Cylichna* Loven 1846

Cylichna gouldi Couthouy　　　　　　　　　　Gould's Barrel-bubble

Plate 26r

Massachusetts Bay to off Cape Cod. Arctic Seas.

⅜ inch (9 mm.) in length, fragile, chubby, with the spire usually sunk in and consisting of 4 or 5 whorls. Color dirty-white with a yellowish periostracum. The whorls are much more globose and the anterior end more constricted than in the much smaller species, *Retusa obtusa*. Formerly placed in the genus *Retusa*. Uncommon from 26 to 34 fathoms.

Cylichna alba Brown　　　　　　　　　　　Brown's Barrel-bubble

Arctic Seas to North Carolina. Bering Sea to San Diego, California.

¼ inch (5 mm.) in length, fragile, narrowly oblong with flat sides. Apex with a dished, shallow depression. Upper ⅔ of aperture narrow; below it is wide. Columella short, rounded, slightly raised. Shell white, smoothish, except for microscopic, spiral scratches. Periostracum thin, shiny, yellowish, but often darkly stained with brown. Commonly dredged from 1 to 1000 fathoms in cold water.

Subgenus *Cylichnella* Gabb 1872

Cylichna bidentata Orbigny　　　　　　　Orbigny's Barrel-bubble

Plate 26q

North Carolina, Florida to Texas and West Indies.

3 mm. in length, somewhat resembling *alba*, but its columella has a spiral, callous fold and an indistinct nodule below. The shell is more oval. Glossy-white. Commonly found from shallow water to 200 fathoms. This is *C. biplicata* of authors, not A. Adams.

Family PHILINIDAE

Related to the Scaphander Canoe-shells, but different in having the mantle reflexed and closed over the shell, in lacking central teeth in the radula, and in having a much more degenerate shell.

GENUS *Philine* Ascanius 1772

Philine quadrata S. Wood　　　　　　　　　　　Quadrate Paper-bubble

Figure 59e

Arctic Seas to North Carolina.

⅓ inch in length, moderately fragile, semi-transparent, white, squarish-oval and more constricted toward the top. Aperture large, flaring, and rounded below. Early whorls very small. Sculpture of numerous spiral rows of microscopic oval punctations. Suture deep. The narrow top of the aperture is slightly higher than the apex. Commonly dredged off the New England states from 20 to 400 fathoms.

Philine lima Brown　　　　　　　　　　　　　File Paper-bubble

Figure 59f

Arctic Seas to Cape Cod, Massachusetts.

⅓ inch in length, much more oblong than *quadrata*, with the top of the aperture well below the apex, and sinuate from a top view. Columella fairly strong. Sculpture of spiral rows of scalloped lines forming chains, between which are a single scalloped line. Moderately common in fairly shallow but cold water. Alias *P. lineolata* Couthouy.

Philine sagra Orbigny　　　　　　　　　　Crenulated Paper-bubble

North Carolina to southeast Florida and the West Indies.

⅛ to ¼ inch in length, oblong, fragile, white, with a large aperture, with numerous spiral lines of small oblong rings placed end to end, and characterized by the finely crenulated lip. Top of the aperture the same height as the apex. Not uncommon from 15 fathoms down.

Family GASTROPTERIDAE
GENUS *Gastropteron* Kosse 1813

Shell entirely internal and consisting of a minute, nautiloid, calcareous spire. Body sack-shaped, with two large, wing-like, fleshy flaps, one on each side of the body. These peculiar, small sea-slugs swim through the water in a bat-like manner.

Gastropteron rubrum Rafinesque　　　　　　　Bat-wing Sea-slug

Figure 60e

West coast of Florida to the West Indies. Mediterranean.

⅓ to 1 inch in length. General color varying from red-purple to pale-rose, sometimes with bluish-white spots. There is a vivid, iridescent blue border on the head disk and the "wings." Found for the first time in the western Atlantic by Harold J. Humm in 1950 at Alligator Harbor, Florida. Rare? This is probably *G. meckeli* "Dall."

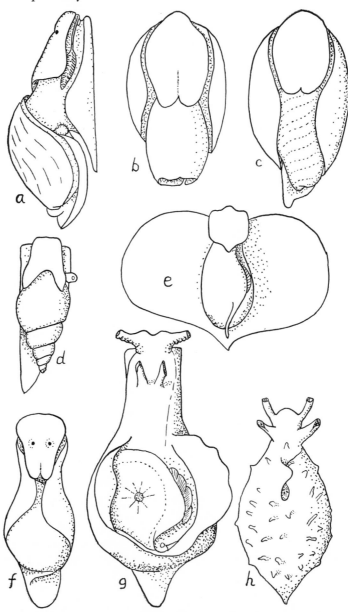

FIGURE 60. Animals of some Tectibranchs. **a,** *Haminoea* (side view, X3); **b,** *Philine* (X3); **c,** *Scaphander* (X2); **d,** *Acteocina* (X3); **e,** *Gastropteron* (X5); **f,** *Bulla* (X3); **g,** *Aplysia* (X½); **h,** *Bursatella* (X½). (After Guiart 1901.)

G. *pacificum* Bergh from the Aleutians is similar, but yellowish with red flecks. There are 16 to 20 gill leaflets. Margin of mantle without a flagellum, as in *rubrum*. Uncommon from 9 to 15 fathoms. *G. cinereum* Dall (British Columbia) is 11 mm. in length, and a uniform dusky-slate color. It also lacks a posterior flagellum on the mantle.

Superfamily *APLYSIACEA*
Family *APLYSIIDAE*
Genus *Aplysia* Linné 1767

Dorsal lobes free, well-separated and used for swimming. Shell internal, thin, flat, horny, with little or no lime, and colored amber. Skin smoothish. They give off a harmless purple ink. *Tethys* is a name which was for a long time applied to this group, but it is now restricted to a nudibranch genus. *Aplysia* is a conserved name (see fig. 60g).

Aplysia willcoxi Heilprin   Willcox's Sea-hare

Cape Cod to both sides of Florida.

5 to 9 inches in length. Mantle under the lobes with a minute perforation or fleshy tube above the area of the shell. Color dark-brown with slight maculations on the swimming lobes, head and neck. There are large, rounded, fairly regular, yellowish scallopings along the inner border of the lobes. Mantle and gills light-purple and yellow. Common. The form *perviridis* Pilsbry is clear green on the head and tentacles, the lobes olive-green with a coarse-meshed reticulation of black, subdivided by fine veins; irregularly maculated all over with light-green, with an occasional clumping of white dots.

Aplysia dactylomela Rang   Spotted Sea-hare

South half of Florida and the West Indies.

4 to 5 inches in length, characterized by its pale-yellow to yellowish-green color and the fairly large, usually irregular circles of violet-black scattered over the body. Common in some grassy localities. *A. protea* Rang of the West Indies is very similar, but the circles are more numerous, and often with smaller circles or large spots within the larger ones.

Aplysia floridensis Pilsbry   Sooty Sea-hare

Lower Florida Keys. The West Indies?

4 inches in length. Color deep purple-black, the inside of the swimming

lobes slightly lighter, and with blotches of black at the edges. Mantle purple-black, spotted irregularly with lighter purple. Uncommon?

Subgenus *Metaplysia* Pilsbry 1951

Aplysia badistes Pilsbry 1951   Walking Sea-hare

Biscayne Bay, Florida. And south?

2 to 4 inches in length. Mantle under the lobes with a large perforation. Sole of foot with a characteristic, muscular disk at each end. Exterior of animal dark-olive, indistinctly mottled with irregular spots of dusky buff, and having small, sparsely scattered, ragged black spots. Sole of foot yellowish olive. Found recently along the Venetian Causeway under rock ledges at low tide. (For details see *Notulae Naturae*, Philadelphia, no. 240, pp. 1 to 6, illustrated.)

GENUS *Bursatella* Blainville 1817

Bursatella leachi plei Rang   Ragged Sea-hare

West and northwest Florida and the West Indies.

4 inches in length, elongate-oval, plump, soft and flabby. Greenish gray to olive in color, sometimes with white flecks. Surface covered with numerous, ragged filaments. Shell absent in adults. Commonly found in grassy, mud-bottom areas at low tide. The east side of Sanibel Island is a good collecting spot. This is the only western Atlantic species known in this genus. Formerly placed in another genus, *Notarchus* Cuvier 1817.

Family PLEUROBRANCHIDAE
GENUS *Pleurobranchus* Cuvier 1805
Subgenus *Susania* Gray 1857

Pleurobranchus atlanticus Abbott   Atlantic Pleurobranch
                                              Figure 61

Southeast Florida (and the West Indies?).

1½ to 2 inches in length. Mantle with U-shaped notch in front where two tube-like rhinophores protrude up. Dorsum or back with numerous small rounded warts. Color yellowish orange with irregular splotches of deep maroon-brown. Largest warts translucent pale-yellow with a chocolate ring around the base. Some tipped with chalk-white. Foot pale-yellow to orangish. Gill plume on right side of body, with 20 to 22 primary leaflets on each side, with a nodule on the main stem where they originate. Primary leaflets with 15 smaller leaflets, each of which has 5 to 10 microscopic plates. Shell

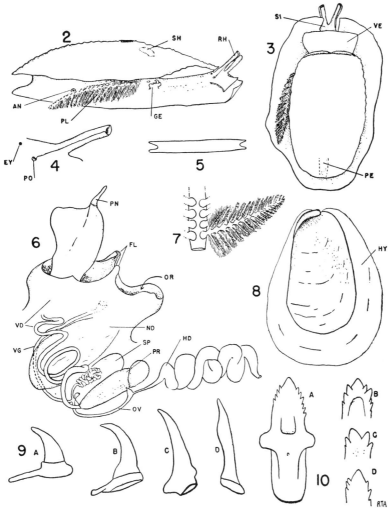

FIGURE 61. *Pleurobranchus atlanticus* Abbott, 1½ inches. 2 and 3, entire animal; 4, rhinophore; 5, cross-section of velum; 6, genitalia; 7, details of gill plume; 8, shell; 9, radula; 10, platelets in mandibles. (From R. T. Abbott 1949, *Nautilus* vol. 62.)

small, calcareous, pinkish white, flat with a small spire, and located under the dorsum. Moderately common in shallow water in winter on Soldier Key, near Miami. Originally collected by F. M. Bayer. *P. gardineri* White is a synonym.

Superfamily PYRAMIDELLACEA
Family PYRAMIDELLIDAE

This is a well-known family of very small gastropods which are extremely baffling to novices attempting to identify any one of the several hundred so-called species. Even among the experts there is not always agreement on what constitutes a species, subgenus or genus in this group. It would be impossible to present in a book this size even an account of only the most common species. Those interested in delving into this interesting maze of species are referred to the works of Bartsch, W. H. Dall and K. Bush.

The recent work of Fretter and Graham has shown that the Pyrams are ectoparasites, with each species feeding on a particular host, usually a tubicolous polychaete worm or a bivalve mollusk. The Pyrams attach themselves

---

ATLANTIC COAST PYRAMIDELLIDAE
(FIGURE 62)

(The names in parentheses are subgenera. Illustrations are from P. Bartsch, 1909, Proc. Boston Soc. Nat. Hist., vol. 34, no. 4. The measurements refer to the length of an average specimen).

a, *Odostomia (Chrysallida) willisi* Bartsch. Willis' Odostome. 3 mm.; milky white. Prince Edward Island, Canada. Uncommon.

b, *Turbonilla (Pyrgiscus) interrupta* Totten. Interrupted Turbonille. 6 mm.; pale waxy yellow. Casco Bay, Maine, to the West Indies. 2 to 107 fathoms. Common.

c, *Turbonilla (Turbonilla) stricta* Verrill. Varied Turbonille. 5 mm.; milky white. Massachusetts to North Carolina. 3 to 8 fathoms. Moderately common.

d, *Turbonilla (Turbonilla) nivea* Stimpson. Snowy Turbonille. 5 mm.; milky white. Maine to Connecticut. 40 to 400 fathoms. Uncommon.

e, *Pyramidella (Syrnola) fusca* C. B. Adams. Brown Pyram. 6 mm.; light brown. Gulf of St. Lawrence to Florida. Common.

f, *Odostomia (Menestho) trifida* Totten. Three-lined Odostome. 4 mm.; shiny white. Maine to New Jersey. Shore. Common.

g, *Odostomia (Iolaea) hendersoni* Bartsch. Henderson's Odostome. 3 mm.; glossy white. Woods Hole, Massachusetts. Uncommon.

h, *Odostomia (Menestho) bisuturalis* Say. Double-sutured Odostome. 5 mm.; milky white. Nova Scotia to Delaware Bay. Shore to 2 fathoms. Common.

i, *Odostomia (Menestho) impressa* Say. Impressed Odostome. 5 mm.; milky white. Massachusetts Bay to Gulf of Mexico. Common in shallow water.

j, *Odostomia (Chrysallida) seminuda* C. B. Adams. Half-smooth Odostome. 4 mm.; whitish. Nova Scotia to Gulf of Mexico. Shore to 12 fathoms. Common.

k, *Odostomia (Odostomia) gibbosa* Bush. Fat Odostome. 3 mm.; shiny, yellowish white. Maine to southern Massachusetts. Uncommon. (= *modesta* Bartsch, not Stimpson).

to the host by means of an oral sucker, and pierce the body wall of the host with a buccal stylet. They suck the host's blood by means of a buccal pump. Embryological and other data have shown that this family of mollusks is closely related to the tectibranch mollusks, rather than to the prosobranchs with which they have been formerly placed. (See *Journal of the Marine Biological Assoc.*, vol. 28, pp. 493-532, 1949.)

Genus *Pyramidella* Lamarck 1799

Pyramidella dolabrata Lamarck   Giant Atlantic Pyram
Plate 4q

Bahamas and the West Indies. Florida Keys?

¾ to 1 inch in length, solid and glossy smooth. Columella large, and with 2 or 3 strong, spiral plicae. Color opaque-white with 3 fine, spiral lines of brown, 1 of which is just above the suture. Common in the West Indies

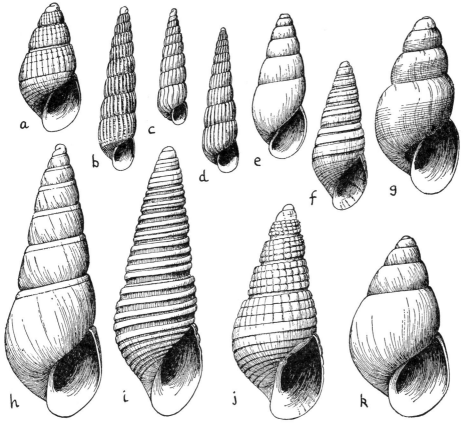

FIGURE 62. Atlantic Coast Pyramidellidae. (See opposite page.)

(The names in parentheses are subgenera. "D. and B." is the abbreviation for Dall and Bartsch 1909, *Bull. 68*, U. S. Nat. Mus., pls. 1-30, from which these drawings are taken. The measurements refer to the length of an average specimen.)

a, *Pyramidella (Lonchaeus) adamsi* Carpenter. Adams' Pyram. 15 mm. White to dark-brown, spotted or banded. San Pedro to Mexico. Common.

b, *Turbonilla (Chemnitzia) kelseyi* D. and B. Kelsey's Turbonille. 5 mm. Semi-transparent; ribs not on base of shell. Santa Barbara to Mexico. Shore to 30 fathoms. Moderately common.

c, *Turbonilla (Turbonilla) acra* D. and B. Acra Turbonille. 10 mm. Milk-white; 15 whorls; ribs extend down over the base of the shell. California. Rare.

d, *Turbonilla (Strioturbonilla) buttoni* D. and B. Button's Turbonille. 6 mm.; yellowish white; spire and base of shell with microscopic, wavy, spiral lines not shown in drawing). Southern California to Mexico. Shore to 25 fathoms. Common at many places.

e, *Turbonilla (Pyrgolampros) chocolata* Carpenter. Chocolate Turbonille. 12 mm.; shiny, golden brown, with 2 or 3 spiral bands of lighter color. Monterey to San Diego. Shore to 25 fathoms. Uncommon.

f, *Turbonilla (Mormula) tridentata* Carpenter. Three-toothed Turbonille. 10 mm.; chestnut; obscurely banded; minutely reticulated; 3 folds on white columella. Monterey to Lower California. Shore to 40 fathoms. Common.

g, *Odostomia (Evalina) americana* D. and B. American Odostome. 3 mm.; milky white. San Pedro to Lower California. Shore to 10 fathoms. Uncommon.

h, *Odostomia (Miralda) aepynota* D. and B. Tower Odostome. 2 mm.; translucent. San Pedro to Lower California. Shore to 12 fathoms. Uncommon.

i, *Odostomia (Odostomia) farella* D. and B. Farelle Odostome. 3 mm.; white; fine growth lines only. Off Long Beach. Rare.

j, *Odostomia (Evalea) phanea* D. and B. Phanea Odostome. 5 mm.; milky white. Monterey to San Diego. On rocks and abalones. Common.

k, *Turbonilla (Pyrgiscus) aragoni* D. and B. Aragon Turbonille. 7 mm.; milky white; lower half of whorls brown, upper half flesh-colored. Base with 15 spiral lines. Monterey to Redondo Beach. 10 to 40 fathoms. Uncommon.

l, *Turbonilla (Bartschella) laminata* Carpenter. Laminate Turbonille. 7 mm.; apex waxy yellow; last whorl brown; columella white. Redondo Beach to Lower California. Shore to 25 fathoms. Common.

m, *Odostomia (Ividella) pedroana* D. and B. San Pedro Odostome. 7 mm.; chocolate-brown. San Pedro to Lower California. Shore to 12 fathoms. Common.

n, *Odostomia (Chrysallida) helga* D. and B. Helga Odostome. 5 mm.; milky white. Redondo Beach to Gulf of California. Shore to 25 fathoms. Common.

o, *Odostomia (Ivara) terricula* D. and B. Earth Odostome. 4 mm.; milky white. Monterey to Lower California. Shore to 25 fathoms. Common.

p, *Odostomia (Iolaea) amianta* D. and B. Pure Odostome. 5 mm.; yellowish white. San Mateo to Lower California. Shore to 75 fathoms. Common.

q, *Odostomia (Amaura) nota* D. and B. Nota Odostome. 7 mm.; light yellow. Santa Rosa Island to San Diego. Among weeds. Common.

r, *Odostomia (Evalea) donilla* D. and B. Donille Odostome. 5 mm.; bluish white. Santa Monica to Lower California. Shore to 10 fathoms. Common.

s, *Odostomia (Menestho) fetella* D. and B. Fetelle Odostome. 4 mm.; milky white. Santa Monica to Lower California. Shore to 6 feet. Common.

t, *Odostomia (Salasiella) laxa* D. and B. Lax Odostome. 4 mm.; milky white. Catalina Island to Lower California. Shore to 70 fathoms. Common.

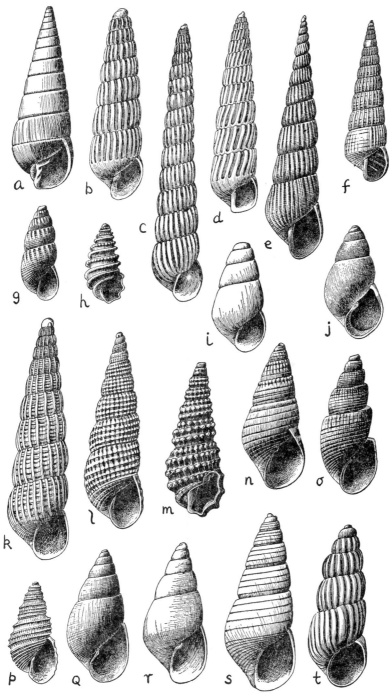

FIGURE 63. Pacific Coast Pyramidellidae. (See opposite page.)

and possibly present in the Lower Florida Keys. This species is a sand-dweller.

Order PTEROPODA
(Sea Butterflies or Pteropods)

These small, pelagic gastropods are very abundant in the open seas in nearly every part of the world. They are occasionally washed ashore, but more commonly their shells are found in dredge hauls. The identification of pteropods is important to many types of oceanographic studies. There are two suborders, *Thecosomata* or those having shells, and the *Gymnosomata* or those without shells. We have omitted the latter group, and refer interested workers to our bibliography. Every known American species (eastern Pacific and western Atlantic) of the shelled *Thecosomata* has been included and figured.

Suborder THECOSOMATA
Family SPIRATELLIDAE
GENUS *Spiratella* Blainville 1817
(*Limacina*)

Spiratella helicina Phipps   Helicid Pteropod
Figure 64a

Arctic Seas to the Gulf of Maine. Arctic Seas to northern California.

Up to 8 mm. in length, spire short, shell wider than long. Surface with relatively large, axial threads. Adults (over 3 mm.) without an operculum. Abundant enough in the Arctic Seas to serve as an important source of food for certain whales. *S. pacifica* Dall is the same.

Spiratella retroversa Fleming   Retrovert Pteropod
Figure 64c

Arctic Seas to Cape Cod, Massachusetts.

Up to 5 or 6 mm. in length, spire slightly elevated, umbilicus distinct, shell higher than wide. Entire surface covered with fine, spiral lines. 10 whorls. *Limacina balea* Möller and *Spirialis gouldi* Stimpson are this species.

Spiratella trochiformis Orbigny   Trochiform Pteropod
Figure 64e

Cape Cod, Massachusetts, to Brazil. (N. Lat. 42° to S. Lat. 28°).

1 mm. in length, very close in characters, except shape, to *S. retroversa*, and thought by some workers to be a warm-water subspecies of that species.

Spiratella lesueuri Orbigny Lesueur's Pteropod
Figure 64b

Cape Cod, Massachusetts, south to Brazil. Indo-Pacific. (N. Lat. 42°
to S. Lat. 40°).

1.5 mm. in length, spire elevated somewhat; umbilicus distinct. Shell as
long as wide. Spiral lines only around the umbilicus.

Spiratella bulimoides Orbigny Bulimoid Pteropod
Figure 64d

New York to southern Brazil. (N. Lat. 39° to S. Lat. 40°).

2 mm. in length, spire high, shell twice as long as wide. Umbilicus very
indistinct. Lip fragile and often broken. 6 to 7 whorls.

Spiratella inflata Orbigny Planorbid Pteropod
Figure 64h

Cape Cod, Massachusetts, to Argentina. (N. Lat. 42° to S. Lat. 40°).

1.5 mm. in length, spire depressed, with the globose whorls in one plane
to give a planorboid shape. *Limacina scaphoidea* Gould is this species.

Family PERACLIDAE
GENUS *Peracle* Forbes 1844

Shell fragile, with sinistral or left-handed whorls (resembling the fresh-
water pond snail, *Physa*); aperture very large and elongated; columella prolonged into an elongate rostrum; no umbilicus. Operculum thin, paucispiral,
sinistral and subcircular in outline. There are only two species in the genus.
Peraclis Pelseneer is the same genus.

Peracle reticulata Orbigny Reticulate Pteropod
Figure 64g

Worldwide, pelagic. (40° N. to 20° S.).

4 mm. in length, brownish yellow, sinistral and with 4 whorls. Suture
deep. The surface exhibits a raised hexagonal reticulation, the sides of the
hexagons bearing a regular row of minute teeth. *P. physoides* Forbes and
P. clathrata Eyd. and Soul. are the same.

Peracle bispinosa Pelseneer Two-spined Pteropod
Figure 64f

Atlantic, pelagic. (38° N. to 28° S.).

7 mm. in length, milky-white, similar to *reticulata*, but with a wide,

shallow suture bearing axial ridges, and with the shoulder of the outer lip bearing a small, triangular projection. Uncommonly collected.

Family *CAVOLINIDAE*

Shell symmetrical (not coiled), fragile, white to brown, and of various shapes—needle-like, cylinder-shaped, flattened triangular or bulbous.

GENUS *Creseis* Rang 1828

Shell a long cone, almost circular in cross-section, needle-like.

Creseis acicula Rang   Straight Needle-pteropod
                                                              Figure 64n
Atlantic and Pacific, pelagic. (N. Lat. 48° to S. Lat. 40°).

20 to 33 mm. (about an inch) in length. A long, straight, slender cone tapering to a sharp point. *Styliola vitrea* Verrill and *conica* Esch. are this species.

Creseis virgula Rang   Curved Needle-pteropod
                                                              Figure 64p
Atlantic and Pacific, pelagic. (N. Lat. 41° to S. Lat. 35°).

8 to 10 mm. in length. A drawn-out, slender shell similar to *acicula*, but with its narrow end hooked to one side. The amount of bend of hook is variable. *Hyalaea coniformis* Orb. and *Cleodora virgula* Soul. and Eyd. are the same.

GENUS *Styliola* Lesueur 1825

Styliola subula Quoy and Gaimard   Keeled Clio
                                                              Figure 64-o
Worldwide in warm seas, pelagic.

10 mm. in length, conical, straight, considerably elongated. The surface is smooth, and with a dorsal groove not parallel to the axis of the shell, but slightly oblique, turning from left to right, with only the anterior extremity (which ends in a rostrum) in the median line. There is only one species in the genus and it is world-wide in distribution.

GENUS *Hyalocylis* Fol 1875

Hyalocylis striata Rang   Striate Clio
                                                              Figure 64q
Worldwide in warm seas, pelagic.

8 mm. in length, conical, slightly compressed dorso-ventrally (oval in cross-section); apex slightly recurved dorsally; surface with transverse grooves; embryonic shell small, smooth, bulbous and separated from the main shell by a constriction. This is the only species in the genus.

GENUS *Clio* Linné 1767

Shell of a somewhat angular form, colorless, compressed dorso-ventrally, and with lateral keels. A cross-section of the anterior or open portion is thus always angular at the sides. There is generally a crest or ridge extending longitudinally along the back. Embryonic shell varies in form, but is always definitely separated from the rest of the shell. *Cleodora* Peron and Lesueur 1810 is the same genus.

Clio pyramidata Linné   Pyramid Clio
Figure 64k

Worldwide, pelagic.

16 to 21 mm. in length. No lateral keels on the posterior portion; without lateral spines. Lateral margins very divergent. No posterior transverse grooves. Dorsal ribs undivided. Common. *C. lanceolata* Lesueur and *Cleodora exacuta* Gould are this species. The shell exhibits considerable variation in form.

Clio cuspidata Bosc   Cuspidate Clio
Figure 64i

Atlantic and Indo-Pacific, pelagic.

Without lateral keels on the posterior portion. Lateral spines very long. Common.

Clio recurva Children   Wavy Clio
Figure 64j

Atlantic and Indian Ocean. Warm water, pelagic.

1 inch in length, with lateral keels over its entire length. 3 dorsal ribs markedly projecting. A large, fragile, transparent and very exquisite species. This is *C. balantium* Rang.

Clio polita Pelseneer   Two-keeled Clio
Figure 64l

Atlantic, pelagic.

With lateral keels over its entire length. Dorsal ribs very slightly projecting. The posterior portion of the shell is narrow. *C. falcata* Pfeffer is the same species.

Genus *Cavolina* Abildgaard 1791

Shell squat, bulbous, horny-brown in color, characterized by a much constricted aperture, which is, however, very broad transversely. Sides of shell often prolonged into spine-like projections. *Cavolinia* is an alternate, incorrect spelling, and *Hyalaea* Lamarck is a synonym of this genus.

Cavolina longirostris Lesueur Long-snout Cavoline

Figure 64v

Worldwide, pelagic. (47° N. to 40° S.).

5 to 9 mm. in length. Dorsal lip with a thin margin. Posterior portion of the ventral lip markedly projecting laterally. Common. *Hyalaea limbata* Orb. and *H. angulata* Souleyet are synonyms.

Cavolina gibbosa Rang Gibbose Cavoline

Figure 64w

Worldwide, pelagic. (43° N. to 38° S.).

About 10 mm. in length. Dorsal lip with a thin margin. Shell without appreciable lateral points. Ventral lip not more developed than the dorsal. Ventral surface with an anterior transverse keel. Common.

Cavolina tridentata Forskal Three-toothed Cavoline

Figure 64u

Worldwide, pelagic. (40° N. to 40° S.).

10 to 20 mm. in length. Dorsal lip with a thin margin. Ventral lip not more developed than the dorsal one. Shell without appreciable lateral points. The shell is as broad at the end of the lips as it is at the anterior end. *C. gibbosa* is narrower at the ends of the lips. *Hyalaea affinis* Orb. is merely a form of this species. *C. telemus* Linné might possibly be this species.

FIGURE 64. The Pteropods or Sea-butterflies of American Waters. **a**, *Spiratella helicina* Phipps; **b**, *S. lesueuri* Orb.; **c**, *S. retroversa* Fleming; **d**, *S. bulimoides* Orb.; **e**, *S. trochiformis* Orb.; **f**, *Peracle bispinosa* Pelseneer; **g**, *Peracle reticulata* Orb.; **h**, *Spiratella inflata* Orb.; **i**, *Clio cuspidata* Bosc; **j**, *Clio recurva* Children; **k**, *Clio pyramidata* Linné; **l**, *Clio polita* Pelseneer; **m**, *Cuvierina columnella* Rang; **n**, *Creseis acicula* Rang; **o**, *Styliola subula* Q & G; **p**, *Creseis virgula* Rang; **q**, *Hyalocylis striata* Rang; **r**, *Cavolina inflexa* Lesueur; **s**, *Cavolina quadridentata* Lesueur; **t**, *C. trispinosa* Lesueur; **u**, *C. tridentata* Forskal; **v**, *C. longirostris* Lesueur; **w**, *C. gibbosa* Rang; **x**, *C. uncinata* Rang.

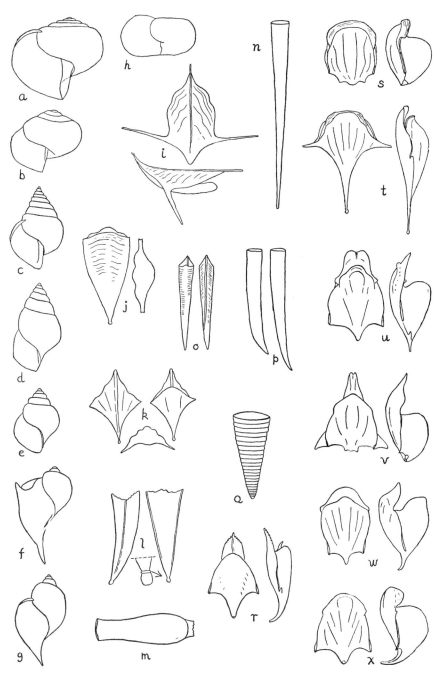

FIGURE 64. Explanations on opposite page.

Cavolina uncinata Rang   Uncinate Cavoline
                                                            Figure 64x

Worldwide, pelagic. (40° N. to 40° S.).

6 to 7 mm. in length. Dorsal lip with a thin margin. Ventral lip not
more developed than the dorsal one. Shell with distinct lateral points. Upper
lip flattened posteriorly. *C. uncinatiformis* Pfeffer is a synonym.

Cavolina inflexa Lesueur   Inflexed Cavoline
                                                            Figure 64r

Atlantic and Indo-Pacific. (41° N. to 42° S.).

6 to 7 mm. in length, similar to *uncinata*, but the upper lip is directed
straight forward, instead of flattened posteriorly; and the ventral side of the
shell is weakly, instead of strongly, convex. *C. labiata* Orb., *C. imitans* Pfeffer
and *C. elongata* Blainville are this species.

Subgenus *Diacria* Gray 1842

Similar to *Cavolina s.s.*, but the dorsal lip of the shell is thickened into
a pad, and not thin as the true Cavoline is. Some workers use this as a genus.

Cavolina trispinosa Lesueur   Three-spined Cavoline
                                                            Figure 64t

Worldwide, pelagic. (60° N. to 41° S.).

About 11 mm. in length. Dorsal lip thickened into a pad. Shell with a
long lateral spine on each side, and a very long terminal one. Aperture
scarcely discernible. Ventral side of shell very slightly convex. *C. mucro-
nata* Q. and G., *C. cuspidata* Delle Chiaje and *C. reeviana* Dunker are this
species. Very common.

Cavolina quadridentata Lesueur   Four-toothed Cavoline
                                                            Figure 64s

Worldwide, pelagic. (36° N. to 28° S.).

2 to 4 mm. in length. Dorsal lip thickened into a pad. Without promi-
nent lateral spines. Aperture well-developed. Ventral side greatly inflated.
Upper lip longer than the bottom one. *C. inermis* Gould, *C. minuta* Sowerby,
C. intermedia Sowerby and *C. costata* Pfeffer are synonyms. Quite common.

Genus *Cuvierina* Boas 1886

Shell cylindrical, shaped somewhat like a fat cigar. Surface smooth. A
cross-section is almost circular. Behind the aperture the shell is slightly con-

stricted. There is only one species in the genus. The genera *Cuvieria* Rang and *Herse* Gistel 1848 (non Oken 1815) are synonyms.

Cuvierina columnella Rang   Cigar Pteropod
                                                             Figure 64m

Worldwide, pelagic. (43° N. to 42° S.).

10 to 14 mm. in length. See generic description and figure. The shell varies somewhat in shape. *C. oryza* Benson, *C. urceolaris* Mörch and *cancellata* Pfeffer are the same. Common.

Suborder GYMNOSOMATA

Pteropods characterized by the absence of shell, pallial cavity and mantle-skirt; by the presence of a well-developed head, bearing two pairs of tentacles, of which the two posterior bear rudimentary eyes. Jaws and radula present. Found pelagic in all seas, and sometimes in great abundance. Rarely exceed one inch in length. They are carnivorous. Ascend to the surface at night, and sink to a lower level in the daytime. They are not treated in this book. The group contains such genera as *Pneumodermopsis* Bronn 1862, *Pneumoderma* Cuvier 1805 (= *Pneumonoderma* Agassiz), *Cliopsis* Troschel 1854 (= *Clionopsis* Bronn), *Notobranchaea* Pelseneer 1886, *Clione* Pallas 1774, *Paedoclione* Danforth 1907, *Anopsia* Gistel 1848 (= *Halopsyche* Bronn and *Euribia* Rang).

Order NUDIBRANCHIA
(Nudibranchs and Sea-slugs)
Superfamily DORIDACEA
Family DORIDIDAE

Branchial plumes in an arc or circle usually joined together at their bases, usually retractile into a cavity. Rhinophores always with a perfoliate club. Pharyngeal bulb never suctorial.

Genus *Archidoris* Bergh 1878

Body not hard, dorsum granular or tubular; tentacles short, thick, with an external, longitudinal sulcus. No labial armature. Branchial plumes not numerous, 3- to 4-pinnate. Center of radula naked, marginal teeth hooked and bearing minute denticles. Penis and vagina unarmed (without hooks).

Archidoris montereyensis Cooper   Monterey Doris
                                                             Plate 16h

California.

1 to 2 inches in length. Rhinophore stalks conical, the clavus slightly

dilated, conical, perfoliate with 24 to 30 leaves on each side. Each of the 7 branchial plumes large, spreading and 3- to 4-pinnate. Radula with 33 rows; center naked; with 42 to 49 strongly hooked, denticulate pleural teeth. Common in tide pools.

Subgenus *Anisodoris* Bergh 1898

Archidoris nobilis MacFarland   Noble Pacific Doris
Plate 16c

California.

4 inches in length. Rhinophore stalk stout, conical, the clavus perfoliate, with about 24 leaves, and the stalk deeply retractile within low sheaths, the margins of which are tuberculate. Each of the 6 branchial plumes large and spreading, 3- to 4-pinnate. A thin, membrane-like expansion joins the bases of the plumes. Radula with 26 rows; center naked; with 55 to 62 strongly hooked pleural teeth. Moderately common in tide pools.

GENUS *Discodoris* Bergh 1898

Body rather soft, oval in outline; branchial aperture slightly crenulate, stellate or bilabiate; anterior margin of the foot bilabiate, the upper lip more or less notched.

Discodoris heathi MacFarland   Heath's Doris
Plate 16i

California.

1 inch in length. Mantle thick, densely spiculate. Rhinophores cylindroconical, the stalk stout, the clavus with 10 to 15 leaves, and wholly retractile. Each of the 8 to 10 branchial plumes are tripinnate. Radula colorless, with 20 rows of teeth; center naked. 36 to 42 strongly hooked pleural teeth. Rather rare in rock pools in the summer.

GENUS *Rostanga* Bergh 1879

Back covered with minute, spiculose or stiff papillae; branchiae of simple-pinnate leaves.

Rostanga pulchra MacFarland   MacFarland's Pretty Doris
Plate 16g

California.

¾ inch in length. Rhinophores short, stout, translucent-pink, stalk stout, prolonged beyond the 20- to 24-leaved clavus as a blunt, cylindrical process which is ¼ the length of the entire rhinophore. 10 to 12 erect, separate, retractile branchial plumes. Radula with 65 to 80 rows of 80 denticulate pleural teeth. Lives on red sponges. The egg-ribbon is orange-red and often laid on the sponge. Common.

Genus *Diaulula* Bergh 1880

Body fairly soft; back silky finish; branchial aperture round and crenulate; branchial plumes tripinnate.

Diaulula sandiegensis Cooper   San Diego Doris
Plate 16d

Alaska to San Diego, California.

2 to 3 inches in length. Body soft; back velvety. Rings of black varying greatly in number and clarity (2 or 3 to 30). Rhinophores conical, the clavus with 20 to 30 leaves, deeply retractile into a conspicuous sheath with a crenulate margin. 6 branchial plumes tripinnate. Radula broad, with 19 to 22 rows, each row with 26 to 30 falcate teeth on each side of the naked center. Moderately common in rock pools of the fucoid zone at all seasons. The broad, white, spiral egg bands are commonly laid from June to August.

Genus *Aldisa* Bergh 1878

Aldisa sanguinea Cooper   Blood-red Doris

Monterey Bay to Point Lobos, California.

½ inch in length. In form, superficially resembling our figure of *Archidoris montereyensis* (pl. 16h), but bright-scarlet to light-red, sprinkled everywhere with very minute, black spots. Characterized by 2 or 3 very large, oval spots of black on the back. Rhinophores similar in form, with 12 to 15 leaves in the clavus. Branchial plumes 8 to 10, simply or irregularly bipinnate. Radula with 70 rows of teeth, each row with 70 to 100 teeth which are long, slender and with small, swollen bases. Not uncommon in rock pools.

Subfamily *CADLININAE*

Labial armature lamelliform, almost annulate, of extremely small hooks. Middle of radula with a denticulated tooth. External margin of pleural teeth serrate.

Genus *Cadlina* Bergh 1879

Characters of the subfamily. The glans penis is armed with a series of hooks. Usually the animal is from 1 to 2 inches in length.

Cadlina laevis Linné　　　　　　　　　　　　White Atlantic Doris

Arctic Seas to Massachusetts. Europe.

1 inch in length, similar to our figure of *A. nobilis* (pl. 16c), but a pure, waxy, semi-transparent white. Back with numerous very small, obtuse, opaque-white tubercles. An irregular row of white or sulfur-yellow, angular spots located down each side near the margin of the back. Rhinophores opaque-white or yellowish, with 12 or 13 leaflets, surmounted by a short, blunt point. Branchial plumes of 5 imperfectly tripinnate, transparent white plumes. Radula with 50 to 70 rows of teeth. 29 to 30 pleural teeth on each side of the central tooth, the latter with 3 to 4 denticles on each side of the center hook. Locally uncommon. *C. repanda* Alder and Hancock, *C. obvelata* Müller and *C. planulata* Gould are this species.

Cadlina flavomaculata MacFarland　　　　　　　Yellow-spotted Doris

Pacific Grove to San Diego, California.

¾ inch in length. Characterized by the 2 rows of lemon-yellow spots borne upon low tubercles. Rhinophores with 10 to 12 leaves in its club. Branchial plumes small, 10 to 11, either simple pinnate or bipinnate. Radula with about 77 rows of teeth, with 23 pleural teeth on each side of the central tooth which has 4 to 6 equal-sized denticles. All times of the year in small numbers in rocky tide pools.

Cadlina marginata MacFarland　　　　　　　　Yellow-rimmed Doris

British Columbia to Monterey Bay, California.

1½ inches in length, similar to our figure of *A. nobilis* (pl. 16c), but covered everywhere with low, yellow-tipped tubercles surrounded by a narrow ring of white and forming the center of a clearly marked polygonal area. Ground color a translucent yellowish white. There is a distinct narrow band of lemon-yellow around the margins of the mantle and the lateral and posterior edges of the foot. Rhinophores with 16 to 18 leaves in the clavus. Branchial plumes 6, bipinnate, sheath with yellow-tipped tubercles on the margin. 90 rows of teeth, with about 47 pleural teeth on each side of the central tooth which has 4 to 6 even-sized denticles. Not uncommon in rock pools.

Subfamily *GLOSSODORIDINAE*

Brilliantly blue-colored; back smooth, body elongate. Labial armature strong, of very minute hooks. Center of radula very narrow, often with minute, compressed spurious teeth. Penis unarmed. *Chromodoris* Alder and Hancock is a synonym of the following genus.

GENUS *Glossodoris* Ehrenberg 1831

Glossodoris porterae Cockerell   Porter's Blue Doris

Plate 161

Monterey to San Diego, California.

½ inch in length, characterized by its deep ultramarine blue (dissolves out at death) and by the two orange stripes. Foot without orange marks. Fairly common in rocky tide pools. This might be the young or a form of the next species.

Glossodoris californiensis Bergh   Californian Blue Doris

Monterey to San Diego, California.

2 inches in length. Like *G. porterae*, but with numerous, bright, orange, oblong spots in two rows on the mantle, another row down each side of the foot, and a group of round spots on the anterior end. Common in tide pools. *G. universitatis* Ckll. is this species.

Glossodoris macfarlandi Cockerell   MacFarland's Blue Doris

La Jolla to San Pedro, California.

¼ inch in length, like a small *porterae,* but with a ground color of reddish purple (not dissolving out at death); mantle with a yellow-orange margin and 3 longitudinal yellow stripes. End of foot with an orange stripe. Rare.

Family *DENDRODORIDIDAE*

Body soft, Doris-shaped. Pharyngeal bulb and elongated sucking tube, destitute of mandibles and radulae. Penis armed with a series of hooks. *Doriopsis* Pease is a synonym of the following genus.

GENUS *Dendrodoris* Ehrenberg 1831

Dendrodoris fulva MacFarland   Common Yellow Doris

California.

2 inches in length. Back soft, with low, papilla-like elevations tipped with white. Rhinophores with 18 to 20 leaves in the clavus which is ⅔ the

length of the entire rhinophore. It is completely retractile. 5 branchial plumes are tripinnate. No mandibles or radula. One of the commonest Pacific Coast species. In tide pools at all times of the year, especially common in summer. Coiled egg band is yellow.

Family *POLYCERIDAE*

Body limaciform (slug-like); branchial plumes not retractile.

GENUS *Laila* MacFarland 1905

Laila cockerelli MacFarland   Laila Doris
Plate 16j

Monterey to San Diego, California.

¾ inch in length. Rhinophores with 13 leaves in the clavus. 5 branchial plumes tripinnate, non-retractile into the cavity. 76 to 82 rows of radula; center with a series of rectangular, flattened plates; on the side are 2 pleural teeth, then 10 to 13 closely set pavement-like uncinal teeth. Glans penis long, armed with 10 to 12 irregular rows of minute, thorn-like hooks. Not very common. Found under shelving rocks in tide pools.

GENUS *Triopha* Bergh 1880

Triopha carpenteri Stearns   Carpenter's Doris
Plate 16k

Monterey to Point Lobos, California.

1 inch in length. Rhinophores with 20 to 30 leaves in the club. 5 branchial plumes, large, tripinnate. 30 to 33 rows of radulae, with 4 teeth on the center part (the rhachis); pleural teeth 9 to 18, strongly hooked. Uncinal teeth 9 to 18, quadrangular in outline. Very common in rock pools.

Triopha maculata MacFarland   Maculated Doris
Plate 16f

Monterey to Point Lobos, California.

1 inch in length. Rhinophore stalk and club same length, the latter with 18 leaves. Branchial plumes 5, tripinnate 14 rows of teeth, each row with 4 flattened plates, 4 to 5 pleurals, and 7 to 8 uncinal teeth. Blunt glans penis armed with minute hooks. Abundant in summer in rock pools, in winter un-common.

Triopha grandis MacFarland   MacFarland's Grand Doris
Plate 16b

California.

2 to 3 inches in length. With 8 to 12 tuberculate processes in front of head, and 6 to 7 more down the sides of the back. Back yellowish-brown, often flecked with bluish spots. Tips of yellow processes, tip of tail and tips of branchial plumes with yellowish red. Rhinophores set in conspicuous sheaths, club yellow with 20 leaves. Branchial plumes 5, bushy, tri- and quadri-pinnate. 18 rows of radular teeth; 4 centrals, 8 pleurals and 8 uncinal teeth. Found on brown kelp. Fairly common.

Genus *Polycera* Cuvier 1817

Frontal margin with finger-like processes. Finger-like processes bordering branchial plumes. Center of radula naked, flanked by 2 lateral teeth and several uncini.

Polycera atra MacFarland   Orange-spiked Doris
Plate 16e

Monterey to San Diego, California.

½ to 1 inch in length. The blue-black lines shown in our figure are usually thinner and less conspicuous. 8 gill plumes. Common on brown algae. 9 to 10 rows of radular teeth, dark-amber; 2 pleurals, 3 to 4 uncinal teeth.

Family *ONCHIDORIDAE*
Genus *Acanthodoris* Gray 1850

Body Dorid-like with a furry back. Labial disk armed with minute hooks. Center of radula naked; first pleural tooth large, external pleurals 4 to 8, small. Glans penis armed. Vagina very long.

Acanthodoris pilosa Abildgard   Pilose Doris
Plate 15b

Arctic Seas to New Haven, Connecticut. Alaska.

½ to 1¼ inches in length. Semi-transparent. Color variable, ranging from pure white to yellowish white, canary-yellow, yellowish brown, gray-speckled, purple-brown and black. Back covered with soft, slender, conical, pointed papillae, which give it a hairy appearance. Rhinophores long, its club bent backwards and with 19 to 20 leaves. Sheath denticulate. Branchial plumes 7 to 9, large and spreading, tripinnate, transparent. A number of color forms have been described from Alaska by Bergh and from New England by A. E. Verrill. Radula with about 27 rows. No central tooth, 4 pleurals on each side. Moderately common at low tide, sometimes found out of water.

Acanthodoris brunnea MacFarland   Pacific Brown Doris

Monterey Harbor, California.

¾ inch in length, somewhat like our figure of *A. pilosa* (pl. 15b). Somewhat broader at the anterior end. Brown tubercles on back rounder, fewer, not as pointed. Back brown with flecks of black and with small spots of lemon-yellow between the tubercles. Rhinophores deep blue-black, tipped with yellowish white. Club with 20 to 28 obliquely slanting leaves. 7 branchial plumes, wide-spreading, bipinnate. About 10 tubercles are included within the rosette, 4 or 5 of them large and enclosing the anal papilla. Anterior margin of back is yellow. 24 to 28 rows of radular teeth. No centrals. First pleural large, with 14 to 19 denticles on the inner border. 6 to 7 other smaller pleural teeth. Dredged 30 to 60 fathoms.

GENUS *Adalaria* Bergh 1878

Adalaria proxima Alder and Hancock   Yellow False Doris
Plate 15i

Arctic Seas to Eastport, Maine. Europe.

½ inch in length; deep yellow, white or yellow-orange. Back covered with stout, subclavate, or elliptical bluntly pointed tubercles, set at a little distance apart, and mixed with smaller ones. Calcareous spicules appear through the skin, radiating from the tubercles. Rhinophores with 15 leaves reaching almost to the base. Margin of sheath smooth. Branchial plumes 11. 40 rows of radular teeth. No central tooth. First pleural large, sickle-shaped, other 11 small and plate-like. Uncommon (?) in New England.

Family OKENIIDAE
GENUS *Ancula* Loven 1846

Ancula cristata Alder   Atlantic Ancula
Plate 15f

Arctic Seas to Massachusetts. Europe.

½ inch in length, of a transparent watery white, smooth. Rhinophores with 8 to 10 leaves. 3 branchial plumes, tripinnate. Labial armature of rows of imbricated hooks. Radula narrow, center naked, 25 to 27 rows of teeth; inner pleural large, denticulate on the inner margin. Outer pleural tooth small, smooth. *A. sulphurea* Stimpson is probably this species. On the northwest coast of Florida there is a *Polycera* (*hummi* Abbott) which superficially resembles this species in external features, but its radula indicates its true relationships.

Ancula pacifica MacFarland   Pacific Ancula

California.

½ inch in length, very similar to our figure of *A. cristata* (pl. 15f). Color translucent-yellow with 3 narrow, orange lines on the anterior half of the back, and one down the center of the back half. Rhinophores with 9 yellowish leaves. 3 branchial plumes. 4 (not 6) finger-like processes on each side of the plumes. 35 rows of teeth in the radula. Center with a small quadrangular plate, flanked by one large and one small pleural tooth.

GENUS *Hopkinsia* MacFarland 1905

Hopkinsia rosacea MacFarland   Hopkins' Doris
                                                                 Plate 16a
Monterey to San Pedro, California.

1 inch in length. Rhinophores long and tapering, the anterior side smooth along the entire length. ¾ of the posterior side bears about 20 pairs of oblique plates. Branchial plumes 7 to 14, entirely narrow and naked. 1 large pleural tooth on each side, flanked by a tiny, triangular pleural. Spiral egg ribbon rosy. Moderately common at all times of year under shelving rock between tide marks.

Superfamily *AEOLIDIACEA*
Family *DENDRONOTIDAE*
GENUS *Dendronotus* Alder and Hancock 1845

Body compressed; 2 tentacles laminated, with arborescent sheaths; numerous branchiae ramose. Arrow-shaped central tooth with a denticulate margin; about 9 elongate laterals on each side. About 40 rows of teeth.

Dendronotus frondosus Ascanius   Frond Eolis
                                                                 Plate 15e
Arctic Seas to Rhode Island. Alaska to Vancouver Island.

2 inches in length. Rhinophores with 5 or 6 large leaves, interspaced by about 15 smaller ones. Other characters as shown in our figure and the generic descriptions. *D. arborescens* Müller is this species. Common from shore to 60 fathoms.

Dendronotus giganteus O'Donoghue   Giant Frond Eolis

Northwest United States.

5 to 8 inches in length. Similar to our figure of *frondosus*. 16 to 18

leaves, all told, in the club of the rhinophore. Distinguished from *frondosus* by the 3 to 5 small but well-marked dendriform papillae on the posterior edge of the rhinophore sheath. Usually dredged down to 25 fathoms. Probably the largest of the American nudibranchs.

Family SCYLLAEIDAE
GENUS *Scyllaea* Linné 1758

Scyllaea pelagica Linné   Sargassum Nudibranch

Southeast United States. Other warm seas.

1 to 2 inches in length. Translucent cream-brown to orange-brown. With numerous flecks of red-brown. Body elongate. Oral tentacles absent. Two slender long rhinophores. Sides of body with 2 pairs of large, club-like, foliaceous gill plumes or cerata. Common in floating sargassum weed in the Gulf Stream.

Family AEOLIDIIDAE
GENUS *Aeolidia* Cuvier 1798

Body depressed, rather broad; branchiae a little flattened, set in numerous, close, transverse rows; 4 tentacles simple; foot broad, anterior angles acute. Radula of a single, broad, pectinate plate.

Aeolidia papillosa Linné   Papillose Eolis
                                                             Plate 15g

Arctic Seas to Rhode Island. Europe. Arctic Seas to Santa Barbara, California.

1 to 3 inches in length. Color variable: brown, gray or yellowish, always more or less spotted and freckled with lilac, gray or brown and opaque-white. Number of papillae fewer in young specimens. 30 rows in radula of a single, broad, arched tooth bearing about 46 denticles.

Family TERGIPEDIDAE
GENUS *Catriona* Winckworth 1941
(*Cratena* of authors)

Catriona aurantia Alder and Hancock   Orange-tipped Eolis
                                                             Plate 15j

Arctic Seas to Connecticut. Europe.

½ inch in length. Branchiae numerous, occurring in 10 or 11 close, transverse rows, anteriorly with 5 to 6 papillae per row, posteriorly with 2 to 4. Radula of 80 plates which are horseshoe-shaped and with 6 strong, straight denticles. Not too common. *C. aurantiaca* A. and H. is the same species.

GENUS *Tergipes* Cuvier 1805

Body slender; tentacles simple, the oral pair very short. Branchiae not very numerous, fusiform, inflated, set in a single series on each side of the back; foot narrow, anterior angles rounded. Egg mass kidney-shaped. Radula with a single row of plates, each with a stout central denticle and numerous delicate marginal denticles.

Tergipes despectus Johnston — Johnston's Balloon Eolis

Plate 15d

Arctic Seas to New York. Europe.

⅓ inch in length, characters as shown in our figure and in the generic description. Gregarious on hydroids. Shore to 8 fathoms. Common (?).

Family *GLAUCIDAE*
GENUS *Glaucus* Forster 1777

Glaucus marinus Du Pont 1763 — Blue Glaucus

Worldwide, pelagic in warm waters.

2 inches in length, body elongate, head small. Tentacles and rhinophores very small. 4 clumps of vivid-blue frills on each side of the body. Dorsal side smooth, striped with dark-blue, light-blue and white. Underside pale grayish blue. With strong jaws and a radula of a center row of about 10 denticulated teeth. Moderately common at certain seasons. Washed ashore with *Janthina*, the Purple Sea-snail. *G. atlanticus* Forster, *G. radiata* Gmelin and *G. forsteri* Lamarck are all this species.

Family *FLABELLINIDAE*
GENUS *Eubranchus* Forbes 1838

Resembling *Tergipes*. Radula with central plate with several stout denticles. A triangular lateral tooth is on each side of the central plate.

Eubranchus exiguus Alder and Hancock — Dwarf Balloon Eolis

Plate 15c

Arctic Seas to Boston, Massachusetts. Europe.

⅕ inch in length, characters as shown in our figure and the generic description. Seasonally uncommon.

Eubranchus pallidus Alder and Hancock

Painted Balloon Eolis
Plate 15h

Arctic Seas to Boston, Massachusetts. Europe.

½ inch in length, characters as shown in our figure and the generic description. Uncommon (?).

GENUS *Coryphella* Gray 1850

Branchiae numerous, clustered, elongate or fusiform. Foot narrow, with the anterior angles much produced. Radula with a single longitudinal series of central teeth which bear a large central denticle and several marginal denticles. There is a denticulated lateral tooth on each side of the central.

Coryphella rufibranchialis Johnston

Red-fingered Eolis
Plate 15a

Arctic Seas to New York. Europe.

1 inch in length, characters as shown in figure and in the generic description. Common (?) in New England.

Subclass *PULMONATA* Order *BASOMMATOPHORA*
Family *SIPHONARIIDAE*
GENUS *Siphonaria* Sowerby 1824

Shells closely resembling the true limpets, *Acmaea*, but at once distinguished by the nature of the muscle scars on the inside. In both, the long, narrow scar is horseshoe-shaped, but in *Siphonaria* the gap between the ends is located on one side of the shell, while in *Acmaea* it is located at the front end. In some *Siphonaria*, the area near the gap is trough-shaped. These are air-breathers and are more closely related to the land garden snails than to the gill-bearing, water-breathing limpets.

Siphonaria pectinata Linné

Striped False Limpet
Figure 65b

Eastern Florida, Texas, Mexico and St. Thomas.

1 inch in length, rather high, with an elliptical base. Exterior with numerous, fine, radial threads or rather smoothish. Color whitish with numerous, brown, bifurcating, radial lines. Interior glossy, similarly striped. Center cream to brown. Muscle scar with 3 swellings, the gap occurring between the two at the side. Do not confuse with *Acmaea leucopleura* which commonly has a blackish owl-shaped figure inside. Common along the shores on rocks. This is *S. naufragum* Stearns and *S. lineolata* Orbigny.

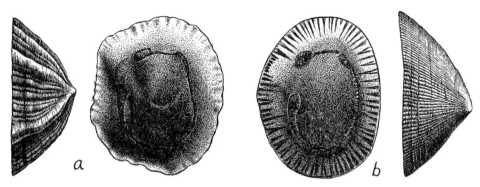

FIGURE 65. Atlantic Coast False Limpets. Side and interior views of a, *Siphonaria alternata* Say; b, *S. pectinata* Linné. Natural size.

Siphonaria alternata Say   Say's False Limpet
Figure 65a

Southeast Florida (and Sarasota), the Bahamas and Bermuda.

½ to ¾ inch in length, with about 20 to 25 small, white, radial ribs between which are smaller riblets. Background gray to cream. Interior glossy-tan, sometimes striped or mottled with dark-brown. Fairly common on rocks near the shore line.

CHAPTER VIII

Coat-of-Mails
and Other Chitons

Class AMPHINEURA

Order *LEPIDOPLEURIDA*
Family *LEPIDOPLEURIDAE*
GENUS *Lepidopleurus* Risso 1826

Lepidopleurus cancellatus Sowerby   Arctic Cancellate Chiton

Greenland to the Gulf of Maine. Bering Sea to Oregon.

½ inch in length, arched; color of exterior an orange-gray to whitish gray; interior white. Anterior valve microscopically granulated in radial rows. Central areas of the intermediate valves very finely granulated with densely placed, round pimples. Posterior valve with a smooth, slightly elevated central apex. Girdle narrow, same color as the valves and densely packed with tiny, split-pea scales. Some scales are commonly club-shaped, especially at the margins of the girdle, or sometimes so irregular and crowded as to give the appearance of fine moss. Moderately common on gravel bottoms from 20 to 100 fathoms.

Key to the Small Red Chitons (⅓ to 1 inch)
Girdle naked:

    Valves dull, color pattern maculated . . *Tonicella marmorea* Fabr.
    Valves glossy, with bright, black and white lines . . . . .
    . . . . *Tonicella lineata* Wood

Girdle with scales:

    With overlapping, split-pea scales . . *Lepidopleurus cancellatus* Sby.

With tiny, granular scales:

Interior of valves bright pink	*Ischnochiton ruber* L.
Interior of valves white	*Ischnochiton albus* L.

Order *CHITONIDA*
Family *LEPIDOCHITONIDAE*
GENUS *Tonicella* Carpenter 1873

Tonicella marmorea Fabricius   Mottled Red Chiton

Greenland to Massachusetts. Japan and the Aleutian Islands.

About 1 inch in length, oblong to oval, elevated and rather acutely angular. Colored a light-tan over which is a heavy suffusion of dark-red maculations and specks. Upper surface appears smooth, although under high magnification it is seen to be granulated. Lateral areas of intermediate valves not very distinctly outlined. Interior of valves tinted with rose. Posterior valve with 8 to 9 slits. Girdle is leathery and without scales or bristles. Superficially this species resembles *Ischnochiton ruber* which, however, has scales on its girdle. Common from 1 to 50 fathoms.

Tonicella lineata Wood   Lined Red Chiton

Japan to the Aleutians to San Diego, California.

About an inch in length, similar to *T. marmorea*, but with its valves smooth and shiny, and it is brightly painted with black-brown lines bordered with white which run obliquely backwards on the intermediate valves. The end valves have these same color lines concentrically arranged. Common on the rocky shores of Alaska. The young live in waters off the shore from 10 to 30 fathoms, but as they mature they migrate toward shore.

GENUS *Lepidochitona* Gray 1821
Subgenus *Cyanoplax* Pilsbry 1892

Lepidochitona dentiens Gould   Gould's Baby Chiton

Alaska to Monterey County, California.

½ inch or slightly more in length, oval, slightly elevated. Color tawny, olivaceous, slaty or brownish, usually covered with specklings of a darker hue. Upper surface of valves covered with microscopic, sharp granulations which are rarely aligned in any direction. Lateral areas may be slightly raised, and may be bounded in front by a very low rib. The apex of the posterior valve is near the center and is raised; behind the apex, the valve is concave. Girdle very narrow, same color as the valves, and with very minute, gritty granules.

L. (C.) dentiens is a common intertidal form found north of Monterey County, California. It is replaced to the south by the very similar and common *L. (L.) keepiana* Berry (Keep's Baby Chiton). In the former, the insertion teeth are prominently developed, their bounding slits in general *widely V-shaped,* and the teeth of the posterior valve very acute on the sides; the eaves are wide and conspicuously porous or "spongy." In the latter species, there are numerous short, *narrowly slitted* teeth in the terminal valves, and extremely thin, narrow, less openly porous eaves. (See S. S. Berry, 1948, *Leaflets in Malacology*, vol. 1, no. 4.)

Lepidochitona hartwegi Carpenter   Hartweg's Baby Chiton

Washington to Lower California.

1 to 1½ inches in length, oval, rather flattened. Similarly colored to *L. dentiens*. Girdle narrow and finely granulated. Sculpture of the end valves and the lateral areas of the middle valves differs from the microscopic granulations of *dentiens* in bearing easily seen, but very tiny, warts. It also differs in having the area behind the apex of the posterior valve convex instead of concave. Moderately common in intertidal areas.

GENUS *Nuttallina* Carpenter 1879

Nuttallina californica Reeve   Californian Nuttall Chiton

Vancouver Island to San Diego, California.

About 1 inch in length, almost 3 times as long as wide. Color dark-brown to olive-brown. Upper surface of valves finely granulated and with a shallow furrow on each side of the smooth dorsal ridge. Apex, or mucro, of posterior valve so far back that it extends beyond the posterior margin of the eaves. Interior of valves bluish. Posterior valve about as wide as long and with 8 to 9 slits. Girdle with short, rigid spinelets mostly brown in color and with a few white ones intermingled. The girdle looks mossy. Moderately common.

Nuttallina scabra Reeve   Rough Nuttall Chiton

San Diego to Lower California.

Very similar to *N. californica,* but the posterior valve twice as wide as long. Color of valves lighter. Girdle spines much less numerous. This is Carpenter's *Acanthopleura flexa*.

Family *MOPALIIDAE*
Genus *Mopalia* Gray 1847

Mopalia ciliata Sowerby   Hairy Mopalia

Alaska to Monterey, California.

1 to 1½ inches in length, oblong, usually colored with splotches of black and emerald-green, although sometimes having cream-orange bands on the sides of the valves. Sometimes grayish green with grayish black or white mottlings. Girdle colored yellowish brown to blackish brown. Valves slightly beaked; lateral area separated from the central area by a prominent, raised row of beads. Central areas with many coarse, wavy, longitudinal riblets, which are sometimes pitted between. Lateral areas coarsely granulated or wrinkled. Posterior valve small, with a deep slit on each side and a broad, deep notch at the very posterior end. Girdle fairly wide, generally notched at the posterior end and clothed with curly, strap-like brown hairs between which are much smaller, glassy white hairs or spicules. Interior of valves greenish white. Anterior valve granulated and with 8 to 9 coarse, raised rays of beads. A common intertidal species. The subspecies *wosnessenski* Middendorff (Alaska to Puget Sound) is supposed to be without the tiny white spicules in the girdle.

Mopalia muscosa Gould   Mossy Mopalia

Alaska to Lower California.

1 to 2 inches in length, oblong to oval. Very similar to *M. ciliata*, but differing in having a very shallow and small notch at the very posterior end. Color usually a dull-brown, blackish olive or grayish. Interior of valves blue-green, rarely stained with pinkish. Girdle with stiff hairs resembling a fringe of moss. The following species have been considered by some workers as varieties of *muscosa*, and perhaps with some justification: *lignosa* Gould, *hindsi* Reeve, *acuta* Carpenter, the latter having also been named *plumosa* and *fissa* by Carpenter. A common intertidal species.

Mopalia lignosa Gould   Woody Mopalia

Alaska to Lower California.

1 to 2½ inches in length, oblong. Color a grayish green or blackish green, rarely with whitish cream and brown, feathery markings. The sculpturing on the valves is very delicate and may consist only of numerous small pittings near the center. Concentric growth lines in smoother specimens are quite easily seen. Radial ribs absent on the end valves. Girdle solid or macu-

lated with browns and yellows. Strap-like, brown hairs not numerous. Interior of valves greenish white to white. Moderately common.

Mopalia hindsi Reeve   Hinds' Mopalia

Alaska to the Gulf of California.

2 to 3 inches in length, oblong, flattened and resembling *M. ciliata* Sowerby, but generally smoother. Girdle brown, rather thin and fairly wide, and almost naked except for a few short hairs. Interior of valves white with short crimson rays under the beaks. Moderately common.

Genus *Placiphorella* Dall 1878

Placiphorella velata Pilsbry   Veiled Pacific Chiton
                                                         Figure 66a
Monterey, California, to Lower California.

1 to 2 inches in length, readily recognized by its flat, oval shape and wide girdle which is very broad in front. There are a few hairs on the girdle which, if viewed under a lens, will be seen to be covered by a coat of diamond-shaped scales. Girdle reddish yellow. Valves colored a dull olivaceous brown with streaks of buff, blue, pink or chestnut. Interior of valves white with a slight bluish tint. Posterior valve with 1 to 2 slits and with very large sutural plates. Fairly common intertidally.

Genus *Katharina* Gray 1847

Katharina tunicata Wood   Black Katy Chiton
                                                         Figure 66b
Aleutian Islands to southern California.

2 to 3 inches in length, oblong and elevated. Characterized by its shiny, naked, black girdle which covers ⅔ of each gray valve. Valves usually eroded. Anterior valve densely punctate. Interior of valves white. Foot salmon to reddish. Very common between tides, especially in the north.

Genus *Symmetrogephyrus* Middendorff 1847
(*Amicula* Gray 1847, not 1842)

Symmetrogephyrus vestitus Broderip and Sowerby   Concealed Arctic Chiton
                                                         Figure 66c
Arctic Seas to Massachusetts Bay. Arctic Seas to the Aleutian Islands.

1 to 2 inches in length, oval, rather elevated. Valves covered with a thin, smooth, brown girdle except for a small, heart-shaped exposure at the

center of each valve. Girdle may have widely scattered tufts of hair. Interior of valves white. Common from 5 to 30 fathoms.

S. pallasi Middendorff (Concealed Pacific Chiton) is very similar to *vestitus*, but the girdle is much thicker and the bunches of reddish hairs more numerous. Uncommon from 3 to 10 fathoms.

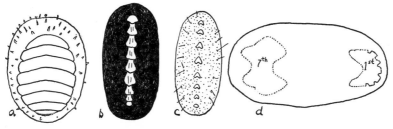

FIGURE 66. Pacific Coast chitons. **a,** *Placiphorella velata* Pilsbry, 1 to 2 inches (California); **b,** *Katharina tunicata* Wood, 2 to 3 inches (Pacific Coast); **c,** *Symmetrogephyrus vestitus* Brod. and Sby., 1 to 2 inches (Arctic waters to Massachusetts and Alaska); **d,** *Amicula stelleri* Midd., 6 to 12 inches (Pacific Coast), showing position of the 1st and 7th valves only.

GENUS *Ceratozona* Dall 1882

Ceratozona rugosa Sowerby   Rough Girdled Chiton

Figure 67a

East Florida to the West Indies.

1 to 2 inches in length, oblong, slightly beaked. Surface commonly eroded, whitish gray with blue-green to moss-green mottlings on the sides of the calves. Surface roughly sculptured. Anterior valve with 10 to 11 strong, rugose, radiating ribs. Lateral areas bounded in front and behind by a large, rugose rib. Central area with low, rough, longitudinal ribs. Interior of valves bluish green. Girdle leathery, yellowish brown and with numerous, yellowish brown clusters of strap-like hairs. Posterior valve rather small and with 8 to 10 slits. 35 to 36 gill lamellae. The gills extend the length of the foot, but do not go as far as the posterior end. Very common, especially in the Greater Antilles.

Family CRYPTOPLACIDAE
GENUS *Cryptoconchus* Burrow 1815

Cryptoconchus floridanus Dall   White-barred Chiton

Figure 67b

Southeast Florida to Puerto Rico.

Rarely over an inch in length, long and narrow, and characterized by its thin, black, naked girdle which extends on to the valves except over the narrow, beaded dorsal area. These exposed bands in the valves make it appear as if a streak of white paint had been applied along the top of the animal. The side of the girdle at each valve-suture has a minute pore bearing short bristles, but these 2 features are commonly difficult to see. A variety is found with a brown-colored girdle. 16 gill lamellae. They begin halfway back along the side of the foot. Uncommon.

Genus *Acanthochitona* Gray 1821

Acanthochitona spiculosa Reeve   Glass-haired Chiton

Southeast Florida and the West Indies.

1 to 1½ inches in length, elongate, with the girdle covering most of the valves. There are 4 clumps of long, glassy bristles near the anterior valve and one on each side of the other valves. The clumps are set in cup-like collars of the girdle skin. End valves and lateral areas of middle valves covered with tiny, round, sharply raised pustules. The dorsal, longitudinal ridge is raised, narrow, distinct and smoothish except for microscopic pin-points. Lower edge of girdle with a dense fringe of brown or bluish bristles. 32 gill lamellae. The gills begin about ⅓ back along the side of the foot and do not extend quite so far back as the posterior mantle margin. A moderately common species in shallow water. *A. astriger* Reeve is the same species.

Acanthochitona pygmaea Pilsbry   Dwarf Glass-haired Chiton

West Coast of Florida to the West Indies.

½ to ¾ inch in length, moderately elongate and colored cream, green, brown or variegated with these colors. Similar to *A. spiculosa*, but smaller, and with its dorsal ridge triangular, less elevated and cut by longitudinal grooves. The pustules on the lateral areas and the end valves are round or oval. The clumps of bristles are the same. Not uncommonly found among rocks and dead shells at low tide.

Acanthochitona balesae Pilsbry 1940 from the Lower Keys is very elongate, only ⅓ inch in length; the pustules on the lateral areas are proportionately larger and fewer, and the dorsal ridge is rounded and covered with small, granulose pustules. Rare at Bonefish Key, Florida.

Genus *Amicula* Gray 1842
(*Cryptochiton* Middendorff 1847)

Amicula stelleri Middendorff   Giant Pacific Chiton
Figure 66d

Japan and Alaska to California.

6 to 12 inches in length, oblong and flattened. The large, white, butter-

fly-shaped valves are completely covered by the large, leathery, firm girdle which is reddish brown to yellowish brown. Minute red spicules make the girdle feel gritty. Common in the northern part of its range. Formerly called *Cryptochiton stelleri*.

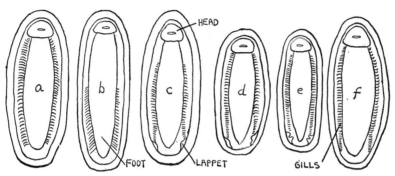

FIGURE 67. Atlantic Coast chitons, showing from the underside the position and length of gills and the nature of the lappets. **a**, *Ceratozona rugosa* Sby., 1 to 2 inches; **b**, *Cryptoconchus floridanus* Dall, 1 inch; **c**, *Calloplax janeirensis* Gray, ¾ inch; **d**, *Chaetopleura apiculata* Say, ½ inch; **e**, *Tonicia schrammi* Shuttleworth, 1 inch; **f**, *Chiton tuberculata* Linné, 3 inches.

Family *ISCHNOCHITONIDAE*
GENUS *Calloplax* Thiele 1909

Calloplax janeirensis Gray   Rio Janeiro Chiton

Figure 67c

Lower Florida Keys and the West Indies. Brazil.

½ to ¾ inch in length, oblong, gray to greenish brown, or speckled with red. Very strongly sculptured. Lateral areas strongly elevated by 3 to 4 very coarse, large, beaded ribs; anterior valve with 12 to 18 such ribs. Central ridge (or jugal tract) with longitudinal rows of fine beads; apex elevated, smooth and rounded. Central area with about 12 very sharp, granulose, longitudinal ribs. Interior white. Anterior valve with 10, middle valves with 1 and posterior valve with 9 slits. Girdle with very fine "sugary" scales and an occasional single hair. Gills start ⅓ the way back from the head and extend posteriorly to a large, fleshy lappet on the posterior margin of the girdle. An uncommon species.

GENUS *Chaetopleura* Shuttleworth 1853

Chaetopleura apiculata Say   Common Eastern Chiton

Figures 67d; 68

Cape Cod, Massachusetts, to both sides of Florida.

⅓ to ¾ inch in length, oblong to oval. Valves slightly carinate. Central areas with 15 to 20 longitudinal rows of raised, neat beads. Lateral areas distinctly defined, raised and bear numerous, larger, more distantly spaced beads which may or may not be present on the more dorsal region. Interior white or grayish. Slits of anterior valve 11, central or middle valves 1, posterior valve 9 to 11. Girdle narrow, mottled cream and brown, microscopically granulose and with sparsely scattered, transparent, short hairs. 22 to 24 gill lamellae in each gill which start just behind the juncture of the head and foot and extend all the way back to the posterior margin of the mantle where there is located a small, single-lobed lappet. Common from 1 to 15 fathoms.

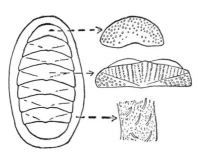

FIGURE 68. Common Eastern Chiton, *Chaetopleura apiculata* Say, ½ inch.

In the north, the exterior color is buff to ashen, rarely reddish. On the west coast of Florida, where they are commonly found attached to *Pinna* shells, the colors vary from light-gray, mauve, yellow to white, and are commonly with a darker or lighter streak down the center or rarely with longitudinal blue stripes.

GENUS *Ischnochiton* Gray 1847
Subgenus *Stenoplax* Carpenter 1878

Ischnochiton floridanus Pilsbry   Florida Slender Chiton

Miami to Dry Tortugas, Florida.

1 to 1½ inches in length, about 3 times as long as wide, elevated, with the valves roundly arched, not carinate. Color whitish to whitish green with markings of olive, blackish olive or gray. Lateral areas raised and with wavy, longitudinal riblets which are commonly strongly beaded. Central areas with wavy, longitudinal ribs. Interior of valves mixed with white, blue and pink, rarely all pink or all white. End valves concentrically (or rarely axially) beaded. Intermediate valves with 1 slit, posterior valve with 9. Girdle marbled with bluish and gray, and densely covered with round, solid, finely striated scales. Moderately common.

Ischnochiton purpurascens C. B. Adams (Purplish Slender Chiton) from the West Indies is very similar, but the end valves and lateral areas have smooth, instead of beaded, wavy, concentric riblets. Common.

Ischnochiton papillosus C. B. Adams   Mesh-pitted Chiton

Tampa to the Lower Keys and the West Indies.

⅓ to ½ inch in length, oval. Moderately sculptured and without very distinct lateral areas. It has microscopic, even, quincunx pittings on the upper surfaces of the valves. End valves with concentric rows of fine, low beads. Lateral areas with fine, wavy, longitudinal, incised lines. Posterior slope of posterior valve is concave and with 9 slits. Color whitish with heavy mottlings of olive-green; rarely with white spots. Girdle narrow, colored with alternating bars of white and greenish brown. Scales like microscopic split peas which are finely striated. A fairly common shallow-water species.

Ischnochiton magdalenensis Hinds   Magdalena Chiton

Coos Bay, Oregon, to Lower California.

2 to 3 inches in length, elongate. Color a drab greenish, commonly whitish due to wear. Central areas with fine, irregular, longitudinal cuts and with diamond-shaped pits near the lateral areas. Lateral areas prominently raised and with 10 to 12 coarse, radial ribs of gross, low beads. Front slope of anterior valve straight. Interior of valves bluish with the posterior end of each whitish. Girdle rather narrow and with alternating, faint bars of brown and yellowish brown. Scales round and so small that the girdle has the texture of fine sandpaper. Common.

Ischnochiton conspicuus Pilsbry   Conspicuous Chiton

San Miguel Island, California, to the Gulf of California.

2 to 6 inches in length, very similar to *magdalenensis*, but the front slope of the anterior valve is very concave; the central areas are practically smooth and with flecks of green. Scales are elongate, hard, and so densely packed that the girdle feels velvety. Interior of valves pinkish and blue. Moderately common between tides.

Ischnochiton acrior Pilsbry   Acrior Chiton

San Pedro, California, to Lower California.

2 to 6 inches in length, very similar to *magdalenensis*. Instead of diamond-shaped pittings on the sides of the central areas, there are wavy ribs. The front slope of the anterior valve is very concave, as in *conspicuus*. The scales are similar to those in *magdalenensis*, but much larger. Interior of valves pinkish with a blue spot at the anterior end of each valve. Moderately common.

Ischnochiton regularis Carpenter   Regular Chiton

Southern California.

1 to 1½ inches in length, oblong, appears smooth to the naked eye. Color an even slate-blue or uniform olive-blue. Valves slightly carinate. Central areas with very fine, longitudinal threads. Lateral areas slightly raised and with radial threads. Interior of valves gray-blue. Girdle with very tiny, closely packed, low, round scales. Moderately common between tides.

Subgenus *Lepidopleuroides* Thiele 1928

Ischnochiton albus Linné   White Northern Chiton

Arctic Seas to Massachusetts. Europe. Arctic Seas to off San Diego, California.

About ½ inch in length, oblong, moderately elevated. Upper surfaces smoothish except for irregular, concentric growth ridges and a microscopic, sandpapery effect. Color whitish, cream, light-orange or rarely marked with brown. Interior of valves white. Posterior valve with 12 to 13 weak slits. 17 to 19 gill lamellae on each side, beginning about halfway alongside the foot. Girdle sandpapery, with tiny, closely packed, gravelly scales. Common from shore to several fathoms in cold water. Distinguished from *ruber* by the anterior slope of the anterior valve which is straight to slightly concave in *albus*, but convex in *ruber*.

Ischnochiton ruber Linné   Red Northern Chiton

Arctic Seas to Connecticut. Europe. Alaska to Monterey, California.

½ to 1 inch in length, oblong, moderately elevated and with the valves rather rounded. Upper surfaces smooth except for growth wrinkles. Colored a light-tan over which is a heavy suffusion of orange-red marblings, or entirely suffused with red. Interior of valves bright pink. Posterior valve with 7 to 11 slits. Girdle reddish brown with weak maculations; covered with minute, elongate scales which do not overlap each other. 15 to 18 gill lamellae, similar to those in *albus*. Common from 1 to 80 fathoms. Do not confuse with *Tonicella marmorea* Fabr. whose girdle is naked.

Subgenus *Lepidozona* Pilsbry 1892

Ischnochiton mertensi Middendorff   Merten's Chiton

Aleutians to Lower California.

1 to 1½ inches in length, rather oval in shape. Color variable: commonly yellowish with dark reddish brown streaks and maculations. Central areas with strong, longitudinal ribs and smaller, lower cross ridges which give a netted appearance. Jugal area V-shaped and with 5 to 6 smooth longitudinal ribs. Lateral areas raised, smoothish and with a few prominent warts. Anterior valves with 30 or more radial rows of warts which are largest near the girdle. Interior whitish, or rarely tinged with pink. Girdle with alternating yellowish and reddish bands; covered with tiny, low, smooth, split-pea scales. Very abundant just offshore, especially in the northern part of its range.

Ischnochiton californiensis Berry   Trellised Chiton

Southern California to Lower California.

1 to 1½ inches in length, oval to oblong, heavily sculptured. Color a dull-greenish with yellowish splotches and with a dark-brown area on the top of each valve. Central area with longitudinal and cross ribs which give a strong netted appearance. Lateral areas raised and with 4 rows of prominent beads. Posterior edge of valves serrated with about 20 small tooth-shaped beads. Anterior valve with 20 to 27 strongly granular ribs. Girdle closely packed with convex, tiny, split-pea scales. Moderately common and formerly thought to be *I. clathratus* Reeve which, however, is only from the Panamic Province to the south.

Ischnochiton cooperi Pilsbry   Cooper's Chiton

Southern California.

1 to 1½ inches in length, rather oval in shape. Color olive-green to olive-brown and clouded with light-blue. Central area with closely packed, sharp, longitudinal ribs which are finely striated. Jugal area with the same type of ribs and with its anterior end having about 10 notches. Lateral areas raised and with 4 to 8 irregular rows of prominent, rounded warts. Interior of valves bluish. Girdle covered with tiny, flat, striated, split-pea scales. Uncommon.

Ischnochiton palmulatus Pilsbry   Big-end Chiton

Southern California.

½ inch in length, oblong and with the posterior valve massive and greatly swollen. Color of valves yellowish brown to light grayish green with dark blackish green in the areas just above the girdle. Central areas carinate at the top, with 20 to 30 strong, rounded, longitudinal ribs. Lateral areas greatly raised and with 2 convex, strong ribs of coarse beads. Anterior valve with 9 convex, beaded ribs. Posterior valve very high, convex and with 7 to 8 pairs of strong ribs. Interior bluish white. Girdle yellowish with reddish bands; thin, narrow and with microscopic, striated scales. Uncommon.

Family CHITONIDAE
Subfamily CHITONINAE
GENUS *Chiton* Linné 1758

Outer edges of insertion plates with tiny, sharp teeth or pectinations; girdle with small, hard scales that look like overlapping split peas; girdle colored with alternating bars of grayish green and black; interior of valves blue-green.

Key to the American *Chiton*

Valves entirely smooth:

Girdle scales glassy; sinus of valves narrow . . *laevigatus* (Pacific)
Girdle scales dull; sinus of valves wide . . *marmoratus* (Atlantic)

Valves with sculpturing:

Central area with longitudinal ribs:

Posterior valve with round pimples . . *tuberculatus* (Atlantic)
Posterior valve with radiating ribs . . . *virgulatus* (Pacific)

Central area smoothish and dull . . . . *squamosus* (Atlantic)

Chiton tuberculatus Linné   Common West Indian Chiton

Plate 1d; figure 67f

Florida to Texas and the West Indies.

2 to 3 inches in length. Color a dull grayish green or greenish brown. Some or all of the valves may have a smooth, dark-brown, arrow-shaped patch on the very top. Girdle with alternating zones of whitish, green and black. The scales are placed on the girdle, so that they appear slightly higher than broad, while in *squamosus* they appear to be broader than high. Lateral areas with 5 irregular, radiating cords. Central areas smooth at the top and with 8 to 9 long, strong, wavy, longitudinal ribs on the sides. End valves with irregular, wavy radial cords. Gills beginning at the juncture of the head and foot and with 46 to 48 lamellae. A very common species on wave-dashed, rocky shores.

Chiton squamosus Linné   Squamose Chiton

Southeast Florida and the West Indies.

2 to 3 inches in length. Color a dull, ashen-gray with wide, irregular, dull-brown, longitudinal stripes. Posterior edge of first 7 valves marked with 4 or 5 squares of blackish brown. Girdle with alternating pale stripes of grayish green and grayish white. Posterior valve with minutely pimpled ribs. Lateral areas of middle valves with 6 to 8 rows of small beads between which are microscopic pinholes. Central areas smoothish, with fine, transverse scratches. Common.

Chiton viridis Spengler from the West Indies is very similar, but the margins of the central areas have 6 to 11 very short, wavy ribs, and the lateral areas have 3 or 4 strong ribs of rounded pustules. Uncommon.

Chiton marmoratus Linné   Marbled Chiton

South Florida to Texas and the West Indies.

2 to 3 inches in length. Color variable: (1) entirely blackish brown; (2) olive with flecks, patches and lines of whitish merging together towards the middle; or (3) purplish brown or light-olive with zebra-like stripes on the sides. Entire surface of valves smooth except for a microscopic, silky texture. Lateral areas a little raised. Underside of posterior valve with 14 to 16 slits. A common West Indian littoral species.

Chiton laevigatus Sowerby   Smooth Panama Chiton

Gulf of California to Panama.

2 to 3 inches in length, similar to *marmoratus*. Girdle scales broader than high and not very glossy. Valves smooth, with a silky sheen and colored a grayish green over which are radiating rays of dark-brown. Similar coloration in the other valves. Underside of posterior valve with 21 narrow slits. Sutural plates on underside of middle valves with a dark blotch at the base. Common.

Chiton albolineatus Sowerby (western Mexico, uncommon) is also a smooth species, but differs in having only 16 to 17 slits in the posterior valve and has snow-white, radiating lines on the lateral areas and on the end valves. The girdle scales are light blue-green and edged with white.

Chiton virgulatus Sowerby   Virgulate Chiton

Magdalena Bay, Lower California, to Panama.

2 to 3 inches in length, similar to *tuberculatus*. Girdle scales glassy. End valves with rather even, raised, radiating threads. Middle valves with the lateral areas bearing about 8 raised threads which split in two at the margin of the valve. Central area with about 60 to 70 even, longitudinal threads. Posterior valve with 19 to 20 prominent slits. Common. The closely resembling species, *C. stokesi* Broderip, from Mexico to West Colombia has only 15 to 16 slits in the posterior valve.

Subfamily *ACANTHOPLEURINAE*
Genus *Acanthopleura* Guilding 1829
Subgenus *Maugeria* Gray 1857

Acanthopleura granulata Gmelin   Fuzzy Chiton

South half of Florida and the West Indies.

2 to 3 inches in length, usually so worn and eroded as to eliminate the brown color and granulated sculpturing. Girdle thick, ashy white with an occasional black band, and matted with coarse, hair-like spines. Underside of valves colored a light-green, with the middle valves having a rather large, black splotch behind the sinus. Posterior valve with about 9 slits. Compare with *Ceratozona rugosa* whose gills do not extend to the very posterior end as they do in this species. Common.

Genus *Tonicia* Gray 1847

Resembling *Chiton* in having pectinate or toothed sutural plates, but the girdle is naked and the upper surface of the valves have microscopic eyes.

Tonicia schrammi Shuttleworth   Schramm's Chiton
                                                    Figure 67e

Southeast Florida and the West Indies.

About an inch in length, colored a brownish red to buff and with darker mottlings and speckles. Upper surface of valves glossy; interiors white with a crimson stain in the center. Lateral areas separated from the smooth central area by a strong, rounded rib. The central area has a peppering of about 75 tiny, black eyes. Head valve smooth except for 8 to 10 broad rays of tiny, black eyes. Girdle naked, leathery and brownish to flesh-colored. Posterior valve with 14 slits. 36 lamellae in each of the 2 gills. They begin just behind the juncture of the head and the foot and extend back almost to the posterior end where there is a bilobed, small, fleshy lappet. A moderately common, intertidal species.

CHAPTER IX

Dentaliums and Other
Tusk-Shells

Class *SCAPHOPODA*

Family *SIPHONODENTALIIDAE*

The tusk-like shell is generally swollen in the middle and is entirely smooth. The foot is worm-like and can be expanded at the end into a round disk. The median tooth of the radula is almost as long as wide.

GENUS *Cadulus* Philippi 1844

Shell small, white, without sculpture and swollen in the middle somewhat like a cucumber. Aperture constricted and very oblique. The genus is divided into four subgenera as follows:

Apex with 2 deep slits . . . . . *Dischides* Jeffreys 1867
Apex with 4 deep slits . . . *Polyschides* Pilsbry and Sharp 1897
Apex with 2 or 4 shallow slits . . *Platyschides* Henderson 1920
Apex without slits:

Obese, convex on both sides . . . . . .*Cadulus s. str.*
Slender, almost flat on one side . . . . *Gadila* Gray 1847

Subgenus *Polyschides* Pilsbry and Sharp 1897

Cadulus carolinensis Bush

Carolina Cadulus
Figure 69a

North Carolina to Florida and to Texas.

10 mm. in length. Slightly swollen. Apex with 4 shallow slits. In cross-section the shell is roundish. Commonly dredged from 3 to 100 fathoms.

327

Cadulus quadridentatus Dall   Four-tooth Cadulus

Figure 69b

North Carolina to both sides of Florida and the West Indies.

5 to 10 mm. in length, swollen behind the aperture. Apex with 4 well-defined slits. In cross-section the shell is roundish. Commonly dredged from 3 to 50 fathoms.

Subgenus *Gadila* Gray 1847

Cadulus mayori Henderson   Mayor's Cadulus

Southeast Florida.

3 to 4 mm. in length, swollen just anterior to the middle of the shell. Apical opening ⅔ the size of the aperture and usually has 1 or 2 callus rings within the opening. Fairly common from 16 to 100 fathoms.

Family *DENTALIIDAE*

Shell with the greatest diameter at the aperture. Foot conical and with epipodial processes. Median tooth of the radula twice as wide as long.

Genus *Dentalium* Linné 1758

The shell is an elongate, curved tube open at both ends, and somewhat resembles an elephant's tusk. The diagnostic characters are the type of sculpturing (ribs, riblets and circular threads or incised lines), the form of the apex, the degree of curvature, the size and thickness of shell and the position and form of the apical slit. The ten or so subgenera are nebulous in character and definition and one should consult the works of J. B. Henderson, H. A. Pilsbry and W. H. Dall.

Subgenus *Dentalium s. str.*

Dentalium laqueatum Verrill   Panelled Tusk

North Carolina to south Florida and the West Indies.

1 to 2½ inches in length, thick-shelled and dull-white in color. Apex sharply curved; anterior ⅔ of shell slightly arched. 9 to 12 strong, elevated, primary longitudinal ribs with equally spaced, concave intercostal (space between ribs) spaces. Ribs fade out at the anterior third. There are fine reticulations over the entire shell. A supplemental tube is present in the young shells. Abundant in sandy mud from 4 to 200 fathoms.

Dentalium texasianum Philippi

Texas Tusk
Figure 69c

North Carolina and the Gulf States.

¾ to 1½ inches in length, thick-shelled, well-curved, hexagonal in cross-section and dull, grayish white in color. The broad spaces between the ribs are flat. Common from 3 to 10 fathoms. The subspecies *cestum* Henderson from Texas has numerous, cord-like riblets between the six main ribs.

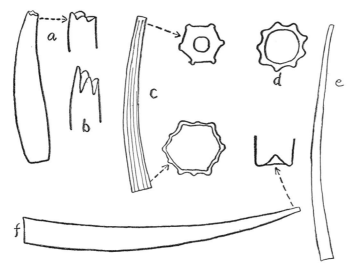

FIGURE 69. Tusk Shells. **a**, *Cadulus carolinensis* Bush, ⅓ inch (southeast United States), showing an enlargement of the apical end; **b**, apical end of *Cadulus quadridentatus* Dall, ⅓ inch (southeast United States); **c**, *Dentalium texasianum* Philippi, 1 inch (southeast United States), showing cross-section at each end; **d**, cross-section of *D. pilsbryi* Rehder, 1 inch (west Florida); **e**, *D. eboreum* Conrad, 2 inches (southeast United States); **f**, *D. pretiosum* Sowerby, 2 inches (Pacific Coast).

Subgenus *Dentale* Da Costa 1778

Dentalium entale stimpsoni Henderson

Stimpson's Tusk

Nova Scotia to Cape Cod, Massachusetts.

1 to 2 inches in length, round in cross-section and dull, ivory-white in color. Region of the apex always very eroded and chalky. Surface uneven and with some longitudinal wrinkles in better preserved specimens. A poor subspecies of the north European *D. entale* Linné. Common from 8 to 1200 fathoms. The subgenus *Antalis* H. and A. Adams is the same as *Dentale*.

Dentalium occidentale Stimpson

Western Atlantic Tusk

Newfoundland to off Cape Hatteras, North Carolina.

1 to 1½ inches in length. Primary ribs 16 to 18, fairly distinct in the young stages; sculptureless in the senile stage. Round in cross-section. Chalky-white when eroded. Common from 20 to 1,000 fathoms.

Dentalium antillarum Orbigny　　　　　　　　　　　　Antillean Tusk

South half of Florida and the West Indies.

About 1 inch in length, roundish in cross-section. Primary ribs 9, but increasing to 12 near the middle and finally with 24 near the aperture. Microscopic, transverse lines between the ribs. Color opaque-white, rarely reflecting a greenish tint. Encircled with weak, zigzag bands or splotches of translucent gray.

Dentalium pilsbryi Rehder　　　　　　　　　　　　Pilsbry's Tusk
　　　　　　　　　　　　　　　　　　　　　　　　　Figure 69d

West Florida and Brazil only.

¾ to 1½ inches in length, roundish in cross-section. Primary ribs 9, with a smaller, weaker, rounded, secondary rib appearing between each. All ribs fade out toward the anterior end. Intercostal spaces flat, crossed by coarse growth lines. No transverse, microscopic sculpture. Color opaque-white; without gray splotches. Formerly known as *D. pseudohexagonum* Henderson 1920, not Arnold 1903. Uncommon from 2 to 5 fathoms.

Dentalium pretiosum Sowerby　　　　　　　　　　　Indian Money Tusk
　　　　　　　　　　　　　　　　　　　　　　　　　Figure 69f

Alaska to Lower California.

About 2 inches in length, moderately curved and solid; opaque-white, ivory-like, commonly with faint dirty-buff rings of growth. Apex with a short notch on the convex side. A common offshore species which was used extensively by the northwest Indians for money.

Subgenus *Fissidentalium* Fischer 1895

Dentalium floridense Henderson　　　　　　　　　　　Florida Tusk

Southeast Florida and the West Indies.

2 to 3 inches in length, roundish in cross-section. Shell hard and yellowish white. Apex hexagonal with concave spaces between. Ribs increase to 24 anteriorly and are rounded, equal-sized and crowded. There is a long, narrow apical slit on the convex side. Rare from 35 to 100 fathoms.

Subgenus *Graptacme* Pilsbry and Sharp 1897

Dentalium eboreum Conrad — Ivory Tusk

Figure 69e

North Carolina to both sides of Florida and the West Indies.

1 to 2½ inches in length, glossy, ivory-white to pinkish. Apical slit deep, narrow and on the convex side. Apical end with about 20 very fine longitudinal scratches. Common in sandy, shallow areas.

Dentalium semistriolatum Guilding — Half-scratched Tusk

South Florida and the West Indies.

About 1 inch in length. Similar to *eboreum*, but curved more, with apical slits on the side, and its color translucent-white with milky patches. Some specimens may be reddish near the apical end. Common from 1 to 90 fathoms.

Dentalium calamus Dall — Reed Tusk

North Carolina to east Florida and the Greater Antilles.

¾ to 1 inch in length, almost straight and glassy-white. Most of the shell has minute, longitudinal scratches (about 16 per mm.). The apical end is sealed over by a bulbous cap which bears a small slit. Uncommon.

Subgenus *Episiphon* Pilsbry and Sharp 1897

Shells very small, needle-like, wholly lacking longitudinal sculpture and, as in some other subgenera, having a projecting, thin tube at the posterior end after the tip is broken or lost. Only one species in the Western Atlantic.

Dentalium sowerbyi Guilding — Sowerby's Tusk

North Carolina and Texas to Florida and the Lesser Antilles.

10 to 15 mm. in length. Needle-like, not fragile, curved, glossy-white. Crowded rings of growth microscopic on tip. Apex without slit and from it projects a very thin inner tube. Erroneously known previously as *D. filum* Sowerby. Commonly dredged from 17 to 180 fathoms.

CHAPTER X

Scallops, Oysters and Other Clams

Class PELECYPODA

The bivalves or clams are dwellers of fresh, marine or brackish waters. They lack a head and are without jaws or radular teeth; they are protected by a pair of shelly valves which are connected or hinged by a horny ligament and which are moved by the contraction of one to three muscles attached to the inner sides of the valves; feeding is usually done with the aid of their ciliated or hair-covered gills. Further details have been presented in the chapter on "The Life of the Clam." The class is also known as *Lamellibranchia*, *Bivalvia* or *Acephala*. The class may be divided into the following orders and suborders:

Order PALAEOCONCHA

    Suborder SOLEMYACEA **(Awning Clams)**

Order PROTOBRANCHIA

    Suborder NUCULACEA (Nut Clams)

Order FILIBRANCHIA

    Suborder TAXODONTA (Ark Shells)
    Suborder ANISOMYARIA (Scallops, Oysters, Sea Mussels)

Order EULAMELLIBRANCHIA

    Suborder SCHIZODONTA (River Mussels)
    Suborder HETERODONTA (Cockles, Lucines, Venus)
    Suborder ADAPEDONTA (Mya and Razor Clams, Teredos)
    Suborder ANOMALODESMACEA (Pandora Clams)

Order SEPTIBRANCHIA

    Suborder POROMYACEA (Dipper Clams, Meat Eaters)

Order *PALAEOCONCHA*
Family *SOLEMYACIDAE*
GENUS *Solemya* Lamarck 1818

The Awning Clams are very primitive in their characters and they have no near relatives. Their shells are fragile, with a weak, toothless hinge, gaping at both ends, and covered by a polished, horny, brown periostracum which extends well beyond the margins of the valves.

Subgenus *Petrasma* Dall 1908

Solemya velum Say   Common Atlantic Awning Clam
Plate 27a

Nova Scotia to Florida.

½ to 1 inch in length, very fragile, and with a delicate, shiny, brown periostracum covering the entire shell and extending beyond the edges. Light radial bands of yellowish brown are present in some specimens. Chondrophore supported by 2 curved arms. Commonly dredged in shallow water in mud bottom. Compare Florida specimens with *occidentalis*.

Solemya borealis Totten   Boreal Awning Clam

Nova Scotia to Connecticut.

2 to 3 inches in length, very similar to *velum*, but more compressed, heavier, and colored grayish blue or lead on the inside of the valves (instead of purplish white). The striking difference is in the siphonal opening of the animal. In *velum*, there are 2 small, median, low tubercles above the opening and 5 or 6 pairs of short tentacles at the lower end of the opening. In *borealis*, there are 3 pairs (one of which is large and long) of tentacles above the opening and about 15 smaller ones bordering the lower half. *S. borealis* is moderately common offshore.

Solemya occidentalis Deshayes   West Indian Awning Clam

West coast of Florida and the West Indies.

¼ inch in length, similar to *S. velum*, but much smaller, and has only one slender ridge or rib bordering the chondrophore. Uncommon just offshore. Described first by Deshayes in 1857, later by Fischer in 1858.

Solemya valvulus Carpenter   Pacific Awning Clam

San Pedro, California, to the Gulf of California.

¾ inch in length, thin, translucent. Periostracum shiny, light-brown, with slender, radial lines of darker brown which are finely striate posteriorly. Ligament bounded by a single, arched prop or rib. Uncommonly dredged offshore.

Order *PROTOBRANCHIA*
Suborder *NUCULACEA*

Key to Families
a. Chondrophore below hinge present:

    b. External ligament absent; shell ovate . . . . . *Nuculidae.*
    bb. External ligament present; shell elongate . . . . *Nuculanidae.*

aa. Chondrophore absent; shell ovate . . . . *Malletiidae.*

Family *NUCULIDAE*
Genus *Nucula* Lamarck 1799

Shell ovate, usually less than ⅓ inch in size; interior pearly; ventral margins usually with fine denticulations.

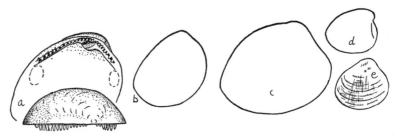

FIGURE 70. Atlantic Coast Nut Clams. **a** and **b**, *Nucula proxima* Say, ¼ inch; **c**, *N. tenuis* Montagu, ⅓ inch (both coasts); **d**, *N. delphinodonta* Mighels, ¹⁄₁₆ inch; **e**, *N. atacellana* Schenck, ⅛ inch.

Subgenus *Nucula s. str.*

Nucula proxima Say   Atlantic Nut Clam
                                            Figure 70a, b

Nova Scotia to Florida and Texas.

¼ inch in length, obliquely ovate, smooth. Color greenish gray with microscopic, embedded, axial, gray lines and prominent, irregular, brownish, concentric rings. Outer shell overcast with oily iridescence. Anterior end often with microscopic, axial lines. Ventral edge minutely crenulate. Common just offshore in mud.

Nucula atacellana Schenck — Cancellate Nut Clam
Figure 70e

Massachusetts to Maryland.

⅛ inch in length and oval in outline; the cancellate sculpturing is due to the crossing of numerous radial and concentric threads. Interior scarcely pearly and with a thick, transparent glaze. Color yellowish brown to light tan. Commonly dredged offshore down to 500 fathoms. Formerly known as *N. reticulata* Jeffreys and *cancellata* Jeffreys.

Nucula crenulata A. Adams — Atlantic Crenulate Nut Clam

South Carolina to Key West, Florida.

¼ inch in length, ovate, internal margin finely crenulate. With numerous concentric, fine ribs which have numerous microscopic crenulations between them. Interior scarcely pearly; overlaid by a thick, transparent layer of shell matter, through which radial fractures or lines are discernible; color yellowish. Dredged in shallow water. Similar in outline to *N. tenuis* (see fig. 70c).

Nucula exigua Sowerby — Pacific Crenulate Nut Clam

Southern California to Ecuador.

³⁄₁₆ inch in length (5 mm.), shaped like *tenuis* and the preceding species. Concentric rings strong with radial crenulations between. Strongly projecting lunular area just under the beaks. Color yellowish. Dredged in shallow water. Very similar to *N. crenulata* from the Western Atlantic.

Nucula delphinodonta Mighels — Delphinula Nut Clam
Figure 70d

Nova Scotia to Maryland.

¹⁄₁₆ inch (3 mm.) in length, ovate, fat and smooth except for coarse concentric growth lines. Anterior end slightly pushed in under beaks and bordered by slight carination. Ventral edge smooth. Color olive-brown. 9 teeth posterior to and 4 teeth anterior to chondrophore.

Nucula tenuis Montagu — Smooth Nut Clam
Figure 70c

Labrador to Maryland. Alaska to Lower California.

Usually ³⁄₁₆ inch in length (up to ⅜ inch in Alaska), ovate, smooth except for irregular growth lines. Color a shiny olive-green, sometimes with darker lines of growth. No radial lines. Ventral edge smooth.

GENUS *Acila* H. and A. Adams 1858

Similar to *Nucula* but characterized by the presence of divaricate sculpture on the outside of the shell. One common species in North American waters.

Subgenus *Truncacila* Schenck 1931

Shell without the shallow sinus as seen in true *Acila*, and the posterior end of the shell nearly at right angles.

Acila castrensis Hinds　　　　　　　　　　　　Divaricate Nut Clam

Figure 72c

Bering Sea to Lower California.

½ inch in length, abruptly truncate at the anterior end. Divaricate, radiating ribs plainly visible. Commonly dredged from 4 to 100 fathoms in sandy mud.

Family NUCULANIDAE
GENUS *Nuculana* Link 1807
(*Leda* Schumacher 1817)

Nuculana pernula Müller　　　　　　　　　　Müller's Nut Clam

Arctic Ocean to Cape Cod. Northern Alaska.

½ to 1 inch in length, elongate and truncate posteriorly, moderately fat, slightly gaping at the rounded anterior end. Numerous raised, concentric growth lines. Periostracum light-brown to dark green-brown, semiglossy. Shell dull-white, interior shiny-white. Interior of rostrum (posterior end of shell) reinforced by a strong radial roundish low rib. Lunule long, prominent, with sharp edge. Commonly dredged offshore in cold water. *N. conceptionis* Dall is much more elongate, smoother and glistening brown.

Nuculana minuta Fabricius (Arctic to off San Diego; and to Nova Scotia) is ½ inch in length, ⅔ as high, rather plump, and with a short rostrum whose smoothish lunule is bounded by a rather coarse rib. Concentric, raised threads are numerous and crowded. Beaks are one third to almost one half the way back from the rounded anterior end. Uncommon offshore.

Nuculana tenuisulcata Couthouy　　　　　　　Thin Nut Clam

Figure 71a

Arctic Seas to Cape Cod.

Up to ¾ inch in length, elongate, moderately compressed; rostrum moderately long with a sharp, high keel down the dorsal center (margin of the valves). Concentric ribs fairly even, well-developed, numerous. Periostracum light- to dark-brown. Commonly found in mud just below low-tide mark.

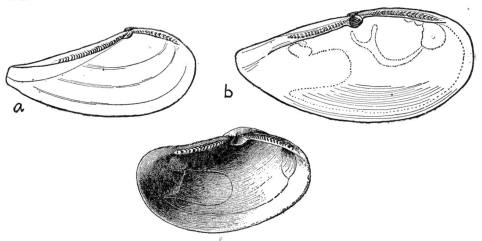

FIGURE 71. Nut and Yoldia Clams. a, *Nuculana tenuisulcata* Couthouy, ¾ inch (Atlantic); b, *Yoldia limatula* Say, 2 inches (Arctic waters, both coasts); c, *Yoldia montereyensis* Dall, 1 inch (California).

Nuculana carpenteri Dall   Carpenter's Nut Clam

North Carolina to West Indies.

About ¼ inch in length, compressed, thin, translucent yellow-brown, with a long, slightly upturning rostrum. Anterior end round. Umbones very small, close together. Almost smooth except for minute, concentric growth lines and microscopic axial scratches which are absent in dead, white valves. Commonly dredged offshore from 10 to 100 or more fathoms.

Nuculana fossa Baird   Fossa Nut Clam

Alaska to Puget Sound, Washington.

¾ to 1 inch in length, elongate, moderately fat and smoothish except for small, pronounced, concentric ribs at the anterior end and on the beaks. Dorsal area of rostrum smoothish, depressed and bounded by 2 weak radial ribs. Periostracum dark- to light-brown. Dredged offshore in shallow water. Some workers consider the following forms or variations as subspecies: *sculpta* Dall, *vaginata* Dall and *curtulosa* Dall.

Subgenus *Ledella* Verrill and Bush 1897

Nuculana messanensis Seguenza   Messanean Nut Clam

Cape Cod to the West Indies.

⅛ to ¼ inch in length, moderately elongate, with a very short, slightly pinched rostrum. Almost smooth except for a few very small concentric growth ridges near the base of the shell. When alive, glistening light-brown with a slight oily iridescence. When dead, grayish white with concentric chalk streaks. Commonly dredged in moderately deep water. One of our smallest species.

Subgenus *Saccella* Woodring 1925

Nuculana acuta Conrad   Pointed Nut Clam

Cape Cod to the West Indies.

¼ to ⅜ inch in length, moderately elongate, with a sharp-pointed posterior rostrum. Concentric ribs evenly sized and evenly spaced and extending over the rib which borders the dorsal surface of the rostrum. Shell usually dredged dead in a white condition. Periostracum thin, very light yellowish. Common offshore.

Nuculana concentrica Say   Concentric Nut Clam

Northwest Florida to Texas.

½ to ¾ inch in length, strong, rather obese and moderately rostrate. Yellow-white, semi-glossy and with very fine, concentric grooves which are evident in adults on the ventral half of the valves. Beaks and the area just below smooth. Radial ridge on rostrum smoothish, not crossed by strong threads. Differing from *acuta* in being more obese, in having a smooth beak area, smooth rostral ridge and in having much finer, more numerous, concentric threads or cut lines. Moderately common in 1 to 3 fathoms.

FIGURE 72. Pacific Coast Nut Clams. **a** and **b**, *Nuculana taphria* Dall, ½ inch (California); **c**, *Acila castrensis* Hinds, ½ inch (Pacific Coast).

Nuculana taphria Dall   Taphria Nut Clam
                                                             Figure 72a, b
Bodego Bay, California, to Lower California.

About ⅓ to ¾ inch in length, shiny green-brown, with prominent concentric sculpture and characterized by the nearly central umbones. Rostrum bluntly pointed, slightly upturned at the end. Concentric ribs disappear just anterior to the carinate border of the dorsal area which is strongly wrinkled. Adults over ½ inch become quite fat. Commonly dredged off southern California in shallow water. Sorensen reports that this species is found commonly in fish stomachs off Monterey.

Nuculana penderi Dall and Bartsch   Pender's Nut Clam

Forrester Island, Alaska, to Santa Barbara, California.

¼ to ⅜ inch in length, moderately elongate, very fat; rostrum short and pointed; concentric ribs prominent and evenly developed. Dorsal area of rostrum oval, finely ribbed and bounded by a sharp, smooth, large rib. Periostracum light-brown. Moderately common offshore.

Nuculana hindsi Hanley   Hinds' Nut Clam

Nazan Bay, Alaska, to Costa Rica.

¼ inch in length, moderately elongate (example: 7.8 mm. long; 4.4 mm. high; both valves 3.0 mm. wide); posterior end rostrate, slightly turned up at the end. Dorsal area of rostrum smoothish except for faint axial threads bounded by smooth carinate rib. Sculpture of evenly sized, closely spaced, distinct, concentric ribs which become obsolete just before the rostral rib. Exterior light yellowish brown. Interior white with faint pearly sheen. *N. acuta* Conrad, a name often given to this Pacific Coast species, has its rostral rib crossed by concentric ribs. *N. penderi* Dall and Bartsch is twice as fat, with a very ovate lunule and is more rounded at the ventral margin. Hanley in 1860 first reported this species from "Gulf of Nicoya, Costa Rica." This is probably "*N. redondoensis*" "Burch" Woodring 1951. It is dredged commonly off the West Coast from 15 to 600 fathoms.

Subgenus *Thestyleda* Iredale 1929

Nuculana hamata Carpenter   Hamate Nut Clam

Figure 26d

Puget Sound to Panama City, California.

Under ½ inch in length, moderately compressed, exterior with strong concentric ribs; characterized by the squarely truncated posterior end of the long rostrum. Fairly commonly dredged off Californian shores from 20 to 200 fathoms.

Genus *Yoldia* Möller 1842

Somewhat similar to *Nuculana,* but the valves are much thinner and fragile, rarely with a long rostrum, usually gaping at both ends, much smoother and glistening.

Subgenus *Yoldia s. str.*

Yoldia limatula Say　　　　　　　　　　　　　　　　　File Yoldia
　　　　　　　　　　　　　　　　　　　　　　　　　　Figure 71b

Maine to Cape May, New Jersey. Northern Alaska.

1 to 2½ inches in length, elongate, narrowing at the posterior end. Umbones very small, halfway between the ends of the shell. Exterior glistening greenish tan to light chestnut-brown, with only faint concentric growth lines. Interior glossy white. A rather common species just below low-water mark. Distinguished from *Y. sapotilla* by its more elongate shape. It is present in northern Alaska, but it is replaced to the south by the following subspecies.

Yoldia limatula gardneri Oldroyd　　　　　　　　　Gardner's Yoldia

Southern Alaska to off San Diego, California.

Very similar to the true *limatula,* but always having the anterior ventral margin with a small concave depression. In general shape it falls within the variations of the Atlantic specimens. Moderately common.

Yoldia sapotilla Gould　　　　　　　　　　　　　　Short Yoldia
　　　　　　　　　　　　　　　　　　　　　　　　　Plate 27b

Arctic Seas to North Carolina.

¾ to 1½ inches in length, oblong, smooth, with a moderately extended posterior end. Periostracum yellowish to greenish brown. Differing from *limatula* in being shorter and less extended and more truncate at the posterior end. Commonly dredged off New England in shallow water; often found in fish stomachs. This species can be confused with the uncommon *Y. myalis* Couthouy (pl. 27d) which is found from Labrador to Cape Cod and Alaska and which, however, is shorter and more pointed at the posterior end.

Subgenus *Megayoldia* Verrill and Bush 1897

Yoldia thraciaeformis Storer　　　　　　　　　　　Broad Yoldia
　　　　　　　　　　　　　　　　　　　　　　　　　Plate 27e

Arctic Seas to Cape Hatteras, North Carolina. Alaska to Puget Sound.

1½ to 2 inches in length, oblong. Characterized by its squarish, up-turned posterior end; coarse, dull, flaky periostracum; large circular chondrophore; and the coarse, oblique rib running from beak to posterior ventral margin. Moderately common from shallow to deep water. Found in fish stomachs.

Family MALLETIIDAE

Shell not pearly inside, oval, compressed, gaping at both ends; ligament external, elongated, resting on nymphs; numerous teeth; no resilium. A linear depression extends from the umbonal cavity to the anterior muscle scar. Worldwide, usually deep water. Includes several genera and subgenera including *Tindaria* Bellardi 1875, *Neilonella* Dall 1881, *Malletia* Desmoulins 1832 and *Protonucula* Cotton 1930.

GENUS *Tindaria* Bellardi 1875

Shell small, resembling a tiny Venus clam; fat; beaks facing slightly forward; ligament minute, external; hinge smooth, continuous just below beaks. Generally deep water and rare.

Tindaria brunnea Dall   Brown Tindaria

Bering Sea, Alaska, to Tillamook, Oregon.

¼ inch in length, fat, moderately pointed at posterior end. Very fine concentric scratches. Exterior dark olive-brown. Interior glossy cream. Has been dredged abundantly in a few places in deep water. There are 8 other rare species on the West Coast of America.

Order FILIBRANCHIA
Suborder TAXODONTA
Superfamily ARCACEA

Key to Families

a. Shell elliptical, hinge straight . . . . *Arcidae*
aa. Shell circular or lopsidedly circular, hinge curved:

    b. Ligament partly sunk into shell . . . . *Limopsidae*
    bb. Ligament external . . . . *Glycymeridae*

Family ARCIDAE

The ark shells have undergone intensive study in the last few years, and the nomenclature is still not settled. It is obvious, though, that not all of the arks can be placed under the single genus *Arca*. The geological history and morphological studies force us to recognize three subfamilies. Many of the subgenera listed here are considered by some authorities as full genera. I have refrained from defining their limits.

Subfamily *ARCINAE*
Genus *Arca* Linné 1758

Characterized by the long, narrow hinge line with numerous small teeth, by the large byssal notch on the ventral side, and the wide ligamental area between the beaks.

Arca zebra Swainson   Turkey Wing
                                                          Plate 27n

North Carolina to Lesser Antilles. Bermuda.

2 to 3 inches in length, about twice as long as deep. Color tan with flecks and zebra-stripe markings of reddish brown. Periostracum brown, matted. Ribs of irregular sizes. No concentric riblets. Do not confuse with *A. umbonata*. A common species which attaches itself to rocks with its byssus. Used extensively in the shellcraft industry. Formerly *A. occidentalis* Philippi.

Arca umbonata Lamarck   Mossy Ark
                                                          Plate 27j

North Carolina to the West Indies.

$1\frac{1}{2}$ to $2\frac{1}{2}$ inches in length. Similar to *A. zebra*, but differing in having beaded ribs and a very large byssal opening, usually having the posterior end much larger, and in lacking the zebra stripes. Periostracum sometimes quite heavy and foliated. Commonly attached to underside of rocks in shallow water.

Genus *Barbatia* Gray 1847
Subgenus *Barbatia* s. str.

Barbatia candida Helbling   White Bearded Ark
                                                          Plate 27r

North Carolina to Brazil.

$1\frac{1}{2}$ to $2\frac{1}{2}$ inches in length; fairly thin, not heavy. Irregular in shape. Byssal opening at base of shell. Numerous weak, slightly beaded ribs; those on the posterior dorsal area being very strongly beaded. Periostracum brown, longest at posterior end. Exterior and interior of shell white. Ligament moderately developed. This species was also named *candida* Gmelin, *jamaicensis* Gmelin and *helblingi* Brug. Common, attached under stones.

Barbatia cancellaria Lamarck   Red-brown Ark
                                              Plate 27q

    Southern Florida and the West Indies.

    1 to 1½ inches in length, similar to *B. candida*, but with low, cancellate sculpture and colored a dark, purplish brown. This is a common species which is erroneously called *B. barbata* Linné (a Mediterranean species).

Subgenus *Acar* Gray 1857

Barbatia domingensis Lamarck   White Miniature Ark
                                              Plate 27u

    North Carolina to Florida and the Lesser Antilles.

    ½ to ¾ inch in length, somewhat box-shaped, whitish in color and with no appreciable periostracum. Similar in shape and sculpture to *Arcopsis adamsi*, but instead of having a small, triangular ligament between the beaks, *domingensis* has a very narrow, long ligament posterior to the beaks. The posterior end is usually larger than the anterior end and characteristically dips slightly downward. Common at low tide under rocks. Erroneously called Arca *reticulata* Gmelin by Dall and others (see Lamy and Woodring).

Barbatia bailyi Bartsch   Baily's Miniature Ark

    Santa Monica, California, to Gulf of California.

    A little over ¼ inch in length, oblong to squarish, fat; cancellate sculpture in which the beads become foliate at the posterior end. Ligament small, narrow and placed well posterior to the fairly close beaks; about 15 teeth. Color white to brownish white.

    Very common in certain localities under stones at low tide. *A. pernoides* Carpenter was thought to be this shell but is apparently some other much larger species of unknown identity.

Subgenus *Fugleria* Reinhart 1937

Barbatia tenera C. B. Adams   Doc Bales' Ark
                                                Plate 27k

    Southern half of Florida to Texas and the Caribbean.

    1 to 1½ inches in length, thin-shelled, rather fat and evenly trapezoidal in shape and with numerous rather evenly and finely beaded, thread-like ribs. Ligamental area fairly wide at the beak end, becoming narrow at the other.

Small byssal gape present on the ventral margin. Moderately common. *Arca balesi* Pilsbry and McLean is this species.

GENUS *Arcopsis* von Koenen 1885

Ligament limited to a very small, triangular, or bar-like area between the umbones.

Arcopsis adamsi E. A. Smith   Adams' Miniature Ark

Figure 26b

Cape Hatteras, North Carolina, to Brazil.

¼ to ⅓ inch in length, oblong in shape, moderately fat, flattened sides; white to cream in color. Periostracum very thin. Sculpture cancellate. Ligament limited to a very small, triangular, black patch between the umbones. The muscle scars are usually bordered by a calcareous ridge. Inner margin of valves smooth. Common under rocks.

Subfamily *ANADARINAE*
GENUS *Anadara* Gray 1847
Subgenus *Larkinia* Reinhart 1935

Anadara multicostata Sowerby   Many-ribbed Ark

Newport Bay, California, to Panama.

Shell large, 3 to 4 inches in length, very thick and squarish. 31 to 36 radial ribs. The left valve slightly overlaps the right valve. Found in sandy areas by dredging in depths over 12 feet. *A. grandis* Broderip and Sowerby in Mexican and Panamic waters is larger, heavier and has 25 to 27 ribs.

Anadara notabilis Röding   Eared Ark

Plate 27p

Northern Florida to the Caribbean and Brazil. Bermuda.

1½ to 3½ inches in length. 25 to 27 ribs per valve. Fine concentric threads cross the ribs and are prominent between the ribs. The ribs never split. Rare in Florida; common in the West Indies.

Formerly called *auriculata* Lamarck which is from the Red Sea. *A. deshayesi* Hanley is a synonym of *notabilis*.

Anadara lienosa floridana Conrad   Cut-ribbed Ark

Plate 27-o

North Carolina, Florida to Texas and the Greater Antilles.

2½ to 5 inches in length, elongate. Ribs 30 to 38 in number, square, faintly divided by a fine-cut line. Fine, raised, concentric lines seen between weakly beaded ribs. Left valve very slightly larger than right valve. Periostracum light- to dark-brown. Not very common.

Typical *lienosa* Say is fossil and very close in characters to *floridana*. This species has often been called *A. secticostata* Reeve which is not so elongate and whose origin is unknown.

Anadara baughmani Hertlein　　　　　　　　　　　Baughman's Ark

Off the Texas Coast.

1½ inches in length, similar to *A. lienosa floridana*, but much fatter, with 28 to 30 weakly noduled ribs which are not split, and with a strongly posterior-sloping anterior ventral margin. Common offshore down to 50 fathoms. *A. springeri* Rehder and Abbott, published a month later, is this species.

Anadara transversa Say　　　　　　　　　　　　Transverse Ark
　　　　　　　　　　　　　　　　　　　　　　Plate 27s

South of Cape Cod to Florida and Texas.

½ to 1½ inches in length. Left valve overlaps right valve. Ligament fairly long, moderately narrow, rough or pustulose. Ribs on left valve usually beaded, rarely so on right valve; 30 to 35 ribs per valve. Periostracum grayish brown, usually wears off except along base of valves. Fairly common in mud below low water. The smallest of the Atlantic Anadaras. Distinguished from *ovalis* by its longer, wider, more distinct external ligament. *A. sulcosa* van Hyning 1946 is this species.

Subgenus *Lunarca* Gray 1847

The subgenera *Argina* Gray and *Arginarca* McLean 1951 are probably the same.

Anadara ovalis Bruguière　　　　　　　　　　　Blood Ark
　　　　　　　　　　　　　　　　　　　　　　Plate 27t

Cape Cod to the West Indies and the Gulf States.

1½ to 2⅓ inches in length, not very thick, roundish to ovate; square, smooth ribs; ligament very narrow and depressed; beaks close together. Periostracum black-brown, hairy. Ribs 26 to 35 in number.

Dall considered the forms "*pexata* Say" and "*americana* Wood" too indistinct for recognition. This species was known for a long time as *campechiensis* Gmelin and is common.

Subgenus *Cunearca* Dall 1898

Anadara brasiliana Lamarck

Incongruous Ark

Plate 27y

North Carolina to West Florida to Texas and the West Indies.

1 to 2½ inches in length; almost as high as long. Beaks facing each other at center of short, transversely striate ligamental area. Left valve overlaps right valve considerably. Ribs 26 to 28, square with strong bar-like beads. Periostracum thin, light-brown. *A. incongrua* Say is this species.

A. chemnitzi Philippi from the Greater Antilles to Brazil is similar, but thick-shelled, less than 1 inch in length; the beaks are slightly forward of the center of the ligamental area.

Subfamily *NOETIINAE*
Genus *Noetia* Gray 1857

Beaks point posteriorly; valves the same size; ligament transversely striate; posterior muscle scar raised to form a weak flange.

Subgenus *Eontia* MacNeil 1938

The subgenus *Eontia* is an Atlantic group only. *Noetia s. str.* differs in having decidedly more regular sculpture, the ribs smoother and never divided; deeper and longer crenulations on the inner margin. There is only one Recent American true *Noetia* (*reversa* H. and A. Adams) which occurs from the Gulf of California to Peru.

Noetia ponderosa Say

Ponderous Ark

Plate 27z; figure 28a

Virginia to Key West, Louisiana and the Gulf of Mexico.

2 to 2½ inches in length, almost as high as long. Ribs raised, square and split down the center by a fine incised line; 27 to 31 ribs per valve. Posterior muscle scar raised to form a weak flange. Periostracum thick, black, but wears off at the beaks. A common shallow-water sand-dweller. Fossil specimens are rarely found on Nantucket, Massachusetts, beaches.

Family *LIMOPSIDAE*
Genus *Limopsis* Sasso 1827

Rather small, obliquely oval, clams with tufted, velvety brown periostracum. Hinge line curved, with a series of oblique teeth. The hinge resembles that of the *Glycymeridae*. Ligament external, small, central, triangular. Mostly deep water. Four species on the Pacific Coast, about six on the Atlantic side.

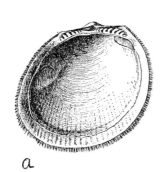
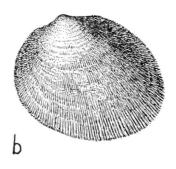

FIGURE 73. *Limopsis diegensis* Dall, ½ inch (California). **a,** interior of valve; **b,** exterior, showing the fur-like periostracum.

Limopsis diegensis Dall   San Diego Limopsis
                                                            Figure 73

Santa Barbara Islands to Coronado Island, California.

⅓ to ½ inch in length, obliquely oval. Shell white; exterior glossy white with concentric striae, often studded by tiny pinpoint holes. Radial scratches present. Periostracum heavy, tufted with hairs, and often with a cancellate pattern. Uncommonly dredged below 20 fathoms.

Limopsis cristata Jeffreys   Cristate Limopsis

Cape Cod to southeastern Florida.

¼ inch in length, similar to *sulcata* but much smaller, less tufted with periostracum, with the inner margin of the valves having a series of strong, pimple-like nobs or teeth, and the outside of the shell having its radial sculpture stronger than its faint concentric sculpture. Commonly dredged off Florida.

Limopsis minuta Philippi (Newfoundland to both sides of Florida) is very close to this species but has cancellate or beaded sculpture and attains a length of ½ inch. The shells of *L. antillensis* Dall (Florida to the Lesser Antilles) are ¼ inch in size and unique in being brightly colored with pink, orange or yellow.

Limopsis sulcata Verrill and Bush   Sulcate Limopsis
                                                            Plate 27f

Cape Cod to Florida, the Gulf States and the West Indies.

½ inch in length, strongly oblique, with prominent, rounded ribs which are finely cut on the upper edge by short radial grooves. Inner margin of valves smooth. Shell dull-white. Periostracum thick, tufted, extending beyond the ventral edge of the shell. Commonly dredged in moderately shallow water.

Family GLYCYMERIDAE
GENUS *Glycymeris* DaCosta 1778

Shell heavy, usually orbicular, equivalve, porcellaneous, usually with a soft, velvety periostracum. Beaks slightly curved inward. Hinge heavy, with numerous, small, similar teeth. Ligament external, its area distinct and with diverging grooves. The largest muscle scar is at the anterior end. Often misspelled *Glycimeris* or *Glicymeris*.

Glycymeris pectinata Gmelin   Comb Bittersweet
Plate 27i
North Carolina to both sides of Florida and the West Indies.

½ to 1 inch in size; characterized by 20 to 40 raised, radial ribs which have no fine radial striae or scratches on them. Color grayish and commonly splotched with brown. A common shallow-water species.

Glycymeris undata Linné   Atlantic Bittersweet
Plate 27g
North Carolina to east Florida and the West Indies.

2 inches in length, heavy, smoothish, except for microscopic radial scratches and somewhat larger concentric scratches, giving a silky appearance. There are numerous very weak and hardly discernible radial ribs separated by lines of white. Beaks at about the middle of the ligamental area. Color cream to white with bold splotches of nut-brown. Interior all white or well-stained with brown. This is *G. lineata* Reeve.

In the region of the Carolinas, an inch-long species (*spectralis* Nicol 1952) is found which is more oval, its beaks face slightly toward the rear and the color is almost a uniform light-brown. Both common.

Glycymeris decussata Linné   Decussate Bittersweet
Plate 27h
Southeast Florida and the West Indies.

2 inches in size, very similar to *undata*, but differs in the posteriorly pointing beaks, and in having nearly all of the ligamental area in front of the beaks. The radial scratches are stronger. This is *G. pennacea* Lamarck. Moderately common.

Glycymeris americana DeFrance   Giant American Bittersweet

North Carolina to north Florida to Texas.

Up to 5 inches in length, rather compressed, always much flatter than *undata*. The dorsal or hinge side of large specimens is quite long. Beaks point toward each other and are located at the midpoint of the hinge. Color drab-gray or tan, rarely with weak mottlings. Rare.

Glycymeris subobsoleta Carpenter   West Coast Bittersweet

Plate 31e

Aleutian Islands to Lower California.

1 inch in size, subtrigonal, texture chalky. Periostracum velvety, but usually worn away. Ligament area short. Radial ribs flat, with narrow interspaces; usually white, but may be with light- to medium-brown markings. A rather common shallow- to rather deep-water species.

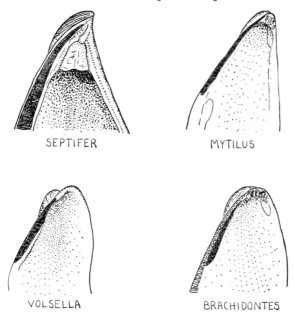

SEPTIFER   MYTILUS

VOLSELLA   BRACHIDONTES

FIGURE 74. Hinges of some genera of mussels.

Suborder ANISOMYARIA
Superfamily MYTILACEA
Family MYTILIDAE
GENUS *Crenella* Brown 1827

Shells small, oval to oblong-oval, thin, brownish periostracum and with fine decussate radial ribs. Ligament weak, internal. Margins crenulated. Interior of shell glossy-white with a faint trace of iridescence. Mantle open in front, and folded at the posterior end into a sessile excurrent siphon. Foot worm-shaped with a disk-shaped end. Hinge finely dentate.

Crenella faba O. F. Müller   Faba Crenella

Figure 75a

Arctic Seas to Nova Scotia.

¼ to ½ inch in length, oval-oblong, with numerous radial ribs. Color reddish brown. Thin periostracum varnish-like. Byssus golden-brown. Common offshore.

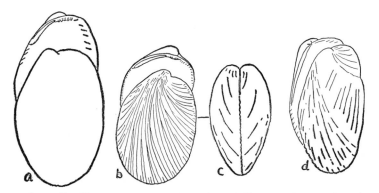

FIGURE 75. Crenella Clams. **a**, *Crenella faba* Müller, ⅓ inch (Arctic waters); **b** and **c**, *Crenella glandula* Totten, ⅓ inch (New England); **d**, *Musculus lateralis* Say, ⅜ inch (Atlantic Coast)

Crenella glandula Totten   Glandular Crenella

Plate 28j; figure 75b, c

Labrador to North Carolina.

¼ to ½ inch in length, squarish, with the beaks near one corner. Radial ribs are fine, numerous, slightly beaded and often crossed by much finer, concentric threads. Color olive-brown. A very common offshore, cold-water species. The smaller *decussata* has its beaks at the center of its more symmetrical shell.

Crenella decussata Montagu   Decussate Crenella

Bering Sea to San Pedro, California. Greenland to North Carolina.

Less than ⅛ inch in size, oval, with numerous fine, decussated radial ribs. Color tan to yellowish gray. Dredged from 3 to 150 fathoms. A food of many marine fishes. Compare with *glandula*.

Crenella divaricata Orbigny (North Carolina to southeast Florida and the West Indies) is even smaller than *decussata*, is pure white, and very inflated.

Crenella columbiana Dall   British Columbia Crenella
Figure 26g, h

Aleutian Islands to San Diego.

A little over ½ inch in length, oval-oblong, inflated, with numerous very fine, decussate radial ribs. Color greenish yellow-brown. 10 to 100 fathoms.

GENUS *Modiolus* Lamarck 1799

This group of mussels have shells of various forms in which the hinge is without teeth. The anterior end of the shell extends in front of the beaks, while in *Mytilus* 3 to 5 tiny teeth are present on the hinge and the beaks are at the very anterior end of the shell. *Volsella* Scopoli has been rejected.

Modiolus modiolus Linné   Northern Horse Mussel
Figure 26j

Arctic Seas to northeast Florida. Arctic Seas to San Pedro, California.

2 to 6 inches in length, heavy, with a coarse, rather thick, black-brown periostracum which in dried specimens flakes off to reveal a mauve-white, chalky shell. One of the largest and commonest mussels found in cooler waters below low-water mark. Do not confuse with *M. americanus*.

Modiolus americanus Leach   Tulip Mussel
Plate 35l

North Carolina to the West Indies.

1 to 4 inches in length, smooth, except for the periostracum, which is commonly hairy and sometimes resembles a beard. Color light-brown flushed with deep rose and sometimes with several radial streaks of light-purple. Anterior ventral area with a deep chestnut splotch. Interior dull white, sometimes stained with bluish, rose or brownish.

Recently killed specimens are commonly washed ashore in large numbers. A very common species. Formerly known as *Volsella* or *Modiolus tulipa* Linné or Lamarck. The subspecies found in Charleston Bay, North Carolina, and Tampa, Florida, is more compressed and a soft brown in color.

Modiolus demissus Dillwyn   Atlantic Ribbed Mussel
Plate 28h

Gulf of St. Lawrence to South Carolina. Introduced to California.

2 to 4 inches in length, black-brown in color, often shiny, and with strong, rough, radial, bifurcating ribs. Interior bluish white with the posterior end flushed with purple or purplish red. This is the only ribbed *Modiolus* in our waters, but do not confuse it with *Brachidontes recurvus* which

has a strongly curved beak, tiny teeth at the umbo, and is a solid rosy-brown on the inside. *Mytilus plicatulus* Lamarck is this species.

The subspecies *demissus granosissimus* Sowerby (both sides of Florida to Texas and Yucatan) is very similar but with almost twice as many ribs which are finely and neatly beaded. Common.

Modiolus fornicatus Carpenter   California Horse Mussel
Plate 29-o

Monterey to San Pedro, California.

About 1 inch in length, smoothish, inflated, light-brown periostracum which wears white at the beak end. Beaks curved strongly forward. Interior dull white. Found in moderately deep water, and rarely cast ashore. Associated with *Haliotis rufescens*.

Modiolus capax Conrad   Capax Horse Mussel

Santa Cruz, California, to Peru.

2 to 6 inches in size. Periostracum thick, often with coarse hairs, chestnut-brown in color. Worn shell brick-red with bluish mottlings. Interior half white, half (ventral) brownish purple. Resembles figure of *Modiolus americanus* (Pl. 351).

GENUS *Brachidontes* Swainson 1840
Subgenus *Brachidontes s. str.*

Brachidontes citrinus Röding   Yellow Mussel
Plate 35i

Southern Florida and West Indies.

1¼ inches in length, elongate, with numerous wavy, fine axial ribs, colored a light brownish yellow outside, and inside mottled a metallic purplish and white. Anterior end has four very tiny white teeth. Bordering the ligament are about 30 very small, equal-sized teeth on the edge of the shell. Compare with *B. exustus* Linné which is wider. The genus is commonly misspelled *Brachydontes*.

Subgenus *Hormomya* Mörch 1853

Brachidontes exustus Linné   Scorched Mussel
Plate 35j

North Carolina to the West Indies.

¾ inch in length, rather elongate with numerous fine axial ribs; colored a yellowish brown to dark-brown outside, and inside mottled with a metallic purplish and white. Anterior end has two very tiny purplish teeth. Beyond the ligament (posterior end) there are 5 to 6 very tiny, equal-sized teeth on the edge of the shell. Compare with *B. citrinus,* which is more elongate.

Brachidontes stearnsi Pilsbry and Raymond — Stearns' Mussel

Santa Barbara, California, to Oaxaca, Mexico.

½ to 1 inch in length, obtusely carinate, with numerous coarse, beaded, radial ribs which bifurcate. Color brownish purple on the dorsal half, straw-yellow to brownish yellow on the flattened ventral half. Hinge on dorsal edge with about a dozen very tiny bar-like teeth. Usually found in colonies in crevices of stones. Two small clams, *Lasaea cistula* Keen and *L. subviridis* Dall, attach themselves to the byssus of this species. Do not confuse *B. stearnsi* with *Septifer bifurcatus,* with which it often lives. *B. multiformis* Carpenter and *B. adamsianus* Dunker are closely related species, if not mere forms, found in the Panamic province.

Subgenus *Ischadium* Jukes-Brown 1905

Brachidontes recurvus Rafinesque — Hooked Mussel
Plate 35n

Cape Cod to the West Indies.

1 to 2½ inches in length, flattish, rather wide, with numerous wavy axial ribs. Color outside a dark grayish black, inside a purplish to rosy brown with a narrow blue-gray border. At the umbonal end there are 3 to 4 extremely small, elongate teeth on the edge of the shell. The anterior end of the shell is strongly hooked. This was known as *M. hamatus* Say and has sometimes been placed in the genus *Mytilus.*

GENUS *Amygdalum* Megerle von Mühlfeld 1811

Shell thin, very smooth, often with colored, cobwebby designs. These clams build nests for themselves with a copious supply of byssal threads.

Amygdalum papyria Conrad — Paper Mussel
Plate 28i

Texas and Maryland to Florida.

1 to 1¼ inches in length, elongate, smooth, glistening, fragile, and colored a delicate two-tone of bluish green and soft yellowish brown. Interior iridescent-white. The ligament is very weak and thin. *A. sagittata* Rehder, sometimes dredged off Florida and Mississippi, is very shiny, ivory-white, half of each valve with fine, gray, cobwebby streaks. The umbo is reinforced inside by a very small, smooth column or rib.

Genus *Septifer* Recluz 1848

Septifer bifurcatus Reeve   Bifurcate Mussel

Crescent City, California, to Gulf of California.

1 to 2 inches in size, subtriangular in outline, inflated. With a couple of dozen strong, wavy radial ribs. Inner margin crenulated. Periostracum black, although often worn white between the ribs. Interior pearly-white, often stained bluish brown on one half of the inner side. The subspecies *obsoletus* Dall from San Diego is mostly black on the interior and is a quite elongate form.

Genus *Mytilus* Linné 1758

Mytilus edulis Linné   Blue Mussel
                                                      Plate 35m
Arctic Ocean to South Carolina. Alaska to California.

1 to 3 inches in length, no ribs but often with coarse growth lines. Ventral margin often curved. Color blue-black with eroded areas of chalky purplish. Periostracum varnish-like. Interior slightly pearly-white with deep purple-blue border. Occasionally, specimens have radial rays of brown-yellow. Very common in New England. Sometimes found in more southerly waters attached to floating wood.

Mytilus edulis diegensis Coe 1946 (Northern California to Lower California) is indistinguishable from specimens of *edulis* found in Alaska and New England, and probably only represents an ecological or physiological race (see W. R. Coe, 1946).

Mytilus californianus Conrad   Californian Mussel
                                                      Plate 29p
Aleutian Islands to Socorro Island, Mexico.

2 to 10 inches in length, thick, inflated; ventral margin nearly straight; with less than a dozen or so, fairly broad, weak radial ribs which are best seen on the middle part of the shell. Growth lines very coarse. An abundant species found between tides attached to rocks.

Genus *Musculus* Röding 1798

Mussel-like shells with the sculpturing divided into three oblique areas, the center one being smooth or almost so, and the two end areas having radial ribs. The ligament is much longer than that in *Crenella*. These are moderately deep-water clams. Mantle folded in front into a wide, incurrent siphon and behind into a conical excurrent siphon. Foot strap-shaped. This genus was formerly known as *Modiolaria* Beck 1838. Hinge finely dentate.

Musculus niger Gray Black Musculus
Plate 28g

Arctic Seas to North Carolina. Alaska to Puget Sound.

About 2 to 3 inches in length. Similar to *M. discors,* but much more compressed and with strongly developed axial, decussated ribs on the posterior and anterior thirds. Center section with microscopic concentric wavy threads and pimples. Often pinkish on the inside. Common.

Musculus laevigatus Gray Smooth Musculus
Plate 28f

Arctic Ocean to Puget Sound, and North Atlantic.

1 to 1¾ inches in length. Distinguished from *discors* by its larger size, and in having no pronounced radial riblet or depression separating the posterior third from the middle area. The periostracum is more often black in this species. The posterior area very often has numerous microscopic concentric scratches which, to the naked eye, give this area a dull finish. Like *discors,* this species is much fatter than *niger.*

Musculus lateralis Say Lateral Musculus
Figure 75d

North Carolina to Florida and the West Indies.

⅜ inch in length, oblong, fragile, with a center area on the valve with concentric growth lines only. Remainder of shell with radial ribs. Color light-brown with a strong blush of blue-green. Interior slightly iridescent. Common offshore.

Musculus discors Linné Discord Musculus
Plate 28e

Atlantic: Arctic Seas to Long Island Sound. Pacific: Arctic Seas to Puget Sound.

1 inch in length, oblong, fairly fragile. Anterior and posterior thirds of outer shell with very weak radial ribs; center section smooth except for irregular growth lines. Periostracum shiny and either dark black-brown or light-brown. Interior bluish white with slight iridescence. Commonly dredged.

GENUS *Botula* Mörch 1853
Subgenus *Adula* H. and A. Adams 1857

Botula falcata Gould   Falcate Date Mussel
                                                              Plate 29k

   Coos Bay, Oregon, to Lower California.

   2 to 4 inches in length, very elongate, slightly curved. Beaks rounded
and about one-eighth the length from the anterior end; a strongly marked
angle occurs from the beaks to the base of the posterior extremity; numerous
vertical, wavy ribs over all the shell. Color a shiny chestnut-brown. Common.

Botula californiensis Philippi   Californian Date Mussel
                                                              Plate 29h

   1 to 1¼ inches in length, elongate, curved and smooth, except for a
velvety, hair-like covering over the posterior end. Shiny, chocolate-brown
in color. Moderately common.

GENUS *Lioberus* Dall 1898

Lioberus castaneus Say   Say's Chestnut Mussel

   Both sides of Florida and the West Indies.

   ¾ inch in length, oval-elongate, well-inflated and thin-shelled. Exterior
chestnut- to dark-brown, the anterior half glossy, the posterior half dull
and commonly with a fine grayish matting of periostracum. Interior bluish
white and with an irregular surface. Hinge simple with a slight swelling or
pad under the beaks. Moderately common in shallow water.
   Botula fusca Gmelin from North Carolina to southeast Florida (rare)
and the West Indies (common) is similar, but distinguished by its longer,
hooked or arcuate shape, by the thick, concentric ridges on the outside, by
the more anteriorly placed beaks, and by the tiny, vertical threads on the
hinge just posterior to the ligament. Attached in clusters to wood and
rocks.

GENUS *Lithophaga* Röding 1798

Lithophaga nigra Orbigny   Black Date Mussel
                                                              Plate 28m

   Southeast Florida and the West Indies.

   1 to 2 inches in length, elongate and cylindrical. Black-brown outside
and an iridescent bluish white inside. Anterior lower third of each valve

with strong, vertical, smooth ribs; remainder of shell smoothish with only irregular growth lines. Commonly found boring into soft coral blocks.

Lithophaga antillarum Orbigny   Giant Date Mussel
                                                              Plate 28k

   Gulf of Mexico and the West Indies.

   2 to 4 inches in length, elongate, cylindrical and colored a light yellowish brown on the outside and iridescent cream inside. Sides of valves marked with numerous, irregular, vertical riblets. Fairly common in soft rocks in moderately deep water.

Subgenus *Myoforceps* P. Fischer 1886

Lithophaga aristata Dillwyn   Scissor Date Mussel
                                                              Plate 29j

   Southern Florida and the West Indies. La Jolla, California, to Peru.

   ½ to 1 inch in length. Characterized by the pointed tips at the posterior end being crossed like fingers. Color yellowish brown, but generally covered by a smooth, gray, calcareous encrustation. Moderately common in soft rock.

Subgenus *Diberus* Dall 1898

Lithophaga bisulcata Orbigny   Mahogany Date Mussel
                                                              Plate 28n

   North Carolina, the Gulf of Mexico, and the West Indies.

   1 to 1½ inches in length, elongate, cylindrical and coming to a point at the posterior end. A sharp, oblique, indented line divides each valve into two sections. Anterior half of valve smooth, mahogany-brown, but commonly encrusted with porous, gray, calcium deposits. Posterior end more heavily encrusted with a gray, porous covering which projects beyond the edge of the shell. A fairly common rock-boring species.

Lithophaga plumula kelseyi Hertlein and Strong   Kelsey's Date Mussel
                                                              Plate 29i

   San Diego north to Mendocino County, California.

   1 to 2 inches in length, similar to *L. bisulcata*, but the calcareous matter on the posterior end is strongly pitted and furrowed to look like a wet, ruffled feather. Typical *plumula* Hanley ranges from Lower California to Peru. Both fairly common in rocks.

Superfamily PTERIACEA
Family ISOGNOMONIDAE
Genus *Isognomon* Solander 1786

Shell thin and greatly compressed; interior pearly; anterior margin with a narrow byssal gape near the dorsal margin. Hinge with numerous parallel grooves perpendicular to the dorsal margin of the valve. *Perna* Bruguière and *Pedalion* Dillwyn are synonyms. *Pedalion* Solander 1770 is invalid.

Isognomon alatus Gmelin

Flat Tree Oyster
Plate 35b

South half of Florida and the West Indies.

2 to 3 inches in length. Hinge has 8 to 12 oblong grooves or sockets into which are set small, brown resiliums. Exterior with rough or smoothish growth lines. External color drab purplish gray to dirty-gray. Interior moderately pearly with stains of purplish brown or mottlings of blackish purple. This very flat, oval bivalve is commonly found in compact clumps on mangrove tree roots. Distinguished from *I. radiatus* by its flat, more regularly fan shape and darker color.

Isognomon radiatus Anton 1839

Lister's Tree Oyster
Plate 35a

Southeast Florida and the West Indies.

½ to 2 inches in size, very irregular in shape, commonly elongate. Sometimes twisted and irregular. Hinge short, straight and with 4 to 8 very small, squarish sockets. Exterior rough with weak, flaky lamellations. Color a solid, translucent yellowish, but commonly with a few wavy, radial stripes of light purplish brown. Common on rocks at low tide. Formerly *I. listeri* Hanley 1843.

Isognomon bicolor C. B. Adams (Lower Florida Keys, Bermuda and Caribbean) is heavier, more oval, and commonly with strong lamellations on the outside. It is usually darkly and heavily splotched with purple inside and out. Common. According to Lamy, *I. vulsella* Lamarck is a different species which is limited to the Red Sea.

The Western Tree Oyster, *Isognomon chemnitzianus* Orbigny, from the Coronado Islands to Chili, lives in crowded colonies under stones in shallow water. It resembles the above two species, is about 1 to 2 inches in size; its right valve flattish, left valve slightly swollen. It is the only Californian *Isognomon*.

Family PTERIIDAE
GENUS *Pteria* Scopoli 1777

Fairly thin-shelled, moderately fat, and with the hinge ends considerably drawn out. Pearly inside. The right and left valves bear 1 or 2 small denticles which fit into shallow sockets in the opposite valve.

Pteria colymbus Röding   Atlantic Wing Oyster
Plate 35d

North Carolina to southeast Florida and the West Indies.

1½ to 3 inches in length, obliquely oval with a long extension of the hinge line toward the posterior end. Left valve inflated. Right valve somewhat flatter and with a strong anterior notch for the byssus. Periostracum matted, brown and with cancellate fimbrications. Exterior color variable: brown, black or brownish purple with broken, radial lines of cream or white. Interior pearly with a wide, non-pearly margin of purplish black with irregular cream rays. Common from low water to several fathoms.

The Western Wing Oyster, *Pteria sterna* Gould is very similar, 3 to 4 inches in length, and deep purplish brown with occasional paler rays. Anchored in mud; from San Diego to Panama. Common.

GENUS *Pinctada* Röding 1798

This is the famous genus of pearl oysters. The byssal gape is in the right valve below the small, triangular auricle. *Margaritifera* Schumacher is a synonym.

Pinctada radiata Leach   Atlantic Pearl Oyster
Plate 35c

South half of Florida and the West Indies.

1½ to 3 inches in length, moderately inflated to flattish, thin-shelled and brittle. There is a small, thin, flat ligament at the center of the hinge. Exterior tan with mottlings or rays of purplish brown or black. Rarely tinted with dull-rose or greenish. In quiet waters, thin scaly and very delicate, periostracal spines may be developed. Interior a beautiful mother-of-pearl. Common in shallow water attached to rocks.

Family PINNIDAE
GENUS *Pinna* Linné 1758

The Pen Shells are large, fragile, fan-shaped clams which live in sandy or mud-sand areas, usually in colonies. The apex or pointed end is deeply buried, and there is a mass of byssal threads attached to small stones or fragments of shells. The broad end of the shell projects above the surface of the sand an inch or so. In the genus *Pinna*, there is a weak groove running down the middle of each valve. In *Atrina*, this character is absent.

Pinna carnea Gmelin   Amber Pen Shell
                                                          Plate 27w
    Southeast Florida and the West Indies.

4 to 9 inches in length, relatively narrow, thin-shelled and with a central, radial ridge in the middle of the valve which is more conspicuous at the pointed or hinge end. With or without 10 radial rows of moderately large, scale-like spines. Color usually a light orangish to translucent-amber. Rare in Florida but common in the Bahamas. This is the only *Pinna* in the western Atlantic. *P. rudis* Linné is apparently a Mediterranean and West African species which is heavier and darker red. The name *P. haudignobilis* Karsten is invalid, as are all this author's names.

Genus *Atrina* Gray 1842

Atrina rigida Solander   Stiff Pen Shell
                                                          Plate 27x
    North Carolina to south half of Florida and the Caribbean.

5 to 9 inches in length, relatively wide, moderately thick-shelled and with 15 to 25 radial rows of tube-like spines; rarely smoothish. Color dark to light brown. Commonly washed ashore. A small, commensal crab lives inside the mantle cavity. A number of unusual snails and chitons are found in or on dead or live *Pinna* shells.

Atrina serrata Sowerby   Saw-toothed Pen Shell
                                                          Plate 27v
    North Carolina and south half of Florida.

Similar in size and shape to *rigida*, but covered with many more, much smaller, sharp spines. It is usually thinner-shelled and lighter in color. Commonly washed ashore with *rigida*.

Superfamily *PECTINACEA*
Family *PLICATULIDAE*
Genus *Plicatula* Lamarck 1801

Shell trigonal or spathate, thick-shelled and attached by either valve to rocks or other shells. Sculpture of broad, radial ribs. Hinge with a narrow, elongate chondrophore which is flanked on each side by a fluted tooth and a socket.

Plicatula gibbosa Lamarck
Kitten's Paw
Plate 35e

North Carolina to Florida, the Gulf States and West Indies.

About 1 inch in length, somewhat cat's-paw-shaped. Shell strong, heavy, with 5 to 7 high ribs which give the valves a wavy, interlocking margin. Hinge in upper valve with 2 strong, equally sized teeth; lower attached valve with 2 sockets in the hinge with 2 smaller teeth set rather close together. Color dirty-white to gray with red-brown or purplish lines on the ribs. A common intertidal to offshore species.

Family *PECTINIDAE*

Because of the great number of fossil and living species of scallops and the almost limitless modifications exhibited by them, there have been no less than 50 genera and subgenera proposed in this family by various authors. Doubtlessly, many more will be invented. Most, if not all, of these genera are closely integrated by connecting species. Workers have a choice of using the single genus, *Pecten*, or employing a genus for nearly every species. We are arbitrarily employing only six genera—*Pecten, Aequipecten, Chlamys, Placopecten, Lyropecten* and *Hinnites*—and we cannot justify these on biological grounds. It may be noted that we have moved the glassy, thin-shelled *Propeamussium* from the Pectinidae into a family of its own on anatomical grounds. This new family refers to what was once called "Amussiidae." True *Amusium*, however, is merely a subgenus of *Pecten* connected to it by a series of species in the *Euvola* group.

GENUS *Pecten* Müller 1776
Subgenus *Pecten* s. str.

Pecten diegensis Dall
San Diego Scallop
Plate 33e

Cordell Bank, California, to Lower California.

2 to 3 inches in size. Right valve convex with 22 or 23 flat-topped ribs which are generally longitudinally ridged on top. Left valve much flatter, with 21 to 22 narrow, rounded ribs. Dredged from 10 to 75 fathoms.

Subgenus *Patinopecten* Dall 1898

Pecten caurinus Gould
Giant Pacific Scallop
Plate 29b

Wrangell, Alaska, to Humboldt Bay, California.

6 to 8 inches in size, roughly circular; upper valve almost flat, reddish gray and with about 17 low, rounded ribs; lower valve deeper, whitish and with a few more, stronger, rather flat-topped ribs. This is the common, edible deep-sea scallop of Alaska.

Subgenus *Amusium* Röding 1798

Pecten papyraceus Gabb   Paper Scallop

The Gulf of Mexico and the West Indies.

About 2 inches in size, oily smooth, glossy, exterior without ribs, but internally with about 22 very fine ribs which are commonly arranged in pairs. Both valves moderately convex to flattish. Upper valve light-mauve to reddish brown with darker flecks. Lower valve whitish at the center with yellow to cream margins or all white. Hinge line strongly arched. Rather rare in collections, but commonly brought up from several fathoms by shrimp fishermen.

Pecten laurenti Gmelin (pl. 33f) from the Greater Antilles is larger, with a straight hinge line and with the lighter-colored valve more convex than the darker valve. Rare.

Section *Euvola* Dall 1897

Pecten ziczac Linné   Zigzag Scallop
                                                             Plate 33d

North Carolina to Florida and the West Indies.

2 to 4 inches in size. Upper (left) valve flat; lower valve very deep and convex. There are 18 to 20 broad, very low, rather indistinct ribs on the deep valve which is generally colored a brownish red (rarely orange). The ribs fade out or are not present near the side margins of the valve. Flat valve with a bright mozaic of whites and browns. A fairly common species. Do not confuse with *raveneli.*

Pecten raveneli Dall   Ravenel's Scallop
                                                             Plate 33g

North Carolina, the Gulf of Mexico, and the West Indies.

1 to 2 inches in size, similar to *ziczac,* but the deep valve has about 25 very distinct ribs which are commonly whitish in color. Between them are fairly wide, tan or pinkish grooves. In the flat valve, the 25 or so ribs are rounded in cross-section whereas in *ziczac* they are flat-topped and much closer together. A rather uncommon species.

Pecten tereinus Dall   Tereinus Scallop

Southern Florida and the Gulf of Mexico.

1 inch in size, quite fragile. Upper (left) valve flat, with about 20 small, narrow ribs; lower valve deep to moderately deep and with low, irregularly defined, roundish ribs. Color whitish tan, slightly translucent, with faint mottlings of pink near the beaks. Rarely, the flat valve may be flecked with brown, zigzag, fine lines. A rare species uncommonly dredged by private collectors in 10 to 40 fathoms.

Genus *Chlamys* Röding 1798

Chlamys sentis Reeve   Sentis Scallop

Plate 34a

North Carolina to southeast Florida and the West Indies.

1 to 1½ inches in length, but not so wide (like a fan opened only 80 degrees). Valves rather flat. One hinge ear small, the other twice as large. With about 50 ribs of varying sizes, each with tiny, closely set scales. There are 2 to 4 smaller ribs between the slightly larger ones. Color commonly brilliant: purple, red, vermilion, orange-red, brownish, white or mottled (especially near the beaks). Common under rocks below low-tide mark. Do not confuse with *ornata, mildredae* or *benedicti.*

Chlamys mildredae F. M. Bayer   Mildred's Scallop

Plate 34c

Southeast Florida and Bermuda.

1 to 1¼ inches in length, similar to *sentis* and *ornata,* but the ribs of the upper valve (one without the byssal notch) 30 in number and every third or fourth one larger. Sculpture of rather large, erect scales set about 1 mm. apart. Ribs of lower valve about 30, in groups of 2 or 3. Exterior color much like *sentis;* interior yellowish with purple stains near the margins. Rare under rocks at low tide.

Chlamys ornata Lamarck   Ornate Scallop

Plate 34b

Southeast Florida to the West Indies.

1 to 1¼ inches in length, similar to *sentis,* but with about 18 high, major, slightly scaled ribs separated by 2 small, scaly cords on the upper valve. Ribs of lower valve are in 18 groups of 3 closely spaced riblets. Exterior ivory to yellowish cream with strong maculations of maroon or purplish. Interior usually white. Compare with *mildredae.* An uncommon and favorite collector's item.

Chlamys benedicti Verrill and Bush   Benedict's Scallop

South half of Florida and the Gulf of Mexico.

Rarely over ½ inch in length. Very similar to *sentis*, but with a greater range of colors and having 2 color variations not found in *sentis* (pure lemon-yellow or mottled with chalk-white zigzag stripes). With about 22 strong ribs alternating with weaker ribs, total about 45. Shorter ear has a sharp, 90-degree corner and bears prominent spines, while in *sentis* it is more rounded or considerably more than 90 degrees and is smoother. Hinge margin of longer ear has small projecting scales. Color pink, pinkish red, light purple or yellow, and commonly with pronounced whitish zigzag markings. A moderately common species usually misidentified as young *sentis* or *muscosa*.

Chlamys imbricata Gmelin   Little Knobby Scallop
                                              Plate 34f
Southeast Florida and the West Indies.

1 to 1¾ inches in length, but not quite so wide. Lower valve (the one with the byssal notch) slightly convex. Upper valve almost flat and fairly thin. Ribs 8 to 10, uncommonly with smaller cords between. They have prominent, cup-shaped, delicate, distantly spaced scales. Color dirty-white or pinkish with small, squarish, red or purplish blotches. Interior yellowish, commonly with purplish stains. Moderately common.

Chlamys hastata hastata Sowerby   Pacific Spear Scallop
                                                     Plate 34j
Monterey to Newport Bay, California.

2 to 2½ inches in size; without microscopic reticulations; right valve (with byssal notch) with about 18 to 21 primary, strongly spined ribs which have 5 to 7 much smaller, weakly spined, secondary ribs in between left valve with 10 to 11 distantly spaced, strongly scaled primary ribs, with 12 to 16 very weak, beaded secondary ribs in between. This is not so common as the subspecies *hericia,* and is much more colorful, commonly being bright orange, red or lemon.

Chlamys hastata hericia Gould   Pacific Pink Scallop
                                                   Plate 34k
Alaska to San Diego, California.

2 to 2¾ inches in size; without microscopic reticulations; right valve (with byssal notch) with about 18 to 21 primary, moderately scaled ribs which have 5 to 7 much smaller spined ribs between; left valve with about 10 to 11 primary, spined ribs which have a single, rounded, almost as large

secondary rib in between. Between these large ribs there are 15 to 18 tiny, spined ribs, 3 of which are on the large secondary rib. Color variable: solid rose, pink, white, light yellowish and blends of all these. Commonly dredged in shallow waters.

Chlamys hindsi Dall   Hinds' Scallop
Plate 34l

    Alaska to off San Diego, California.

2 to 2½ inches in size; with microscopic reticulations between the ribs either near the beaks or the margins of the valves. Left valve (without the byssal notch) with numerous primary ribs, each bearing 3 rows of spines, and with a secondary spined rib between. Right valve flattish, usually lighter-colored, and with fewer ribs which are smoothish, rounded and inclined to be grouped in pairs. The reticulate sculpturing is best seen on this side. Color variable: light-rose, mauve, lemon-yellow, pale-orange and blends of these. A rather common species dredged in shallow water down to 822 fathoms.

Chlamys islandica Müller   Iceland Scallop
Plate 27l

    Arctic Seas to Buzzards Bay, Massachusetts. Alaska to Puget Sound, Washington.

3 to 4 inches in length, not quite so wide. Long hinge ear is twice the length of the short one. Valves moderately convex to flattish. With about 50 coarse, irregular ribs which split in two near the margin of the valve. Rarely, the ribs are grouped more or less in groups of twos, threes or fours. Color usually a dirty-gray or cream, but some are quite attractively tinged with peach, yellow or purplish both inside and out. A very common species offshore on the continental shelf.

GENUS *Leptopecten* Verrill 1897

Leptopecten latiauratus Conrad   Kelp-weed Scallop
Plate 34i

    Point Reyes, California, to Lower California.

About 1 inch in size, thin, lightweight, with 12 to 16 squarish ribs. Ears strongly pointed at the ends. Color varies from translucent yellowish to chestnut-brown; commonly mottled with white. The subspecies *monotimeris* Conrad has rounded ribs which form broad corrugations on the shell and it has less acutely pointed ears. This is a common species found attached to kelp weeds, stones and bottoms of boats. Sometimes spelled *latiauritus*.

GENUS *Placopecten* Verrill 1897

Placopecten magellanicus Gmelin   Atlantic Deep-sea Scallop
                                        Plates 33c; 27m

Labrador to Cape Hatteras, North Carolina.

5 to 8 inches in size, almost circular. Valves almost flat to slightly convex. Interior flaky-white. Exterior rough with numerous very small, raised threads. Exterior yellowish gray to purplish gray or dirty-white. This is the common, edible, deep-sea scallop fished off our New England coasts. The name *grandis* Solander is nude and cannot be used.

GENUS *Lyropecten* Conrad 1862
Subgenus *Lyropecten s. str.*

Lyropecten antillarum Recluz   Antillean Scallop
                                                   Plate 34g

Southeast Florida and the West Indies.

½ to ¾ inch in length and width. Valves fragile, both nearly flat. Only about 15 moderately rounded, low ribs. Growth lines exceedingly fine (seen with the aid of a strong lens). Color either pastel-yellow, tawny-orange or light-brown, commonly with chalk-white mottlings, flecks or stripes. Found uncommonly in shallow water.

Subgenus *Nodipecten* Dall 1898

Lyropecten nodosus Linné   Lion's Paw
                                                        Plate 33b

North Carolina to Florida and the West Indies.

3 to 6 inches in size, rather heavy and strong-shelled. Characterized by the 7 to 9 large, coarse ribs which have large, hollow nodules. The entire shell also has numerous, much smaller, but distinct, riblets. The color is commonly dark maroon-red, but may be bright-red or orange. Fairly common offshore, especially on the west coast of Florida.

GENUS *Aequipecten* P. Fischer 1887
Subgenus *Aequipecten s. str.*

Aequipecten glyptus Verrill   Tyron's Scallop
                                                      Plate 33a

South of Cape Cod to the Gulf of Mexico.

1 to 2½ inches in size. Both valves rather flat. Shell somewhat lopsided and spathate in shape. About 17 ribs which start out as fine, sharp, slightly prickled ribs, but become flattened and indistinct or absent near the

margin of the valve. One valve pure-white, the other with broad, rose rays corresponding to the ribs. Internally white and with weak, fine ribs. Rare, but has been brought in by commercial trawlers. This is *P. tryoni* Dall.

Aequipecten phrygius Dall   Spathate Scallop

Off Cape Cod to east Florida and the West Indies.

About 1 inch in size. Characterized by its peculiar spathate or open-fan shape. With 17 sharp ribs. On closer inspection, it will be seen that each rib is composed of 3 rows of very fine, closely packed scales which are welded together to form a single rib. In cross-section, this would give the rib the shape of the letter M. Hinge-line straight with one ear slightly shorter than the other. Color dull-gray with indistinct blotches of dull-pink. Uncommonly dredged off Miama and the Lower Keys.

Aequipecten lineolaris Lamarck   Wavy-lined Scallop

Florida Keys to the Lesser Antilles.

1 to 2 inches in size, ears about equal. Valves moderately inflated. Surface highly glossy, the colored valve with about 18 very low, rounded ribs. Bottom valve white. Top valve rosy-tan with characteristic, numerous small, wavy, thin lines of pink-brown running concentrically. A few brown mottlings may be present. A very gorgeous and rare species dredged from 7 to 50 fathoms. *A. mayaguezensis* Dall and Simpson is this species.

Aequipecten muscosus Wood   Rough Scallop
Plate 34d, e

North Carolina to both sides of Florida and the West Indies.

¾ to 1¼ inches in size, both valves inflated and fairly deep. Hinge-ears equal to the width of the main part of the shell. 18 to 20 ribs, the center part of each bearing prominent, erect, concave scales, and on each side 2 rows of much smaller scales. Color orange-brown, red, lemon-yellow, orange, or commonly mottled with purple. Beach-worn specimens may lose most of their scaliness. Moderately common just offshore to 90 fathoms. Formerly called *exasperatus* Sby. and *fusco-purpureus* Conrad.

Subgenus *Plagioctenium* Dall 1898

Aequipecten irradians Lamarck   Atlantic Bay Scallop
Plate 33i

Nova Scotia to northern half of Florida and Texas.

2 to 3 inches in size. This is the common edible scallop of our east

coast. It is not a very colorful species, although its drab browns and grays are rarely enlivened with yellow. There are 3 distinct subspecies which previously have been little understood. Each has a distinct geographical range and peculiar habitat.

A. irradians irradians Lamarck. Nova Scotia to Long Island, N.Y. 17 to 18 ribs which are low and roundish in cross-section. Each valve is about the same fatness, and the lower one is only slightly lighter in color. Drab gray-brown with indistinct, darker-brown mottlings. The most compressed of the 3 subspecies. This is *borealis* Say.

A. irradians concentricus Say. New Jersey (rare), Virginia to Georgia and Louisiana to Tampa, Florida. 19 to 21 ribs which are squarish in cross-section. Lower valve (the lightest in color and commonly all white) is much fatter than the dull bluish gray to brown upper valve. Common.

A. irradians amplicostatus Dall (fig. 26f). Central Texas to Mexico and Colombia. Similar to *concentricus*, but with only 12 to 17 ribs; more gibbose; lower valve commonly white and with high, squarish to slightly rounded ribs. Common in Texas.

Aequipecten gibbus Linné   Calico Scallop
                                                                    Plate 33j

North Carolina to Florida, the Gulf of Mexico, and West Indies.

1 to 2 inches. A common, colorful scallop found abundantly in southern Florida a little offshore. Both valves quite fat. Ribs usually 20 (19 to 21), quite square in cross-section. Bottom valve commonly whitish with a little color; upper valve can be of many bright hues (lavender-rose, red, whitish with purple or reddish mottlings, etc.). This is *dislocatus* Say. If collecting in southeast Florida, do not confuse with *A. nucleus.*

Aequipecten gibbus nucleus Born   Nucleus Scallop
                                                                    Plate 34h

Southeast Florida and the West Indies.

1 to 1½ inches in size. This is a difficult subspecies to identify, and it is possible that it is only a form. It is rarely over an inch in size, has 1 to 3 more ribs than *gibbus*, is usually fatter, and is characteristically colored with small, chestnut mottlings on a cream background and commonly with snow-white specklings. Both or only one valve may be heavily colored. Never with the bright shades of orange, red, etc. Not uncommon in the Keys from low tide to a few fathoms on grass.

GENUS *Hinnites* Defrance 1821

Biologically speaking, this genus is really a *Chlamys* in which the adults are attached to rocks and become quite massive like *Spondylus*. For convenience, we are considering it a full genus.

Hinnites multirugosus Gale   Giant Rock Scallop
Plate 29a

Aleutian Island to Lower California.

Up to 8 inches in length. A heavy massive shell characterized by the early "Chlamys-like" shell at the beaks. Interior white with a purplish hinge area. Attached to rocks by the right valve. The ½-inch long young are almost impossible to separate from some species of *Chlamys*, except when they show a mauve spot on the inside of the hinge line on each side of the resilium pit, or if they show signs of distortion or a mottling pattern of color on the outside of the valves. Some young are bright-orange. A common species. Formerly known as *Hinnites giganteus* Gray. This is a regrettable name change which I have followed, since leading workers on the Pacific Coast have adopted it.

Family PROPEAMUSSIIDAE
GENUS *Propeamussium* Gregorio 1883

Propeamussium pourtalesianum Dall   Pourtales' Glass Scallop
Plate 27c

Southeast Florida and the West Indies.

½ inch in length. Valves very slightly convex. Shell extremely thin and transparent (like thin mica flakes). Each valve reinforced inside with about 9 rod-like, opaque white ribs. Exterior of one valve is smoothish, the other valve with numerous, microscopic, concentric threads. Common offshore. Frequently dredged off Miama by amateurs. There have been a number of other species described, some of which may only be forms of this variable species.

Family SPONDYLIDAE
GENUS *Spondylus* Linné 1758

Spondylus americanus Hermann   Atlantic Thorny Oyster
Plate 36b

South half of Florida and the West Indies.

3 to 4 inches in size. Spines 2 or less inches in length, usually standing fairly erect. Color variable: white with yellow unbones, red or purple; sometimes all rose, all cream or all pink. The young are much less spinose, and might be confused with *Chama* which, however, does not have the ball-and-socket type of hinge. Beautiful and large specimens are found clinging

to old wrecks in fairly deep water. Perfect specimens have recently sold for over $40. Formerly called *americanus* Lamarck, *echinatus* Martyn and *dominicensis* Röding. Sometimes called the Chrysanthemum Shell. Not uncommon.

Spondylus pictorum Schreiber   Pacific Thorny Oyster
Plate 36a

Gulf of California to Panama.

Up to 5 inches in size. The spines are 1½ inches or less in length and usually bent over. Color variable, and usually more brilliant than the Atlantic species. A popular, and now fairly high-priced collector's item. Often found on beaches with their spines worn off. They live in fairly deep water attached to rocks and wrecks.

Family LIMIDAE
Genus *Lima* Bruguière 1797

Lima lima Linné   Spiny Lima
Plate 35g

Southeast Florida and the West Indies.

1 to 1½ inches in height and pure-white in color. Sculpture of numerous even, radial ribs bearing many erect, sharp spines. The posterior ear is much smaller than the anterior one. No large posterior byssal gape as in *scabra*. Moderately common under coral stones in shallow water. This species and its various forms or subspecies (*squamosa* Lamarck, *multicostata* Sowerby, *caribaea* Orbigny and *tetrica* Gould) are found all over the world in tropical waters.

Lima pellucida C. B. Adams   Antillean Lima

North Carolina to both sides of Florida and the West Indies.

¾ to 1 inch in height, elongate, fragile, semi-translucent, white, with a large posterior gape and with a long, narrow anterior gape. Radial ribs small, fine, uneven in size and distribution. Hinge-ears almost equal in length. Closely related to *L. hians* Gmelin from Europe. A fairly common species which is often misidentified in collections as *L. inflata* Lamarck (not Gmelin). *L. antillensis* Dall is the same. In thicker and older specimens there is a small, pinhole depression in the hinge just off to one side of the ligamental area.

Subgenus *Ctenoides* Mörch 1853

Lima scabra Born   Rough Lima
Plate 35f, o

Southeast Florida and the West Indies.

1 to 3 inches in height, half as long. Sculpture coarse, consisting of irregular, radial rows of short, bar-like ribs, somewhat giving the appearance of shingles on a roof. Periostracum thin, dark- to light-brown. A common variation of this species (form *tenera* Sowerby) is startlingly different, in that the small radial ribs are much more numerous and much smaller (pl. 35h). Common under rocks in shallow water at low tide.

Subgenus *Mantellum* Röding 1798

Lima hemphilli Hertlein and Strong — Hemphill's Lima
Plate 29c

Monterey, California, to Mexico.

1 inch in length, white, obliquely elliptical in shape. With fine, irregular, radial ribs which are crossed by very fine, rough threads. Anterior and posterior margins smooth. This fairly common species has been erroneously called *dehiscens* Conrad and *L. orientalis* Adams and Reeve.

GENUS *Limatula* Wood 1839

Limatula subauriculata Montagu — Small-eared Lima

Greenland to Puerto Rico. Alaska to Mexico.

½ inch in height, ovate-oblong, greatly inflated (having the shape of the shell of a pistachio nut), and sculptured with numerous small, longitudinal riblets. On the inside of the valves there are 2 prominent, longitudinal riblets at the center of the shell. Periostracum over the white shell is yellowish brown. Moderately common in cooler waters from just offshore to 1000 fathoms.

GENUS *Limea* Bronn 1831

Limea bronniana Dall — Bronn's Dwarf Lima

North Carolina to Florida and the West Indies.

Very small, 5.0 mm. in height, ovate, superficially resembling a small *Cardium*. With about 25 to 30 strong, smooth, rounded, radial ribs. Microscopic, concentric scratches between the ribs. Inner margin of valves serrated and reinforced by small, round teeth. Shell pure-white in color. Hinge-ears with an internal set of 3 or 4 small teeth.

Superfamily *ANOMIACEA*
Family *ANOMIIDAE*
GENUS *Anomia* Linné 1758

The valve without the hole has 1 large and 2 small muscle scars. The shell is attached to a rock or wood surface by means of a calcified byssus which passes through a large notch in the right valve. The genus *Pododesmus* differs in having only 2 muscle scars in the top or holeless valve.

Anomia simplex Orbigny   Common Jingle Shell
                                                              Plate 35k

Cape Cod to Florida, the Gulf of Mexico, and the West Indies.

1 to 2 inches in size, irregularly oval, smoothish, thin but strong. The upper or free valve is usually quite convex; the lower valve is flattish and with a hole near the apex. Color either translucent-yellow or dull-orange. Some with a silvery sheen. Specimens buried in mud become blackened. Very commonly attached to logs, wharfs and boats.

Anomia aculeata Gmelin   Prickly Jingle Shell

Nova Scotia to North Carolina.

Rarely exceeding ¾ inch in size, irregularly rounded, moderately fragile. Upper valve convex, rough, often with small prickles. Lower valve flat and with a small hole near the hinge end. Color drab, opaque whitish tan. A common cold-water species attaching itself to rocks and broken shells.

Anomia peruviana Orbigny   Peruvian Jingle Shell
                                                              Plate 29e

San Pedro, California, to Peru.

1 to 2 inches in size, variable in shape, thin, partially translucent, smooth or with irregular sculpture; colored orange or yellowish green. Occurs between tides attached to rocks, other shells and waterlogged wood. Common.

GENUS *Pododesmus* Philippi 1837

The valve without the hole has 1 large and 1 small muscle scar.

Pododesmus macroschismus Deshayes   False Jingle Shell
                                                              Plate 29d

Alaska to Lower California.

1 to 4 inches in size. Radiating ribs very irregular and coarse. Color yellowish or greenish white, inner surface green and somewhat pearly. Lower valve with a large opening for the byssus. This is a very common species which is found attached to stones and wharf pilings from low-tide mark to about 35 fathoms. Often found on *Haliotis*. *P. cepio* Gray is a synonym.

Pododesmus rudis Broderip from Florida and the West Indies is very similar to the Pacific Coast species. Inch-long, brownish specimens are found in the crevices of coral boulders below low-water mark to several fathoms. Larger, more whitish specimens are found clinging to iron wrecks. Moderately common. *P. decipiens* Philippi. See plate 38b.

Superfamily OSTREACEA
Family OSTREIDAE
GENUS *Ostrea* Linné 1758

This genus used to include all of the oysters, but today several valid genera are recognized, so that only three American species are included in true *Ostrea*. These are *O. equestris* Say and *O. permollis* Sowerby from the Atlantic Coast and *O. lurida* Carpenter from the Pacific Coast. The European oyster, *O. edulis* Linné is also in this group. All of these oysters are relatively small. The eggs are fertilized and developed within the mantle chamber and gills. Usually around one million eggs are produced at one spawning. The prodissoconch hinge is long, the valves symmetrical. In the adults, the muscle scar is near the center of the shell and is not colored.

Ostrea equestris Say   Crested Oyster
                                                              Plate 28c
North Carolina, Florida, the Gulf States and West Indies.

1 to 2 inches in length, more or less oval, and with raised margins which are crenulated. The attached valve has a flat interior with a rather high, vertical margin on one side. Interior dull grayish with a greenish or opalescent-brown stain. Margin sometimes stained a weak-violet. Not very abundant except in some Florida bays. It lives in water that is much saltier than that in which *virginica* lives. Also named *spreta* Orbigny. *O. cristata* Born is quite different and is limited to South America.

Ostrea frons Linné   'Coon Oyster
                                                             Plate 28d
Florida, Louisiana and the West Indies.

1 to 2 inches in size. The radial plicate sculpture and corresponding sharply folded valve margins are characteristic of this intertidal species. Inner margins of valves closely dotted with minute pimples for nearly the entire circumference of the valves. Muscle scars located well up toward the hinge. Beaks somewhat curved. Interior translucent-white, exterior usually purplish red. Frequently elongate and attached to stems of trees by a series of clasping projections of the shell, but may be also oval in shape. *O. rubella* and *O. limacella* Lamarck are this species. *O. folium* Linné is a Philippine species.

Ostrea permollis Sowerby

Sponge Oyster
Plate 28b

North Carolina to Florida and the West Indies.

Rarely over 3 inches in size. Lives embedded in sponges with only the margins of the valves showing. The surface of the valves has a soft, silky appearance. Beak twisted back into a strong spiral. Exterior light-orange to tan; interior white. Inner margins with numerous small, round denticles. Common.

Another flat, but larger and light-shelled oyster, *Pycnodonta hyotis* Linné, is found in deep water attached to old wrecks off Florida and in the West Indies. It is immediately recognized by the peculiar structure of the shell which under a lens appears to be filled with numerous bubbles or empty cells, much like a bath sponge. It reaches a diameter of 3 or 4 inches, is generally circular in outline and may be colored whitish cream, brownish or even lavender. *Ostrea thomasi* McLean is this species according to the French worker, Gilbert Ranson.

Ostrea lurida Carpenter

Native Pacific Oyster
Plate 29f

Alaska to Lower California.

2 to 3 inches in length, of various shapes; generally rough with coarse concentric growth lines, but sometimes smoothish. Interior usually stained with various shades of olive-green, and sometimes with a slight metallic sheen. It occasionally has purplish brown to brown axial color bands on the exterior. This is the common intertidal native species of the Pacific Coast. A number of ecological forms have been described: *expansa* Carpenter, *rufoides* Cpr. and possibly *conchaphila* Cpr.

Genus *Crassostrea* Sacco 1897

This genus includes the commercially important American Oyster, *C. virginica* Gmelin, which was formerly placed in the genus *Ostrea*. In *Crassostrea*, the left or attached valve is larger than the right. The inner margin is smooth. The eggs are small, produced in large numbers at one spawning (over 50 million), and are fertilized and develop in the open waters outside of the parents. The muscle scar is usually colored. The prodissoconch hinge is short, and the valves asymmetrical. The Japanese Oyster (*C. gigas*), introduced to west American shores, the Portuguese Oyster (*C. angulata* Lamarck), and *C. rhizophorae* Guilding from Cuba also belong to this genus. *Gryphaea* Lamarck is a fossil genus which should not be associated with this genus.

Crassostrea virginica Gmelin   Eastern Oyster
                                                  Plate 28a

Gulf of St. Lawrence to the Gulf of Mexico and the West Indies.

2 to 6 inches in length. This is the familiar edible oyster which varies greatly in size and shape. The valve margins are only slightly undulating or are straight. The muscle scar is usually colored a deep purple, the rest of the shell being white inside and dirty-gray exteriorly. Beaks usually long and strongly curved. "Blue Points," a form originally harvested at Blue Point, Long Island, are rounded in shape and with a rather deep, lower valve. "Lynnhavens" are broad, elongate forms originally harvested at Lynnhaven Bay, Virginia. These variations are due to environmental differences. *C. brasiliana* Lamarck and *C. floridensis* Sowerby are this species.

Crassostrea rhizophorae Guilding (*brasiliana* of authors) is found in the Caribbean region, and it is a lightweight shell, deep-cupped, with a flat upper valve small and fitting well down into the lower valve. The inner margin of the lower, attached valve is splotched with bluish purple. Common.

Crassostrea gigas Thunberg   Giant Pacific Oyster
                                                 Plate 29g

British Columbia to California. Japan.

3 to 12 inches in length, of various shapes, but generally characterized by its large size, its coarse, widely spaced, concentric lamellae or very coarse longitudinal flutings or ridges on the outside. Interior enamel white, often with a faint purplish stain on the muscle scar or near the edges of the shell. Very rarely with a greenish stain. A common, large and marketable oyster introduced yearly into Canada and the United States from Japan. The form *laperousi* Schrenck is round. The typical *gigas* is the long, strap-like form. *O. gigas* Meuschen is an invalid name and does not preoccupy that of Thunberg's. Also known as the Japanese Oyster.

Order *EULAMELLIBRANCHIA*
Suborder *HETERODONTA*
Superfamily *ASTARTACEA*
Family *ASTARTIDAE*
Genus *Astarte* Sowerby 1816

Astarte borealis Schumacher   Boreal Astarte
                                                 Plate 28q

Arctic Seas to Massachusetts Bay. Alaska.

1 to 2 inches in length, ovate, moderately compressed. External ligament large. Concentric ridges strong near the beaks but disappearing near

the margins of the valves. Differing from *subequilatera* in being more elliptical in side view, in having the beaks near the middle, with weaker concentric ribs, and with the inner surface of the valve margins smooth. A common shallow-water species.

Astarte subequilatera Sowerby   Lentil Astarte
Plate 28-o

Arctic Seas to off Florida.

1 to 1½ inches in length, ovate, moderately compressed. External ligament small. Concentric ridges strong, rounded, evenly spaced. Internal margin of valves finely crenulate. Beaks turned slightly forward, often eroded. Color dull light- to dark-brown. Found in shallow water in the north and below 50 fathoms in the south. Common. Compare with *borealis*.

Astarte undata Gould   Waved Astarte
Plate 28r

New Brunswick to Maine.

Similar to *subequilatera*, but less elliptical, with its beaks near the center and with fewer and stronger concentric ridges. Probably the commonest Astarte in New England.

Astarte castanea Say   Smooth Astarte
Plate 28s

Nova Scotia to Cape Cod.

1 inch in length, as high, trigonal in shape, quite compressed. Beaks pointed and hooked anteriorly; external ligament small. Shell almost smooth, except for weak, low concentric lines. Color a glossy light-brown. Inner margin of valves finely crenulate. A commonly dredged species.

Astarte nana Dall   Southern Dwarf Astarte

North Carolina to Florida and the Gulf States.

¼ inch in length, slightly trigonal in shape, compressed. With or without about 25 well-developed, evenly spaced, rounded, concentric ridges. Ventral and inner edge of valves usually with 40 to 50 distinct small pits or crenulations. Shell cream, tan, brown or rose-brown in color with the beaks usually whitish. A very abundant species dredged in moderately shallow water, especially off eastern Florida.

Family CRASSATELLIDAE
Genus *Eucrassatella* Iredale 1924

Shell large, thick, equivalve, posteriorly rostrate; ligament and resilium adjacent and internal in a triangular resilifer; left valve with 2 diverging cardinal teeth; right valve with 3, of which the posterior one is more or less obsolete. 3 laterals in each valve. *Crassatella* Lamarck is fossil and not this genus. *Crassatellites* Krueger is believed to be invalid.

Eucrassatella speciosa A. Adams 1852

Gibb's Clam

Plate 30z

North Carolina to both sides of Florida and the West Indies.

1½ to 2½ inches in length, ⅔ as high, heavy, beaks at the center, and the shell somewhat diamond-shaped. Concentric sculpture of neat, rather heavy, closely packed ridges (about 15 per half inch). Lunule and escutcheon sunken, lanceolate in shape and about the same size as each other. Exterior with a thin, persistent, nut-brown periostracum. Interior glossy ivory with either a tan or pink blush. Moderately common just offshore in sand. *C. floridana* Dall is the same, being based on a young specimen. *E. gibbesi* Tuomey and Holmes 1856 is a synonym.

GENUS *Crassinella* Guppy 1874

Shell small, compressed, subtriangular, and slightly inequivalve. 2 cardinals in each valve. 1 anterior lateral in the right valve, 1 posterior lateral in the left valve.

Crassinella lunulata Conrad

Lunate Crassinella

Figure 28k

North Carolina to both sides of Florida and the West Indies.

¼ to ⅓ inch in length, as high, quite compressed, solid, with the tiny, closely pressed-together beaks at the middle or slightly toward the anterior end. Dorsal margins straight and about 90 degrees to each other, the anterior margin slightly longer and with a wider sunken area. The valves are peculiarly askew, so that the posterior dorsal margin of the left valve is more obvious than that of the right valve. Concentric sculpture of coarse but well-developed ribs (about 15 to 17 plainly visible). Color whitish or pinkish, interior commonly brown. Sometimes faintly rayed. A common shell from beach to 60 fathoms.

Crassinella mactracea Lindsley

Lindsley's Crassinella

Plate 30b

Massachusetts Bay to Long Island, New York.

Almost identical with *lunulata* from more southern waters, but more obese, with a more oval lunule, and generally with a chalky texture to the shell. Occasionally the ribs are less strongly developed. Common from just offshore to 30 fathoms.

Superfamily *CARDITACEA*
Family *CARDITIDAE*
Genus *Cardita* Bruguière, 1792

Shell small, thick, radially ribbed, quadrate, with a slight ventral gape and having a byssus. The animal has a marsupium to contain its eggs. Posterior right cardinal usually absent or almost so. This appears to be the accepted use of *Cardita* according to Winckworth, Chavan, Lamy and Dall.

Subgenus *Carditamera* Conrad 1838

Carditamera has shells which are more elongate and have strong lateral teeth.

Cardita floridana Conrad　　　　　　　　　　　Broad-ribbed Cardita

Plate 30*a*

Southern half of Florida and Mexico.

1 to 1½ inches in length, about half as high, elongate, inflated, solid and heavy. Surface with about 20 strong, rounded, raised, beaded, radial ribs. In live material, the gray periostracum obscures the color of the shell. Exterior whitish to gray with small bars of chestnut color on the ribs arranged in concentric series. Interior white with a small light-brown patch above the two muscle scars. Beaks close together. Lunule small, very deeply indented under the beaks. Ligament moderately large, visible from the outside. Very common on the west coast of Florida where it is washed ashore. Used extensively in the jewelry business.

Cardita gracilis Shuttleworth is doubtfully recorded from Florida but is known from Mexico to Puerto Rico. It is quite elongate, narrow at the anterior end, with larger, smoothish ribs, and the posterior lateral tooth is stained dark-brown. Uncommon.

The Pacific Coast species is *Cardita carpenteri* Lamy (pl. 29r) which is ½ inch long and ranges from British Columbia to Lower California in shallow to deep water. Its color is brownish gray with a purplish interior.

Subgenus *Glans* Mühlfeld 1811

Cardita dominguensis Orbigny　　　　　　　　　　　Domingo Cardita

North Carolina to southeastern Florida.

¼ inch in length, ovate, inflated; beaks close together, pointing toward each other, located nearer the anterior end. Lunule narrow, rough, ill-defined. Numerous strong radial ribs are weakly beaded. Color whitish with a rose tint. Moderately common from 1 foot to 70 fathoms on sandy bottoms. Compare with the commoner and closely resembling *Venericardia tridentata*.

GENUS *Venericardia* Lamarck 1801

Shell rounded-trigonal, with strong radial ribs which are commonly beaded; internal margins crenulate; right anterior cardinal and laterals absent. No byssus made.

Subgenus *Cyclocardia* Conrad 1867

Cyclocardia has whitish shells and a rough periostracum.

Venericardia borealis Conrad · · · · · · · · · · · · Northern Cardita

Plate 28t

Labrador to Cape Hatteras.

1 to 1½ inches in height, rounded, obliquely heart-shaped, thick and strong; beaks elevated and turned forward. Surface with about 20 rounded, moderately rough or beaded, radial ribs. Shell white, usually covered by a fairly thick, velvety, rusty-brown periostracum. Lunule small but very deeply sunk. Hinge strong; in the left valve the central tooth under the beak is large, triangular and curved. Very common on the Grand Banks where it serves as a food for fish.

V. novangliae Morse (Nova Scotia to New York) is similar, but is ovate, the length being slightly greater than the height of the shell. It is sometimes considered a variety of *borealis*.

Venericardia ventricosa Gould · · · · · · · · · · · · Stout Cardita

Plate 29l

Puget Sound to Santa Barbara Islands.

About ¾ inch in length, rounded-trigonal, moderately fat, lunule small; with about 13 rather wide, radial ribs which are bluntly beaded. Inner margins of the valves have prominent, squarish, widely spaced crenulations which correspond to the external ribs. There are two other forms, one from Monterey (*stearnsi* Dall), the other from Redondo Beach, which are very close, but their distinctiveness and proper names are yet to be decided. The latter form is *V. redondoensis* "Burch" P. Morris 1952. *C. ventricosa* is dredged fairly commonly.

Subgenus *Pleuromeris* Conrad 1867

Venericardia tridentata Say   Three-toothed Cardita

North Carolina to all of Florida.

¼ inch in length and height, trigonal in shape, inflated, with 15 to 18 heavily beaded strong radial ribs. Beaks close together, pointing slightly forward. Lunule oval, sharply impressed, smoothish. Escutcheon small, narrow. External color grayish brown, sometimes with red-brown mottlings. Hinge-teeth often purplish blue. Interior of valve stained with light-brown on white background. A common, moderately shallow-water species, usually confused with *Cardita dominguensis* which, however, lacks the strong tridentate hinge, is ovate in shape, whose ribs are weakly beaded and whose beaks point toward each other.

Venericardia perplana Conrad   Flattened Cardita
                                                  Plate 281
North Carolina to southern half of Florida.

¼ inch in size, similar to *V. borealis* but much smaller, without a periostracum, pinkish or mottled brown, and more oblique. The ribs are wider, and close to each other. The subspecies *flabella* Conrad from Tampa Bay, Florida, has fewer ribs which are squarish and separated by furrows almost equal in size to the ribs themselves. *V. perplana* is common, *flabella* only locally found at certain seasons in few numbers.

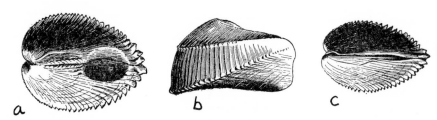

FIGURE 76. Kelsey's Milner Clam, *Milneria kelseyi* Dall, ⅛ to ¼ inch (southern California).

GENUS *Milneria* Dall 1881

Milneria kelseyi Dall   Kelsey's Milner Clam
                                           Figure 76
Monterey to Lower California.

⅛ to ¼ inch in length. An extraordinary clam which resembles a tiny Brazil nut. The bottom margins of the valves are pushed in to form a small cup-shaped hollow. Into this, the females put the 50 or so young whose shells are smooth and round. The hollow is covered over by a sheath of peri-ostracum. Hinge of adult with large triangular tooth in left valve which fits snugly between two smaller ones in the right valve. Found in shallow water under stones. External sculpture of scaled ribs and concentric ridges. Color light-brown. Shell thick, translucent glaze inside.

Family CORBICULIDAE
Genus *Polymesoda* Rafinesque 1820

Polymesoda caroliniana Bosc   Carolina Marsh Clam
Plate 30bb

Virginia to north half of Florida and Texas.

1 to 1½ inches in length, about as high, subtriangular in outline, rather obese and with a strong shell. Exterior of smoothish shell is covered with a very fuzzy or minutely scaled periostracum which is mostly glossy-brown and rather thin. Interior white, rarely stained with purple. Each hinge with 3 small, almost vertical, equally sized teeth below the beaks and each hinge with 1 anterior and posterior lateral. Ligament external, long, narrow and dark-brown. Common at the mouths of rivers where the influence of the tides is felt.

Genus *Pseudocyrena* Bourguignat 1854

Pseudocyrena floridana Conrad   Florida Marsh Clam
Plate 30y

Key West to northern Florida and to Texas.

1 inch in length, quite similar to *Polymesoda caroliniana*, but more vari-able in shape (ovalish to elongate), without the fuzzy periostracum, with its beaks never eroded away and with 2 long, slender anterior and posterior laterals. Exterior with irregular growth lines, dull dirty-white, commonly flushed with purple or pink. Interior white with a wide margin of deep purple or entirely purple. Brackish warm water in mud. Common.

Family ARCTICIDAE
Genus *Arctica* Schumacher 1817
(*Cyprina* Lamarck)

Arctica islandica Linné   Ocean Quahog
Plate 32f

Newfoundland to off Cape Hatteras, North Carolina.

3 to 5 inches in length, almost circular in outline, rather strong, porcel-laneous, but commonly chalky. Exterior covered with a brown to black, rather thick periostracum. The posterior laterals and the absence of a pallial

sinus will distinguish this clam from the true Quahogs (see *Mercenaria mercenaria*). A common, commercially dredged species found in sandy mud from 5 to 80 fathoms. This is the only living species in this family. There are numerous fossil species. Also called the Black Clam and Mahogany Clam.

Family TRAPEZIIDAE
GENUS *Coralliophaga* Blainville 1824

Shell cigar-shaped, with the beaks at the anterior end. 3 cardinals in each valve, the posterior one extending along the hinge line like a lateral. Posterior muscle scar considerably larger than the anterior one. Some workers have placed this genus in the *Petricolidae*.

Coralliophaga coralliophaga Gmelin　　　　　　　　Coral-boring Clam

Plate 28p

West coast of Florida to Texas and the West Indies.

½ to 1½ inches in length, oblong to elongate, and quite thin. Very finely sculptured with radial threads. Concentric lamellations present at the posterior end. Exterior yellowish white; interior white. This shell is very similar in appearance to *Lithophaga antillarum*, but may be told from it by the presence of distinct teeth in the hinge. This is an uncommon species which lives in the burrows of other rock-boring mollusks.

Superfamily DREISSENACEA
Family DREISSENIDAE
GENUS *Congeria* Partsch 1835
Subgenus *Mytilopsis* Conrad 1857

Congeria leucophaeata Conrad　　　　　　　　Conrad's False Mussel

New York to Florida to Texas and Mexico.

½ to ¾ inch in length, superficially resembling a *Mytilus* or *Septifer* because of its mussel-like shape. The Septifer-like shelf at the beak end has a tiny, downwardly projecting, triangular tooth on the side facing the long, internal ligament. The hinge has a long thin bar under the ligament. Exterior bluish brown to tan with a thin, somewhat glossy periostracum. Interior dirty bluish tan. This common bivalve attaches itself by its short byssus to rocks and twigs in clumps which resemble colonies of *Mytilus*. Found in brackish to fresh water near rivers.

Superfamily *LUCINACEA*
Family *DIPLODONTIDAE*
Genus *Diplodonta* Bronn 1831

Shell thin, orbicular and strongly inflated. There are 2 cardinal teeth in each valve. The left anterior and right posterior ones are split or bifid. Laterals obscure or absent. *Taras* Risso, commonly used in place of the name *Diplodonta*, is a doubtful name which has been recently abandoned.

Subgenus *Diplodonta* s. str.

Diplodonta punctata Say   Common Atlantic Diplodon

North Carolina to both sides of Florida and the West Indies.

⅓ to ¾ inch in length, moderately strong, almost orbicular, well-inflated and pure-white in color. Smooth near the beaks, elsewhere very finely scratched with concentric lines and commonly with distantly spaced, coarse growth lines. Fairly common in shallow to deep water.

Diplodonta orbella Gould   Pacific Orb Diplodon

Alaska to Panama.

¾ to 1 inch in length, almost circular in outline, quite inflated and smoothish except for moderately coarse growth lines. Beaks small, pointing slightly forward. Ligament posterior to beaks is long, raised and conspicuous. 2 rather large teeth in each valve below the beaks. Left anterior and right posterior teeth split. In many shallow-water localities, this clam builds a compact nest of periostracal material and detritus. In its more southerly range, specimens are usually more compressed, less orbicular in shape and more glossy externally (subspecies *subquadrata* Carpenter). Alias *Taras orbella*.

Subgenus *Phlyctiderma* Dall 1899

Diplodonta semiaspera Philippi   Pimpled Diplodon

North Carolina to Florida, Texas and the West Indies.

Rarely over ½ inch in length, similar to *D. punctata*, but chalky-white externally and with numerous concentric rows of microscopic pimples. Moderately common in sand below low-water mark to 40 fathoms. Alias *D. granulosa* C. B. Adams.

Subfamily *THYASIRINAE*
Genus *Thyasira* Lamarck 1818

Shell subglobular and of an earthy texture; umbones directed forward; posterior region of valve deeply furrowed; lunule absent; ligament in a groove and partly external; hinge without teeth and indented in front of the umbo; pallial line without a sinus.

Thyasira trisinuata Orbigny   Atlantic Cleft Clam

Nova Scotia to south half of Florida and the West Indies.

¼ to ½ inch in length, oblong, fragile and translucent-white. Hinge weak and with only a very long, weak posterior lateral. Posterior slope of shell with 2 strong, radial waves or rounded grooves. Moderately common in dredgings from 15 to 90 fathoms on sandy bottom.

Thyasira gouldi Philippi (Labrador to off North Carolina) is similar, but only ¼ inch in size, almost round but slightly higher, and with a weak yellowish periostracum. The hinge lacks teeth. Common offshore to 60 fathoms. Called Gould's Cleft Clam.

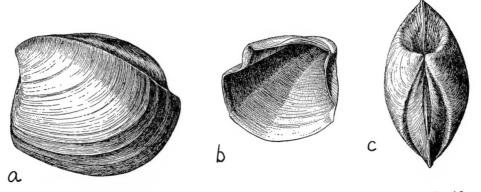

FIGURE 77. Pacific Cleft Clams. **a**, *Thyasira bisecta* Conrad, 1 inch (Pacific Coast); **b** and **c**, *Thyasira excavata* Dall, ½ inch (Gulf of California).

Thyasira bisecta Conrad   Pacific Cleft Clam
                                                        Figure 77a

Alaska to Oregon.

1 to 1½ inches in length, almost square in side view and moderately obese. Characterized by the almost vertical, straight, anterior end which is 90 degrees to the dorsal margin. Ligament long and narrow and flush with the dorsal margin of the shell. There is a deep, prominent radial furrow on the exterior running posteriorly from the beaks. Shell chalky-white, com-

monly with a thin, yellowish gray periostracum. Irregular coarse growth lines present. Uncommon from 4 to 139 fathoms. Closely related to *T. disjuncta* Gabb, if not that species.

Family *LUCINIDAE*
GENUS *Lucina* Bruguière 1797

Shell orbicular, strong and laterally compressed. Cardinal teeth small, obscure in the adults, but the laterals are well-developed. The use of *Lucina* here is based on Anton's designation of *pensylvanica* Linné as the genotype. The genus *Linga* Gregorio is this genus.

Subgenus *Lucina s. str.*

Lucina pensylvanica Linné   Pennsylvania Lucina
Plate 38h
North Carolina to south Florida and the West Indies.

1 to 2 inches in length, ovate, usually quite inflated. Concentric ridges very delicate and distinct. Color pure-white with a thin yellowish periostracum. Lunule heart-shaped, well-marked and raised at the center. The furrow from the beak to the posterior ventral edge of the valve is very pronounced. Beachworm specimens become smooth and shiny-white. The species name was incorrectly spelled by Linné. Moderately common in shallow water.

Subgenus *Here* Gabb 1866

Lucina sombrerensis Dall   Sombrero Lucina
Figure 78b
Southern Florida.

¼ inch in length, oval, greatly inflated and pure white in color. No radial sculpture. Concentric riblets numerous, sharp and irregularly crowded. Concentric growth irregularities commonly make the outer surface wavy. Commonly dredged off Miami from 20 to 90 fathoms.

Subgenus *Bellucina* Dall 1901

Lucina amiantus Dall   Lovely Miniature Lucina
Figure 78c
North Carolina to both sides of Florida.

¼ to ⅜ inch in length, not quite so high, quite obese, thick-shelled, pure-white in color and beautifully sculptured with 8 to 9 wide, rounded, radial ribs across which run numerous, small concentric riblets. Near the posterior upper margin of the shell there is a radial row of about 8 to 11 small, scale-like nodes. Behind the tiny, curved beaks there is an ovalish, heart-shaped depression. Internal margin of valves strongly crenulated with tiny teeth. Adults are commonly misshapen by concentric growth stops. Common from shallow water to 68 fathoms. Compare with *L. multilineata*.

FIGURE 78. American Lucinas. ATLANTIC: **a,** *Phacoides filosus* Stimpson, 1 to 3 inches; **b,** *Lucina sombrerensis* Dall, ¼ inch; **c,** *Lucina amiantus* Dall, ⅜ inch; **d** and **e,** *Lucina leucocyma* Dall, ¼ inch; **f,** *Lucina multilineata* Toumey and Holmes, ⅜ inch. PACIFIC: **g,** *Lucina approximata* Dall, ¼ inch; **h,** *Lucina tenuisculpta* Cpr., ½ inch.

Subgenus *Parvilucina* Dall 1901

Lucina multilineata Tuomey and Holmes   Many-lined Lucina

Figure 78f

North Carolina to both sides of Florida.

⅜ to ¼ inch in length, almost circular in shape, very obese, moderately thick-shelled, white, and very finely sculptured. Somewhat like *L. amiantus*, but without radial ribs, except for exceedingly fine threads seen best near the beaks. Concentric sculpture of numerous, rather irregular, growth threads. The shell commonly continues growth after a long rest, thus causing an irregular, concentric hump in the shell. Inner margin very finely denticulate. *P. crenella* Dall is the same species. Common from beach to 120 fathoms.

Lucina tenuisculpta Carpenter
Fine-lined Lucina
Figure 78h

Bering Sea to Lower California.

½ inch in length, slightly less in height, oval in outline, chalky-white and with a thin, grayish or yellowish green periostracum. Sculpture of numerous, small, weak, raised, radial threads. Concentric growth lines fine and irregularly placed. Beaks fairly prominent and pressed closely together. Behind them, the narrow, depressed ligament is visible from the outside. In front is the small, heart-shaped, depressed lunule. Inner margin of valves finely toothed. Common just offshore.

Lucina approximata Dall
Approximate Lucina
Figure 78g

Monterey, California, to Panama.

¼ inch or less in size. Very similar to *tenuisculpta*, but smaller, almost round in outline, more inflated and with fewer and quite strong, radial riblets. Periostracum very thin, commonly worn off. Shell texture less chalky. Common in sandy mud just offshore to 48 fathoms.

Subgenus *Pleurolucina* Dall 1901

Lucina leucocyma Dall
Four-ribbed Lucina
Figure 78d, e

North Carolina to southeast Florida and the Bahamas.

¼ inch in length, roughly oval, fairly thick-shelled, inflated and white in color. With 4 conspicuous, large, rounded, radial ribs, and with numerous, small, crowded, squarish, concentric riblets. The inner margins of the valves are finely denticulate. A common, bizarrely sculptured species found from low water to several fathoms.

Subgenus *Pseudomiltha* P. Fischer 1885

Lucina floridana Conrad
Florida Lucina
Plates 38i, 30aa

West coast of Florida to Texas.

1½ inches in length, almost circular, compressed, smoothish, except for a few weak, irregular growth lines. Pure-white with a dull-whitish, flaky periostracum. The beaks point forward, and in front of them there is a deep, small pit. Hinge plate fairly wide and strong, but the teeth are weakly defined. Moderately common in shallow water to a few fathoms.

Genus *Phacoides* Gray 1847

Shell orbicular, quite compressed. Sculpture mostly concentric. Cardinal teeth obsolete in adults, but the laterals are well-developed. *Phacoides* Blainville is the same but is not considered valid. *Dentilucina* Fischer is the same.

Subgenus *Phacoides s. str.*

Phacoides pectinatus Gmelin   Thick Lucina
                                                                 Plate 38g
North Carolina to Florida, Texas and the West Indies.

1 to 2½ inches in length, ovate, compressed, white or flushed with bright-orange. Concentric ridges moderately sharp, usually unequally spaced. Ligament partially visible from the outside. Lunule strongly raised into a rather thin, rough blade. Anterior and posterior lateral tooth strong. Cardinals very weak. Moderately common in shallow water. Alias *Lucina jamaicensis* Lamarck. Do not confuse with *P. filosus*.

Subgenus *Lucinisca* Dall 1901

Phacoides nassula Conrad   Woven Lucina

North Carolina to Florida, Texas and the Bahamas.

½ inch in length, almost circular, inflated, strong and pure white. Sculpture of strong, closely spaced, concentric and radial ribs. These form a reticulate, rough surface. Where the ribs cross each other there is a tiny, raised scale. The ventral margin of the valve is strongly beaded by the distal ends of the axial riblets. Common in shallow water to 100 fathoms.

Phacoides nuttalli Conrad   Nuttall's Lucina
                                                                 Plate 31g
Santa Barbara, California, to Manzanillo, Mexico.

1 inch in length, circular, moderately inflated and with a fine, sharp, cancellate sculpturing. The shell is divided off at the anterior and upper portion into a slightly more compressed region which is less sculptured concentrically. Lunule very deep, short and larger in the left valve. Moderately common offshore in sand.

The subspecies *centrifuga* Dall, from Lower California, has stronger and distantly spaced, concentric, raised lines.

Subgenus *Lucinoma* Dall 1901

Phacoides filosus Stimpson

Northeast Lucina
Plate 38j; figure 78a

Newfoundland to north Florida and the Gulf States.

1 to 3 inches in length (south of North Carolina rarely over 1½ inches), almost circular, compressed, white, with a thin, yellowish periostracum. Beaks small, close together and centrally located. Sculpture of sharp, raised, thin, concentric ridges each about ⅛ inch apart. The young commonly lack these ridges. No anterior lateral tooth present. Common offshore. Do not confuse with *pectinatus* which has a strong anterior lateral tooth, is tinted inside with orange and whose concentric ridges are unevenly spaced.

Phacoides annulatus Reeve

Western Ringed Lucina
Figure 28f

Alaska to southern California.

2 to 2½ inches in length, oval to circular and slightly inflated. With strongly raised, concentric threads about 1/16 inch apart. Shell chalky-gray to white, overlaid by a thin, greenish-brown periostracum. Fairly commonly dredged from 8 to 75 fathoms. The Tertiary fossil species *acutilineatus* Conrad may be the same.

Genus *Anodontia* Link 1807

Shell large, obese, fairly thin and subcircular in outline. Hinge without distinct teeth. Anterior muscle scar long and parallels the pallial line.

Anodontia alba Link

Buttercup Lucina
Plate 38f

North Carolina to Florida, the Gulf States and West Indies.

1½ to 2 inches in length, oval to circular, inflated and fairly strong. Hinge with very weak teeth, the posterior lateral being the most distinct. Exterior dull-white with weak, irregular concentric growth lines. Interior with a strong blush of yellowish orange. A common species used in the shellcraft business. This is *Lucina chrysostoma* Philippi.

Anodontia philippiana Reeve

Chalky Buttercup
Plate 38e

North Carolina to east Florida, Cuba and Bermuda.

2 to 4 inches in length, very similar to *A. alba*, but with a more chalky shell, never with orange color, interior usually pustulose, and the long, anterior muscle scar juts away from the pallial line at an angle of about 30 degrees instead of paralleling it as in *alba*. An uncommon species, commonly confused with *alba*. It lives down to 50 fathoms but at times is washed ashore. *A. schrammi* Crosse is this species.

Genus *Codakia* Scopoli 1777

Shell large, orbicular, moderately compressed. Hinge of right valve with a prominent anterior lateral which is typically close to the cardinals (an anterior, a posterior and a middle cardinal). Hinge of left valve with a large double anterior lateral, only 2 cardinals, and with a small, double posterior lateral.

Subgenus *Codakia* s. str.

Codakia orbicularis Linné   Tiger Lucina
                                                                Plate 38d

Florida to Texas and the West Indies.

2½ to 3½ inches in length, slightly less in height, well-compressed, more or less orbicular in outline, thick and strong. Beaks and ¼ inch of subsequent growth smoothish. Remainder of the shell roughly sculptured by numerous coarse radial threads which are crossed by finer concentric threads. This commonly gives the radial ribs a beaded appearance. Exterior white. Interior white to pale-lemon, commonly with a rose tinge on the ends of the hinge or along the margins of the valves. Lunule just in front of the beaks is deep, heart-shaped, small and nearly all on the right valve. A common tropical species. Do not confuse with *C. orbiculata*.

Codakia costata Orbigny   Costate Lucina

North Carolina to southeast Florida and the West Indies.

½ inch in length, variable in shape, but usually orbicular, quite obese, white to yellowish in color. With fine radial ribs, usually in pairs which are crossed by very fine concentric threads. Beaks also· with this sculpturing. Lunule small, indistinct, lanceolate, slightly more on the right valve. Compare its poorly defined lunule with those of *orbicularis* and *orbiculata*. Moderately common offshore on sandy bottoms.

Subgenus *Epilucina* Dall 1901

Codakia californica Conrad   Californian Lucina
                                                                Plate 31c

Crescent City, California, to Lower California.

1 to 1½ inches in length, oval to circular, moderately inflated. Exterior dull-white with numerous, crowded, rather distinct, but small, concentric threads. Lunule of right valve like a small, depressed, lanceolate shield which fits snugly into a similarly shaped recess in the left valve. A common littoral species in southern California and down to 78 fathoms. Do not confuse with large specimens of *Diplodonta*.

Subgenus *Ctena* Mörch 1860

Codakia orbiculata Montagu   Dwarf Tiger Lucina
                                                      Plate 30l

North Carolina to the south half of Florida and the West Indies.

1 inch or less in length, very similar to *orbicularis*, but with a large, elongate lunule in front of the beaks (instead of small and heart-shaped), and with stronger, less numerous, commonly divaricate ribs which are noticeable right up to the ends of the beaks. This species is much fatter and never has pink coloring inside. Common in sand from low water to 100 fathoms.

The form *filiata* Dall has finer sculpturing much like *orbicularis*, is often yellowish in color, but can be readily distinguished from the latter by its elongate lunule. Common in the Gulf of Mexico.

Genus *Divaricella* von Martens 1880

Divaricella quadrisulcata Orbigny   Cross-hatched Lucina
                                                      Plate 30m

Massachusetts to south half of Florida and the West Indies.

½ to 1 inch in length, almost circular, moderately inflated, and glossy-white in color. Sculpture of fine, criss-cross or divaricate, impressed lines. Inner margin minutely impressed. A very common species washed ashore on sandy beaches. It is used extensively in the shellcraft business. *D. dentata* Wood from the West Indies is very similar, but its inner margin is smooth.

Family *CHAMIDAE*

a. Shell equivalve, with a distinct lunule; radial rows of spines   .   .   .
                                                         . *Echinochama*
aa. Shell very inequivalve; no lunule:

  b. Umbones turning from right to left; attached by left valve   . *Chama*
  bb. Umbones turning from left to right; attached by the right valve.   .
                                                          . *Pseudochama*

Genus *Chama* Linné 1758

Chama macerophylla Gmelin   Leafy Jewel Box

Plate 37b; figure 79b

North Carolina to southeast Florida and the West Indies.

This is the most common and most brightly hued Atlantic species. In quiet waters it may develop spine-like foliations to such an extent that it resembles the Spiny Oyster, *Spondylus*. Exterior variously colored: lemon-yellow, reddish brown, deep- to dull-purple, orange, white, or a combination of these colors. Inner edges of the valves have tiny, axial ridges or crenulations. The scale-like fronds have minute radial lines. Compare with *sinuosa*.

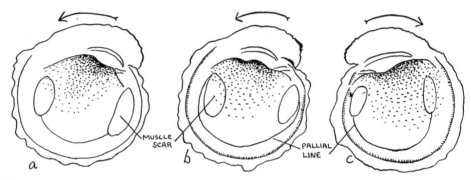

FIGURE 79. Atlantic Chamas. Diagrammatic drawings of the deep valves, showing direction of growth and the juncture of the pallial line and muscle scars. a, *Chama sinuosa* Broderip; b, *Chama macerophylla* Gmelin; c, *Pseudochama radians* Lamarck.

Chama congregata Conrad   Little Corrugated Jewel Box

Plate 37d

North Carolina to Florida and the West Indies.

Rarely over 1 inch in size. This species closely resembles the common *macerophylla*, but in place of numerous foliations there are low axial corrugations or wavy cords. The unattached valve may have a few short, flat spines. There are fine crenulations on the inner margins of the valves. The color is usually dull with darker specklings. In rocky areas they live in crevices and under stones. Commonly found attached to pen and ark shells.

Chama sinuosa Broderip   White Smooth-edged Jewel Box

Figure 79a

South half of Florida and the West Indies.

1 to 3 inches in size. The color is always whitish, although the interior may be stained with dull-green. There are no crenulations on the inner edges

of the valves. The pallial line runs directly to the anterior muscle scar and not past the end as in the other species. This is a reef species. An ecological variety of heavy shell has been named *firma* Pilsbry and McGinty 1938.

Chama pellucida Broderip   Clear Jewel Box
Plate 37a

Oregon to Chile.

1½ to 3 inches in size, with frond-like, smoothish foliations. Color opaque to translucent-white. Interior chalk-white, the margins minutely toothed or crenulate. Commonly found attached to pilings, breakwaters and floating wood. Also dredged down to 25 fathoms.

GENUS *Pseudochama* Odhner 1917

These are mirror images of the chamas. According to Odhner, the anatomy and prodissoconchs differ in the two genera.

Pseudochama radians Lamarck   Atlantic Left-handed Jewel Box
Plate 37c; figure 79c

Southern Florida and the West Indies.

1 to 3 inches in size. This is the only species of *Pseudochama* in eastern America. It is not very colorful, and ranges from a dull-white to a dull purplish red. The interior is commonly stained with mahogany-brown. Crenulations are present on the inner edges of the valves. In shape, it is a mirror image of *sinuosa*. *P. ferruginea* Reeve is considered a synonym. Common.

Pseudochama exogyra Conrad   Pacific Left-handed Jewel Box

Oregon to Panama.

Similar to *pellucida*, but attached by the right valve which, when viewed from the inside, is arched counterclockwise. The opaque whitish area inside is generally not bordered by tiny crenulations. A common intertidal species.

Pseudochama echinata Broderip in the Gulf of California is a popular shell which is characterized by a watermelon-red hinge and purple-stained interior.

Pseudochama granti Strong (Grant's Chama), dredged off central California and Catalina Island, is about 1 inch in size, with prickly spines on the underside of the attached, cup-formed valve. One end of the valve is tinted with rose inside and out. Not common.

Genus *Echinochama* P. Fischer 1887

Echinochama cornuta Conrad   Florida Spiny Jewel Box

Plate 37g

North Carolina to both sides of Florida to Texas.

1 to 1½ inches in length, quadrate in outline and rather obese and heavy. Lunule distinct and broadly heart-shaped. With 7 to 9 rows of moderately long, stoutish spines, between which the shell is grossly pitted. Exterior creamy-white; interior white or flushed with bright pinkish mauve. Attached to a small pebble or broken shell by the right valve. Common from 3 to 40 fathoms, and commonly washed ashore.

Echinochama arcinella Linné (True Spiny Jewel Box, pl. 37h) from the West Indies to Brazil has 16 to 35 (commonly 20) radial rows of slender spines. The shell is not as obese nor as heavy as *cornuta*. The subspecies *californica* Dall (pl. 37e) is very similar, with slightly longer spines and with a more compressed shell. It ranges from the Gulf of California to Panama in offshore water.

Superfamily LEPTONACEA
Family LEPTONIDAE

A group of small, fragile, inflated, translucent clams which are parasitic or commensal on other marine creatures or are active crawlers like the gastropods. Most species brood their young inside the mantle cavity. The family is also named *Erycinidae* and *Kelliidae*.

Genus *Kellia* Turton 1822

Shell unsculptured, inflated and oval-oblong. Lateral teeth present. 2 cardinal teeth in the right valve.

Kellia laperousi Deshayes   La Perouse's Lepton

Alaska to Panama.

¾ to 1 inch in length, oval-oblong, rather obese and with small beaks near the center. Shell fairly strong, chalk-white, but commonly covered with a smooth, glossy, greenish to yellowish-brown periostracum which, however, is commonly worn away in the beak area. Very common. Found attached to wharf pilings among mussels and chama shells.

Genus *Lasaea* Brown 1827

Shell very small, beaks nearer one end. Teeth the same as in *Kellia*.

Lasaea cistula Keen   Little Box Lepton

Southern half of California to Peru.

$\frac{1}{16}$ of an inch in length (one of the smallest of our American clams), oval-oblong to quadrate, with one end slightly more rounded. Beaks slightly nearer the posterior end. Shell very obese to moderately inflated. Color light-tan with dark carmine around the dorsal margin area, and commonly blushed on the sides with light-carmine. Coarse, concentric growth lines, especially in the adults. Periostracum thin and yellowish tan. Found nestled together in great numbers attached to seaweed holdfasts and among mussels.

Lasaea subviridis Dall (British Columbia and south) is much more compressed, smoother, with smaller beaks, but otherwise similar to *cistula*. Common.

GENUS *Pseudopythina* P. Fischer 1884

Shell small and quadrangular. Lateral teeth absent; 1 cardinal tooth in each valve.

Pseudopythina rugifera Carpenter   Wrinkled Lepton
                                                         Figure 80a

Alaska to Lower California.

$\frac{1}{2}$ to $\frac{3}{4}$ inch in length, oval-oblong, moderately obese, fairly fragile, beaks close together and located about the middle of the shell. Shell white, but in live specimens covered with a thin, light-brown, semi-glossy periostracum which is feebly and concentrically wrinkled. The ventral edge of the valves is slightly indented in the middle in some specimens. May be found attached to crustaceans and the sea mouse, *Aphrodita*, or be free.

Pseudopythina compressa Dall (Alaska to Mexico, the Compressed Lepton) is similar in size and outline, but is considerably compressed (thinner), smoothly polished in the beak area and with much less periostracum. Common.

GENUS *Mysella* Angas 1877

Two cardinal teeth in the right valve, none in the left. *Rochefortia* Velain is this genus but a later name by several months.

Mysella planulata Stimpson   Atlantic Flat Lepton

Nova Scotia to Texas and the West Indies.

$\frac{1}{8}$ inch in length, oval-oblong in side view, well-compressed, and fairly fragile. Beaks small, ¾ the distance back from the anterior end. Dorsal margin of valves pushed in, both in front and back of the beaks. There is no thickening of the hinge line directly below the beaks. Color white, with a thin, nut-brown, smoothish periostracum. Moderately common attached to buoys, eel-grass and wharf pilings.

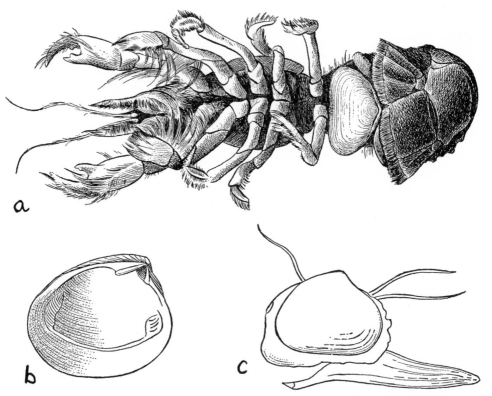

FIGURE 80. Pacific Lepton Clams. **a**, *Pseudopythina rugifera* Cpr., ½ inch, attached to the underside of a crawfish; **b**, *Mysella tumida* Cpr., ⅛ inch; **c**, animal of the clam, *Bornia longipes* Stimpson, ⅓ inch (Carolinas).

Mysella golischi Dall   Golisch's Lepton

Southern third of California.

¼ inch in length, oval-oblong in side view, moderately compressed and rather fragile. Beaks small, ¾ the distance back from the anterior end. The dorsal margin of the valve is pushed in slightly just anterior to the beak. Shell white, semi-transparent, with its glossy exterior having irregular, concentric wrinkles. In live specimens, there is a thin yellowish brown periostracum. These clams are found attached to the gills or legs of the large sand crab, *Blepharopoda occidentalis*. Common. *M. pedroana* Dall, known from a single specimen, is much more oblique in shape, resembling the equally rare *Erycina fernandina* Dall from off Florida.

Mysella tumida Carpenter

Fat Pacific Lepton
Figure 80b

Alaska to Lower California.

⅛ to ³⁄₁₆ of an inch in length, moderately compressed, somewhat triangular in shape. The tiny beaks are almost at the very posterior end. Shell dull-white, but commonly covered with a light-brown, smoothish periostracum which is faintly marked with concentric, microscopic wrinkles. The hinge teeth are large in comparison to those in other species. Common from low water to 99 fathoms. Has been found in duck stomachs.

Superfamily CARDIACEA
Family CARDIIDAE
Subfamily TRACHYCARDIINAE
GENUS *Trachycardium* Mörch 1853

Trachycardium muricatum Linné

Yellow Cockle
Plate 39p

North Carolina to Florida, Texas and the West Indies.

2 inches in height, subcircular, with 30 to 40 moderately scaled, radiating ribs. Externally light-cream with irregular patches of brownish red or shades of yellow. Interior commonly white, rarely yellow-tinted especially in Florida. A very common, shallow-water species. Compare with *egmontianum* and *magnum* which are both more elongate.

Trachycardium egmontianum Shuttleworth

Prickly Cockle
Plate 39-o

North Carolina to south Florida and the West Indies.

2 inches in height, with 27 to 31 strong, prickly, radial ribs. Externally whitish to tawny-gray with odd patches of weak yellow, brown or dull-purple. Interior glossy, commonly brightly hued with salmon, reddish and purple. Do not confuse with *muricatum* which is more oval, has more ribs which are not sharply scaled at the center of the shell and is commonly only yellowish inside. A common shallow-water species, especially on the Gulf side of Florida.

Trachycardium isocardia Linné from the West Indies has larger and slightly different scales, 32 to 37 ribs and has not been recorded from Florida.

Trachycardium magnum Linné

Magnum Cockle

Lower Florida Keys and the West Indies.

2 to 3½ inches in height, elongate, with 32 to 35 mostly smooth ribs. The ribs at the posterior end have small, tooth-like scales. Middle ribs completely smooth and squarish. Externally light-cream with irregular patches of reddish brown. Interior china-white with the deepest part flushed with orange-buff. As a rule, the posterior margin is pale-yellow, merging into pale-purple at the extreme edge. A West Indian species which has been found on the most southerly keys.

Trachycardium quadragenarium Conrad — Giant Pacific Cockle

Plate 31a

Santa Barbara to Lower California.

3 to 6 inches in size, commonly slightly higher than long, inflated, and with 41 to 44 strong, closely set, squarish, radial ribs which bear small, upright, strong, triangular spines, especially at the anterior, posterior and ventral portions of the shell. Ribs on beaks smoothish. Exterior whitish tan, but commonly covered with a thin, opaque-brown periostracum. Interior dull-white. Moderately common from shore to 75 fathoms. Known locally as the Spiny Cockle.

Genus *Papyridea* Swainson 1840

Papyridea soleniformis Bruguière — Spiny Paper Cockle

Plate 39n

North Carolina to south half of Florida and the West Indies.

1 to 1¾ inches in length, fairly fragile, moderately compressed, and gaping posteriorly where the margin of the valve is strongly denticulated by the ends of the dozen radial, finely spinose ribs. Exterior tawny with rose flecks or mottlings. Interior glossy, mottled with violet and white, rarely a solid pastel-orange. Moderately common from low tide to several fathoms. The name *hiatus* Meuschen used for this species in *Johnsonia* is not valid (ruled non-binomial).

A similar species, *P. semisulcata* Sowerby (Frilled Paper Cockle, pl. 32c) found from low water to 40 fathoms from southern Florida to the West Indies, is less than ½ inch in length, white, twice as fat, and with longer denticulations at the end of the 12 to 15 radial ribs. Uncommon, except off Miami where it is commonly dredged.

Subfamily *FRAGINAE*
Genus *Trigoniocardia* Dall 1900

Trigoniocardia media Linné — Atlantic Strawberry Cockle

Plate 39m

North Carolina to southeast Florida and the West Indies.

1 to 2 inches in size, squarish in outline, thick, inflated, with 33 to 36 strong radial ribs which are covered with close-set, chevron-shaped plates. External color whitish with mottlings of reddish brown. Interior usually white, or may be flushed with orange, rose-brown or purple. The posterior slope is pushed in somewhat and is slightly concave. A relatively common species found in shallow to moderately deep water.

The Western Strawberry Cockle, *T. biangulata* Sowerby, is the Californian counterpart of the above species. It is 1½ inches in length, with about 30 strong ribs; exterior yellowish white, interior reddish purple. Moderately common.

Subfamily PROTOCARDIINAE
GENUS *Nemocardium* Meek 1876

Nemocardium centifilosum Carpenter   Hundred-lined Cockle

Alaska to Lower California.

½ to ¾ inch in length, almost circular; posterior third of shell with cancellate sculpturing and separated from the finely ribbed anterior two thirds of the shell by a single raised rib. Edge minutely serrate. Exterior with gray, greenish gray or brownish gray, thin, fuzzy periostracum. Interior dull-white. Fairly common.

GENUS *Microcardium* Thiele 1934

Microcardium peramabile Dall   Eastern Micro-cockle

Rhode Island to southeast Florida and the West Indies.

½ to ¾ inch in length, thin, inflated, subquadrate, white, but may be mottled tan on the anterior slope. Sculpture prominent on the posterior third of the valve. It consists of about 90 closely packed, radial ribs (spinose posteriorly) which are crossed by minute concentric threads. The anterior two thirds is separated from the rest of the shell by a single, crested, spinose, radial rib. Very commonly dredged off eastern Florida.

Microcardium tinctum Dall, found with the above species, is ¾ inch in length, stained with rose-red and has more than 150 minute, radial ribs. Uncommon.

Subfamily LAEVICARDIINAE
GENUS *Laevicardium* Swainson 1840

Laevicardium laevigatum Linné   Common Egg Cockle

Plate 39k

North Carolina to both sides of Florida and the West Indies.

1 to 2 inches in size, higher than long, polished smooth, inflated, fairly thin and obscurely ribbed. Exterior generally whitish, but may be rose-tinted, mottled with brown or flushed with purple, yellow or burnt-orange. Interior similarly colored. With about 60 very fine, subdued radial ribs. A common shallow-water species. The name *serratum* Linné has been erroneously applied to our Atlantic species by some workers.

Laevicardium mortoni Conrad   Morton's Egg Cockle
Plate 39l

Cape Cod, Massachusetts, to Florida and the Gulf of Mexico.

¾ to 1 inch in size, ovate, glossy, similar to *laevigatum*, but commonly with brown, zigzag markings and with fine, concentric ridges which are minutely pimpled. Common in southern New England from shallow water to 2 fathoms. A food of wild ducks.

Laevicardium pictum Ravenel   Ravenel's Egg Cockle

South Carolina to southeast Florida and the West Indies.

½ to 1 inch in height, obliquely triangular in shape, polished and only moderately inflated. Exterior white or cream with delicate shades or rose or brown and with a weak, iridescent sheen. A color form has strong, brown, zigzag streaks. Beaks very low and near the anterior end. Very faint radial and concentric lines present. Dredged from 75 to 85 fathoms. An uncommon and attractive species.
Laevicardium sybariticum Dall 1886 (Dall's Egg Cockle, rare, same range), is more inflated, squarish in shape and with deep-pink breaks.

Laevicardium substriatum Conrad   Common Pacific Egg Cockle

Ventura County, California, to the Gulf of California.

Less than 1 inch in size, obliquely ovate, smooth and slightly compressed. Color tan with closely set, narrow, radial bands of reddish brown. These lines are commonly interrupted. Interior cream with cobwebby mottlings of purplish brown. Very common in such localities as Mission Bay and Newport.

Laevicardium elatum Sowerby   Giant Pacific Egg Cockle

San Pedro, California, to Panama.

3 to 7 inches in height, oval, inflated, slightly oblique, with numerous, shallow, radial grooves, but the posterior and anterior regions are smooth.

Exterior orange-yellow; interior china-white. This is the largest species of recent cockles and is moderately common.

GENUS *Dinocardium* Dall 1900

Dinocardium robustum Solander   Giant Atlantic Cockle
Plate 32a

Virginia to north Florida, Texas and Mexico.

3 to 4 inches in size, ovate, inflated, with 32 to 36 rounded, radial, smoothish ribs. Externally straw-yellow with its posterior slope mahogany-red shading toward purple near the edge. Interior rose, with brownish posteriorly and with a white anterior margin. This is the large, common cockle washed ashore along the Carolina and Georgia strands. It is not found in southwest Florida.

When the Florida Canal project was begun in 1935, President F. D. Roosevelt was presented with a large silver platter on which was set a specimen of *Dinocardium*, encased in gold and containing a portion of the first earth excavated as a result of the blast set off by the President. The canal was never completed.

Dinocardium robustum vanhyningi Clench and L. C. Smith
Vanhyning's Cockle
Plate 32b

Tampa Bay to Cape Sable, Florida.

3½ to 5 inches in size, higher than long, with 32 to 36, smoothish, rounded, radial ribs. Externally straw-yellow with irregular patches and bands of mahogany-red to purplish brown. It is more elongate, glossier and more colorful than *robustum*, and is the common, large cockle on the west coast of Florida. They are popular souvenirs, being used for ash trays, melted-butter dishes, baking dishes and for holding pincushions.

GENUS *Serripes* Gould 1841

There is only one species of this peculiar genus of cockles in North American waters. The hinge is narrow, the cardinal teeth weak, and the ligament is large. The radial ribs are very weak.

Serripes groenlandicus Bruguière   Greenland Cockle
Plate 32d

Arctic Seas to Cape Cod, Massachusetts. Alaska to Puget Sound, Washington.

2 to 4 inches in length, moderately thin but strong, inflated, almost

PLATE 33

LARGE SCALLOPS

a. TRYON'S SCALLOP, *Aequipecten glyptus* Verrill, 2½ inches (Massachusetts to the Gulf of Mexico), p. 366.

b. LION'S PAW, *Lyropecten nodosus* L., 5 inches (North Carolina to the West Indies), p. 366.

c. ATLANTIC DEEPSEA SCALLOP, *Placopecten magellanicus* Gmelin, 8 inches (Labrador to off North Carolina), p. 366.

d. ZIGZAG SCALLOP, *Pecten ziczac* L., 3 inches (North Carolina to the West Indies), p. 362.

e. SAN DIEGO SCALLOP, *Pecten diegensis* Dall, 3 inches (California), p. 361.

f. LAURENTIAN SCALLOP, *Pecten laurenti* Gmelin, 3 inches (West Indies), p. 362.

g. RAVENEL'S SCALLOP, *Pecten raveneli* Dall, 1½ inches (North Carolina to Texas and the West Indies), p. 362.

h. CIRCULAR PACIFIC SCALLOP, *Aequipecten circularis* Sby., 2½ inches (Pacific side of Central America), not in text.

i. ATLANTIC BAY SCALLOP, *Aequipecten irradians irradians* Lam., 3 inches (Atlantic Coast), p. 367.

j. CALICO SCALLOP, *Aequipecten gibbus* L., 1½ inches (North Carolina to West Indies), p. 368.

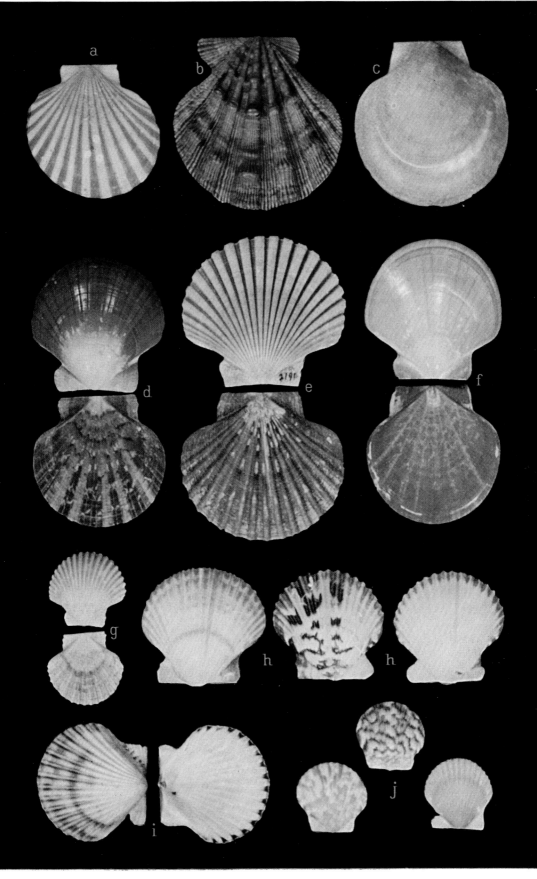

PLATE 34

SMALL SCALLOPS

a. SENTIS SCALLOP, *Chlamys sentis* Reeve, 1 inch (North Carolina to the West Indies), p. 363.

b. ORNATE SCALLOP, *Chlamys ornata* Lam., 1 inch (Southeastern Florida and West Indies), p. 363.

c. MILDRED'S SCALLOP, *Chlamys mildredae* F. M. Bayer, 1 inch. Holotype (Florida and Bermuda), p. 363.

d. and e. ROUGH SCALLOP, *Aequipecten muscosus* Wood, e. is a large, worn specimen. 1 inch (Southeastern United States and the West Indies), p. 367.

f. LITTLE KNOBBY SCALLOP, *Chlamys imbricata* Gmelin, 1 inch (Southeastern Florida and the West Indies), p. 364.

g. ANTILLEAN SCALLOP, *Lyropecten antillarum* Recluz, ½ inch (Southeastern Florida and the West Indies), p. 366.

h. NUCLEUS SCALLOP, *Aequipecten nucleus* Born, 1 inch (Southeastern Florida and the West Indies), p. 368.

i. KELP-WEED SCALLOP, *Leptopecten latiauratus* Conrad, 1 inch (California), p. 365.

j. PACIFIC SPEAR SCALLOP, *Chlamys hastata hastata* Sby., 2 inches (California), p. 364.

k. PACIFIC PINK SCALLOP, *Chlamys hastata hericia* Gould, 2 inches (Alaska to San Diego, California), p. 364.

l. HIND'S SCALLOP, *Chlamys hindsi* Dall, 2 inches (Alaska to San Diego, California), p. 365.

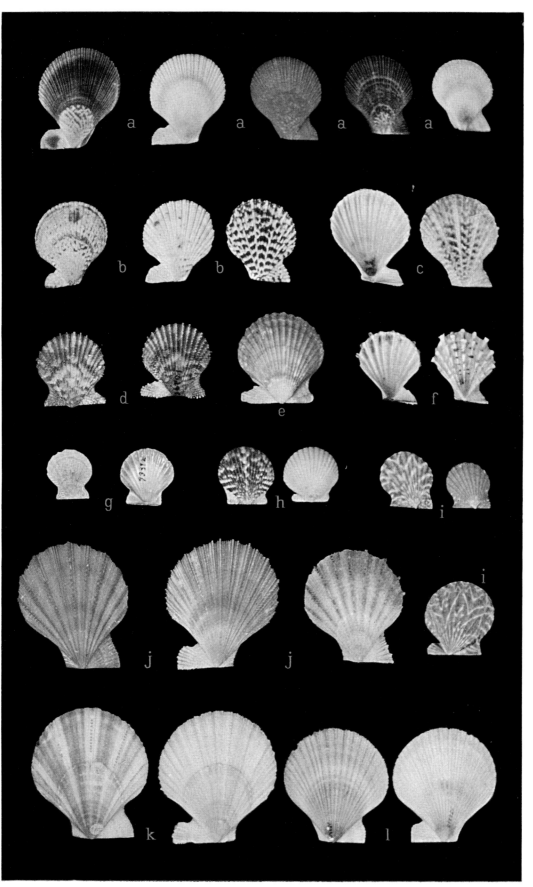

PLATE 35

PEARL OYSTERS AND MUSSELS

a.   LISTER'S TREE OYSTER, *Isognomon radiatus* Anton, 1½ inches (Southeastern Florida and the West Indies), p. 358.

b.   FLAT TREE OYSTER, *Isognomon alatus* Gmelin, 2½ inches (Florida and West Indies), p. 358.

c.   ATLANTIC PEARL OYSTER, *Pinctada radiata* Leach, 2 inches (Florida and West Indies), p. 359.

d.   ATLANTIC WING OYSTER, *Pteria colymbus* Röding, 2 inches (North Carolina to West Indies), p. 359.

e.   KITTEN'S PAW, *Plicatula gibbosa* Lam., 1 inch (North Carolina to Gulf States and south), p. 361.

f.   ROUGH LIMA, *Lima scabra* Born, 2 inches (Southeastern Florida and the West Indies), p. 370.

g.   SPINY LIMA, *Lima lima* L., 1½ inches (Southeastern Florida and the West Indies), p. 370.

h.   ROUGH LIMA, *Lima scabra* Born, smooth form *tenera* Sby., 2 inches (Southeastern Florida and West Indies), p. 371.

i.   YELLOW MUSSEL, *Brachidontes citrinus* Röding, 1 inch (Florida and West Indies), p. 352.

j.   SCORCHED MUSSEL, *Brachidontes exustus* L., ¾ inch (North Carolina to the West Indies), p. 352.

k.   COMMON JINGLE SHELL, *Anomia simplex* Orbigny, 1 inch (Atlantic Coast), p. 372.

l.   TULIP MUSSEL, *Modiolus americanus* Leach, 3 inches (North Carolina to the West Indies), p. 351.

m.   BLUE MUSSEL, *Mytilus edulis* L., 2½ inches (Arctic to South Carolina), p. 354.

n.   HOOKED MUSSEL, *Brachidontes recurvus* Raf., 2 inches (Cape Cod to West Indies), p. 353.

o.   Living LIMA CLAM, *Lima scabra* Born, showing the delicate tentacles along the mantle edge which aid this clam in swimming (Southeastern Florida and West Indies), p. 370.

PLATE 36

THORNY OYSTERS

a. PACIFIC THORNY OYSTER, *Spondylus pictorum* Schreiber, 5 inches (Gulf of California to Panama), p. 370.

b. ATLANTIC THORNY OYSTER, *Spondylus americanus* Hermann, 3 to 4 inches. Comes in many colors. Specimen from the Leo Burry collection. (Florida and the West Indies), p. 369.

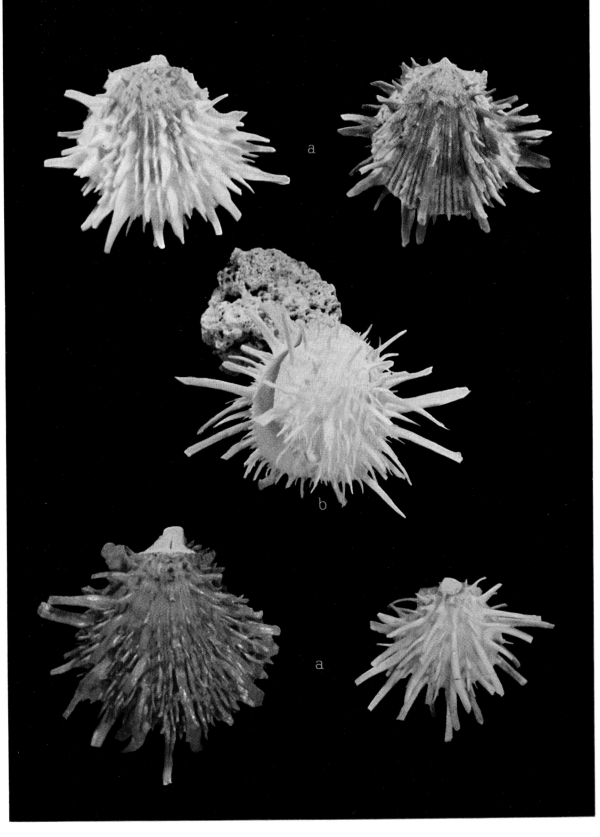

PLATE 37

JEWEL BOXES

a. CLEAR JEWEL BOX, *Chama pellucida* Brod., 2 inches (Oregon to Chili), p. 393.

b. LEAFY JEWEL BOX, *Chama macerophylla* Gmelin, 2½ inches (North Carolina to the West Indies), p. 392.

c. ATLANTIC LEFT-HANDED JEWEL BOX, *Pseudochama radians* Lam., 3 inches (Southern Florida and the West Indies), p. 393.

d. LITTLE CORRUGATED JEWEL BOX, *Chama congregata* Conrad, 1 inch (North Carolina to the West Indies), p. 392.

e. CALIFORNIA SPINY JEWEL BOX, *Echinochama arcinella californica* Dall, Holotype. 2 inches (off Lower California to Panama), p. 394.

f. CHERRY JEWEL BOX, *Chama florida* Lam., 1 inch (West Indies), not in text.

g. FLORIDA SPINY JEWEL BOX, *Echinochama cornuta* Conrad, 1 inch (North Carolina to Texas), p. 394.

h. TRUE SPINY JEWEL BOX, *Echinochama arcinella* L., 1½ inches (West Indies to Brazil), p. 394.

PLATE 38

ATLANTIC OYSTERS, LUCINAS AND VENUS CLAMS

a. HONEYCOMBED OYSTER, *Pycnodonta hyotis* L., 2½ inches (Florida and West Indies), p. 374.

b. ATLANTIC FALSE JINGLE SHELL, *Pododesmus rudis* Broderip, 2 inches (North Carolina, south), p. 372.

c. COON OYSTER, *Ostrea frons* L., 2 inches a, b and c are from a sunken wreck off Miami. Gift of Ethel Townsend (Florida, south), p. 373.

d. TIGER LUCINA, *Codakia orbicularis* L., 3 inches (Gulf coast and West Indies), p. 390.

e. CHALKY BUTTERCUP, *Anodontia philippiana* Reeve, 3 inches (North Carolina to Cuba), p. 389.

f. BUTTERCUP LUCINA, *Anodontia alba* Link, 2 inches (North Carolina to Texas and West Indies), p. 389.

g. THICK LUCINA, *Phacoides pectinatus* Gmelin, 2 inches (North Carolina to Texas and West Indies), p. 388.

h. PENNSYLVANIA LUCINA, *Lucina pennsylvanica* L., 1½ inches (North Carolina to West Indies), p. 385.

i. FLORIDA LUCINA, *Lucina floridana* Conrad, 1½ inches (Western Florida to Texas), p. 387.

j. NORTHEAST LUCINA, *Phacoides filosus* Stimpson, 2 inches (off entire east coast), p. 389.

k. AMETHYST GEM CLAM, *Gemma gemma* Totten, ⅛ inch (east coast; for details see figure 84), p. 418.

l. EMPRESS VENUS, *Antigona strigillina* Dall, 1½ inches (Southeastern Florida, south), p. 404.

m. QUEEN VENUS, *Antigona rugatina* Heilprin, 1 inch (North Carolina to West Indies), p. 405.

n. HEART-SHAPED VENUS, *Pitar cordata* Schwengel, 1½ inches (Gulf of Mexico, offshore), p. 414.

o. DISK DOSINIA, *Dosinia discus* Reeve, 3 inches (Virginia to Gulf of Mexico and Bahamas), p. 417.

PLATE 39

VENUS CLAMS AND COCKLES

a. KING VENUS, *Chione paphia* L., 1½ inches (Florida Keys and West Indies), p. 409.

b. SUNRAY VENUS, *Macrocallista nimbosa* Solander, 5 inches (North Carolina to the Gulf States), p. 416.

c. IMPERIAL VENUS, *Chione latilirata* Conrad, 1 inch (North Carolina to the Gulf States), p. 409.

d. LIGHTING VENUS, *Pitar fulminata* Menke, 1 inch (North Carolina to the West Indies), p. 414.

e. CALICO CLAM, *Macrocallista maculata* L., 2 inches (North Carolina to the West Indies), p. 416.

f. ROYAL COMB VENUS, *Pitar dione* L., 1½ inches (Texas to the Caribbean), p. 415.

g. LADY-IN-WAITING VENUS, *Chione intapurpurea* Conrad, 1½ inches (North Carolina to the Gulf and West Indies), p. 407.

h. CROSS-BARRED VENUS, *Chione cancellata* L., 1 inch (North Carolina to the West Indies), p. 407.

i. GLORY-OF-THE-SEAS VENUS, *Callista eucymata* Dall, 1½ inches (North Carolina to the West Indies), p. 415.

j. POINTED VENUS, *Anomalocardia cuneimeris* Conrad, ¾ inch (Florida to Texas), p. 409.

k. COMMON EGG COCKLE, *Laevicardium laevigatum* Linné, 2 inches (North Carolina to the West Indies), p. 399.

l. MORTON'S EGG COCKLE, *Laevicardium mortoni* Conrad, 1 inch (Massachusetts to Texas), p. 400.

m. ATLANTIC STRAWBERRY COCKLE, *Trigoniocardia media* L., 1½ inches (North Carolina to the West Indies), p. 398.

n. SINY PAPER COCKLE, *Papyridea soleniformis* Brug., 1 inch (North Carolina to the West Indies), p. 398.

o. PRICKLY COCKLE, *Trachycardium egmontianum* Shuttleworth, 2 inches (North Carolina to Cuba), p. 397.

p. YELLOW COCKLE, *Trachycardium muricatum* L., 2 inches (North Carolina to the Gulf States and West Indies), p. 397.

PLATE 40

SEMELES AND TELLINS

a. GAUDY ASAPHIS, *Asaphis deflorata* L., 2 inches (Southeastern Florida and West Indies), p. 439.

b. PURPLISH SEMELE, *Semele purpurascens* Gmelin, 1 inch (North Carolina to West Indies), p. 435.

c. LARGE STRIGILLA, *Strigilla carnaria* L., ¾ inch (North Carolina to West Indies), p. 428.

d. ATLANTIC SANGUIN, *Sanguinolaria cruenta* Solander, 1½ inches (Florida, the Gulf and West Indies), p. 439.

e. SUNRISE TELLIN, *Tellina radiata* L., 3 inches (South Carolina to the West Indies), p. 421.

f. and g. WHITE ATLANTIC SEMELE, *Semele proficua* Pulteney, 1 inch. f is the rayed form, *radiata* Say (North Carolina to the West Indies), p. 434.

h. ROSE PETAL TELLIN, *Tellina lineata* Turton, 1½ inches (Florida and the West Indies), p. 427.

i. GREAT TELLIN, *Tellina magna* Spengler, 4 inches (North Carolina to the West Indies), p. 427.

j. FAUST TELLIN, *Arcopagia fausta* Pulteney, 3 inches (North Carolina to the West Indies), p. 428.

k. SMOOTH TELLIN, *Tellina laevigata* L., 3 inches (Florida and the West Indies), p. 422.

l. SPECKLED TELLIN, *Tellina interrupta* Wood, 3 inches (North Carolina to the West Indies), p. 422.

m. CANDY STICK TELLIN, *Tellina similis* Sby., 1 inch (Florida and the West Indies), p. 426.

n. ALTERNATE TELLIN, *Tellina alternata* Say, 2½ inches (North Carolina to the Gulf States), p. 427.

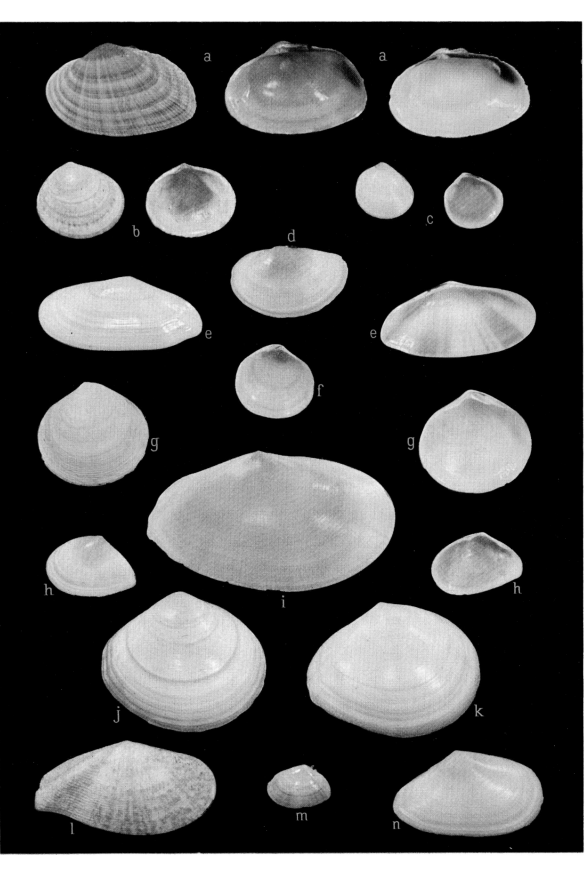

round and slightly gaping at the posterior end. Exterior brownish gray and may be with brown, concentric rings of growth. Interior dull-white. Beaks inflated and high. Ligament large and strong. No lunule or escutcheon. Weak radial ribs seen at both ends only. Concentric growth ridges prominent near the margins. Muscle scars and pallial line deeply impressed. Foot of animal large and suffused with heavy, red mottlings. Very commonly dredged in cold, northern waters.

GENUS *Clinocardium* Keen 1936

Clinocardium ciliatum Fabricius

Iceland Cockle
Plate 32e

Greenland to Massachusetts. Alaska to Puget Sound, Washington.

1½ to 3 inches in size, a little longer than high, with 32 to 38 ridged radial ribs which are crossed by coarse concentric lines of growth. Externally drab grayish yellow with weak, narrow, concentric bands of darker color. Interior ivory. Periostracum gray and conspicuous. Especially abundant from Maine northward in offshore waters.

Clinocardium nuttalli Conrad

Nuttall's Cockle
Plate 31b

Bering Sea to San Diego, California.

2 to 6 inches in length; smaller ones being almost round, adults tending to be higher than long; moderately compressed; commonly with 33 to 37 coarse radial ribs which are creased by half-moon-shaped riblets. Older specimens worn smoothish. Exterior drab-gray, with a brownish yellow, thin periostracum. Common offshore. Once called *C. corbis* Martyn. Known locally as the Basket Cockle.

Clinocardium fucanum Dall

Fucan Cockle

Sitka, Alaska, to off Monterey, California.

1 to 1½ inches in length, longer than high, moderately inflated, and with 45 to 50 low, poorly developed, radial ribs which are crossed by microscopic concentric lines. No wavy, radial furrow on the upper posterior edge of the shell. Color whitish with a grayish-brown periostracum. Common in the Puget Sound area.

Young *C. nuttalli* are distinguished from this species by their 2 first ribs behind the ligament which are large, rounded and make a wavy edge to the shell. In small specimens of *C. ciliatum*, the top edges of the ribs are sharp; in *fucanum* they are rounded.

Genus *Cerastoderma* Poli 1795

Cerastoderma pinnulatum Conrad

Northern Dwarf Cockle

Plate 30c

Labrador to off North Carolina.

¼ to ½ inch in length, thin, with 22 to 28 wide, flat ribs which have delicate, arched scales on the anterior slope of the shell. Scales missing on the central portion of the valve. Externally cream; interior glossy and white, rarely tinted with orange-brown. Commonly dredged from 7 to 100 fathoms.

Superfamily *VENERACEA*
Family *VENERIDAE*

The classification of the family of Venus clams has been one of continual debate and rearranging for some years. Our presentation here is no better than has been suggested before, but at least it is in a form which is conservative and most likely to be accepted by the majority. Rather than accept a separate family Chionidae or consider it a subfamily remotely related to the Venerinae, we have allied it as an artificial group in the subfamily Venerinae. I suspect that an anatomical study of the soft parts will support this course.

Subfamily *VENERINAE*

Sculpture usually both radial and concentric; anterior lateral present, especially in the left valve, but often extraordinarily vestigial.

Genus *Antigona* Schumacher 1817
Subgenus *Dosina* Gray 1835

Antigona listeri Gray

Princess Venus

Plate 32m

Southeast Florida and the West Indies.

2 to 4 inches in length, oblong-oval, obese. Resembling *Mercenaria campechiensis*, but characterized by numerous, fine, radial riblets which cause the sharp, concentric ribs to be serrated or beaded. Each side of the lunule is bounded by a long, deep, narrow furrow. Posterior muscle scar usually stained brown. Moderately common in shallow water in sand.

Subgenus *Circomphalus* Mörch 1853

Antigona strigillina Dall

Empress Venus

Plate 38l; figure 81d

Southeast Florida and the West Indies.

1½ inches in length, externally very much like a small *Mercenaria campechiensis*, but not as elongate and with more distinct, concentric riblets. Internally, it is distinguished easily by the extremely small, if not absent, pallial sinus, by the very thick margin of the shell, and in the left valve by the presence of a button-like anterior lateral "tooth." Exterior whitish. Dredged occasionally from 40 to 70 fathoms. Considered a collector's item.

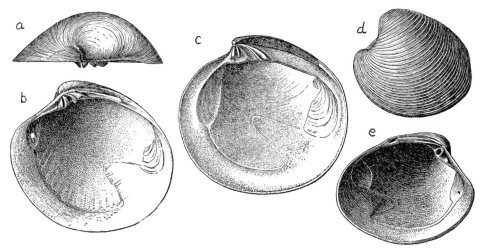

FIGURE 81. Some venerid clams. **a** and **b**, *Compsomyax subdiaphana* Cpr., 2 inches (Pacific Coast); **c**, *Dosinia discus* Reeve, 2 inches (Atlantic Coast); **d**, *Antigona strigillina* Dall, 1½ inches (Florida and West Indies); **e**, left valve of *Pitar morrhuana* Linsley, 1½ inches (Atlantic Coast).

Antigona rugatina Heilprin   **Queen Venus**
Plates 38m; 32n

North Carolina to southeast Florida and the West Indies.

1 to 1½ inches in length, rather circular, inflated, and characterized by strong, raised, lamellate, concentric ribs between which are 5 or 6 smaller, raised concentric ridges. Lunule heart-shaped, well-impressed, bordered by a fine, deep line, and crossed by numerous raised threads. Escutcheon well-formed, smoothish. Color cream to whitish with light-mauve mottlings. Very uncommon.

Antigona rigida Dillwyn from the West Indies is very similar, but not nearly so obese, and its concentric ridges are stronger and smoother.

"Chionid" Group

Ovate-trigonal, inequilateral, sculpture usually cancellate; lunule impressed. Inner margins usually crenulate. Teeth strong, without the tiny, pimple-like anterior lateral. Pallial sinus short. This section or group is considered by some to be of family or subfamily rank. See Frizzell and Myra Keen.

Genus *Mercenaria* Schumacher 1817

The Hard-shell Clams or Quahogs belong to this genus. The shell is large and thick; lunule large, heart-shaped and bounded by an incised line. Inner margin crenulate. 3 cardinals in each valve. Left middle cardinal split. Formerly placed in the genus *Venus* many years ago, but almost universally placed in a genus by itself by modern workers.

Mercenaria mercenaria Linné　　　　　Northern Quahog
Plate 32h

Gulf of St. Lawrence to Florida and the Gulf of Mexico.
Introduced to Humboldt Bay, California.

3 to 5 inches in length, ovate-trigonal, about ⅚ as high, heavy and quite thick. Moderately inflated. Sculpture of numerous, concentric lines of growth or small riblets. Near the beaks these lines are prominent and distantly spaced. The exterior center of the valves has a characteristic smoothish or glossy area. Exterior dirty-gray to whitish; interior white, commonly with purple stainings. The entire lunule is ¾ as wide as long. The form *notata* Say from the same region is externally marked with brown, zigzag mottlings. This species is very common and is used commercially for chowders and as clams-on-the-half-shell or "cherrystones." Also known as the Hard-shelled Clam. Do not confuse with *M. campechiensis*.

Mercenaria mercenaria texana Dall is a subspecies from the northern Gulf of Mexico region. It is characterized by a glossy central area on the outside of the shell, but has large, irregular, coalescing, flat-topped, concentric ribs.

Mercenaria campechiensis Gmelin　　　　　Southern Quahog
Plate 32g

Chesapeake Bay to Florida, Texas and Cuba.

3 to 6 inches in length, very similar to *mercenaria*, but much more obese, a heavier shell, lacks the smooth central area on the outside of the valves, and the entire lunule is usually as wide as long. Always white internally. Rarely it has a purplish stain on the escutcheon and brown mottlings on the side. There have been a number of forms described. In the vicinity of St. Petersburg, Florida, there is a malformed race in which there is a sharp, elevated ridge passing from the umbo obliquely backward toward the pallial sinus on the inside of each valve. The Southern Quahog is common but has not been exploited commercially to any great extent.

GENUS *Chione* Mühlfeld 1811

Shells trigonal or ovate; thick; 3 cardinal teeth in each valve; no anterior laterals; pallial sinus small and triangular; inner margins crenulated; lunule bounded by an indented line; escutcheon smooth and bounded by a small ridge.

Subgenus *Chione s. str.*

Chione cancellata Linné
Cross-barred Venus
Plate 39h

North Carolina to Florida, Texas and the West Indies.

1 to 1¾ inches in length, varying from ovate to subtriangular in shape, thick; with strong, raised, curved, leaf-like, concentric ribs and numerous coarse radial ribs. Escutcheon long, smooth and V-shaped, commonly with 6 to 7 brown, zebra-stripes. Lunule heart-shaped, with minute vertical threads. Color externally is white to gray; internally glossy-white with a suffusion of purplish blue. A very common, shallow-water species in Florida. Beachworn specimens have a cancellate sculpturing. The subspecies *mazycki* Dall, off the Carolinas, Georgia and northeast Florida has a beautiful rosy interior.

Chione intapurpurea Conrad
Lady-in-waiting Venus
Plate 39g

North Carolina, the Gulf States to the West Indies.

1 to 1½ inches both ways, thick, glossy-white to cream; interior white, commonly with a violet, radial band or splotch at the posterior third. Exterior with crowded, smooth, low, rounded, concentric ribs. The lower edge of these ribs bears many small bars which are lined up one below the other to give the shell the impression that it has axial ribs. The concentric ribs become sharp and higher at the shell's extreme ends. Lunule with raised lamellations; escutcheon with very fine, transverse lines. Uncommon. Incorrectly spelled *interpurpurea*.

Chione californiensis Broderip
Common Californian Venus
Plate 31j

San Pedro to northern South America.

2 to 2½ inches high, a little longer, subtrigonal, moderately compressed, with sharp, raised, concentric ribs whose edges turn upwards, and with low, rather wide, rounded, radial riblets. Lunule heart-shaped and striated; escutcheon V-shaped in cross-section, long and smooth. The dorsal posterior end of the right valve is not as smooth and overlaps the left valve. Exterior

creamy-white with faint mauve stripes on the escutcheon. Interior white, commonly with a purple splotch at the posterior end. This is a common shore species, formerly called *C. succincta* Val.

Chione californiensis undatella Sowerby   Frilled Californian Venus
Plate 31i

San Pedro, California, to northern South America.

Differing from *californiensis* in being more inflated, usually with more numerous and more closely spaced, thinner concentric ribs, and retaining mauve-brown color splotches in the adults. Very common. Many workers consider this a full species, and apparently additional field study is necessary.

Chione fluctifraga Sowerby   Smooth Pacific Venus
Plate 31k

San Pedro, California, to the Gulf of California.

2½ inches in height, slightly longer, moderately compressed, subtrigonal; radial grooves or ribs strong at the posterior third and at the anterior quarter of the shell; central area with stronger, low, rather wide, concentric ribs which may have coarse, half-moon-shaped beads. Lunule not well-defined; escutcheon not well seen and not sunken nor smooth as in *californiensis*. Exterior creamy-white, semi-glossy, rarely stained with blue-gray. Interior white with purple splotches near the muscle scars or on the teeth. Not uncommon along the sandy shores in southern localities.

Section Timoclea Brown 1827

Chione grus Holmes   Gray Pygmy Venus
Plate 32i

North Carolina to Key West to Louisiana.

¼ to ⅜ inch in length, oblong, with 30 to 40 fine, radial ribs which are crossed by very fine, concentric threads. The posterior dozen ribs are cut along their length by a very fine groove. Dorsal margin of right valve fimbriated and overlapping the left valve. Lunule narrow, heart-shaped, colored brown. Escutcheon very narrow and sunken. Exterior colored a dull-gray, but some Florida specimens tend to be whitish, pinkish or even orange. Interior glossy-white with purplish brown area at the posterior end. *Purple color on hinge at both ends.* Commonly dredged in shallow water.

Chione pygmaea Lamarck   White Pygmy Venus

Southeast Florida and the West Indies.

¼ to ½ inch in length, similar to *grus*, but with prominent scales, 4 to 5 brown, zebra stripes on the escutcheon, with a white lunule, and the teeth purple only on the posterior half of the hinge. Beaks commonly pink. Interior all white. Fairly common in shallow water.

Subgenus *Lirophora* Conrad 1863

Chione paphia Linné — King Venus

Plate 39a

Lower Florida Keys and the West Indies.

1½ inches in length, similar to *latilirata*, but not so heavy, with 10 to 12 smaller, concentric ribs which are thin at their ends. From a side view, the dorsal margin of the lunule is very concave. Not very common in the United States.

Chione latilirata Conrad — Imperial Venus

Plate 39c

North Carolina to Florida and to Texas.

1 inch in length, very thick and solid, with 5 to 7 large, bulbous, concentric ribs, usually rounded, but may also be sharply shelved on top. The ribs are not thin and flattened at their ends. Lunule heart-shaped, and, from a side view, its dorsal margin is almost straight. Surface of shell glossy, cream with rose and brown mottlings. Rather uncommon offshore in about 20 fathoms.

GENUS *Anomalocardia* Schumacher 1817

Anomalocardia cuneimeris Conrad — Pointed Venus

Plate 39j

South half of Florida to Texas.

½ to ¾ inch in length, about ¾ to ½ as high, pointed into a sharp, wedge-like rostrum at the posterior end. Lunule oval to slightly heart-shaped and faintly impressed. Wide, shallow escutcheon bordered by a weak ridge. Beaks tiny and inrolled. Sculpture of small, but distinct, rounded, concentric ribs which are more prominent near the beaks. Color variable: glossy-cream, white or tan with brown or purple rays of fine specklings. Interior white, purple or brown. Brackish water specimens are dwarfed. A common sandy shore species.

A. brasiliana Gmelin (West Indian Pointed Venus) which is twice as large, less elongate, and with the concentric ribs extending over into the escutcheon area, has been erroneously reported from the United States. Common in the West Indies and south to Brazil.

GENUS *Protothaca* Dall 1902

Protothaca tenerrima Carpenter　　　　　　　　Thin-shelled Littleneck

Vancouver, B.C., to Lower California.

About 4 inches in length and 2¾ inches high, very compressed, relatively thin, with a chalky texture, with a few raised concentric lines and numerous very small radial threads. Lunule fairly defined. Exterior light gray-brown. Interior chalky-white. A fairly common species, commonly washed ashore on Californian beaches.

Protothaca staminea Conrad　　　　　　　Common Pacific Littleneck
Plate 31m, n

Aleutian Islands to Lower California.

1½ to 2 inches in length, subovate, beaks nearer the anterior end; sculpture of concentric and radial ribs which form beads as they cross each other at the anterior end of the shell. Radial ribs stronger on the middle of the valves. Beaks almost smooth. Exterior rusty-brown with a purplish cast. A very abundant, wide-spread species with a number of varieties. Sometimes with a mottled color pattern.

Variety or form: *laciniata* Carpenter reaches 3 inches in length, is coarsely cancellate and beaded, its color rusty-brown to grayish.

Variety or form: *ruderata* Deshayes (typically a northern form) is chalky-white to gray, with concentric ribs large and coarse, commonly lamellate (see pl. 31-o).

Compare with *Tapes philippinarum*, the Japanese Littleneck.

GENUS *Humilaria* Grant and Gale 1931

Humilaria kennerleyi Reeve　　　　　　　　Kennerley's Venus

Alaska to Carmel Bay, California.

2½ to 4 inches in length, ovate-oblong, with the beaks near the anterior end. With sharp, concentric ribs whose edges are bent upwards. Spaces between ribs. Color and texture like gray Portland cement. Interior white. Margin of shell finely crenulate, a feature that will distinguish it from worn specimens of *Saxidomus*. Dredged on mud bottoms from 3 to 20 fathoms. A collector's item, although reasonably common.

GENUS *Tapes* Mühlfeld 1811

Subgenus *Ruditapes* Chiamenti 1900

Tapes philippinarum Adams and Reeve　　　　Japanese Littleneck

Puget Sound southward.

1½ to 2 inches in length, extremely close to *Protothaca staminea* and *grata* Sowerby (the latter's range is from the Gulf of California to Panama), but differing from both in being much more elongate and more compressed. Its lunule and small escutcheon are more distinct and quite smooth as compared to those of *staminea*. The hinges are extremely similar. *P. grata* differs in having tiny, distinct crenulations on the inside of the anterior dorsal margin. *T. semidecussata* Reeve appears to be the same as this introduced species. Its colors are variable and commonly variegated. Alias *T. bifurcata* Quayle 1938.

GENUS *Compsomyax* R. Stewart 1930

Compsomyax subdiaphana Carpenter   Milky Pacific Venus
                                                     Plate 31f; figure 81a, b

Alaska to Lower California.

1½ to 2½ inches in length, elongate-ovate, moderately inflated; beaks anterior and pointing forward. Sculpture of fine, irregular, concentric lines of growth, otherwise rather smoothish. Lunule poorly defined. 3 cardinal teeth in each valve, the most posterior one in the right valve deeply split. Color usually chalky-white, but younger specimens are yellowish white and semi-glossy. Interior white. Dredged in soft mud from 5 to 25 fathoms. Abundant in some Californian localities. Formerly placed in the genus *Clementia*.

FIGURE 82. *Psephidia* of the Pacific Coast. **a-b,** *P. lordi* Baird; **c,** *P. ovalis* Dall. Both ¼ inch.

GENUS *Psephidia* Dall 1902

Psephidia lordi Baird   Lord's Dwarf Venus
                                                     Figure 82a

Alaska to San Diego, California.

¼ inch in length, ovate, compressed or slightly flattened; beaks small; sculpture of microscopic, concentric growth lines. 3 cardinals in hinge of each valve. No laterals. On the dorsal margin there is a microscopic groove parallel to the edge. Color whitish to greenish white, commonly with darker, concentric color bands. Tiny young shells may be found inside the adult clams in the summer and spring months. Common.

Genus *Irus* Oken 1815

Irus lamellifera Conrad

Californian Irus Venus

Plate 31r

Monterey to San Diego, California.

1 to 1½ inches in length, usually oblong, although some specimens may be almost round. Characterized by about a dozen, strongly raised, concentric lamellae or thin ridges. Shell whitish and with a chalky texture. Moderately common. Found burrowing in gray shale from low water to several fathoms.

Subfamily MERETRICINAE
Genus *Tivela* Link 1807

Tivela floridana Rehder

Florida Tivela

Palm Beach County, Florida.

⅜ inch in length, subquadrate, beaks in the center, highly polished and with microscopic growth lines near the margins. Exterior glossy, tan or purplish. Interior mottled with purplish brown. This is the only *Tivela* recorded from eastern United States. Uncommon offshore.

Subgenus *Pachydesma* Conrad 1854

Tivela stultorum Mawe

Pismo Clam

Plate 31h; figure 28d

San Mateo County, California, to Lower California.

3 to 6 inches in length, ovate, heavy, moderately inflated, glossy-smooth, except for weak lines of growth. Ligament large and strong. Color brownish cream with wide, mauve, radial bands. Bands may be absent. Posterior end marked off by a single, sharp thread. Lunule lanceolate and with vertical scratches. Periostracum thin and varnish-like. A common and edible species. This is the only West Coast *Tivela*, but it has received a number of unnecessary names, *T. crassatelloides* Stearns being one of many.

Genus *Transennella* Dall 1883

Left anterior lateral fitting into a socket in the right valve. Internal margins are obliquely grooved with numerous, microscopic lines. These are parallel to the growth lines at the ventral margin of the valves.

Transennella stimpsoni Dall

Stimpson's Transennella

Figure 83a, b

North Carolina to southeast Florida and the Bahamas.

¼ to ½ inch in length, glossy, rounded trigonal in shape, smooth except for fine growth lines. Inner margins of valves creased with microscopic oblique threads. Exterior cream with 2 or 3 wide, radial bands of weak brown. Interior commonly flushed with purple. Pallial sinus long. Fairly common in shallow water.

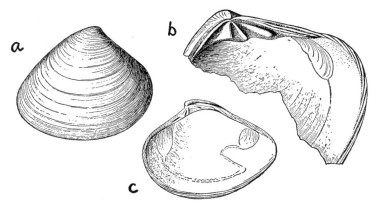

FIGURE 83. Atlantic Transennella Clams. **a** and **b**, *Transennella stimpsoni*, ½ inch; **c**, *T. conradina* Dall, ½ inch.

Transennella conradina Dall   Conrad's Transennella

Figure 83c

South half of Florida and the Bahamas.

Shell very similar to that of *stimpsoni*, but differing in being pointed posteriorly, hence more elongate, and more variable in color. Zigzag brown lines present in some, others are solid cream or solid brownish. Exterior with fine, raised, concentric lines. Pallial sinus short. Common in shallow water.

Transennella tantilla Gould   Tantilla Transennella

Alaska to Lower California.

¼ inch in length, ovate, angle at beaks about 90 degrees, smooth except for weak, concentric lines of growth. Immediately recognized under the hand lens by the tiny grooves running on the inside of the shell margins. Exterior cream with the posterior end stained bluish. Interior white with a wide, radial band of purple-brown at the posterior end. Dredged in large numbers off California and at times found washed ashore.

"Pitarid" Group

Beaks generally nearer the anterior end; cardinal teeth not tending to radiate; anterior laterals well-developed.

Genus *Pitar* Römer 1857

Anterior left lateral fitting into a well-developed socket in the right valve. Middle left cardinal large; posterior right cardinal split.

Subgenus *Pitar s. str.*

Pitar fulminata Menke   Lightning Venus
Plate 39d

North Carolina to Florida and the West Indies.

1 to 1½ inches in length, plump, umbones large and full; lunule very large and outlined by an impressed line. Anterior end broader than the posterior end. Sculpture of crowded, rather heavy lines of growth. Exterior whitish with spots and/or zigzag markings of yellowish brown. Moderately common in shallow water. ⅓-inch young are commonly dredged off Miami.

Pitar albida Gmelin of the West Indies is very similar, but all white in color, more quadrate in shape, has a narrower and more elongate lunule, and is usually more compressed. Common.

Pitar morrhuana Linsley   Morrhua Venus
Plate 32l; figure 81e

Gulf of St. Lawrence to North Carolina.

1 to 1½ inches in length, oval-elongate, moderately plump, with the lunule large and elongate. With numerous, heavy lines of growth. Color dull grayish to brownish red. *P. fulminata* is similar, but is found only to the south of Cape Hatteras, is not so elongate (compare figures), and is marked with brown. Fairly commonly dredged off New England.

Pitar simpsoni Dall   Simpson's Venus

South half of Florida and the West Indies.

¾ inch in length, plump, with fine, irregular, concentric threads; the large, *ovate* lunule is polished smooth. Color white to purplish white, commonly with zigzag, yellow-brown markings. Escutcheon absent. Nearest in shape to *morrhuana*. Uncommon at low tide to 26 fathoms.

Subgenus *Pitarenus* Rehder and Abbott 1951

Pitar cordata Schwengel   Schwengel's Venus
Plate 38n

Off the Florida Keys and the Gulf of Mexico.

1½ inches in length, very similar to *morrhuana*, but much fatter, with more distinct concentric threads on the outside, and with fine crenulations along the inside of the ventral margins of the valves. Interior white, commonly with a pinkish blush. Dredged from 30 to 50 fathoms and brought in by shrimp fishermen. Uncommon.

Subgenus *Hysteroconcha* P. Fischer 1887

Pitar dione Linné Royal Comb Venus

Plate 39f

Texas to Panama and the West Indies.

1 to 1¾ inches in length, characterized by its violet and purple-white colors and 2 radial rows of long spines at the posterior end of the valve. A common species washed ashore in Texas. The closely resembling species, *Pitar lupanaria* Lesson, occurs in the Pacific from Lower California to Peru.

GENUS *Gouldia* C. B. Adams 1845

Shell less than ½ inch in length; beaks minute; lunule long, bounded by an impressed line; no escutcheon. With concentric or reticulate sculpture. Anterior lateral teeth present. This genus is put in the separate subfamily *Circinae* by some workers.

Gouldia cerina C. B. Adams Serene Gould Clam

North Carolina to south half of Florida and the West Indies.

⅓ inch in length, solid, trigonal in shape, beaks in the center, high and very small; lunule long, bounded by an impressed line; no escutcheon. Sculpture reticulate in which the fine, concentric ribs predominate. The radial ribs are stronger anteriorly. Color white, uncommonly with purplish or brownish flecks. A common species from shallow water to 95 fathoms.

GENUS *Callista* Poli 1791
Subgenus *Costacallista* Palmer 1927

Callista eucymata Dall Glory-of-the-Seas Venus

Plate 39i

North Carolina to south half of Florida, Texas and the West Indies.

1 to 1½ inches in length, fairly thin, oval, with about 50 slightly flattened, concentric ribs which have a short dorsal and long ventral slope, and separated by a narrow, sharp groove. Color glossy-white to waxy pale-brown, with clouds and zigzag markings of reddish brown. No escutcheon. Margins rounded. A beautiful and rare species dredged from 25 to 110 fathoms.

Genus *Macrocallista* Meek 1876

Macrocallista nimbosa Solander
Sunray Venus
Plate 39b

North Carolina to Florida and the Gulf States.

4 to 5 inches in length, elongate, compressed, glossy-smooth with a thin varnish-like periostracum. Exterior dull salmon to dull mauve with broken, radial bands of darker color. Interior dull white with a blush of reddish over the central area. Moderately common in shallow, sandy areas and not uncommonly washed ashore after storms.

Macrocallista maculata Linné
Calico Clam
Plates 1b; 39e

North Carolina to south half of Florida and the West Indies.

1½ to 2½ inches in length, ovate, glossy-smooth with a thin varnish-like periostracum. Exterior cream with checkerboard markings of brownish red. Rarely albino or all dark-brown. Moderately common in shallow, sandy areas in certain localities. A popular collector's item. Also known as the Checkerboard or Spotted Clam.

Genus *Callocardia* A. Adams 1864
Subgenus *Agriopoma* Dall 1902

Callocardia texasiana Dall
Texas Venus
Plate 32k; figure 28e

Northwest Florida to Texas.

1½ to 3 inches in length, ¾ as high. Externally resembling *Pitar morrhuana*, but much more elongate, having the beaks rolled in under themselves, and with a more elongate, faint lunule. The posterior cardinal is S-shaped in the right valve. Uncommon, if not rare. Found on the beaches, but its biology and habits are unknown.

Genus *Amiantis* Carpenter 1863

Amiantis callosa Conrad
Pacific White Venus

Santa Monica, California, to south Mexico.

3 to 4½ inches in length, longer than high, beaks pointing anteriorly, shell hard, heavy, glossy and with neat concentric ribs. Lunule small, heart-shaped and pressed in slightly under the beaks. Anterior end round. Color solid ivory. A very attractive, fairly common species living just below tide line on sandy bottoms in the open surf. Commonly washed ashore alive after

storms between Seal Beach and Huntington Beach. Once known as *C. nobilis* Reeve.

GENUS *Saxidomus* Conrad 1837

Shell large, slightly gaping posteriorly, hinge with 4 or 5 cardinal teeth in the right valve, 4 in the left. Pallial sinus long and fairly narrow.

Saxidomus nuttalli Conrad — Common Washington Clam

Plate 31l

Humboldt Bay, California, to Lower California.

3 to 4 inches in length, oblong, with the beaks nearer the anterior end; heavy, with coarse, crowded, concentric ribs. Color a dull, dirty, reddish brown to gray with rust stains. Interior glossy-white, commonly with a flush of purple at the posterior margins. No lunule. Ligament large. Valves slightly gaping posteriorly. Young specimens less than 2 inches are thin-shelled, somewhat glossy and with pretty, mauve, radial streaks on the dorsal edge, both in front and behind the beaks. A very common species which is edible. Also called the Butter Clam.

Saxidomus gigantea Deshayes — Smooth Washington Clam

Aleutian Islands to Monterey, California.

Possibly this is only an ecologic variation or an example of a geographical gradient within a species. It is similar to typical *nuttalli*, but generally lacks the rust-stain color and rarely, if ever, develops the prominent concentric ridges. This is the commonest and best food clam in Alaska.

Subfamily *DOSINIINAE*
GENUS *Dosinia* Scopoli 1777
Subgenus *Dosinidia* Dall 1902

Dosinia elegans Conrad — Elegant Dosinia

West Florida to Texas and south.

2 to 3 inches in length, circular, compressed, glossy, straw-yellow with numerous even, concentric ridges (20 to 25 per inch in adults). Moderately common. Do not confuse with *D. discus*.

Dosinia discus Reeve — Disk Dosinia

Plate 38-o; figure 81c

Virginia to Florida, the Gulf States and the Bahamas.

2 to 3 inches in length, similar to *elegans*, but having more and finer concentric ridges (about 50 per inch in adults), and not so circular. Commonly washed ashore in perfect condition after storms along the Carolina coasts and middle western Florida.

Subfamily EMMINAE

Very small shells, with marginal grooves and denticles simulating lateral teeth. Inner ventral margins crenulate.

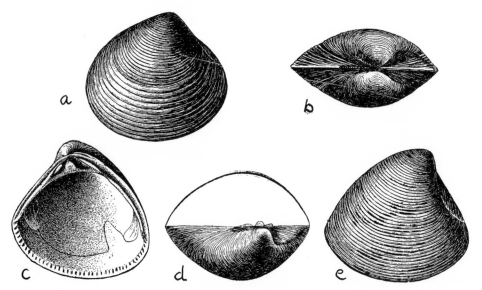

FIGURE 84. Amethyst Gem Clam. a and b, *Gemma gemma* Totten, ⅛ inch; c to e, the form *purpurea* Lea, ⅛ inch.

GENUS *Gemma* Deshayes 1853

Shell the size and shape of a split-pea; lunule large, faintly impressed; no escutcheon; 2 large teeth in the left valve with a large, median socket between the two. A very thin ridge which might be termed a tooth occurs posteriorly beneath the ligament. 3 teeth in right valve. Pallial sinus small and triangular. The shells of the brooded young may be found inside some females.

Gemma gemma Totten   Amethyst Gem Clam
                                        Plate 38k; figure 84

Nova Scotia to Florida, Texas and the Bahamas. Puget Sound, Washington (introduced).

⅛ inch in length, subtrigonal, moderately inflated and rather thin-shelled. Exterior polished and with numerous, fine, concentric furrows or riblets.

Color whitish to tan with purplish over the beak and posterior areas. Pallial sinus commonly, but not always, about the length of the posterior muscle scar. It points upward. This is a very common shallow-water species. A number of subspecies or forms have been described, but their validity needs clarification: *purpurea* Lea (fig. 84c to e), *manhattensis* Prime and *fretensis* Rehder.

GENUS *Parastarte* Conrad 1862

Shell the size of a split-pea, very similar to *Gemma*. In *Parastarte*, the ligament is high and situated beneath the beak, occupying a very high and broad area. In *Gemma*, the ligament is very narrow and elongated, and extending posterior to the beaks. Pallial sinus much smaller in *Parastarte*.

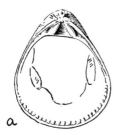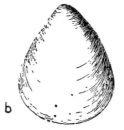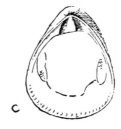

FIGURE 85. Brown Gem Clam, *Parastarte triquetra* Conrad, ⅛ inch (Florida).

Parastarte triquetra Conrad

Brown Gem Clam
Figure 85

Both sides of Florida (to Texas?).

⅛ inch in size, very similar to *Gemma gemma*, but much higher than long, with the beaks larger and elevated. Exterior highly polished and smoothish. Color usually tan to brown, but may be flushed with pink in beachworn specimens. The pallial sinus is almost absent. Moderately common on sand bars and obtained by screening the sand.

Family PETRICOLIDAE
Subfamily PETRICOLINAE
GENUS *Petricola* Lamarck 1801
Subgenus *Naranio* Gray 1853

Petricola lapicida Gmelin

Boring Petricola

South half of Florida and the West Indies.

½ inch in length (up to 1¾ inches in the Lesser Antilles), ovate, inflated, chalk-white, with criss-cross, threaded sculpturing. Beaks swollen and close together. Posterior end with wavy ribs consisting of fine mud particles laid down over the shell by the animal. There is an enclosed, elongate furrow between the beaks and the hinge. Color yellowish white. Found in burrow holes in coral rocks. Not uncommon.

Subgenus *Petricolaria* Stoliczka 1870

Petricola pholadiformis Lamarck   False Angel Wing
Plate 32z; figure 94b

Gulf of St. Lawrence to the Gulf of Mexico and south.

2 inches in length, elongate, rather fragile and chalky-white. With numerous radial ribs. The anterior 10 or so are larger and bear prominent scales. Ligament external, located just posterior to the beaks. Cardinal teeth quite long and pointed. The siphons are translucent-gray, large, tubular and separated from each other almost to their bases. A very common clay and peat-moss borrower.

Genus *Rupellaria* Fleuriau 1802

Rupellaria typica Jonas   Atlantic Rupellaria
Plate 30e
North Carolina to the south half of Florida and the West Indies.

About 1 inch in length, oblong, flattened anteriorly; compressed, usually attenuated and gaping posteriorly. Beaks point anteriorly. Exterior gray or whitish and with numerous, irregularly spaced, coarse radial ribs. Interior uneven and brownish gray. This coral borer is variable in shape and uneven in texture. It may also be truncate at the posterior end. Moderately common.

Rupellaria tellimyalis Carpenter   West Coast Rupellaria
Plate 31t
Santa Monica, California, to Mazatlan, Mexico.

1 to 1⅓ inches in length. Oblong-elongate, variable in shape and outline due to crowding in the rock burrow. Shell fairly thick, white, except for purplish blotches commonly behind the hinge and at the posterior end. Radial threads are coarser at the anterior end. Growth lines are irregular and coarse. Pallial sinus broadly rounded at its anterior end. Early or nepionic shell is shaped somewhat like a *Donax*, smooth, translucent purplish brown and rarely found attached at this early stage to rocks and kelp stalks. *R. californiensis* Pilsbry and Lowe is identical.

Rupellaria denticulata Sowerby known only from Peru has a similar nepionic shell (contrary to other reports), has a narrower, triangular pallial sinus, and (contrary to reports) is a more fragile shell. Its anterior end is pointed and slightly uplifted. Interior blushed with mottlings of chestnut to purplish brown.

Rupellaria carditoides Conrad Hearty Rupellaria

Vancouver, B.C., to Lower California.

1 to 2 inches in length. Very variable in shape, usually oblong; in some, squat and almost orbicular. Shell white to grayish white and very chalky in texture. Concentric growth lines quite coarse and irregular. Radial sculpture of peculiar, fine, scratched lines crowded together, but worn away in some specimens. Fairly common. Found boring into hard rock. Nepionic shell usually oblong. _R. californica_ Conrad is the same.

Subfamily COOPERELLINAE
GENUS _Cooperella_ Carpenter 1864

Hinge plate narrow, with 2 right and 3 left short, divaricating cardinals under the beaks. The left central cardinal is always, and the others commonly, split or bifid. No laterals. Muscle scars small and oval. Pallial line narrow, the sinus long.

Cooperella subdiaphana Carpenter Shiny Cooper's Clam

Southern California to Lower California.

About ½ inch in length, oval-oblong, opaque-white with a brilliant gloss and slight opalescence. Fragile. Outer surface with slightly wavy concentric growth lines. Ligament tiny, short, set just behind the beaks and visible externally. Moderately common offshore to 40 fathoms.

Superfamily TELLINACEA
Family TELLINIDAE
GENUS _Tellina_ Linné 1758
Subgenus _Tellina s. str._

Tellina radiata Linné Sunrise Tellin
Plate 40e

South Carolina to south half of Florida and the West Indies.

2 to 4 inches in length, elongate, moderately inflated. Characterized by its oily smooth, glistening surface and rich display of colors—either creamy-white or rayed with pale-red or yellow. Interior flushed with yellow. The beaks are usually tipped with bright-red. Uncommon in Florida but abundant in the West Indies.

Tellina laevigata Linné   Smooth Tellin
                                                             Plate 40k

Southern Florida and the West Indies.

2 to 3 inches in length, oval to slightly elongate, moderately compressed, strong, with a smooth, glossy surface except for microscopic, radial scratches. Exterior color either whitish or usually faintly rayed, or banded at the ventral margins with soft, creamy-orange. Inside polished white to yellowish. Rare in Florida, fairly common in the West Indies.

Subgenus *Tellinella* Mörch 1853

Tellina interrupta Wood   Speckled Tellin
                                                             Plate 40l

North Carolina to south half of Florida and the West Indies.

2½ to 3½ inches in length, well elongated, moderately inflated, twisted at the posterior end where at the dorsal margin on the right valve there are 2 rough ridges. Concentric threads numerous, evenly spaced. Color whitish with numerous small, prominent, zigzag specklings of purplish brown. Interior yellowish. Not uncommon in southeast Florida, but abundant in some shallow West Indian bays.

Tellina idae Dall   Ida's Tellin
                                                          Figures 87a, b; 28h

Santa Monica to Newport Bay, California.

2 to 2½ inches in length, elongate, compressed. With strong, rather evenly spaced, concentric, lamellate threads. Posterior end narrow, slightly twisted and with a rounded, radial ridge near the dorsal margin (in right valve) or a ridge at the dorsal margin and a furrow below it (left valve). Ligament elongate and sunk deeply into the long, deep dorsal-margin furrow. Color grayish white. Moderately common.

Subgenus *Angulus* Mühlfeld 1811

Tellina agilis Stimpson   Northern Dwarf Tellin
                                                          Plate 30x; figure 86f

Gulf of St. Lawrence to North Carolina.

⅓ to ½ inch in length, moderately elongate, compressed, fairly fragile; glossy-white externally with an opalescent sheen. Interior white. Ligament external and prominent. With a large rounded pallial sinus almost extending to the anterior muscle scar. External sculpture of faint, microscopic, concentric, impressed lines. Commonly found washed on shore from Maryland north. Formerly known as *Tellina tenera* Say and *Angulus tener* Say (not

Schrank 1803 nor Leach 1818). In 1858, W. Stimpson gave this species a new name (Amer. Journ. Sci., vol. 25, p. 125). *T. elucens* Mighels might be this species, although it is very doubtful.

Tellina texana Dall replaces this species in the Gulf of Mexico. It has more distinct, crowded concentric lines, is more inflated and has very small, microscopic striae in most specimens. Compare with *T. versicolor*.

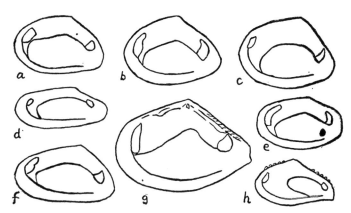

FIGURE 86. Interior views of southeast coast Tellins, showing outline shapes and pallial sinus scars. **a,** *Tellina lineata* Turton, 1½ inches; **b,** *T. tampaensis* Conrad, ½ inch; **c,** *T. mera* Say and *promera* Dall, ¾ inch; **d,** *T. sybaritica* Dall, ⅓ inch; **e,** *T. similis* Sowerby, 1 inch; **f,** *T. agilis* Stimpson and *sayi* Dall, ½ inch; **g,** *Quadrans lintea* Conrad, ¾ inch; **h,** *Phylloda squamifera* Deshayes, ¾ inch.

Tellina versicolor DeKay   DeKay's Dwarf Tellin

New York to the south half of Florida and the West Indies.

½ inch in length, very similar to *T. agilis*, but more elongate, colored white, red, pink or rayed, is more inflated, and has a nearly straight instead of curved ventral margin. The exterior of *versicolor* has a brighter iridescence. The pallial sinus is much closer to the anterior muscle scar.

Tellina mera Say   Mera Tellin
                                                   Figure 86c

Eastern Florida and the Bahamas.

½ to ¾ inch in length, roughly elliptical, moderately inflated, pure opaque-white in color. Fairly thin but strong. Beaks fairly large for a Tellin, touching and pointing toward each other and located nearer the posterior than the center of the shell. The valves show hardly any posterior bend or twist. Exterior smoothish with fine, irregular, concentric lines of growth more evident near the margins. Moderately common in shallow water between tides. Compare with *promera* and *tampaensis*.

Tellina promera Dall   Promera Tellin
                                                          Figure 86c

South half of Florida and the West Indies.

¾ inch in length, very similar to *mera*, but larger, more inflated, thicker-shelled, more oval and with the umbones a little closer to the center of the shell. This may be only a subspecies of *mera*. Quite common on both sides of Florida.

Tellina tampaensis Conrad   Tampa Tellin
                                                          Figure 86b

South half of Florida to Texas, the Bahamas and Cuba.

½ to ¾ inch in length, similar to *mera*, but more pointed posteriorly, whitish with a faint pinkish blush, and with very numerous, microscopic, concentric lines of growth. The pallial sinus line in this species runs forward nearly to the anterior muscle scar and then drops almost vertically toward the ventral margin of the shell before continuing posteriorly. In *mera* and *promera*, the pallial sinus line toward the anterior muscle scar, makes a U-shaped turn, and then runs posteriorly but does not join the lower pallial line until about the middle of the ventral region of the valve. Common in shallow water.

Tellina texana Dall   Say's Tellin

New Jersey to south half of Florida and Cuba.

½ inch in length, white with a faint opalescent sheen. Extremely close to *agilis*, but distinguished by the heavy, enamel-white finish on the inside of the shell and 1 or 2 fairly distinct radial grooves running from the posterior muscle scar to the ventral margin of the valve. The faint pallial sinus just touches the anterior muscle scar. Fairly common. *T. sayi* Dall (Dec. 1900) is a synonym of *T. texana* Dall (Nov. 1900). *T. polita* Say (not Spengler) is also this species.

Tellina sybaritica Dall   Dall's Dwarf Tellin
                                                          Figure 86d

North Carolina, the Gulf of Mexico and Cuba.

¼ to ⅓ inch in length, very elongate, shiny, with quite strong, numerous concentric threads or cut lines. Color varying from translucent-white, yellowish, pinkish to bright watermelon-red. Our smallest and most colorful Tellin, and plentiful from 1 to 60 fathoms. It somewhat resembles young *alternata*, but the latter are smoother and have a short instead of long posterior lateral tooth in the left valve, and has no lateral lamina in the left valve.

Section *Oudardia* Monterosato 1885

Tellina modesta Carpenter   Modest Tellin

Plate 31u

Alaska to the Gulf of California.

¾ to 1 inch in length, elongate, moderately pointed at the posterior lower corner. Surface white with iridescent sheen and with fine concentric threads or grooves. These fade out at the posterior fourth of the shell, but reappear more coarsely on the very posterior slope. There is a well-formed, radial rib inside just behind the anterior muscle scar. Common in certain sandy localities from shore to 25 fathoms. It appears that *T. buttoni* Dall is the same species.

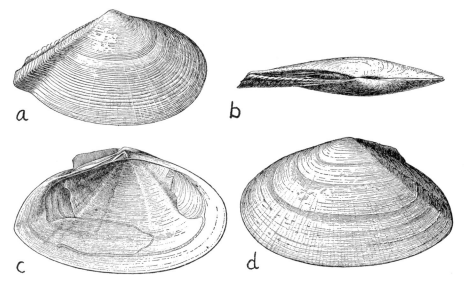

FIGURE 87. Pacific Coast Tellins. **a** and **b**, *Tellina idae* Dall, 2 inches (California); **c** and **d**, *Tellina lutea* Wood, 3 inches (Alaska).

Section *Peronidia* Dall 1900

Tellina lutea Wood   Great Alaskan Tellin

Figure 87c, d

Arctic Ocean to Cook's Inlet, Alaska. Japan.

3 to 4 inches in length, elongate, quite compressed, and with a posterior twist to the right. Worn shells chalky-white, commonly with a pink flush. Periostracum in young is greenish yellow and glossy; in adults dark-brown. Ligament prominent. Commonly found from beach to 23 fathoms. *T. venulosa* Schrenck 1861 is an ecologic form with brownish cracks in the shell.

Subgenus *Moerella* Fischer 1887

Tellina salmonea Carpenter   Salmon Tellin
Plate 31y

Aleutian Islands to San Pedro, California.

½ inch in length, ovalish, with a short, blunt posterior end. Ligament behind the beaks prominent. Dorsal margin in front of beaks almost straight. Color chalky-white, commonly with a pinkish cast. Periostracum smooth, thin, yellowish tan. Characterized by about 4 to 7 prominent, concentric, former growth-stop lines which are usually stained dark-brown. Common from low tide to 34 fathoms in sand. Do not confuse with *meropsis*.

Tellina meropsis Dall   Meropsis Tellin
Plate 30u

San Diego, California, to the Gulf of California.

½ inch in length, ovalish, pure white, smoothish, with exceedingly fine growth lines. Surface silky, but rarely with an iridescent sheen. Beaks slightly toward the posterior end. Ligament not prominent and light-brown. Without growth stoppage lines. See *T. salmonea*. Common from shore to 15 fathoms.

Tellina carpenteri Dall   Carpenter's Tellin

Forrester Island, Alaska, to the Gulf of California.

⅓ inch in length, moderately elongate, with a rounded anterior end and rather truncate posterior end. Ligament short. Color cream, whitish and commonly blushed with watermelon-pink inside and out. It also has a faint iridescent sheen. Found very abundantly in many localities in mud and sand from shore to 369 fathoms.

Subgenus *Scissula* Dall 1900

Tellina similis Sowerby   Candy Stick Tellin
Plate 40m; figure 86e

South half of Florida, the Bahamas and western Caribbean.

1 inch in length, moderately elongate, moderately compressed, thin but fairly strong. Color opaque-white with a yellowish blush and with 6 to 12 short radial rays of red. Interior yellowish with red rays or solid pink or yellow. A red splotch commonly occurs on the hinge in front of the cardinal teeth. Sculpture of concentric growth lines and numerous fine concentric *threads which cross the shell at an oblique angle*. Common on sand flats. *T. decora* Say is the same species.

Tellina iris Say   Iris Tellin

North Carolina to Florida, the Gulf of Mexico, and Bermuda.

½ inch in length, very similar to *similis*, and often as colorful, but very thin-shelled, translucent and more elongate. On the interior of the valves, the wavy oblique lines are evident, and there are 2 radial thickenings or weak, white, internal ribs at the posterior end. Common from intertidal flats to 20 fathoms.

Tellina candeana Orbigny from the Lower Florida Keys and the West Indies commonly yellowish white and is more wedge-shaped (blunter at the anterior end) and, of course, has the peculiar sculpture of this subgenus.

Subgenus *Scrobiculina* Dall 1900

Tellina magna Spengler   Great Tellin
                                                                      Plate 40i

North Carolina to southeast Florida and the West Indies.

3 to 4½ inches in length, half as high, quite compressed and glossy-smooth. Posterior dorsal region dull, bordered by a weak, radial ridge. Left valve glossy white, rarely faintly yellowish; right valve glossy orange to pinkish and microscopically cut with concentric scratches. An uncommon and very lovely species much sought after by collectors. Found just below low tide in sand.

Subgenus *Eurytellina* P. Fischer 1887

Tellina lineata Turton   Rose Petal Tellin
                                                                      Plate 40h

All of Florida and the West Indies.

1½ inches in length, moderately elongate, slightly inflated, solid and with a fairly strong twist to the right at the posterior end. Smoothish and glossy, but under a lens fine, concentric, crowded grooves may be seen. The outer surface has a slight opalescent sheen. Color pure-white or strongly flushed with watermelon-red. Pallial sinus just touches the anterior muscle scar, while in the similar but more elongate *T. alternata* it does not. Common in shallow water.

Tellina alternata Say   Alternate Tellin
                                                                      Plate 40n

North Carolina, Florida, and the Gulf States.

2 to 3 inches in length, elongate and compressed, solid, and with a moderately pointed and slightly twisted posterior end. Sculpture of numer-

ous evenly spaced, fine, concentric grooves. Area near umbones smooth. Color glossy and variable: whitish, yellowish or flushed with pink. Interior glossy-yellow or pinkish. A common shallow-water species which should always be compared with *lineata*.

The similar *Tellina angulosa* Gmelin from the Keys and West Indies is not so elongate, has finer grooves and a more highly glossed surface which is commonly covered with a greenish-yellow, thin periostracum. Common in sand.

T. punicea Born (which Dall called *angulosa* Gmelin) from the Keys (rare) and West Indies (common) is similar, but is always bright watermelon-red internally and purplish red exteriorly. The pallial sinus just touches the anterior muscle scar, which it does not in *alternata* or *angulosa*.

GENUS *Arcopagia* Brown 1827
Subgenus *Cyclotellina* Cossmann 1886

Arcopagia fausta Pulteney 1799

Faust Tellin
Plate 40j

North Carolina to southeast Florida and the West Indies.

2 to 4 inches in length, oval, moderately inflated, fairly heavy, and smoothish, except for small, rough, concentric lines of growth. Hinge strong, the posterior lateral in the right valve being long and strong. Color outside a semi-glossy-white; inside highly glossed and enamel-white with a yellowish flush. Do not confuse with *T. laevigata* which is glossy outside and has orange-tinted margins. Moderately common in the West Indies. Donovan gave this species the same name in 1801.

GENUS *Strigilla* Turton 1822

Tellin-like shells, usually oval in shape and with inconspicuous growth lines crossed by fine, oblique, cut lines. There are only four species in the western Atlantic.

Strigilla carnaria Linné

Large Strigilla
Plate 40c

North Carolina to Florida and western Caribbean.

¾ to 1 inch in length, oval, slightly oblique, moderately compressed, fairly thin but strong. Outer surface finely sculptured by cut lines which are obliquely radial in the central and posterior regions of the valve. At the anterior third of the valve, there are wavy, oblique threads running in the opposite direction. Exterior pinkish white with former, concentric growth stages a deeper pink. Interior bright watermelon-red. The upper line of the pallial sinus runs directly posterior to the anterior muscle scar. Common in shallow water in sand and commonly washed ashore.

West Indian collectors should not confuse this species with *Strigilla rombergi* Mörch (southeast Florida, the Bahamas to Lesser Antilles) which is very similar, except that the upper line of the pallial sinus does not reach the anterior muscle scar. The radial cut lines are more numerous and more curved in *rombergi*. Common.

Strigilla mirabilis Philippi   White Strigilla

North Carolina to Florida, Texas and the West Indies.

⅓ inch in length, oval, inflated, shiny, all white in color and with the peculiar Strigillid sculpturing which is very similar to that in *S. pisiformis* (the radial cut lines meet the ventral margin of the valves at an angle of about 45 degrees). The pallial line runs forward from the posterior muscle scar but does not reach the anterior muscle scar as it does in *pisiformis*. Common. Do not confuse with the larger *Divaricella quadrisulcata* (page 391) which has fine denticulations on the inner margins of the valves. *S. flexuosa* Say 1822 is preoccupied by Montagu 1803 and Turton 1807, and must take the name *mirabilis* Philippi 1841.

Strigilla pisiformis Linné   Pea Strigilla

Florida Keys, the Bahamas and the West Indies.

⅓ inch in length, similar to *S. carnaria*, but always much smaller and more inflated. The pink color inside is concentrated in the deepest part of the valve, and the margins are usually white. The radial, oblique cut lines meet the ventral margin of the valve at about 45 degrees angle, while in *carnaria* the lines are almost vertical. This is a very abundant species, especially in the Bahamas where they are gathered in great numbers and brought to Florida for use in the shellcraft business.

GENUS *Phylloda* Schumacher 1817
Subgenus *Phyllodina* Dall 1900

Phylloda squamifera Deshayes   Crenulate Tellin
                                                                        Figure 86h
North Carolina to the south half of Florida.

½ to 1 inch in length, elongate, concentrically and finely ridged. Characterized by the strong crenulations on the posterior dorsal margin and by the lightly hooked-down posterior ventral margin. Color whitish with a yellow or orangish tint. Moderately common from low water to 60 fathoms.

Compare with *lintea* which lacks the dorsal crenulations. Formerly placed in the genus *Tellina*.

Genus *Quadrans* Bertin 1878

Quadrans lintea Conrad   Lintea Tellin
                                                            Figure 86g

North Carolina to both sides of Florida and the West Indies.

¾ to 1 inch in length, moderately oval, slightly inflated, quite strong and all white in color. Posterior dorsal slope with 2 radial ridges in the right valve, 1 in the left. Concentric lamellae numerous, sharp and minutely raised. Left valve with 2 extremely weak, long laterals, but these are well-developed in the right valve. Dorsal line of the pallial sinus meets the pallial line not far from the anterior muscle scar. Posterior twist to the right is fairly pronounced. Commonly dredged off the Carolinas (9 to 16 fathoms), uncommonly found in a few feet of water on the west coast of Florida. Formerly placed in *Tellina*.

Genus *Tellidora* H. and A. Adams 1856

Tellidora cristata Recluz   White Crested Tellin
                                                            Plate 30-o

North Carolina to west Florida and Texas.

1 to 1½ inches in length, roughly ovate, compressed and all white. The left valve is very flat, the right valve slightly inflated. Dorsal margins of valves with large, saw-tooth crenulations. A bizarre clam found uncommonly in shallow water.

Genus *Macoma* Leach 1819

The Macomas are modified tellins which may be distinguished by (1) no lateral teeth; (2) usually dingy-white in color and of a chalky consistency; (3) there is a strong posterior twist; (4) the pallial sinus is larger in one valve than the other.

Macoma calcarea Gmelin   Chalky Macoma
                                                            Figure 88f

Greenland to Long Island, New York. Bering Sea to off Monterey, California.

1½ to 2 inches in length. Oval-elongate, moderately compressed, but somewhat inflated at the larger, anterior half. Beaks ⅗ the way toward the narrowed, slightly twisted posterior end. Shell dull, chalky-white. Concentric sculpture of fine, irregular threads. Periostracum remaining on the margins is gray. Pallial sinus in left valve runs from the posterior muscle scar anteriorly toward the anterior muscle scar, but does not meet the latter, and then descends posteriorly to meet the pallial line about the middle of the lower margin of the shell. A common cold-water species, distinguished from *balthica* by its larger size, more elongate shape and pattern of the pallial sinus scar (see fig. 88g).

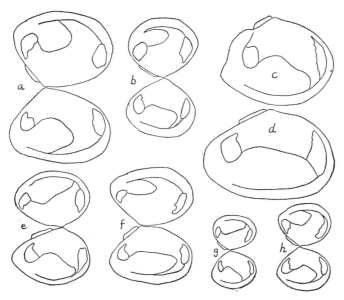

FIGURE 88. American Macomas, showing interior scars. a, *Macoma brota* Dall, 3 inches (Alaska to Washington); b, *M. incongrua* von Martens, 2 inches (Alaska to Washington); c, *M. secta* Conrad, 3 inches (Pacific Coast); d, *M. nasuta* Conrad, 2 inches (Pacific Coast); e, *M. irus* Hanley, 2 inches (Pacific Coast); f, *M. calcarea* Gmelin, 2 inches (Atlantic and Pacific Coasts); g, *M. balthica* Linné, 1 inch (Atlantic and Pacific Coasts); h, *M. planiuscula* Grant and Gale, 1 inch (Alaska to Washington).

Macoma balthica Linné

Balthic Macoma

Figure 88g

Arctic Seas to off Georgia. Bering Sea to off Monterey, California.

½ to 1½ inches in length, oval, moderately compressed. Color dull whitish, in some with a flush of pink, and with a thin, grayish periostracum which readily flakes off. The shape is somewhat variable. A common intertidal and deep-water species. Compare muscle scars with those of *calcarea*.

Macoma tenta Say

Tenta Macoma

Cape Cod to south half of Florida and the West Indies.

½ to ¾ inch in length, fragile, elongate, white in color with a delicate iridescence on the smooth exterior. Posterior and narrower end slightly twisted to the left. This small, tellin-like species is very common in shallow water in sand. *M. souleyetiana* Recluz is the same.

Macoma limula Dall is very similar in size and shape, although somewhat more elongate, and is distinguished by the finely granular external surface of the valves. Commonly dredged from North Carolina to Florida.

Macoma constricta Bruguière   Constricted Macoma

Florida to Texas and the West Indies.

1 to 2½ inches in length, moderately elongate. The posterior end is twisted to the right and is narrowed to a blunt point. Color all white with concentric growth lines stained by the gray periostracum. Common just off-shore.

Macoma nasuta Conrad   Bent-nose Macoma
Figure 88d

Alaska to Lower California.

2 to 3½ inches in length, elongate, rather compressed and strongly twisted to the right at its posterior end. Beaks slightly nearer the anterior end. Can be distinguished from other Pacific Coast species by the pallial sinus in the *left* valve which reaches the anterior muscle scar. One of the commonest species on the west coast and lives about 6 inches below the surface of the mud in quiet waters from shore to 25 fathoms.

Macoma secta Conrad   White Sand Macoma
Figure 88c

Vancouver Island to the Gulf of California.

2 to 4 inches in length. This is the largest Macoma in America and is characterized by the almost flat left valve, rather well-inflated right valve, and by the wide and relatively short ligament which is sunk partially into the shell. There is a large, oblique, rib-like extension just behind the hinge inside each valve. Color cream to white. Common in bays and beaches from shore to 25 fathoms. A small form occurs in protected waters in bays in its more southerly range.

Students of the Pacific Coast fauna consider *M. indentata* Carpenter (same range) as a distinct species in which the shell is 1½ inches in length, a little more elongate, with a more pointed posterior end, and with a slight indentation on the posterior ventral margin. It may possibly be a form of young *secta*. *M. tenuirostris* Dall is even more elongate and may also be a form.

Macoma yoldiformis Carpenter   Yoldia-shaped Macoma

Alaska to San Diego, California.

½ to ¾ inch in length, elongate, moderately rounded at each end and with a small, but distinct, twist to the right at the posterior end. Color a uniform, glossy, porcellaneous white. Rarely translucent with an opalescent sheen. Common from shore to 25 fathoms.

Macoma carlottensis Whiteaves   Queen Charlotte Macoma

Arctic Ocean to Lower California.

About 1 inch in length, extremely fragile, inflated and with a very short, inconspicuous ligament. Color translucent-white with a thin, greenish, glossy periostracum. This species was named *inflatula* Dall at a later date.

Macoma brota Dall   Brota Macoma

Figure 88a

Arctic Ocean to Puget Sound, Washington.

3 inches in length, moderately elongate, moderately inflated and rather thick-shelled. Beaks ⅔ toward the posterior end. Resembles *calcarea* whose pallial sinus in the left valve, however, is more elongate, not as high and generally reaches nearer the anterior muscle scar. *M. brota* is larger and more truncate posteriorly than that species. Common.

Macoma planiuscula Grant and Gale   Grant and Gale Macoma

Plate 30t; figure 88h

Arctic Ocean to Puget Sound, Washington.

About 1 inch in length. Extremely similar to *calcarea*, but porcellaneous, with a glossy, yellowish periostracum and more oval in shape. This species was thought by Dall and others to be "*carlottensis* Whiteaves."

GENUS *Gastrana* Schumacher 1817
Subgenus *Heteromacoma* Habe 1952

Gastrana irus Hanley   Irus Macoma

Figure 88e

Bering Sea to Los Angeles, California. Japan.

Commonly 1½ inches in length (rarely 3); oval-elongate, moderately inflated, very slightly twisted, if at all, at the posterior end. Pallial sinus in left valve almost reaches the bottom of the anterior muscle scar. Beaks slightly anterior. Common in Washington and Oregon. Formerly known as *inquinata* Deshayes.

Macoma incongrua von Martens (fig. 88b) (Alaska to Washington) is similar, but more oval, with a more pointed slope at the beaks, a straighter posterior dorsal edge. Its upper pallial sinus line, after nearly reaching the anterior muscle scar, turns downward and then runs anteriorly before connecting with the pallial sinus. Common in Alaska.

Genus *Apolymetis* Salisbury 1929

Apolymetis intastriata Say   Atlantic Grooved Macoma
Plate 32y

South half of Florida and the Caribbean.

2 to 3 inches in length, elliptical, fairly thin but strong and all white in color. The shell is strongly twisted. At the posterior end, the right valve bears a strong, radial rib, while on the left valve there is a fairly strong, radial groove. Pallial sinus very large. Living specimens are not commonly collected, although shells are commonly washed ashore.

Apolymetis biangulata Carpenter   Pacific Grooved Macoma

Santa Barbara, California, to Ensenada, Mexico.

2 to 3½ inches in length, oval, moderately compressed, strong, and dull grayish white in color. Interior glossy-white with the central portion blushed with pastel-peach. Left valve with a shallow, radial groove near the posterior end. Right valve with a corresponding ridge at the end of which the margin of the shell is shallowly notched. Alias *A. alta* Conrad. Common.

Family SEMELIDAE
Genus *Semele* Schumacher 1817

Resilium supported in a horizontal, chondrophore-like depression which is internal and parallel with the hinge line. 2 cardinal teeth in each valve. Right valve with 2 distinct lateral teeth, but practically absent in the left valve. The ligament is external.

Semele proficua Pulteney   White Atlantic Semele
Plate 40g

North Carolina to south half of Florida and the West Indies.

½ to 1½ inches in length, almost round, beaks almost central. **Lunule** small and pushed in. Fine concentric lines and microscopic radial striations. Externally whitish to yellowish white. Interior glossy, commonly yellowish, rarely speckled a little with purple or pink. Moderately common in shallow water. The color form *radiata* Say, has a few indistinct radial rays of pink (pl. 40f).

Semele purpurascens Gmelin

Purplish Semele
Plate 40b

North Carolina to south half of Florida and the West Indies.

1 to 1¼ inches in length, oblong, thin-shelled, smooth except for very fine concentric growth threads over which run another set of fine, microscopic concentric lines at an oblique angle. External color variable: commonly gray or cream with purple or orangish mottlings. Interior glossy and suffused with purple, brownish or orange. A fairly common, shallow-water species.

Semele bellastriata Conrad

Cancellate Semele
Plate 30j

North Carolina to south half of Florida and the West Indies.

½ to ¾ inch in length, similar in shape to *purpurascens*, but a much smaller species with numerous radial and concentric riblets which cross to give a cancellate appearance. Some specimens well-beaded, others have the radial ribs more prominent. External color yellowish white with reddish flecks or a solid, purplish gray. Interior white, cream or suffused with mauve or violet. *S. cancellata* Orbigny is this species. Fairly common just offshore.

Semele rubropicta Dall

Rose Petal Semele
Plate 29w

Alaska to Mexico.

1 to 1¾ inches in length; beaks ⅔ toward the posterior end; rather thick-shelled, especially in the south. Concentric sculpture of small, irregular growth lines. Radial incised lines numerous. Periostracum thin, smooth and yellowish brown. Exterior of shell dull grayish or tannish white with faint, radial rays of light-mauve. Interior glossy-white, with a small splotch of mauve at both ends of the hinge line. Uncommon from 20 to 50 fathoms.

Semele rupicola Dall

Rock-dwelling Semele
Plate 29t

Santa Cruz, California, to the Gulf of California.

1 to 1½ inches in length. Irregular in shape: ovalish, oval-elongate or obliquely oval. Exterior yellowish cream with numerous, concentric crinkles and a few weak radial threads. Interior glossy, white at the center, bright purplish red at the margins and hinge. John Q. Burch finds this species common in *Chama* and *Mytilus* beds and in rocks and crevices.

Semele decisa Conrad

Bark Semele
Plate 29z

San Pedro, California, to Lower California.

2 to 3 inches in length, equally high. Characterized by its heavy shell, by the coarse, wide, irregular, concentric folds on the outside (resembling rotting bark or wood). Exterior yellowish gray; interior glossy-white, with a purple tinge, especially prominent on the hinge and margins. Commonly found in rocky rubble in shallow water.

Genus Cumingia Sowerby 1833

Shell delicate, with concentric lamellae; slightly gaping behind. Resilium internal and supported by a spoon-shaped chondrophore. One cardinal and 2 elongate lateral teeth in each valve.

Cumingia tellinoides Conrad   Tellin-like Cumingia

Nova Scotia to St. Augustine, Florida.

½ to ¾ inch in length, oblong and fairly thin. Slightly pointed at the posterior end. Exterior chalky-white, with tiny, sharp, concentric lines. This is a moderately common mud-digger which externally resembles a *Tellina*.

The subspecies *vanhyningi* Rehder 1939 replaces the typical species in southern Florida to Texas. It is not so high, is more elongate and more drawn out posteriorly. Common in shallow water. *C. coarctata* Sowerby (Lower Keys and West Indies) has stronger, more widely separated concentric ridges.

Cumingia californica Conrad   Californian Cumingia
Plate 31v

Crescent City, California, to Chile.

1 to 1⅓ inches in length, elongate-oval, moderately compressed. In front of the beaks there is a small, elongate depression. Just posterior to and partially covered by the inrolled beaks is a small, short ligament, posterior to which is a wide, flaring furrow. Concentric sculpture of numerous wavy, rather sharp, fairly large threads. Color grayish white. Pallial sinus very long. Abundant in rock crevices and wharf pilings. Also dredged down to 25 fathoms.

Genus Abra Lamarck 1818

Shell small (¼ inch), fragile, ovalish, smooth, moderately compressed. Translucent-white in color. Resilium internal and supported by a linear chondrophore. Right valve with 2 cardinals and generally with 2 lamellar laterals.

Abra aequalis Say
Common Atlantic Abra
Plate 30v

North Carolina to Texas and the West Indies.

¼ inch in size, orbicular, smooth, glossy and rather inflated. Surface may show a slight iridescence. Periostracum very thin and clear yellowish. Anterior margin of right valve grooved. A very abundant and very simple-looking bivalve. Compare with *lioica*.

Abra lioica Dall
Dall's Little Abra
Plate 30w

Cape Cod to south Florida and the West Indies.

¼ inch in size, similar to *aequalis*, but the beaks are nearer the anterior end, the shell is thinner, and more elongate. Anterior margin of right valve not grooved. The prodissoconch at the beaks is large, tan and more trigonal in shape than the adult. Common from 6 to 200 fathoms.

Family DONACIDAE
GENUS *Donax* Linné 1758

The posterior end is the shortest and the fattest. 2 cardinal and an anterior and a posterior lateral in each valve. Pallial sinus deep.

Donax variabilis Say
Coquina Shell
Plate 30r

Virginia to south Florida and Texas.

½ to ¾ inch in length. Ventral margin of the valves straight and almost parallel with the dorsal margin. The thinner, anterior end is commonly smooth, but may be microscopically scratched with radial lines. From the middle of the valve to the blunt posterior end, small radial threads appear which become increasingly larger posteriorly. Internal margin of valves minutely denticulate. Color variable and commonly very bright, especially inside: white, yellow, pink, purple, bluish, mauve and commonly with rays of darker shades.

The subspecies *roemeri* Philippi (pl. 30q) is very common along the Texas shores. The posterior end is blunter, the whole shell not so elongate, and the ventral margin sags down.

Donax fossor Say
Fossor Donax

Long Island, New York, to Cape May, New Jersey.

Never exceeds ½ inch in length, very similar to *variabilis*, but almost smooth, even at the blunt, posterior end. Only 2 color phases: yellowish white or with weak purplish rays. Subsequent biological studies may show that this common beach species is a subspecies or cold-water form of *variabilis*. Young specimens of *variabilis* from several southern states look suspiciously like this so-called species.

Donax tumidus Philippi   Fat Gulf Donax

Northern shores of the Gulf of Mexico.

⅓ to ½ inch in length, very obese, somewhat trigonal in shape, with its beaks swollen and posterior end strongly truncate. Threads on the blunt posterior end are heavily beaded, sometimes giving a cancellate appearance. Narrow anterior end commonly with distinct, microscopic, distantly spaced, incised, concentric lines. Color whitish with bluish, yellowish or pinkish undertones. Rarely, if ever, rayed. Uncommon. Found in 2 to 3 feet of water.

Donax denticulatus Linné (pl. 30p) from the southwest Caribbean has been erroneously recorded from our shores. It is 1 inch in length and characterized by 2 curved, low ridges on the posterior slope of each valve and by microscopic pin-points on the sides of the valves.

Donax striatus Linné also a lower West Indian species is as large, but characterized by a flat to slightly concave posterior slope which bears numerous fine radial threads.

Donax gouldi Dall   Gould's Donax
Plate 31q

San Luis Obispo, California, to Mexico.

Common form: ¾ inch in length, fairly obese, truncate at the posterior end where the beaks are located. Shell glossy, smooth, except for numerous, microscopic, axial threads at the anterior end. Exterior cream to white with variable color rays of light-tan. Margins of valves commonly flushed with purple. Interior stained with purple or bluish brown. Common on beaches, especially in the north. Known locally as the Bean Clam.

Small form: ½ inch in length, slightly more obese, without the color rays in most cases. Common, especially in the south.

Donax californicus Conrad   California Donax
Plate 31p

Santa Barbara, California, to Panama.

Up to 1 inch in length, narrowly pointed at both ends. Shell glossy or oily-smooth with a tan to greenish-tan periostracum. Interior white or with a purple flush, and with a strong splotch of purple at each end of the dorsal margin. Common in shallow waters of bays along the shore.

Genus *Iphigenia* Schumacher 1817

Iphigenia brasiliensis Lamarck   Giant False Donax
Plate 32u

South half of Florida and the West Indies.

2 to 2½ inches in length, rather heavy, elongate, roughly diamond-shaped in side view and moderately inflated. Posterior dorsal slope flattish. Exterior smoothish, cream with a purple-stained beak area. Commonly entirely covered with a thin, glossy, brown periostracum. Moderately common in shallow water in sand.

Family *SANGUINOLARIIDAE*
Genus *Sanguinolaria* Lamarck 1799

Sanguinolaria cruenta Solander   Atlantic Sanguin
Plate 40d

South Florida, the Gulf States and the West Indies.

1½ to 2 inches in length, moderately compressed, the left valve slightly flatter than the right. With a slight posterior gape. Exterior glossy, smooth, except for minute concentric scratches. Pallial sinus with a U-shaped hump at the top. Color white with the beaks and area below a bright-red which fades ventrally into white. Uncommon in the West Indies, rare in Florida. *S. sanguinolenta* Gmelin is a later name for this species.

Subgenus *Nuttallia* Dall 1898

Sanguinolaria nuttalli Conrad   Nuttall's Mahogany Clam
Plate 29x

Monterey, California, to Lower California.

2½ to 3½ inches in length. A handsome species characterized by its smooth, oval form, glossy nut-brown color, with its right valve almost flat and its left valve inflated. External ligament like a brown leather button. Interior whitish, commonly with rosy or purplish blush. Common near estuaries in 6 to 8 inches of mud.

Genus *Asaphis* Modeer 1793

Asaphis deflorata Linné   Gaudy Asaphis
Plate 40a

Southeast Florida and the West Indies.

2 inches in length, moderately inflated. Sculpture of numerous, coarse, irregularly sized, radial threads. Color variable and brighter on the inside: whitish, yellow, or stained with red, rose or purple. Beaks inflated and rolled in under themselves a little. A moderately common, intertidal species, also known from the Indo-Pacific.

Genus *Tagelus* Gray 1847

Tagelus plebeius Solander   Stout Tagelus
Plate 30d

Cape Cod to south Florida and the Gulf States.

2 to 3½ inches in length, oblong, subcylindrical, rather inflated, rounded posteriorly, obliquely truncate anteriorly. Beaks indistinct, close together and nearer the posterior end of the shell. Hinge with 2 small, projecting cardinal teeth, with a large bulbous callus just behind them. Exterior smoothish, with tiny, irregular, concentric scratches. Periostracum moderately thick, shiny, olive-green to brownish yellow. Moderately common in shallow water in mud-sand intertidal areas. *T. gibbus* Spengler is a later name for this species.

Tagelus californianus Conrad   Californian Tagelus
Plate 29u

Monterey, California, to Panama.

2 to 4 inches in length. Pallial sinus does not extend past a line vertical to the beaks. External color yellowish white under a dark-brown periostracum which is radially striated. Interior white. Common on muddy sand flats near marshes. Lives 8 to 10 inches below the surface of the sand.

Tagelus affinis C. B. Adams   Affinis Tagelus

Southern California to Panama.

1½ to 2½ inches in length. Shell thin. Pallial sinus extends to a line slightly beyond the beaks, that is, about 53 to 60 percent of the total length of the shell. Dredged from 3 to 40 fathoms in mud-sand, but commonly washed ashore.

Subgenus *Mesopleura* Conrad 1867

Tagelus divisus Spengler   Purplish Tagelus
Plate 30g

Cape Cod to south Florida, the Gulf States and the Caribbean.

1 to 1½ inches in length, elongate, subcylindrical, fragile and smooth. The valves are reinforced internally by a very weak, radial rib (commonly

obscure) running across the center of the valve just anterior to the 2 small, projecting cardinal teeth. Color of shell whitish purple, covered externally with a very thin, chestnut-brown, glossy periostracum. A common shallow-water species.

Tagelus subteres Conrad — Purplish Pacific Tagelus
Plate 29s

Santa Barbara, California, to Lower California.

1 to 2 inches in length, subcylindrical, slightly arcuate with the dorsal margins sloping down from the beaks. Color pale purple inside and out. Periostracum yellowish brown and finely wrinkled. Moderately common in shallow water in sandy mud.

Tagelus politus Carpenter from Central America does not slope down from the beaks so strongly, is a thinner shell and much more darkly colored with violet.

GENUS *Heterodonax* Mörch 1853

Shell less than 1 inch in length, resembling a strong, oval *Tellina*. Two cardinals and two lateral teeth in each valve, the laterals usually not very distinct. Pallial sinus extends ⅗ the length of the shell.

Heterodonax bimaculatus Linné — Small False Donax
Plate 30h

South half of Florida and the West Indies. Southern California to Panama.

½ to 1 inch in length, oval, with a truncate anterior end and moderately inflated. Exterior smoothish, with numerous fine growth lines. 2 cardinals in each valve. Anterior to the beaks (which point forward), the hinge is thick for a short distance, then followed by a thinner, concave portion. Color variable: white with 2 oblong crimson spots inside; violet with radial streaks; pink, yellow or mauve; some are speckled with black or brown. This is a common species found with *Donax* on the slopes of sandy beaches. *H. pacificus* Conrad is a synonym.

GENUS *Gari* Schumacher 1817

Shell fairly large, elongate-oval, beaks near the center; hinge thick and with 2 small, bifid teeth just under the beak. Alias *Psammobia* Lam. 1818.

Subgenus *Psammocola* Blainville 1824

Gari californica Conrad — Californian Sunset Clam
Plate 29n

Aleutians to off San Diego, California. Japan.

2 to 4 inches in length, elongate-oval, fairly strong; the low beaks are nearer the anterior end. Sculpture of strong, irregular, concentric growth lines. Periostracum brownish gray, fairly thin and irregularly wrinkled. Exterior shell dirty-white or cream, and may have faint, narrow, radial rays of purple. Common offshore to 25 fathoms. Commonly washed ashore after storms, especially between Sea Beach and Huntington Beach, California. The subgenus *Gobraeus* Leach 1852 is a later name for *Psammocola.*

Suborder *ADAPEDONTA*
Superfamily *SOLENACEA*
Family *SOLENIDAE*
Genus *Siliqua* Mühlfeld 1811

Shell commonly 6 inches in length, oval, compressed laterally; with a rather straight, raised, internal rib ventrally directed. Hinge like *Ensis*.

Siliqua costata Say	Atlantic Razor Clam

Plate 30f

Gulf of St. Lawrence to New Jersey.

2 to 2½ inches in length, ovate-elongate, compressed, fragile, smooth and with a shiny, green periostracum. Interior glossy, purplish white, with a strong, white, raised rib running down from the hinge to the middle of the anterior end. Very common on shallow-water sand-flats along the New England coast.

Siliqua squama Blainville found offshore from Newfoundland to Cape Cod is larger, thicker, white internally, and its internal, supporting rib slanting posteriorly instead of anteriorly as in *costata*. Uncommon.

Siliqua lucida Conrad	Transparent Razor Clam

Bolinas Bay, California, to Lower California.

1 to 1½ inches in length. Very thin, fragile and translucent. Moderately elongate. Shell whitish tan, with broad, indistinct, radial rays of darker tan or rosy purplish. Periostracum thin, olive-green and varnish-like. Moderately common in sand at low tide to 25 fathoms. This species can be distinguished from the young of *S. patula* by its narrower and higher internal rib which crosses the shell at right-angles, and in being more arcuate on its ventral margin.

Siliqua patula Dixon	Pacific Razor Clam

Plate 29y

Alaska to Monterey, California.

5 to 6 inches in length, oval-oblong in shape, laterally compressed and moderately thin. Periostracum varnish-like and olive-green. Interior glossy and whitish with a purplish flush. Internal rib under teeth descending obliquely toward the anterior end. Animal without dark coloration. The variety *nuttalli* Conrad is a synonym. Do not confuse with *S. alta*. An abundant, edible species found in mud and sand on ocean beaches.

Siliqua alta Dall   Dall's Razor Clam

Arctic Ocean to Cook's Inlet, Alaska. Russia.

4 to 5 inches in length, similar to *patula*, but chalky-white inside, more truncate at both ends, a heavier shell, and with a stronger, narrower and vertical (not oblique) rib on the inside. *S. media* Sowerby from the same region may possibly be the young of this species, although it is blushed with purple inside. *S. alta* is common and edible.

GENUS *Ensis* Schumacher 1817

The Jackknife Clams closely resemble *Solen*, but the left valve has 2 vertical, cardinal teeth, and each valve has a long, low posterior tooth.

Ensis directus Conrad   Atlantic Jackknife Clam
Plate 30k

Labrador to South Carolina. Florida?

Up to 10 inches in length, 6 times as long as high, moderately curved and with sharp edges. Shell white, covered with a thin, varnish-like, brownish-green periostracum. Common on sand-flats in New England. Edible.

Ensis minor Dall from both sides of Florida to Texas rarely exceeds 3 inches in length, is more fragile, relatively longer, and is more pointed at the free end (not the end with the teeth). Internally it has purplish stains. Moderately common between tide marks. Some workers consider this a subspecies of *directus*. *E. megistus* Pilsbry and McGinty are probably 5-inch-long specimens of *minor*.

Ensis myrae S. S. Berry   Californian Jackknife Clam

Southern California.

2 inches in length, with much the same characters as in *directus*. This is the only *Ensis* in California and it is not very common. It has been erroneously called *californicus* Dall which, however, is a more southerly species. For a new name for the Californian *Ensis*, consult future works by Pacific Coast students (probably S. S. Berry).[1]

[1] Since appeared Aug. 1953 in Trans. San Diego Soc. Nat. Hist., vol. 11, p. 398.

Genus *Solen* Linné 1758

Similar to *Ensis*, but with only a single tooth at the end of each valve.

Solen sicarius Gould   Blunt Jackknife Clam

Plate 29v

British Columbia to Lower California.

2 to 4 inches in length, 4 times as long as wide. Exterior with a varnish-like, olive to greenish periostracum. Moderately common in certain localities, especially sandy mud flats. Also dredged in 25 fathoms.

Solen viridis Say   Green Jackknife Clam

Plate 30n

Rhode Island to northern Florida and the Gulf States.

About 2 inches in length, about 4 times as long as wide; dorsal edge straight, ventral edge curved. Hinge with a single projecting tooth at the very end of the valve. Color white; periostracum thin, varnish-like, light greenish or brownish. Moderately common in shallow-water sand flats.

Solen rosaceus Carpenter   Rosy Jackknife Clam

Santa Barbara, California, to Mazatlan, Mexico.

1 to 3 inches in length, almost 5 times as long as wide. Shell fragile, with a thin, glossy, olive periostracum. Beachworm specimens are whitish with rosy stains inside and out. It is more cylindrical, the anterior extremity is more rounded and narrower than in *sicarius*. An abundant species along the sandy shores of bays. Also to 25 fathoms.

Genus *Solecurtus* Blainville 1825

Quadrate to rectangular in shape; gaping at both ends. With weak, "clapboard" sculpturing. Ligament prominent, external and posterior to the small beaks. Right valve with 2 strong, horizontally jutting cardinal teeth just under the centrally located beaks. Left valve with 1 cardinal. *Psammosolen* Risso 1826 is a synonym.

Solecurtus cumingianus Dunker   Corrugated Razor Clam

North Carolina to south half of Florida to Texas.

1 to 2 inches in length with characters of the genus. Color all white, with a dull, yellowish-gray periostracum. Outer surface sculptured with

coarse, concentric, irregular lines and with sharp, small, oblique, wavy threads. Uncommon offshore.

Solecurtus sanctaemarthae Orbigny   St. Martha's Razor Clam

Plate 30i

North Carolina to southeast Florida and the West Indies.

½ inch in length, differing from *cumingianus* in being twice as long as high (instead of 2½ times) and in having stronger sculpturing. Uncommon from shallow water to 25 fathoms.

Superfamily *MACTRACEA*
Family *MACTRIDAE*

Key to Some Genera of *Mactridae*

a. Ligament entirely internal . . . . *Mulinia*
aa. Ligament partially external

    b. Chondrophore set off from ligament by a thin shelly plate:

        c. Posterior rostrate; shell thin and fragile . . . . . *Labiosa*
        cc. Both ends rounded; shell thick . . . . *Mactra*

    bb. Chondrophore not set off by a plate . . . . . . *Spisula*

FIGURE 89. Hinge of the Atlantic Fragile Mactra, *Mactra fragilis* Gmelin, 2 inches.

GENUS *Mactra* Linné 1767

Mactra fragilis Gmelin   Fragile Atlantic Mactra

Plate 32s; figure 89

North Carolina to Florida, Texas and the West Indies.

2 to 2½ inches in length, oval, moderately thin, but strong, and smoothish. Posterior slope with 2 radial, small ridges, one of which is very close to the dorsal margin of the valve. With a fairly large posterior gape. Color cream-white, with a thin, silky, grayish periostracum. Moderately common in shallow water. Rarely reaches 4 inches in length.

Mactra californica Conrad   Californian Mactra

Figure 90d

Neeah Bay, Washington, to Panama.

Up to 1½ inches in length, moderately fragile, moderately elongate and with the beaks central. Characterized by peculiar, concentric undulations on the beaks. This small species of *Mactra* is common in lagoons and bays of southern California. It lives 3 to 6 inches below the surface of the sand.

Mactra nasuta Gould   Gould's Pacific Mactra

San Pedro, California, to Mazatlan, Mexico.

Up to 3½ inches in length, similar to *californica*, but more oval at the ventral margin, without concentric undulations on the beaks, and with 2 very distinct, raised, radial ridges on the posterior dorsal margin. The whitish shell is glossy and the periostracum is shiny and yellowish tan. Not very common.

GENUS *Spisula* Gray 1837

Spisula solidissima Dillwyn   Atlantic Surf Clam
Plate 32p

Nova Scotia to South Carolina.

Up to 7 inches in length (usually about 4 or 5 inches), strong, oval and smoothish, except for small, irregular growth lines. The lateral teeth bear very tiny, saw-tooth ridges. Color yellowish white with a thin yellowish brown periostracum. Common below low-water mark on ocean beaches. After violent winter storms, these clams are cast ashore in incredible numbers, some estimates giving an approximate count of 50 million clams along a ten-mile stretch.

The subspecies *similis* Say (Cape Cod to both sides of Florida and to Texas) is more elongate, its anterior slope flatter, and its pallial sinus longer and not sloping slightly upward. In the left valve, the tiny double tooth, just anterior to the spoon-shaped chondrophore, is usually much larger and stronger. Moderately common, and commonly existing with the typical species in the northern part of its range. Compare with *polynyma* which has a larger pallial sinus.

Spisula polynyma Stimpson   Stimpson's Surf Clam
Plate 31w; figure 26k

Arctic Seas to Rhode Island. Arctic Seas to Puget Sound. Also Japan.

3 to 5 inches in length, beaks very near the middle of the valve. Anterior

end smaller than the elliptical posterior end. Shell chalky, dirty-white and with a coarse, varnish-like, yellowish brown periostracum. Worn shells have coarse, concentric, wide growth lines. The pallial sinus is larger in this species than in *solidissima*. Moderately common from low-tide line to 60 fathoms. The form *alaskana* Dall is probably a synonym. The fossil, *S. voyi* Gabb 1868 from the Miocene or Pliocene is possibly only a subspecies.

Spisula falcata Gould   Hooked Surf Clam

Puget Sound, Washington, to California.

2 to 3 inches in length, rather elongate at the narrower anterior end. Exterior chalky with a partially worn-off, light-brown, shiny periostracum. Anterior upper margin of shell slightly concave. Moderately common in sand below low-water line.

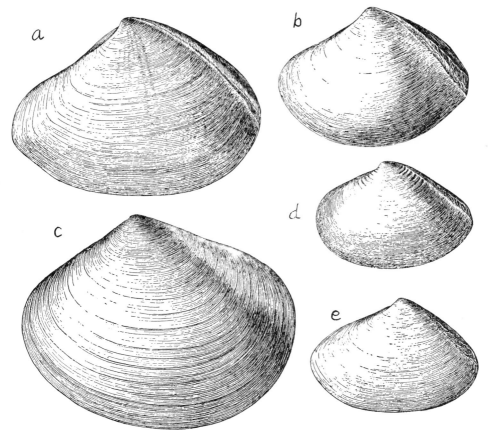

FIGURE 90. Pacific Surf Clams. **a**, *Spisula dolabriformis* Conrad, 3 inches; **b**, *S. hemphilli* Dall, 6 inches; **c**, *Spisula catilliformis* Conrad, 4 inches; **d**, *Mactra californica* Conrad, 1½ inches; **e**, *S. planulata* Conrad, 2 inches.

Spisula planulata Conrad   Flattish Surf Clam
Figure 90e

Monterey, California, to Lower California.

1 to 2 inches in length, ⅓ as high. Beaks almost at the middle. Anterior
upper margin of the shell sharp-edged and straight. Exterior smooth, yellowish, shiny with the edges commonly stained with rusty-brown. Not very
common. Found from low-water line to 36 fathoms.

Spisula catilliformis Conrad   Catilliform Surf Clam
Figure 90c

Washington State to Ensenada, California.

4 to 5 inches in length, almost as high as long. An oval shell with the
beaks slightly nearer the anterior end. Moderately obese. Dull-ivory, commonly stained with reddish brown. With numerous, irregularly sized and
spaced growth lines. Periostracum glossy, thin and usually worn off. Pallial
sinus deep, running anteriorly as far back as the middle of the shell. Rather
uncommonly washed ashore. Live specimens rare.

Spisula dolabriformis Conrad   Hatchet Surf Clam
Figure 90a

Lobitas, California, to Mexico.

3 to 4 inches in length, rather elongate, compressed and smooth. Posterior end shorter, but more expansive than the rather drawn-out anterior end.
Right valve with the posterior lateral tooth separated into 2 teeth lengthwise
by a long, deep channel. Color a smooth, ivory white, with a dull, light-tan,
thin periostracum. Small gape at the posterior end. Do not confuse with
Mactra nasuta which dips down at the ventral margin, has a shiny periostracum and a wide posterior gape, nor with *Spisula falcata* which is similar in
shape, but chalky and with very convex ventral margin to the hinge just
below the chondrophore. Moderately common.

Spisula hemphilli Dall   Hemphill's Surf Clam
Figure 90b

San Pedro, California, to Central America.

Up to 6 inches in length, about ¾ as high. Rather obese. Posterior end
more obese and shorter than the downwardly swept, compressed anterior end.
Periostracum grayish brown, dull, coarsely and concentrically wrinkled. The
pallial sinus is moderately deep and inclined upward. Fairly common along
the southern beaches of California.

Genus *Mulinia* Gray 1837

Mulinia lateralis Say   Dwarf Surf Clam

Plate 32-o

Maine to north Florida and to Texas.

⅓ to ½ inch in length, resembling a young *Spisula* or *Mactra*, moderately obese, beaks quite prominent and near the center of the shell and pointing toward each other. Exterior whitish to cream and smoothish, except for a fairly distinct, radial ridge near the posterior end. Concentric lines plainly seen in the thin, yellowish periostracum. Distinguished from young *Spisula solidissima* which have a proportionately much larger chondrophore in the hinge and which have tiny, saw-tooth denticles on the lower anterior and lateral hinge-teeth. A very abundant species in warm, shallow water in sand.

Genus *Labiosa* Möller 1832

Posterior slightly gaping. Shell fragile. Hinge with a prominent chondrophore. Cardinal teeth small and close to the chondrophore. Ligament submerged, except at the anterior end, and separated from the chondrophore by a shelly plate. *Raeta* Gray 1853 is the same. This is also *Anatina* Schumacher 1817, not Bosc 1816.

Labiosa plicatella Lamarck   Channeled Duck Clam

Plate 32q

North Carolina to Florida, Texas and the West Indies.

2 to 3 inches in length, ⅘ as high, egg-shell thin, but moderately strong. Concentric sculpture of smoothish, distinct ribs which on the inside of the valves show as grooves. Radial sculpture of very fine, crinkly threads. Color pure white. Formerly known as *Raeta canaliculata* Say. *R. campechensis* Gray is also a synonym. Commonly washed ashore, especially along the strands of the Carolinas, but rarely seen alive.

Labiosa lineata Say   Smooth Duck Clam

North Carolina to the north ⅔ of Florida and to Texas.

2 to 3 inches in length, ¾ as high, fairly thin but strong. White to tan in color. Moderately smooth, except for irregular growth lines and tiny, but distinct, concentric ribs near the beaks. Posterior end with a distinct radial rib behind which the shell gapes with flaring edges. Uncommon in most areas of its range.

Genus *Schizothaerus* Conrad 1853

Shell large with a roundish posterior gape. Hinge with small cardinal teeth; lateral teeth very small and close to the cardinals. Ligament external and separated from the cartilage pit by a shelly plate.

Schizothaerus nuttalli nuttalli Conrad   Pacific Gaper

Plate 31z

Washington to Lower California.

Up to 8 inches in length. An oblongish to oval, strong, smoothish shell with a prominent gape at the posterior end. The neat, well-formed beaks are located ¼ to ⅓ from the anterior end. The pallial sinus is very large and deep. Periostracum grayish. Common. Compare with the northern subspecies *capax* Conrad.

Schizothaerus nuttalli capax Gould   Alaskan Gaper

Kodiak Island, Alaska, to Monterey, California.

Up to 10 inches in length, differing from the typical *nuttalli* in being much more oval, more obese, and dipping downward into a well-rounded, ventral margin. This species is very common on most sandy and mud beaches in Puget Sound.

Genus *Rangia* Desmoulins 1832

Rangia cuneata Gray   Common Rangia

Figure 91a, b

Northwest Florida to Texas.

1 to 2½ inches in length, obliquely ovate, very thick and heavy. The beaks which are near the oval, anterior end are high, inrolled and pointing downward and anteriorly. Exterior whitish, but covered with a strong, smoothish, gray-brown periostracum. Interior glossy, white and with a blue-gray tinge. Pallial sinus small, but moderately deep and distinct. A common fresh-water to brackish-water species found in coastal areas. *R. nasuta* Dall is probably only a rostrate form of this species. Compare with *R. flexuosa*.

Subgenus *Rangianella* Conrad 1867

Rangia flexuosa Conrad   Brown Rangia

Figure 91c, d

Louisiana to Texas and Vera Cruz, Mexico.

1 to 1½ inches in length, resembling an elongate *cuneata*, but with no

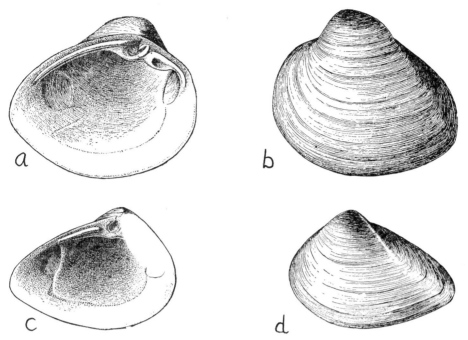

FIGURE 91. Rangia Marsh Clams of the Atlantic Gulf States. **a** and **b**, *Rangia cuneata* Gray, 2 inches; **c** and **d**, *R. flexuosa* Conrad, 1½ inches.

distinct pallial sinus, with much shorter laterals, with a faintly impressed, large lunule, and colored light-brown inside. A rare and elusive species from marsh areas. *R. rostrata* Petit is a synonym.

Family MESODESMATIDAE
GENUS *Mesodesma* Deshayes 1830

Like a large Donax with a prominent chondrophore. Laterals with fine denticles.

Mesodesma arctatum Conrad   Arctic Wedge Clam
Plate 32r

Greenland to Chesapeake Bay.

About 1½ inches in length, somewhat shaped like a *Donax*, fairly thick and compressed. Chondrophore fairly large and spoon-shaped. Left valve with a long anterior and posterior lateral tooth, both of which have fine, comb-like teeth on each side. Pallial sinus small and U-shaped. Interior tan to cream. Exterior with a thin, yellowish, smooth periostracum. Common from low water to 50 fathoms.

GENUS *Ervilia* Turton 1822

Shell small, concentrically striate, and sometimes brightly hued. Ligament absent; resilium small and internal. Laterals small. Left cardinal large and bifid or split.

Ervilia concentrica Gould   Concentric Ervilia

North Carolina to both sides of Florida and the West Indies.

3/16 of an inch in length, 2/3 as high, elliptical in outline, moderately compressed, although some are somewhat inflated. Each end is rounded to the same degree, and the beaks are central. There is a pinpoint depression just behind the glossy, inrolled beaks. Sculpture of fine, numerous concentric ridges. Radial threads may be present to form tiny beads. Color white, yellow or commonly with a pink blush. Common just offshore to 50 fathoms. *E. rostratula* Rehder from Lake Worth, Florida, is similar, but the posterior end is slightly more pointed.

Family HIATELLIDAE
GENUS *Hiatella* Daudin 1801

Shell irregular due to nestling and burrowing habits. Texture chalky. No definite teeth in the thickened hinge of the adults. Pallial line discontinuous; siphons naked and slightly separated at the tips. *Saxicava* Fleuriau 1802 is the same.

Hiatella arctica Linné   Arctic Saxicave
                                             Figure 92a

Arctic Seas to deep water in the West Indies. Arctic Seas to deep water off Panama.

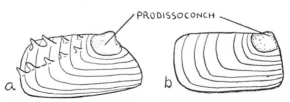

FIGURE 92. The young shells of Hiatella reared under the same artificial conditions. a, *Hiatella arctica* Linné; b, *H. striata* Fleuriau. (After M. Lebour 1938.)

Generally 1 inch in length, but rarely 2 to 3 inches. A very variable species in its shape. The young are rather evenly oblong, but the adults become oblong, oval or twisted and misshapen. The shell is elongate with the dorsal and ventral margins usually parallel to each other. A posterior gape may be present. Beaks close together, about ⅓ back from the anterior end. Just behind them is a conspicuous, bean-shaped, external ligament. Shell chalky, white and with coarse, irregular growth lines. Periostracum gray, thin and usually flakes off in dried specimens. With a weak or fairly strong radial rib at the posterior end. The rib may be scaled. Common in cold water. This is *H. rugosa* Linné and *pholadis* Linné.

Hiatella striata Fleuriau (*H. gallicana* Lamarck and *rugosa* of some authors) is almost indistinguishable in the adult form from *arctica*. The young, however, in *striata* do not have the two radial spinose ribs. This species breeds in winter, while *arctica* breeds in summer. The eggs are pinkish cream, while those of *arctica* are red. It nearly always bores into stone.

Genus *Cyrtodaria* Daudin 1799

Cyrtodaria siliqua Spengler   Northern Propeller Clam

Labrador to Rhode Island.

2 to 3 inches in length, about ½ as high. Gapes at both ends, but more so posteriorly where the large siphonal snout projects out about an inch. Beaks hardly noticeable, placed slightly toward the anterior end. The strong, wide ligament is external at the very anterior end of the dorsal margin. Shell chalky, white with a bluish tint. The valves are thick-shelled, with a coarse callus inside. The valves are slightly twisted in propeller fashion. Hinge a simple bar with a fairly large bulbous swelling under the ligament. In life, the periostracum is light-brown, glossy, smooth and covers the entire exterior. Dried valves soon lose the flaky, blackened periostracum. Moderately common offshore down to 90 fathoms. On occasion, found in fish stomachs.

A similar species, *Cyrtodaria kurriana* Dunker, is found in arctic waters along the shores at low tide. It is 3 times as long as high, hardly twisted and rarely exceeds 1½ inches in length. Uncommon in collections.

Genus *Panomya* Gray 1857

Panomya arctica Lamarck   Arctic Rough Mya

Arctic Seas to Chesapeake Bay.

2 to 3 inches in length, about ½ as high, squarish in outline. Looks somewhat like a misshapen *Mya*, but lacks teeth and a chondrophore in the hinge, and has a coarse, flaky, light-brown periostracum. Characterized by oblong or oval, sunk-in muscle and pallial line scars. There are 2 poorly de-

fined radial ridges near the center of the valves. Common in mud in cold waters offshore.

Panomya ampla Dall   Ample Rough Mya

Aleutian Islands to Puget Sound, Washington.

2 to 3 inches in length. A peculiarly distorted, heavy shell which is much gaping at both ends. Anterior end crudely pointed; posterior broadly truncate. With 3 to 6 depressed scars on the white interior. Exterior concentrically roughened, ash-white in color, with a border of thick, irregular, black periostracum. Hinge without definite teeth. Uncommon offshore in cold water.

GENUS *Panope* Menard 1807

Panope generosa Gould   Geoduck
                                                                (Goo-ee-duck)

Alaska to the Gulf of California.

7 to 9 inches in length. Inflated, slightly elongate and rather thick-shelled. Gaping at both ends. Coarse, concentric, wavy sculpture present, especially noticeable near the small, central, depressed beaks. Periostracum thin and yellowish. Exterior of shell dirty-white to cream; interior semi-glossy and white. Hinge with a single, large, horizontal thickening. The 2 long, united siphons of the animal are half the weight of the entire clam. Common in mud 2 or 3 feet deep in the northwestern states. Edible but tough. Freaks have been named *solida* Dall, *globosa* Dall and *taeniata* Dall. For an interesting and well-illustrated account of this species, see *Natural History Magazine* (N.Y.), April, 1948, on "We Go Gooeyducking" by the Milnes.

Panope bitruncata Conrad from North Carolina to Florida is 5 to 6 inches in length and resembles the Pacific geoduck. Dead valves are rarely found, and I have never seen a live specimen. Possibly extinct.

Superfamily MYACEA
Family MYACIDAE
GENUS *Mya* Linné 1758

These are the soft-shell or "steamer" clams which are so popular in New England. The valves are slightly unequal in size and have a large posterior gape. Resilium internal, placed posterior to the beaks and attached in the left valve to a horizontally projecting chondrophore.

Mya arenaria Linné   Soft-shell Clam

Labrador to off North Carolina. Introduced to western United States.

1 to 6 inches in length. Pallial sinus somewhat V-shaped in contrast to U-shaped in *truncata*. Shell elliptical. Periostracum very thin and light-gray to straw. Chondrophore in left valve long, spoon-shaped and shallow. *M. japonica* Jay of the Pacific coast is probably this speices. This common, delectable clam, also known as the long-necked clam, steamer and nanny nose, is harvested from the readily accessible mud flats of New England in great numbers. In 1935, nearly 12 million pounds, valued at $704,000, were taken from our eastern shores. A hundred pounds of clams furnishes 35 pounds of meat, while the equivalent weight of oysters would give only 13 pounds.

Mya truncata Linné   Truncate Soft-shell Clam

Arctic Seas to Nahant, Massachusetts. Europe. Arctic Seas to Port
        Orchard, Washington. Japan.

1 to 3 inches in length, similar to *arenaria*, but widely gaping at its abruptly truncate, posterior end. The pallial sinus is U-shaped. In Greenland and Iceland this species is fairly common and is considered a delicacy. It is also a food for the walrus, king eider duck, arctic fox and the codfish. It is uncommon in American collections.

GENUS *Sphenia* Turton 1822

Shell small and fragile; surface with concentric ridges; hinge-teeth absent. The elongate, flattened chondrophore in the left valve juts obliquely under the hinge margin of the right valve. These clams are nestlers, and are consequently irregular in shape in many instances.

Sphenia fragilis Carpenter   Fragile Sphenia

Oregon to Mazatlan, Mexico.

¾ inch in length, quite elongate, with a long, narrow, compressed, posterior snout. Anterior half obese and rotund. Beaks fat and close together. Shell fragile, chalky, white and with fine, concentric threads. Periostracum yellowish gray, dull and usually worn off the beak area. Chondrophore in left valve large and with 2 lobes. Socket in right valve large and round. The posterior snout is commonly twisted. Low tide to 46 fathoms in mud. Common.

Sphenia ovoidea Carpenter (Alaska to Panama) is half as large, smoother, and more ovoid in outline without the prominent snout. Uncommon.

Genus *Cryptomya* Conrad 1849

Somewhat like a small, fragile *Mya*, but more fragile, and the right valve is larger and more obese than the left. Large chondrophore in the left valve is thin, flat-topped with an anterior ridge. Posterior gape and pallial sinus almost absent. Siphons very short.

Cryptomya californica Conrad   Californian Glass Mya

Alaska to central Mexico.

1 to 1¼ inches in length, oval, fragile, moderately obese. Right valve fatter. Right beak crowds slightly over the left beak. Posterior gape very small. Chondrophore in left valve large, tucks against a small, concave shelf under the right beak. Exterior chalky and with small growth lines. Periostracum dull-gray, faintly and radially striped at the posterior end. Interior slightly nacreous in fresh specimens. Common in sand where it may live as deep as 20 inches. The short siphons enter the burrows of other marine animals.

Genus *Platyodon* Conrad 1837

Shell somewhat resembling a very fat *Mya*, with a fairly thick shell, rugose sculpturing and fairly small chondrophore.

Platyodon cancellatus Conrad   Chubby Mya

Queen Charlotte Island, B.C., to San Diego, California.

2 to 3 inches in length, rounded rectangular and obese. Gaping widely posteriorly. Shell strong, rather thick and with fine, clapboard-like, concentric growth lines. Rarely with very weak radial grooves. Chondrophore in left valve quite thick and arched. Beak of right valve crowds under beak of left valve. Shell chalky and white; periostracum thin, yellowish brown to rusty, and rugose posteriorly. Moderately common near beds of pholads. Lives in sand.

Family CORBULIDAE
Genus *Varicorbula* Grant and Gale 1931

Varicorbula operculata Philippi 1848   Oval Corbula

North Carolina to the Gulf of Mexico and the West Indies.

⅜ of an inch in length, ¾ as high, moderately thin-shelled and glossy. Beaks high, curled under and pointing anteriorly. Right valve subtrigonal in shape, very obese and with strong, concentric ridges. Left valve more elongate, smaller, less obese and with numerous but weaker ridges. Color white, but some may be tinted with rose near the margins. Uncommonly dredged from 12 to 250 fathoms. Live specimens very rare. This is *C. disparilis* of authors.

GENUS *Corbula* Bruguière 1792

Small, thick shells characterized by one valve (commonly the right) being larger than the other. Posterior end commonly rostrate. Resilium and ligament internal. The genus *Aloidis* Mühlfeld was in current use until recently.

Corbula contracta Say   Contracted Corbula

Cape Cod to Florida and the West Indies.

¼ inch in length, oblong, moderately to strongly obese. Both valves about the same size, except that the posterior, ventral margin of the right valve overlaps that of the left. The numerous, poorly defined, concentric ridges on the outside of the valves extend over the posterior, radial ridge on to the posterior slope. The left valve has a V-shaped notch in the hinge just anterior to the beak. Color dirty-gray. A common shallow-water species.

Corbula dietziana C. B. Adams   Dietz's Corbula

North Carolina to southeast Florida and the West Indies.

⅓ to ½ inch in length, like *contracta*, but larger, thicker-shelled and pinkish inside. The ventral margins are blushed or rayed with carmine-rose. Microscopic threads numerous between the few coarse, concentric ridges. Compare with the smaller and more compressed *barrattiana*. Commonly dredged offshore in the Miami region.

Corbula nasuta Say   Snub-nose Corbula

North Carolina to both sides of Florida and the West Indies.

³⁄₁₆ to ¼ inch in length, oblong, obese and strongly rostrate at the posterior end. The posterior end looks as if it had been severely pinched. Right valve considerably larger than the left. Margins of valves with a thick border

of dark-brown periostracum. Concentric sculpture of distinct ridges. Color yellowish to brownish white. Uncommon in shallow water.

Corbula barrattiana C. B. Adams   Barratt's Corbula
Figure 93a

North Carolina to both sides of Florida and the West Indies.

¼ inch in length, moderately compressed, rostrate at the posterior end, with poorly developed or without concentric ridges in the beak area. Right valve slightly larger than the left. The posterior end of the rostrum in the right valve projects far beyond that of the left valve. Color variable: white, pink, mauve, yellow, orange or reddish. Uncommon in shallow water to 50 fathoms.

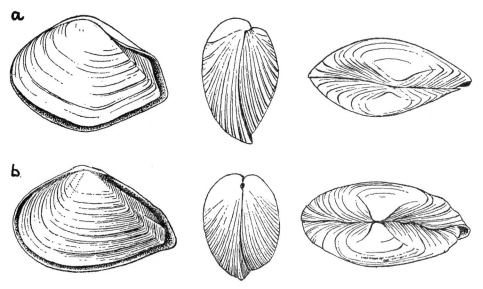

FIGURE 93. Corbula clams. a, *Corbula barrattiana* C. B. Adams, ¼ inch (Atlantic Coast); b, *Corbula swiftiana* C. B. Adams, ¼ inch (Atlantic Coast). Three views of each

Corbula swiftiana C. B. Adams   Swift's Corbula
Figure 93b

Massachusetts to east Florida and the West Indies.

¼ inch in length, oblong, moderately obese; right valve larger, more obese and overlapping the left valve at the ventral posterior region. Posterior slope in the right valve bounded by 2 radial ridges, one of which is close to the margin of the valve. The left valve has only 1 ridge. Shell thick, with the posterior muscle scar on a slightly raised platform. External sculpture the same on each valve, consisting of irregular, concentric ridges. Color dull white with a thin yellowish periostracum. Moderately common from 6 to 100 fathoms.

Corbula luteola Carpenter   Common Western Corbula

Monterey, California, to Panama.

⅓ inch in length, slightly obese, with the right valve more obese and overlapping the left valve on the ventral margin. Anterior end elliptical in outline; posterior end coming to a blunt point. Beaks strong, close together and slightly nearer the posterior end. Shell porcellaneous, whitish gray and may be flushed with pinkish or purplish. Interior whitish, but commonly yellowish with purple-red staining. Sculpture of weak, concentric growth lines which are less noticeable toward the smoothish beaks. Common in some localities in sand and rocky, rubbly beaches to 25 fathoms. *C. rosea* Williamson is merely a pink color phase of this species.

Corbula porcella Dall   Ribbed Western Corbula

Santa Rosa Island, California, to Panama.

¼ to ⅓ inch in length, similar to *luteola*, but much fatter, chalky, gray, and with much stronger concentric riblets. The right valve overlaps the left valve very prominently on the ventral margin. Moderately common from shallow water to 53 fathoms.

Family GASTROCHAENIDAE
Genus ROCELLARIA Blainville 1828

Both valves fairly thin, somewhat chalky in substance, and equal in size and shape. The anterior gape is very large. Beaks very near the anterior end. Hinge-teeth obscure. These clams form flask-shaped excavations in the rocks which they line with calcareous material. When not protected by a burrow, they form a shelly tube to which debris is attached. *Gastrochaena* Spengler is another genus from the Indo-Pacific area (formerly *Fistulana* Brug., not Müller).

Rocellaria hians Gmelin   Atlantic Rocellaria

North Carolina to Texas and the West Indies.

½ to ¾ inch in length; valves rather spathate, with low, indistinct, fine, concentric ridges. Posterior end large and rounded. The entire anterior-ventral end is widely open to accommodate the foot. Color white. Common in soft coral rocks. Erroneously called *cuneiformis* Spengler which is an Indo-Pacific species (see E. Lamy, 1924).

Rocellaria ovata Sowerby (Bermuda and the Atlantic Coast, rare) is very similar, but the beaks are at the very end of the valve, and the shell is more elongate. In *hians*, there is a very small, wing-like projection of the valve in front of the beak.

GENUS *Spengleria* Tryon 1861

Similar to *Rocellaria*, but truncate at the posterior end where there are strong, concentric ribs. Beaks at the anterior third of the shell.

Spengleria rostrata Spengler   Atlantic Spengler Clam

Southeast Florida and the West Indies.

1 inch in length; valve truncate at the posterior end. There is a very characteristic, elevated, triangular area which radiates from the beaks to the large, posterior end. This area is crossed by strong, transverse lamellations resembling a washboard. Commonly found boring in soft coral rock. Uncommon in Florida.

Superfamily *ADESMACEA*
Family *PHOLADIDAE*

Pending publication of the extensive researches on this group by Dr. Ruth D. Turner at Harvard University, we are dividing the groups in this family and treating the species in the conventional manner.

GENUS *Cyrtopleura* Tryon 1862

Barnea costata Linné   Angel Wing
                                                          Figure 94a

Massachusetts to Florida, Texas and the West Indies.

4 to 8 inches in length, moderately fragile; pure white in color, but in life covered by a thin, gray periostracum. With about 30 well-developed, beaded radial ribs which are scale-like at the anterior end of the valve. In fresh material, there is a shelly accessory plate over the hinge area. It is somewhat triangular and with complicated furrows. The internal brace under the beaks is spoon-shaped and with a narrow, strongly hooked attached end. Common in sticky mud about a foot under the surface (consult "Diggin' 'Em Out" by B. R. Bales, *The Nautilus*, vol. 59, pp. 13 to 17). Some colonies have shells which have pink, concentric stains on the inside of

the valves. No scientific study has been made of this variant, but it may be due to environmental or dietetic conditions.

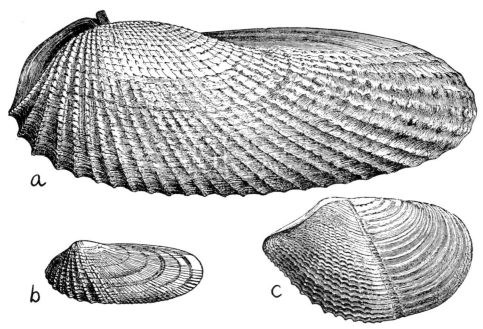

FIGURE 94. Mud, peat and rock borers of the Atlantic Coast. a, Angel Wing, *Barnea costata* Linné; b, *Petricola pholadiformis* Lamarck, False Angel Wing (p. 420); c, Northern Piddock, *Zirfaea crispata* Linné. All slightly reduced. (From Gould and Binney 1870.)

Barnea truncata Say   Fallen Angel Wing

Massachusetts Bay to south Florida.

2 to 2½ inches in length, somewhat resembling the Angel Wing, *B. costata*, but widely gaping at both ends, truncate at the posterior end, smoothish at the posterior half, and with an elongate, narrow accessory plate over the beak area. Shell fragile, white to grayish white. The internal condyle is long and about the thickness of a toothpick. The siphons are encased in a large, rough, gray, tubular sheath which may be extended 2 to 3 times the length of the shell. Bores into clay, soft rock or wood. Common in intertidal zones. Do not confuse with *Zirfaea crispata* which is much more squat and has a radial, indented line dividing each valve into 2 areas.

Barnea subtruncata Sowerby 1834   Pacific Mud Piddock

San Francisco, California, to Lower California.

2 to 2½ inches in length, very elongate, moderately gaping at the bluntly rounded, posterior end, and widely gaping at the foot end. Top margin of shell folded back on itself at, and just anterior to, the beaks. No distinct radial groove on the sides of the valves. Sculpture at anterior end is of prickly, concentric lamellae which fade out toward the smoothish posterior third. Siphon covered with a heavy, brown periostracum. May be dug out of very soft shale at very low tides in a few localities. Moderately common. Miss Turner advises me that this species must take the earlier name of *Barnea spathulata* Deshayes. It was formerly known as *B. pacifica* Stearns.

Genus *Zirfaea* Gray 1842

Zirfaea crispata Linné   Great Piddock
Figure 94c

Newfoundland to New Jersey. Europe.

1 to 2 inches in length, about half as high. Gaping at both ends. Characterized by the radial, indented line which divides the valves into 2 areas: the posterior section which has coarse, irregular growth lines, and the anterior section which has fimbriated or scaled growth lines and a serrated edge. Moderately common in cold water where it burrows into soft rock. This is the only member of the genus on the Atlantic Coast.

Zirfaea pilsbryi Lowe   Pacific Rough Piddock

Bering Sea to San Diego, California.

2 to 4 inches in length, gaping at both ends. Purportedly there are no accessory plates present, except for a small mesoplax. The folded-back dorsal margin at, and in front of, the beaks is covered with a thin, periostracal membrane in fresh material. The yellowish periostracum at the posterior end (siphonal end) is thin, but is not leaf-like as in *Parapholas*. No eggshell-like material over the anterior foot gape. This species appears in old books as *gabbi* Tryon. Dead specimens are commonly washed ashore. Siphon studded with tiny, rounded flecks of chitinous material.

Genus *Pholas* Linné 1758

Pholas campechiensis Gmelin   Campeche Angel Wing
Plate 32t

North Carolina to the Gulf States and Central America.

3 to 4 inches in length, ⅓ as high, fragile, pure-white and closely resembling the False Angel Wing, *Petricola pholadiformis*, but lacking a toothed hinge and ligament, and having a glossy plate rolled over the beaks

which is supported by about a dozen, vertical, shelly plates. Not very common.

Genus *Xylophaga* Turton 1822

Xylophaga washingtona Bartsch   Washington Wood-eater

Washington and Oregon.

⅓ inch in length, globular and fragile. The large anterior gape has ⅓ of its area covered by a triangular callum in each upper and outer corner. Middle of inside of each valve reinforced by a small, strong, radial cord which is welded to the valve. Color yellowish white. Bores into pieces of water-logged wood. Uncommonly collected in dredging hauls.

The Atlantic species X. *dorsalis* Turton has much the same characters as the above species. It occurs from 100 to 300 fathoms from the Arctic Seas to off Cape Cod.

Genus *Penitella* Valenciennes 1846

Penitella penita Conrad   Flap-tipped Piddock

Bering Sea to San Diego, California.

1½ to 3 inches in length, elongate to moderately short, rotund anteriorly, becoming compressed and narrow posteriorly where there is a leathery prolongation protecting the siphons. Anterior gape commonly closed by an eggshell-like, globose, callus plate which continues up on to the beak region in the form of a pair of rather large plates. Just in back of these there is a short, triangular plate or protoplax set on top of the middle of the dorsal region. Otherwise somewhat resembling the large *Parapholas californica*. Commonly found boring in rocks and shale. The form *sagitta* "Stearns" Dall lacks the calcareous covering over the foot gape. It is found with the typical form.

Genus *Pholadidea* Turton 1819

Pholadidea ovoidea Gould   Wart-necked Piddock

Bering Sea to the Gulf of California.

2 to 3 inches in length, very similar to *pilsbryi*, but oval in side view. About half of the anterior foot gape is covered by eggshell-like, smoothish calcareous material. It is not impossible that *pilsbryi* is an ecological form of this species. More studies in their natural history are needed. Moderately

common. John Q. Burch reports that the siphons have characteristic tubercles on their surface. Siphon cream, covered with a brownish periostracum at the base, the remainder studded with small bars of chitinous material, thus giving a warty appearance.

Genus *Nettastomella* Carpenter 1865

Nettastomella rostrata Valenciennes   Rostrate Piddock

Puget Sound to San Diego, California.

¾ inch in length, with a very large, oval gape posteriorly, and anteriorly with a long, pointed, smooth rostrum or snout which bends downward. Shell proper with widely spaced, delicate, high, concentric lamellae. Without accessory plates and apophyses. Found by breaking stones apart. Moderately common.

Genus *Parapholas* Conrad 1849

Parapholas californica Conrad   Scale-sided Piddock

Southern California.

3 to 5 inches in length, oval-oblong, rotund, especially anteriorly. Posterior round in cross-section and gaping. Dorsal margin with 2 long, complicated, accessory plates. Ventral edge covered by one elongate plate. Anterior gape closed over by 2 thin, eggshell-like extensions. Sides of snout with foliated periostracum. Shell proper with a long, radial groove pushing in at the middle. Color grayish white. Interior with 2 long, descending, shelly rods under the middle of the hinge. Common.

Genus *Navea* Gray 1851

Navea subglobosa Gray   Abalone Borer

San Pedro to Monterey, California.

⅓ to ¾ inch in size. Very obese and subglobular. Anterior end widely gaping with the thick edge of the valves coarsely denticulate. Commonly found boring in abalone shells (*Haliotis*).

Genus *Martesia* Blainville 1824

Martesia striata Linné   Striate Martesia
Plate 32w

West Florida to Texas and the West Indies.

¾ to 1¼ inches in length, elongate, moderately fragile, gaping at the posterior end and whitish in color. There is a faint radial groove from beak to ventral margin in each valve. Posterior to this groove, the shell is weakly sculptured with concentric growth lines. Anterior to the groove, there are crowded, finely denticulated riblets. Dorsal margins of valve joined by a long, narrow, fragile plate. Over the umbones there is a broadly rounded, somewhat heart-shaped, thick, calcareous plate. Anterior gape closed over by an egg-shell callus. Commonly found boring in wood.

Martesia cuneiformis Say   Wedge-shaped Martesia

North Carolina to Florida, Texas and the West Indies.

½ to ¾ inch in length, similar to *striata*, but smaller, chubbier, and with a somewhat triangular or lanceolate accessory plate which commonly has a crease running down the center. There is no free flange extending over the beak. Very commonly found burrowing in wood.

Martesia smithi Tryon   Smith's Martesia

New York to north half of Florida and to Texas.

⅓ to ½ inch in length. Bores into rocks and old shells. Very chubby, with a narrow, pointed posterior end. The shelly plate over the beaks is somewhat diamond-shaped, with a pointed anterior end and with no crease on the upper surface. Posterior half of the valves with fine, silken, concentric threads of periostracum. Fairly common in old oyster shells.

Family *TEREDINIDAE*
Genus *Bankia* Gray 1842

This is a genus of shipworms which is very common in warm American waters and which is of great economic importance. The two highly specialized valves of the clam are very similar to those in *Teredo,* and cannot be used as reliable identification characters. The two plume-like pallets at the posterior end of the worm-like animal which can close off the end of the burrow are used to distinguish species. In adult specimens, the shell at the anterior end which scrapes away the moist wood in the tunnel is about ¼₀ the total length of the animal. The mantel secretes a thin, smooth calcareous lining for the burrow as an added protection to the soft body. A much fuller and technical account of this group is given by Clench and Turner in Johnsonia (1946). The life history is explained in detail by Sigerfoos (1908). We have included all three known species in eastern American waters, although five others are found in the West Indies.

Pallets and shells are generally preserved in one part glycerin to four parts alcohol (70% grain) to permit later study of the delicate cones in the pallets. Permanent slides can be made by soaking the pallets in 90% alcohol for 12 hours, then placing on a slide, covering with a few drops of diaphane or euparol, and adding a long slip-cover.

Subgenus *Bankiella* Bartsch 1921

Bankia gouldi Bartsch   Gould's Shipworm
                                                            Figures 16; 95d

New Jersey to Florida, Texas and the West Indies.

Pallets about ½ inch in length. Cones deep-cupped, with smooth, drawn-out edges. Cones not very crowded at the distal end. Do not confuse with *B. caribbea*. Gould's Shipworm is the most widespread and abundant species in this genus on the Atlantic Coast, and hence is the most destructive. It has been found on the Pacific side of the Panama Canal. It is believed that *B. mexicana* Bartsch is the same species.

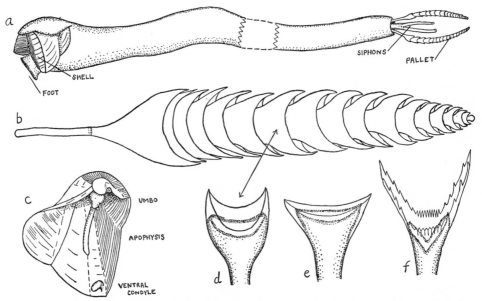

FIGURE 95. Atlantic Bankia Shipworms. a, entire animal; b, a pallet; c, interior view of one of the shell valves; d, *Bankia gouldi* Bartsch; e, *B. caribbea* Clench and Turner; f, *Bankia fimbriatula* Moll and Roch. (From Clench and Turner 1946 in Johnsonia.)

Subgenus *Bankiopsis* Clench and Turner 1946

Bankia caribbea Clench and Turner   Caribbean Shipworm
                                                            Figure 95e

North Carolina to the Caribbean.

Pallets about ⅓ inch in length. Cones shallow-cupped, with smooth, not drawn-out edges. Cones very crowded at the distal end. Do not confuse with *B. gouldi* which is more abundant and a larger species. Uncommon.

Subgenus *Plumulella* Clench and Turner 1946

Bankia fimbriatula Moll and Roch   Fimbriated Shipworm

Figure 95f

South half of Florida and the West Indies.

Pallets ½ to 1 inch in length. Cones deeply cupped, with beautiful, comb-like serrations on the edges. It has been found on the Pacific side of the Panama Canal, and those specimens were named *Bankia canalis* Bartsch.

GENUS *Teredo* Linné 1758

Similar to *Bankia*, but the pallet is made up of a single, paddle-shaped piece.

Teredo navalis Linné   Common Shipworm

Both coasts of the United States. Europe and Africa.

Shell like that in *Bankia* and subject to many minute variations. Each of the 2 calcareous pallets is spathate and compressed, but typically symmetrical. The leathery blade is urn-shaped, widening regularly from a stalk of medium length, then tapering somewhat toward the tip, which is decidedly excavated. The base of the blade is calcareous, but approximately the distal third is normally covered by a yellowish or brownish chitinous epidermis. A very common and destructive species found boring in wood.

Teredo bartschi Clapp   Bartsch's Shipworm

South Carolina to north half of Florida to Texas. Introduced to San Diego, California.

Shell close to *T. navalis*, but with the auricle typically semi-circular rather than sub-triangular in outline. Pallets: stalk long; blade short and deeply excavated at the top. Only the distal half of the blade is invested with periostracum, which is light horn-colored and semi-transparent, permitting the calcareous portion to be seen within as an irregular, hourglass-shaped structure with a deep sinus on either side. A common species.

Teredo diegensis Bartsch   San Diego Shipworm

Southern half of California. **And south?**

Shell similar to that of *T. navalis*, but smaller, more finely sculptured and transparent, and with numerous, closely set ridges. Pallets: blade with an oval, calcareous base, surmounted by a horny cap, amber to black in color. The horny portion is commonly deeply excavated at the tip, but may be cut off bluntly. The two elements of the blade come apart very easily. *T. townsendi* Bartsch is the same according to Kofoid and others.

Suborder *ANOMALODESMACEA*
Superfamily *PANDORACEA*
Family *LYONSIIDAE*
Genus *Lyonsia* Turton 1822

Lyonsia hyalina Conrad   Glassy Lyonsia
                                                            Plate 28u

Nova Scotia to South Carolina.

½ to ¾ inch in length; very thin and fragile. Semi-translucent and whitish to tan. Shell elongate, with the anterior end somewhat obese and the posterior end tapering and laterally compressed. Without teeth in the weak hinge, but with a small, free, elongate, calcareous ossicle inside just under the small, inflated, anteriorly pointing beaks. Periostracum very thin, with numerous raised radial lines. Commonly has tiny sand grains attached. Common from low water to 34 fathoms.

Lyonsia hyalina floridana Conrad, known from the west coast of Florida to Texas, is very similar, differing only in being ⅓ as high as long (instead of ½) and in having a narrower, more rostrate posterior end. Common.

Lyonsia arenosa Möller   Sanded Lyonsia

Greenland to Maine. Alaska to Vancouver.

½ to ¾ inch in length, resembling *hyalina*, but much less obese, with a heavier, greenish-yellow periostracum, and with its posterior end more oval and higher than the anterior end. The dorsal margin of the right valve behind the beak overlaps that of the left valve considerably. There is no posterior gape as in *hyalina*. Like other species in the genus, it glues sand grains to itself. Moderately common from low water to 60 fathoms.

Lyonsia californica Conrad   California Lyonsia

Puget Sound to Lower California.

1 inch in length, very thin, fragile and almost transparent. Quite elongate and moderately obese. Beak area swollen. Posterior end tapering and later-

ally compressed. Outer surface whitish (opalescent when worn), commonly with numerous, weak, radial, dark lines of periostracum. Interior glossy and with an opalescent sheen. Ossicle inside, under hinge, is opaque-white. Very similar to our figure of the Atlantic *hyalina*. Common in sandy mud bottoms of many California sloughs and bays down to 40 fathoms.

GENUS *Entodesma* Philippi 1845
Subgenus *Agriodesma* Dall 1909

Entodesma saxicola Baird   Northwest Ugly Clam
Plate 29m

Alaska to San Pedro, California.

2 to 5 inches in length. A very peculiar, ugly and misshapen clam found along the shore burrowing into rocks. Generally oblong in shape, with the posterior end flaring and gaping. Covered with a thick, rough, brown periostracum which partially flakes off when dry. Interior brownish tan to whitish with a slight opalescence. Hinge without teeth, but with a rather large, oblong, whitish ossicle lying under the internally placed ligament. Moderately common from Washington to southwest Alaska.

GENUS *Mytilimeria* Conrad 1837

A peculiar, bladder-shaped, very thin shell found embedded in compound ascidians or sea squirts.

Mytilimeria nuttalli Conrad   Nuttall's Bladder Clam
Plate 29q

Alaska to Lower California.

1 to 2 inches in length, obliquely oval, inflated, very fragile, opaque with a thin, brownish periostracum. Beaks small and spiral. No teeth in the weak hinge, but a small, calcareous ossicle is present. Color white with underlayers of slightly pearly material. Common under rocks at low tide to 10 fathoms, always embedded in compound ascidians or sea squirts.

Family PANDORIDAE
GENUS *Pandora* Chemnitz 1795 (Opinion 184)

Pandora trilineata Say   Say's Pandora
Figure 96b

Cape Hatteras, North Carolina, to Florida and Texas.

¾ to 1 inch in length, almost half as high; half-moon-shaped in outline, and with a strong, squarish ridge along the hinge margin which extends posteriorly into a fairly long rostrum. Valves very flat, the entire shell very compressed. Beaks tiny, quite near the rounded anterior end. Right valve smoothish and translucent cream. Left valve more prominently divided into 2 portions by a slight radial groove, anterior to which the shell is dull cream, and posterior to which the shell is more glossy but with microscopic concentric growth lines. Sometimes slightly iridescent. Interior pearly. Moderately common in sand below low water to 60 fathoms.

Pandora arenosa Conrad, the Sand Pandora, (Carolinas to southeast Florida) is commonly dredged by amateurs. It is very much like *trilineata*, but never over ½ inch in length, completely lacking the rostrum and with a quite convex left valve which has an external radial rib below the hinge margin ridge. Common. *P. carolinensis* Bush is this species.

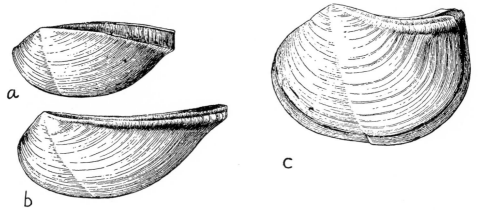

FIGURE 96. Atlantic Pandoras. **a**, *Pandora bushiana* Dall, ⅓ inch (Florida); **b**, *P. trilineata* Say, ¾ inch (Atlantic Coast); **c**, *P. gouldiana* Dall, 1 inch (northern Atlantic Coast).

Pandora gouldiana Dall   Gould's Pandora
                                                  Figure 96c

Gulf of St. Lawrence to Cape May, New Jersey.

¾ to 1⅓ inches in length, similar to *trilineata*, but less elongate with the height slightly more than half the length. The posterior rostrum on the hinge line is very short, stubby and turned up. The shell is opaque, chalky and commonly worn away, showing the pearly underlayers. Margin of valves bordered with blackish brown periostracum. Common from intertidal areas to 20 fathoms.

Subgenus *Kennerlia* Carpenter 1864

Pandora filosa Carpenter   Western Pandora

Alaska to Ensenada, Lower California.

1 inch in length, thin but not too fragile, opalescent-white with a brownish border of periostracum. Interior opalescent. Valve semi-circular in outline. Right valve almost flat and with a single, fairly large tooth which juts laterally. Left valve moderately convex. The posterior dorsal margin is almost straight, the posterior end somewhat drawn out into a rostrum. Moderately common from 10 to 75 fathoms.

Pandora bilirata Conrad (Alaska to Lower California. Dredged) is half as large, not rostrate posteriorly, and with 2 strong radial ribs on the posterior dorsal margin of the left valve.

Pandora granulata Dall (Santa Barbara south to La Paz, Mexico) is much like *bilirata*, but half its size, more elongate, and the 2 radial ribs are granulated.

Family THRACIIDAE
GENUS *Thracia* Blainville 1824

Shells up to 4 inches in size, thin, chalky in texture, beaks so close that the *right one becomes punctured* by the left beak; ligament external; shell commonly moderately rostrate at the posterior end; right valve fatter than the left.

Thracia conradi Couthouy   Conrad's Thracia
                                                        Plate 28y

Nova Scotia to Long Island Sound, New York.

3 to 4 inches in length, about ⅘ as high. Valves obese and chalky-white. Hinge without teeth, only thickened considerably behind the beak and below the large, wide, external ligament. Right beak always punctured by the beak in the left valve. Pallial sinus U-shaped but not very deep. Posterior end of valves slightly rostrate and with a weak radial ridge. Rarely washed ashore. Not uncommon offshore and down to 150 fathoms.

Thracia trapezoides Conrad   Common Pacific Thracia

Alaska to Redondo Beach, California.

2 inches in length and not quite so high; thin, chalky, and with the posterior end broadly rostrate. The beak of the right valve has a hole punctured in it by the beak of the left valve. The posterior rostrated part of the valve is set off by a broad, radial, depressed furrow which is bordered by a low, rounded, radial ridge. Color drab, grayish white. Commonly dredged off the west coast. Compare with *T. curta* Conrad.

Thracia curta Conrad   Short Western Thracia

Alaska to Lower California.

1 to 1½ inches in length, very similar to *trapezoides*, but suboval and lacking the prominence of rostration. It is very close in shape to our illustration of the Atlantic *T. conradi*. A moderately common species. John Q. Burch reports that it is relatively abundant at San Onofre, California, in the rubbly reef at extreme low tides. It has also been taken from wharf pilings, and it is commonly dredged in over 20 fathoms on shale bottoms.

Genus *Cyathodonta* Conrad

Similar to *Thracia*, but the right beak is without a round hole, and the ligament is internal on a definite chondrophore, and the valves are with oblique, concentric undulations. This is sometimes considered a subgenus of *Thracia*, not without justification. There is only one species on the Pacific Coast of America.

Cyathodonta undulata Conrad   Wavy Pacific Thracia
Plate 31s

Monterey, California, to Tres Marias Islands, Mexico.

1½ inches in length, subovate, very thin and fragile, white, and with obliquely concentric undulation which are largest at the anterior end, but disappear toward the posterior end of the shell. Minute, crowded, granulated, radial lines are also present. Uncommon. *C. dubiosa* Dall and *C. pedroana* Dall appear to be this species.

Family PERIPLOMATIDAE
Genus *Periploma* Schumacher 1817

Shell small, oval, right valve fatter than the left, with a slight pearly sheen, hinge with a narrow, oblique spoon and a small, free, triangular lithodesma; ligament absent; anterior muscle scar long and narrow, the posterior one small and ovate.

Periploma papyratium Say   Paper Spoon Clam
Plate 28w

Labrador to Rhode Island.

½ to 1 inch in length, oval, moderately compressed, thin-shelled, and dull-white with a thin, yellowish-gray periostracum. Beaks slit or broken by a short, radial break. Spoon-like chondrophore faces downward and is reinforced by a sharp, curved rib which runs to the inner surface of the valve in a ventral direction. Sculpture of irregular, fine, concentric growth lines. A weak radial groove runs from the beak to the anterior part of the ventral edge. Moderately common in dredge hauls from 1 to 200 fathoms. *P. papyraceum* is an incorrect spelling for this species.

Periploma fragile Totten, the Fragile Spoon Clam (Labrador to New Jersey, 4 to 40 fathoms), differs in being more rostrate anteriorly, the beaks pointing more forward and placed more anteriorly. Its chondrophore is more horizontal to the hinge line.

Periploma inequale C. B. Adams   Unequal Spoon Clam
                                                       Plate 28x

South Carolina to Florida and to Texas.

¾ to 1 inch in length, oblong, the left valve more inflated and slightly overlapping the right valve. Fragile and pure white. Beaks close together, each with a short, radial break or slit in the surface. An oblique, low keel runs from the beaks to the anterior ventral margin of the valve. The keel is bounded posteriorly by a groove. Sculpture consists of microscopic, concentric scratches. Hinge with a single, large, spoon-shaped tooth or chondrophore, above which is a deep slit where the small, free, triangular lithodesma fits. This species is especially abundant along certain Texas beaches. *P. inaequivalvis* Schumacher from the West Indies has not been found in the United States, despite several erroneous records.

Periploma planiusculum Sowerby   Common Western Spoon Clam
                                                       Plate 31x

Point Conception, California, to Peru.

1 to 1¾ inches in length, ovate, thin, with weak, concentric lines of growth. Right valve fatter than the left; chondrophore ovate-trigonal, its longer diameter directed forward and reinforced posteriorly by an elongate, rib-like buttress. Commonly washed ashore on southern Californian beaches.

Periploma discus Stearns   Round Spoon Clam
                                                    Figures 97a; 28l

Monterey, California, to La Paz, Lower California.

1 to 1½ inches in length, similar to *planiusculum*, but almost circular in outline, except that the posterior end is slightly lengthened into a short, broad, blunt rostrum. Uncommonly dredged in mud bottom at several fathoms; rarely washed ashore after storms.

Subgenus *Cochlodesma* Couthouy 1839

Periploma leanum Conrad   Lea's Spoon Clam
Plate 28v

Nova Scotia to off North Carolina.

1 to 1½ inches in length, ovate, quite compressed, fairly fragile and white in color. Smoothish. The beaks located near the center of the dorsal edge of the valves have a natural, radial crease at the anterior end. Chondrophore large, points ventrally and is reinforced anteriorly by a low, sturdy ridge. The muscle scar above the pallial sinus is commonly quite silvery. Periostracum thin and yellowish. Uncommon just offshore. This was put in a separate subgenus, *Aperiploma*, by Habe in 1952.

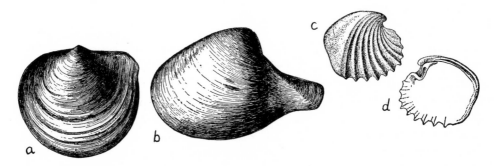

FIGURE 97. Spoon and Dipper Clams. **a**, *Periploma discus* Stearns, 1½ inches (California); **b**, *Cuspidaria glacialis* Sars, 1 inch (Atlantic Coast); **c** and **d**, *Verticordia ornata* Orbigny, ¼ inch (Atlantic Coast).

Order *SEPTIBRANCHIA*
Family *VERTICORDIIDAE*
GENUS *Verticordia* Sowerby 1844
Subgenus *Trigonulina* Orbigny 1846

Verticordia ornata Orbigny   Ornate Verticord
Figure 97c, d

Massachusetts to south half of Florida and the West Indies.

¼ inch in length, oval to round, compressed and with about a dozen strong, sharp, curved radial ribs on the anterior ¾ of the valve. The ribs extend beyond the ventral margin to give a strongly crenulate margin. Exterior dull and cream-white; interior very silvery. Commonly dredged off our east coast from 5 to 200 fathoms.

Verticordia (*Haliris*) *fischeriana* Dall is similar, but much fatter, ⅓ inch in length, and with about 28 small, finely beaded, radial ribs over the entire surface of the valve. Uncommon, 10 to 100 fathoms, from North Carolina to the West Indies. Called Fischer's Verticord.

Family POROMYIDAE
GENUS *Poromya* Forbes 1844

Shell small, fragile; sculpture of fine granules in radial series. Hinge of right valve with a strong cardinal tooth in front of a wide chondrophore; hinge of left valve with a small cardinal tooth behind and above the chondrophore.

Poromya granulata Nyst   Granular Poromya
                                                        Plate 30s
Cape Cod, Massachusetts, to east Florida and Cuba.

¼ to ⅓ inch in length, ovate, inflated and fragile. Beaks inflated and turned forward. Exterior cream-white, with an irregular coating of fine granules which resemble sugar-coating. In fresh material, this granular deposit is also found on the inner margins of the valves. Slightly gaping at the posterior end. Interior of valves silvery white. Commonly dredged in a few fathoms of water off eastern Florida.

Poromya rostrata Rehder (Cape Hatteras, North Carolina, to east Florida and the West Indies, 60 to 100 fathoms) is distinctly rostrate posteriorly and the granules are larger, more evenly spaced and generally cover the entire outer shell. It is relatively uncommon.

Family CUSPIDARIIDAE
GENUS *Cuspidaria* Nardo 1840

Shell small, globose in front, rostrate behind. Hinge with a posterior lateral tooth in the right valve. External ligament elongated. Resilium in a small, spoon-shaped fossette. Lithodesma distinct and semi-circular.

Cuspidaria glacialis Sars   Glacial Cuspidaria
                                                        Figure 97b
Nova Scotia to Maryland. Gulf of Mexico? Alaska.

1 to 1½ inches in length, rostrum moderately long and compressed laterally. Main part of valves fat and round. Sculpture consists of small, irregular growth lines. Periostracum grayish white. Shell cream to white. A common species dredged from 64 to over 1400 fathoms. *C. jeffreysi* in the south is very similar, but smaller, with a rostrum which in cross-section is much more oval and less compressed, and with hardly any periostracum.

Cuspidaria rostrata Spengler   Rostrate Cuspidaria
                                                    Plate 32j

Cape Cod, Massachusetts, to the West Indies.

½ to 1 inch in length, with a tube-like rostrum which is ½ the length
of the entire shell. The rostrum points slightly downward. Shell fairly
smooth with moderately coarse, concentric growth lines. Whitish in color
and sometimes with granular lumps of gray mud attached to the rostrum.
Moderately common in deep water (65 to over 1600 fathoms).

Cuspidaria jeffreysi Dall   Jeffrey's Cuspidaria

Southern Florida and the West Indies.

⅓ inch in length, smoothish with only fine lines of growth. Rostrum
moderately long; main part of shell round and fat. Similar to *glacialis* in the
north which, however, is larger, more compressed, and whose rostrum points
slightly downward instead of directly posteriorly as in this species. Creamy
white in color. Uncommonly dredged in waters over 100 fathoms off Miami.

Subgenus *Leiomya* A. Adams 1864

Cuspidaria granulata Dall   Granulated Cuspidaria

Off Miami, Florida, and the West Indies.

½ inch in length, similar to *jeffreysi*, but snow-white and covered
with numerous, small, opaque-white granules. Uncommon from 30 to 100
fathoms.

Genus *Cardiomya* A. Adams 1864

With strong, sharp radial ribs; fossette more vertical and prominent,
otherwise like *Cuspidaria*.

Cardiomya costellata Deshayes   Costate Cuspidaria

North Carolina to south Florida and the West Indies.

⅓ inch in length, fragile, with a short rostrum, and with a few prominent radial ribs just in front of the rostrum which gives the ventral margin
of the valve in that area a scalloped edge. Anteriorly, the radial ribs are closer
together, but weaker, and are rarely present at the anterior end of the shell.
Additional ribs may develop in older specimens and become more even in
size. (*C. multicostata* Verrill and Smith may be the old form and *C. gemma*
Verrill and Bush the young form.) Commonly dredged off eastern Florida.

Cardiomya pectinata Carpenter   Pectinate Cuspidaria

Puget Sound, Washington, to Panama.

¼ inch in length; anterior end globular, bearing 8 to 12 strong radial ribs (there may be smaller ribs between the main ones). Posterior end drawn-out like a short handle and bearing 2 to 4 weak, longitudinal riblets. Color dull-gray with a glossy, grayish white interior. Commonly dredged offshore.

Consider the case of the oyster
Which passes its time in the moisture.
Of sex alternate,
It chases no mate,
But lives in a self-contained cloister.

JOEL W. HEDGPETH
(*Maryland Tidewater News*
September 1950)

CHAPTER XI

Squid, Octopus and Cuttlefish

Class *CEPHALOPODA*

Subclass *TETRABRANCHIA*

This subclass, which includes the Chambered Nautilus and about 5000 species of fossil and extinct Ammonites, is not represented in American waters. The living species of *Nautilus* from the Indo-Pacific are characterized by a large, chambered, external shell, by two pairs of gills, and by their numerous suckerless arms.

Subclass *DIBRANCHIA*

All of the living cephalopods with the exception of *Nautilus* belong to this subclass which is characterized by animals that have one pair of gills, 8 or 10 arms which bear rows of suckers, and whose shell is internal or entirely absent.

Order *DECAPODA*
(Spirula and Squid)

With 10 arms, 2 of which are the long tentacular arms; body long and cylinder-shaped. An internal shell is present in most cases, and may be calcareous (the cuttlebone) or thin and horny (squid pen). The small suckers on the arms are usually set on small stalks or peduncles and their apertures are armed with horny rings or hooks.

Family *SPIRULIDAE*
Genus *Spirula* Lamarck 1799

Spirula spirula Linné

Cape Cod to the West Indies. Worldwide.

Common Spirula
Figure 98

478

The rather fragile, white shell is a chambered cone coiled in a flat spiral, usually less than 1 inch in diameter and with the coils not in contact. Each small chamber in the shell is divided from its neighbor by a nacreous-white,

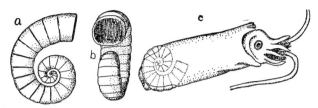

FIGURE 98. The white, inch-long shells (a and b) of *Spirula spirula* Linné are commonly washed ashore on southern beaches, but the squid-like animal (c) lives at great depths and is very seldom captured alive.

concave, fragile septum or wall. There is a small siphonal tube running back into the shell and piercing the septa. These shells are cast up on the beaches quite commonly. The body is short and cylindrical, and surrounds the shell almost completely except for two small areas. The 8 sessile arms and 2 pedunculated tentacular arms are very short.

Myopsid Squid—transparent cornea over eyes; pupils crescent-shaped

Family SEPIOLIDAE
GENUS *Rossia* Owen 1828

Short, "tubby" animals whose bodies are rounded at the end. The mantle edge is free all around. 8 arms short with 2 to 4 rows of spherical suckers which have smooth, horny rims. The two tentacular arms can be almost entirely withdrawn. The internal pen is slender, lanceolate and very thin and delicate. The rather large, semi-circular fins are on the middle of the sides of the body. Eye with small eyelid on the lower side, none above. No sulcus or notch on front of the eye.

Rossia pacifica Berry   Pacific Bob-tailed Squid

Alaska to San Diego, California.

Total length, not including the tentacles, 3 to 4 inches. Body smooth, mantle flattened above and below, rounded behind. Fins large, semi-circular or subcordate, with a free anterior lobe, their attachment more or less oblique to the general plane of the body. Color in life unknown; in alcohol, reduced to brownish buff, heavily spotted above and in less degree below with purplish chromatophore dots, which extend even over the fins, although fewer on under surfaces and margins. This is the only *Rossia* recorded on the Pacific Coast, and it is rather abundant from 9 to 300 fathoms.

Subgenus *Semirossia* Steenstrup 1887

Rossia tenera Verrill　　　　　　　　　　　Atlantic Bob-tailed Squid

Figure 99d

　　Nova Scotia to southeastern Florida.

3 to 4 inches in length, including mantle and longest arm. A small and delicate species, very soft, translucent, and delicately rose-colored when living. Internal pen small, very thin and soft. Length of each side fin is about ⅔ of the body, and the base of attachment of the fin is about ½ the body-length. Arms unequal, the dorsal ones considerably shorter. This species is characterized by the larger size of the suckers located along the middle of the lateral arms. Commonly dredged from 18 to 233 fathoms. Formerly listed as *Heteroteuthis tenera* Verrill. *R. equalis* Voss, dredged off southeast Florida, differs in having the suckers on the lateral arms about equal in size.

Family SEPIIDAE
Genus *Sepia* Linné 1758

The common cuttlefish squid of Europe is not represented in our waters, although in rare instances the cuttlefish bone or internal shell has been found in western Atlantic waters. The cuttlefish bone is an oblong, six-inch or so, very light slab of chalky material, rounded at one end, pointed at the other. It is used in the manufacture of toothpaste, and is tied to the bars of canary cages for the birds to peck at as a source of lime. Ink from *Sepia* was at one time a main source for durable, black writing ink.

Family LOLIGINIDAE
Genus *Loligo* Schneider 1784

Ten-armed, with elongate, tapering, cylindrical body and large, terminal, triangular fins. Arms with two rows of suckers provided with horny, dentated rings; fourth left arm hectocotylized in the males. Tentacular arms with four rows of suckers on their clubs. Internal pen horny, lanceolate with its shaft keeled on the under side. The female receives the sperm sacs of the male upon a specially developed pad below the mouth. In the genus *Lolliguncula*, the sperm sacs are received upon a callused patch within the mantle near the left gill.

Subgenus *Loligo* s. str.

Loligo pealei Lesueur　　　　　　　　　　Atlantic Long-finned Squid

Figure 99a

　　Nova Scotia to both sides of Florida.

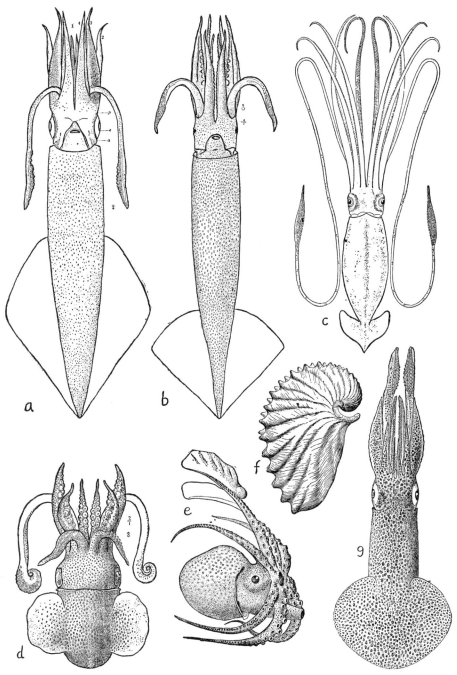

FIGURE 99. Atlantic Cephalopods. **a**, *Loligo pealei* Lesueur (length: 1 to 2 feet); **b**, *Illex illecebrosus* Lesueur (1 to 1½ feet); **c**, *Architeuthis harveyi* Kent (40 to 55 feet); **d**, *Rossia tenera* Verrill (3 to 4 inches); **e**, *Argonauta hians* Solander (2 to 3 inches); **f**, shell of same; **g**, *Lolliguncula brevis* Blainville (5 to 10 inches). (From A. E. Verrill 1879.)

Total length, including tentacular arms, 1 to 2 feet. Easily recognized by the accompanying illustration which shows the rather long, triangular fins. The proportion of fin-length to mantle-length varies from 1 to 1.8 and down to a ratio of 1 to 1.5. Adult males have the left ventral arm conspicuously hectocotylized (see fig. 99a). A very abundant species caught commercially for fish bait in New England. Living specimens are very beautifully speckled with red, purplish and pink.

Loligo opalescens Berry　　　　　　　　　　Common Pacific Squid

Puget Sound, Washington, to San Diego, California.

Total length, not including tentacles, 6 to 8 inches. This is the common squid of the Pacific Coast and can be readily recognized by the accompanying illustration. At certain seasons, they occur in great schools by the thousands.

Genus *Lolliguncula* Steenstrup 1881

Lolliguncula brevis Blainville　　　　　　　　Brief Squid

Figure 99g

Delaware Bay to Florida and to Brazil.

Total length, including the tentacular arms, 5 to 10 inches. Characterized by its short, rounded fins, very short upper arms, and large color spots. Underside of fins white. Consult the figure. Common in warm waters. This is *L. brevipinna* Lesueur and *L. hemiptera* Howell.

Genus *Sepioteuthis* Blainville 1824

Similar to *Loligo*, but with large, triangular fins that extend along the entire length of the mantle, thus giving the animal an oval outline. Siphonal funnel attached to the head by muscular bands. There is a strong wrinkle behind the eye.

Sepioteuthis sepioidea Blainville　　　　　　　Atlantic Oval Squid

Bermuda, Florida and the West Indies.

Total length, including tentacular arms, 4 to 5 inches. Characterized by the long fins which commence a short distance behind the mantle edge (¼ to ⅓ inch). Internal pen thin, lanceolate and without any marginal thickenings. Skin regularly and closely spotted with purple dots. The eggs are large, 5 to 8 mm. in diameter, and laid in long jelly tubes. A rather common, warm-water species.

GENUS *Doryteuthis* Naef 1912

Doryteuthis plei Blainville   Plee's Striped Squid

Florida and the West Indies.

Up to 8 inches in length, including the tentacular arms. Characterized by the long, narrow, slightly wavy, dark-colored bands running back along the side of the mantle. The rest of the mantle is moderately covered with small round dots. The body is long and slender, the triangular fins on the last third of the mantle, and the arm suckers do not have pointed teeth on the horny circles. A common surface-living species of the Caribbean region.

Oigopsid Squid—eyes naked in front, pupils circular; eyelids

Family ARCHITEUTHIDAE
GENUS *Architeuthis* Steenstrup 1857

Architeuthis harveyi Kent   Harvey's Giant Squid
                                                                 Figure 99c
Newfoundland Fishing Banks.

Total length 40 to 55 feet. Body stout, nearly round, swollen in the middle. Arms nearly equal in length, all bearing sharply, serrated suckers. Tentacular arms 4 times as long as the 8 sessile arms. The peculiar backward pointing tail fins separate this species from *A. princeps* Verrill, another giant squid found in the same area. A large well-preserved specimen of any giant squid is worth its weight in gold. No large specimens have been brought back from the fishing banks in many years. They may occasionally be washed ashore from Nova Scotia north. If you find one, take photographs if possible, and notify one of the leading museums. Giant squid of unknown identity have been seen in the Gulf of Mexico.

Family OMMASTREPHIDAE
GENUS *Illex* Steenstrup 1880

Resembling *Loligo* somewhat, but with half-hidden eyes, the lids free and with a distinct notch or sinus in front. Internal pen narrow along the middle portion, and with three ribs. There are 8 rows of tiny suckers on the end section of the 2 long, tentacular arms. Further study may show that the genus *Illex* is the same as *Ommastrephes* Orbigny 1835.

Illex illecebrosus Lesueur   Common Short-finned Squid
                                                                 Figure 99b
Greenland to North Carolina and the Gulf of Mexico.

Total length, including tentacular arms, 12 to 18 inches. A common squid characterized by the small opening to the eyes and the small, narrow sinus or notch in front of the eyes, and by the proportion of fin-length to mantle-length which is roughly 1 to 3. The sides of the head, back of the eyes, have a rather prominent, transverse ridge, back of which the head suddenly narrows to the neck. Under surface of head with a deep, smoothish excavation to receive the dorsal half of the siphonal tube. In males, either the left or right ventral arm is hectocotylized. A very common species used for fish bait. It may be seen in large schools near shore, especially in summer in New England.

Genus *Sthenoteuthis* Verrill 1880

Very similar to *Illex* in almost every way, but the sucker-bearing area includes less than one half the total length of the tentacular arms. The larger suckers on the tentacular club are strongly toothed, with an additional large tooth in each of the four quadrants.

Sthenoteuthis bartrami Lesueur   Flying Squid

Worldwide.

2 to 3 feet in total length, resembling the common *Illex*, but more slender, with shorter fins, and with 4, not 8, rows of tiny suckers on the end of the 2 long tentacular arms. Preserved specimens show a distinct dark, purple-brown dorsal stripe. In life, the colors are very brilliant and are continually changing. Along the middle dorsal line there is a broad violet stripe with a stripe of reddish yellow on each side of it. Body elsewhere bluish; fins rosy. Skin covered with small, red-violet chromatophore dots. On the eyes there are two elongated spots of brilliant blue, and below a bright spot of red. Color of ink reported to be a coffee-and-milk color. A common ocean-going species which swims with great speed, and not infrequently jumps out of water and lands on the decks of ships. Like most squid, it is attracted by artificial light.

Order *OCTOPODA*

The octopods have only 8 arms and are without the 2 long tentacular arms that are characteristic of the squid. The suckers on the arms are without stalks and are not equipped with horny rings. No internal pen or shell. This order includes the many forms of *Octopus* and the Paper Nautilus, *Argonauta*. The female *Argonauta* secretes a shell to hold her eggs.

Family ARGONAUTIDAE
GENUS *Argonauta* Linné 1758

Pelagic octopods in which the dorsal arms of the female are broadly expanded into glandular membranes that secrete and hold a delicate, calcareous shell for containing the eggs. The males are considerably smaller than the females, do not have a shell, and the third right arm is modified into a detachable copulatory organ which persists separately for a certain length of time in the mantle cavity of the female.

Argonauta argo Linné   Common Paper Nautilus

Plates 1c; 26y

Worldwide in warm waters.

4 to 8 inches in length, quite fragile, laterally compressed with a narrow keel, numerous sharp nodules which in the early part of the shell are stained with dark purplish brown. Rest of shell opaque, milky-white. Occasionally washed ashore. *A. americana* Dall is the same.

Argonauta hians Solander   Brown Paper Nautilus

Figure 99e, f

Worldwide in warm waters.

Similar to *A. argo*, but smaller, much "fatter" with a rapidly broadening keel that bears larger and fewer nodules. Color brownish white with darker stains on the early part of the keel. Uncommonly washed ashore.

Family TREMOCTOPODIDAE
GENUS *Tremoctopus* Delle Chiaje 1829

Tremoctopus violaceus Delle Chiaje   Common Umbrella Octopus

Pelagic in warm waters. Worldwide.

Total length, including the arms, 3 to 6 feet. Deep purplish red in color. Characterized by the long skin webs between the four dorsal arms, and the two large holes in the body near the base of the third arm and in front of the eyes. The species is gregarious, and is occasionally washed ashore on the east coast of Florida.

Family OCTOPODIDAE
GENUS *Octopus* Lamarck 1798

There are only five valid species of littoral *Octopus* so far recorded along the Atlantic Coast. There are a few deep water ones, some of which belong to closely related genera. The characters most relied upon in distinguishing species are relative length of the arms, the skin surface, the nature and relative length of the small ligula (the tiny pad-like extension on the end of the third right arm in the males, i.e., the hectocotylized arm). The number of gill plates and color pattern are used to a less extent.

The eight arms have each been given a number, in order that comparisons may be made. This is done by setting the octopus down with the body up, and the arms spread-eagle out in all directions. Turn the octopus so that the two eyes are on the side away from you. By going from the eyes out to the mantle edge away from you, and choosing the first arm to the right, you have located the first arm. Further clockwise are the second, third and fourth right arms. Instead of counting further (fifth arm, etc.), return to the center again, and count to the left—hence, the first, second, third and fourth left arms. When giving an arm formula, only the right ones are generally given, and they are set down in order of large to smaller size. Hence, 4.3.1.2 means the fourth arm is the largest, the second one the smallest in length. It may be pointed out, that on rare occasions an octopus may accidentally lose an arm.

There are two simple sets of measurements (all in millimeters) which are important in distinguishing the species of Octopus. The first is the mantle-arm index which simply means the comparison of the length of the mantle (measure from the round, bulbous "head" end to a point just between the eyes) with the length of the longest arm (turn the octopus over, measure from the mouth to the tip of the longest stretched-out arm). An index is obtained by multiplying the mantle-length by 100 and then dividing the result by the arm-length.

The ligula index is obtained only from males and from the third right arm which is a modified sex organ. The ligula is measured from tip to the last sucker. The arm length is obtained as explained in the preceding paragraph. The index is: length of ligula, multiplied by 100, the result divided by the total arm-length. The number of gill plates and the size of eggs are determined by cutting a deep slit in the body.

Octopus vulgaris Lamarck   Common Atlantic Octopus
Figure 100a

Connecticut to Florida and the West Indies. Europe.

Length, including the longest arm, 1 to 3 feet (the latter would give a radial spread of about 7 feet). Mantle-arm index in Florida and North Carolina is about 25 (that is, the arms are 4 times as long as the mantle). Ligula-index below 2.5. Gill plates 7 to 9 (in Bermuda, usually 10 or 11). In life, skin smoothish; preserved, it is rugose with variously shaped warts. Eggs 3 mm. or less in length. A common harmless species found hidden away under large rocks and crevices near shore. This is *O. rugosus* of authors, *O. americanus* Blainville and *O. carolinensis* Verrill.

Octopus macropus Risso of the West Indies (and possibly Key West) has its first arm the largest and longest, has a ligula-index up to 14, and a wart over one side of the eye. The skin in preserved material has small reddish warts. The eggs are less than 2 mm. in length (see fig. 100e).

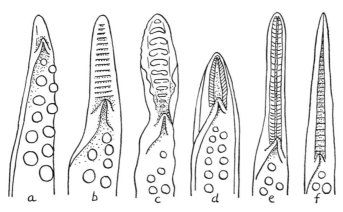

FIGURE 100. Atlantic and Pacific Octopus. The tip end or *ligula* of the third right arm in the male. **a**, *Octopus vulgaris* Lamarck; **b**, *O. burryi* Voss; **c**, *O. briareus* Robson; **d**, *O. joubini* Robson; **e**, *O. macropus* Risso; **f**, *O. hongkongensis* Hoyle (Pacific).

Octopus briareus Robson   Briar Octopus
                                                    Figure 100c

Southern Florida and the West Indies.

Length, including longest arm, 1 to 1½ feet. Arms fairly thick at the bases, quite long, especially the third and sometimes the second. Mantle-index 13 to 30, but usually about 17. Ligula-index about 4. Gill plates 7, rarely 8. Skin smoothish, or finely granular in preserved material; in life, pinkish brown to red-mottled. Eggs elongate, translucent-white, 10 to 12 mm. in length and with equally long attachment stalks. Fairly common between tides under large coral blocks on the Lower Florida Keys.

Octopus burryi Voss   Burry's Octopus
                                                    Figure 100b

Southern half of Florida and the Gulf of Mexico.

Length, including the longest arm, 6 to 10 inches. Characterized by a broad band of dark purple on the top surface of the arms, and, in preserved specimens, by the skin which is covered with closely set, *round* papillae or warts. Gill plates 8 to 10 in number. Ligula-index 4 to 5. This is a recently described species named after a famous Florida collector, Leo L. Burry of

Pompano Beach, Florida. It is a moderately common, fairly shallow-water species.

Octopus joubini Robson   Joubin's Octopus
Figure 100d

Southern half of Florida, and the West Indies.

A small species with a length, including the longest arm, of from 4 to 6, rarely 7 inches. The arms are short, with a mantle-arm index of about 40 to 50. Ligula-index about 6 to 7. Gill plates 5 or 6 usually. Skin smoothish, except for little pimples at scattered intervals. In this species, the longest arm is only 2 or 3 times the mantle-length, while in *O. briareus* the longest arm is 5 or 6 times as long as the mantle. Eggs large, amber-colored, and about 7 to 10 mm. in length. Occasionally cast ashore in fair numbers on the west coast of Florida. *O. mercatoris* Adam 1937 is the same. Formerly placed in the genus *Paroctopus* which is now considered of no value.

Octopus hongkongensis Hoyle   Common Pacific Octopus
Figure 100f

Alaska to Lower California. Japan to south China.

Length, including longest arm, ½ to 3 feet (possibly with a radial spread of nearly 28 feet in Alaskan waters). Skin in preserved specimens covered everywhere by numerous small, pimple-like tubercles with star-shaped bases, and by many heavy, much interrupted, longitudinal wrinkles. Above each eye there is a rather small, conical wart and with a very large, pinnacle-like protuberance behind it. Ligula index 4 to 7. The web between the second and third arms usually extends out to a quarter of the arm's length. Elsewhere the webs are shorter. The commonest littoral Octopus on the Pacific Coast found from shore to 100 fathoms. This is *O. punctatus* Gabb.

O. californicus Berry, an off-shore species, has a large lingula with an index of 14 to 17. The skin in preserved material is covered with numerous, large stellate warts. The Californian Deep-water Octopus.

Octopus bimaculatus Verrill   Two-spotted Octopus

Los Angeles, California, to Lower California.

Total length ½ to 2 feet. Characterized by a large, distinct, round, dark spot in front of each eye near the base of each third arm. Eggs small, 1.8 to 4.0 mm. in length with long stalks, attached in festoons. Mantle-arm index usually 22, but ranging from 14 to 29. Ligula-index 2.0, not significant in separating this species from *bimaculoides*. Fairly common. Lives in the lower part of the intertidal zone down to several feet where there is rock bottom.

Aside from egg-size and egg-clusters and mantle-arm index, there is great difficulty in separating this species from *O. bimaculoides* Pickford which lives nearby in shallower water where there is mud present.

Octopus bimaculoides Pickford   Mud-flat Octopus

Los Angeles, California, to Lower California.

Almost identical with *O. bimaculatus* Verrill. Eggs large, 9.5 to 17.5 mm. in length, with shorter stalks, attached in small clusters. Mantle-arm index 34, but ranging from 29 to 39. Fairly common in shallow water among rocks where mud is present. Adults are somewhat smaller than *bimaculatus*.

CHAPTER XII

Guide to the
Molluscan Literature

THE literature dealing with mollusks is very extensive and widely scattered in many journals and books. Since 1900, approximately 60,000 separate articles on mollusks have appeared. Most of these are listed according to author and subject matter in the *Zoological Record* (printed for the Zoological Society of London and in its 86th volume). This valuable journal may be found in any large museum or university library. The bibliography included here is intended only as a guide or lead to the more important books and articles dealing primarily with American marine mollusks.

GENERAL TEXTS

BRONN, H. G. 1892-1940: *Klassen und Ordnungen des Tierreichs*. Leipzig. Several large volumes on biology and anatomy of mollusks. Technical. Monumental work in German by H. Simroth, H. Hoffmann, F. Haas and others. Large bibliographies.

BULLOUGH, W. S. 1950: *Practical Invertebrate Anatomy*. 463 pp. Macmillan, N.Y. Large section on anatomy and dissecting techniques (pp. 317-391, 34 figs.) on mollusks.

COOKE, A. H. 1895: *Mollusca*. Volume 3 of the Cambridge Natural History Series. 459 pp., 311 figs. Macmillan, N.Y. Very good, but out-of-date, general introduction to mollusks.

JOHNSTON, G. 1850: *An Introduction to Conchology*. 614 pp., 102 figs. J. Van Voorst, London. Interesting reading, history and lore, but very much out-of-date.

MACGINITIE, G. E. and N. 1949: *Natural History of Marine Animals*. 473 pp. McGraw-Hill, N.Y. Ecology and habits of marine mollusks, pp. 327-401.

PELSENEER, PAUL 1906: *Mollusca*. A Treatise on Zoology (vol. 5). 355 pp., 301 figs. Adam and C. Black, London. Excellent college-level text. Somewhat out-of-date.

THIELE, J. 1929-35: *Handbuch der Systematischen Weichtierkunde*. 4 vols. 1154 pp. Jena, Germany. Standard text on classification and arrangement of mollusks.

TRYON, G. W. 1882-84: *Structural and Systematic Conchology*. Philadelphia. 3 vols. Vol. 1, 312 pp., 22 pls., contains introductory matter, biology, history and anatomy. Vol. 2, 430 pp., 68 pls., contains general systematic account of marine forms. Vol. 3 deals with land forms. Useful, but considerably out-of-date.

POPULAR BOOKS

We have listed most of the recently published popular books, but do not necessarily recommend all that are included. The prices listed here are only approximate, especially for those available only second-hand.

General and Foreign

ALLAN, JOYCE 1950: *Australian Shells*. 470 pp., 12 colored pls., 28 halftones, 110 figs. Georgian House, Melbourne. $7.50.

BARTSCH, P. 1931: *Mollusks*. In Smithsonian Scientific Series, vol. 10, pt. 3, pp. 251-357, 36 pls. Series Publishers, N.Y. $8.25. Good general account of mollusks.

COTTON, B. C. and F. K. GODFREY 1938-40: *The Molluscs of South Australia*. Pt. 1, The Pelecypoda; pt. 2, The Scaphopoda, Cephalopoda and Crepipoda. 600 pp., 589 figs. Government Publ., Adelaide. About $3.00. Excellent and well illustrated.

EDMONDSON, C. H. 1946: *Reef and Shore Fauna of Hawaii*. 381 pp., 223 figs. B. P. Bishop Museum, Honolulu. 100 pp. and 75 figs. on mollusks.

HIRASE, S. and ISAO TAKI 1951: *A Handbook of Illustrated Shells from the Japanese Islands and their Adjacent Terr*. 134 pls. (130 in color). About $5.00. Bunkyokaku Publ., Tokyo. Excellent.

PLATT, R. 1949: *Shells Take You Over World Horizons*. 50 pp., 32 color plates. National Geographic Magazine, Wash., D.C. July 1949 issue. Separates 50 cents. Out of print (1952). Excellent illustrations.

POWELL, A. W. B. 1946: *The Shellfish of New Zealand*. 2nd ed. 106 pp., 26 pls. (1 in color). Whitcombe and Tombs Ltd., Auckland. Complete checklist and illustrations of common species. Excellent.

ROGERS, JULIA 1951: *The Shell Book*. 485 pp., 87 pls. (8 in color). C. T. Branford, Boston. A reprint of the 1908 edition with the names brought up-to-date in an appendix by Harald A. Rehder. $6.50. Excellent for beginners.

SMITH, MAXWELL 1940: *World-wide Sea Shells*. 139 pp., many drawings. $4.50. Obtained from author, Box 65, Winter Park, Florida.

TINKER, S. W. 1952: *Pacific Sea Shells*. 237 pp., illust. Paper covered: $2.75. Honolulu, T.H. Mainly gastropods of Hawaii.

VERRILL, A. H. 1936: *Strange Sea Shells and their Stories*. 211 pp., figs., 5 pls. (1 in color). L. C. Page, Boston. $2.50. For children.

VERRILL, A. H. 1950: *The Shell Collector's Handbook*. 228 pp., illus. Putnam's, N.Y. $4.00.

WEBB, W. F. 1948: *Handbook for Shell Collectors*. 8th ed., 236 pp., about 1000 species figured. Interesting notes. $5.00. Obtained from author: 2515 Second Ave. N., St. Petersburg, Florida.

East Coast of America

ALDRICH, B. D. E. and E. SNYDER 1936: *Florida Sea Shells*. 126 pp., 11 pls. Houghton Mifflin, Boston. $1.25. About 150 species included.

ARNOLD, AUGUSTA 1903: *The Sea Beach at Ebb-Tide*. 470 pp., 600 figs. Century, N.Y. Second-hand, $3.00 to $5.00. Section on mollusks included, but out-of-date.

MORRIS, P. A. 1939: *What Shell Is That?* 198 pp., 175 figs. Appleton-Century, N.Y. $2.25. Small pocket guide for New England collectors.

MORRIS, P. A. 1951: *A Field Guide to the Shells of Our Atlantic and Gulf Coasts*. 2nd ed. 236 pp., 45 pls. (8 in color). Houghton Mifflin, Boston. $3.75. Names somewhat out-of-date.

PERRY, LOUISE 1940: *Marine Shells of the Southwest Coast of Florida*. 260 pp., 39 pls. Paleontological Research Inst., Ithaca, N.Y. $3.50. Mostly shells of Sanibel Island. Well illustrated, good descriptions. For amateurs and advanced students.

SMITH, MAXWELL 1937: *East Coast Marine Shells*. 308 pp., illus. Obtained from author, Box 65, Winter Park, Fla. $5.00. Names out-of-date.

VILAS, C. N. and N. R. VILAS 1945: *Florida Marine Shells*. 151 pp., 12 color plates. C. N. Vilas Publ., Box 108, Sarasota, Florida. $2.75. For beginners; names somewhat out-of-date.

West Coast of America

KEEP, J. and J. L. BAILY, JR. 1935: *West Coast Shells*. 350 pp., 334 text figs. Stanford University Press, Calif. $3.75.

MORRIS, P. A. 1952: *A Field Guide to Shells of the Pacific Coast and Hawaii*. 220 pp., illus. Houghton Mifflin, Boston. $3.75. Recommended only for the part concerning our Pacific Coast.

SMITH, MAXWELL 1944: *Panama Marine Shells*. 127 pp., illus. Obtained from author, Box 65, Winter Park, Fla. $6.00.

Directories and U.S. Journals

Directory of Conchologists. An American and international list of over 900 people interested in mollusks. Gives addresses, interests and exchange activities. Mimeographed. Obtained from: Mr. John Q. Burch, 1584 West Vernon Ave., Los Angeles 37, Calif. $1.50.

Annual Report of the American Malacological Union. Names and addresses of over 350 active members. Obtained from: Secretary, The American Malacological Union, Buffalo Museum of Science, Buffalo 11, N.Y.

The Nautilus, a quarterly devoted to the interests of conchologists. Technical and semi-popular articles, notes and news. $2.50 per year; quarterly. Dr. H. B. Baker, Bus. Mgr., Zool. Lab., University of Pennsylvania, Philadelphia 4, Pa.

Johnsonia. Monographs of the Marine Mollusca of the Western Atlantic (quarto size). W. J. Clench, ed. Museum Comparative Zoölogy, Harvard Univ., Cambridge 38, Mass. $4.00 per year or per 100 pp. Excellent illus., descriptions, ranges, collecting localities, book reviews, etc.

Occasional Papers on Mollusks. Depart. of Mollusks, Museum Comparative Zoology, Harvard Univ., Cambridge 38, Mass. Revisions, bio-bibliographies, medical snails, catalogs and other useful articles. Price list from W. J. Clench, ed.

Leaflets in Malacology. S. Stillman Berry, ed., Redlands, Calif. Useful and interesting articles, dealing mainly with the Eastern Pacific fauna.

Revista de la Sociedad Malacologica "Carlos de la Torre," Univ. Habana, Cuba. C. G. Aguayo, ed. Technical articles of importance in English and Spanish.

Current foreign journals are: *Proceedings of the Malacological Society of London; Journal of Conchology* (Great Britain); *Archiv für Molluskenkunde* (Frankfurt-am-Main); *Journal de Conchyliologie* (Paris); *Proceedings Malacological Society of Japan* (formerly the *Venus*); *Basteria* (Leiden).

CHAPTER I—MAN AND MOLLUSKS

Mollusks and Medicine in World War II. R. Tucker Abbott. 1948. *Smithsonian Ann. Report for 1947*, pp. 325-338, 3 pls. (no. 3933).

Snail Invaders. R. Tucker Abbott. *Natural History Magazine* (N.Y.), Feb. 1950, vol. 59, pp. 80-85, 13 figs.

The Venomous Cone Shells. R. Tucker Abbott. 1950. *The Science Counselor* (Duquesne Univ.), Dec. 1950, pp. 125-126; 153, 3 pls.

Molluscan Species in Californian Shell Middens. R. E. Greengo. 1951. *Report 13*, Univ. Calif. Archaeol. Survey, pp. 1-23. Good bibliography.

The Geographical Distribution of the Shell-Purple Industry. J. W. Jackson. 1916. *Memoirs and Proc.* Manchester Literary and Philos. Soc., vol. 60, pt. 2, no. 7, 29 pp., map.

The Use of Cowry-Shells for the Purposes of Currency, Amulets and Charms. J. W. Jackson. 1916. *Ibid.*, vol. 60, pt. 3, 72 pp., maps.

Shell Trumpets and Their Distribution in the Old and New World. J. W. Jackson. 1916. *Ibid.*, vol. 60, pt. 2, no. 8, 22 pp.

The Geographic Distribution of the Use of Pearls and Pearl-Shell. J. W. Jackson. 1916. *Ibid.*, vol. 60, pt. 3, no. 12, 53 pp.

Mémoire sur la Pourpre. H. Lacaze-Duthiers. 1859. *Annales des Science Naturelles (Zoologique)*, 4th series, vol. 12, pp. 5-84, 2 pls.

Ethno-Conchology: A Study of Primitive Money. R. E. C. Stearns. 1889. *Annual Report* U.S. Nat. Mus. for 1886-87, pp. 297-334, 9 pls. (Out of print).

The Use of Marine Mollusca and Their Value in Reconstructing Prehistoric Trade Routes in the American Southwest. D. B. Tower. 1945. Papers Excavators Club (Cambridge, Mass.), vol. 2, no. 3, pp. 1-54. Illus.

European Ballast Shells in Staten Island, N.Y. M. P. Weingartner. 1951. *The Nautilus*, vol. 65, p. 132.

Trade Marks. W. W. Wigginton. 1946. *The Shell Magazine* (Asiatic Petroleum Co.), London, vol. 26, no. 480, Dec., pp. 322-324. Also see the Jubilee Pamphlet of the "Shell" Transport and Trading Co., Ltd. (London), 1947, 26 pp., illus.

CHAPTER II—LIFE OF THE SNAIL

Habits, Life Histories and Ecology

A Preliminary Investigation of the Importance of Desiccation, Temperature and Salinity as Factors Controlling the Vertical Distribution of Certain Intertidal Marine Gastropods in False Bay, South Africa. G. J. BROEKHUYSEN. 1940. *Trans*. Royal Soc. South Africa, vol. 28, pt. 3, pp. 278-292, 2 pls.

Notes on the Louisiana Conch, Thais haemastoma in its Relation to the Oyster, Ostrea virginica. M. D. BURKENROAD. 1931. *Ecology*, vol. 12, no. 4, pp. 656-664, 2 figs.

Observations on the Local Movements of Littorina littorea (L.) and Thais lapillus (L.). R. W. DEXTER. *The Nautilus*, vol. 57, pp. 6-8.

The Mud Snail: Nassa obsoleta. A. C. DIMON. 1905. *Cold Spring Harbor Monogr.* 5, pp. 1-48, 2 pls. (Habits and anatomy).

The Mechanism of Locomotion in Gastropod Molluscs. H. W. LISSMANN. 1945. *Journ. Exper. Biol.*, vol. 21, pp. 58-69. Illus.

The Biology of Purpura lapillus. H. B. MOORE. 1936-39. Pts. I, II, and III. *Journ. Marine Biol. Assoc.*, vol. 21, pp. 61-89; vol. 23, pp. 57-74.

The Leaping of the Stromb (Strombus gigas Linn.). G. H. PARKER. 1922. *Journ. Exper. Zool.*, vol. 36, no. 2, pp. 205-209, 2 figs.

Biology of Acmaea testudinalis Muller. M. A. WILLCOX. 1905. *Amer. Nat.*, vol. 39, no. 461, pp. 325-333.

Growth and Feeding

Influence of Natural and Experimental Conditions in Determining Shape of Shell and Rate of Growth in Gastropods of the Genus Crepidula. W. R. COE. 1942. *Journ. Morph.*, vol. 71, no. 1, pp. 35-47.

A Statistical Test of the Species Concept in Littorina. J. COLMAN. 1932. *Biol. Bull.*, vol. 62, no. 3, pp. 223-243, 11 figs., 8 tables.

How Fulgar and Sycotypus Eat Oysters, Mussels and Clams (Busycon). H. S. COLTON. 1908. *Proc.* Acad. Nat. Sci. Phila., 1908, pp. 3-10, 5 pls.

Growth of the Oyster Drill, Urosalpinx cinerea, Feeding on Four Different Food Animals. J. B. ENGLE. 1942. *Anatom. Records*, vol. 84, p. 505ff.

Observations on Some of the Probable Factors Controlling the Size of Certain Tide Pool Snails. R. R. HUMPHREY and R. W. MACEY. 1930. *Publ.* Puget Sound Biol. Sta., vol. 7, pp. 205-208.

The Mode of Feeding of Crepidula . . . J. H. ORTON. 1912. *Journ. Mar. Biol. Assoc.*, vol. 9, no. 3, pp. 444-478. Illus.

Anatomy

The Structure and Function of the Alimentary Canal of Some Tectibranch Molluscs with a Note on Excretion. VERA FRETTER. 1939. *Trans*. Royal Soc. Edinburgh, vol. 59, pt. 3, no. 22, pp. 599-646. Excellent figs.; also same subject on Chitons vol. 59, p. 119ff; on Limpets vol. 57, p. 287ff; on Aeolids vol. 59, p. 267ff.

The Anatomy of the Gastropod Crepidula adunca Sowerby. C. E. MORITZ. 1938. *Univ. Calif. Publ. Zool.*, vol. 43, no. 5, pp. 83-92, 6 figs.

The Anatomy of Acmaea testudinalis Muller. M. A. WILLCOX. 1906. *Amer. Nat.*, vol. 40, pp. 171-187, 4 figs. (External anatomy.)

The Prosobranchiate Mollusca; A Functional Interpretation of Their Structure and Evolution. C. M. YONGE. 1939. *Philos. Trans.* Royal Soc. London, ser. B, no. 566, vol. 230, pp. 79-147. Illus.

Reproduction and Development

Sexual Differentiation in Mollusks. W. R. COE. 1943-44. *Quart. Review Biology*, vol. 18, pp. 154-164; vol. 19, pp. 85-97. Large bibliography.

Sur le Dimorphism Sexual des Coquilles. ED. LAMY. 1937. *Journ. de Conchyl.*, vol. 81, pp. 283-301. Illus.

Reproduction and Larval Development of Danish Marine Bottom Invertebrates. . . . GUNNAR THORSON. 1946. Medd. Komm. Danmarks Fisk.-Havund. *Series on Plankton*, vol. 4, no. 1, pp. 1-523. (34 pp. of bibliography.)

The Embryology of Fulgur: A Study of the Influence of Yolk on Development. E. G. CONKLIN. 1907. *Proc.* Acad. Nat. Sci. Phila. 1907, pp. 321-358, 6 pls.

The Development of a Mollusk. B. E. DAHLGREN. 1906. *Amer. Mus. Journ.*, vol. 6, pp. 28-53. (Illus. of models.)

Early Development of Haminea. R. E. LEONARD. 1918. *Publ.* Puget Sound Biol. Sta., vol. 2, no. 34, pp. 45-63, 5 pls.

The Embryology of Patella. W. PATTEN. 1885. *Arbeit.* Zoolog. Inst., Wien, vol. 6, pp. 149-174, 5 pls.

Egg-Cases and Larvae

The Egg Capsules of Certain Neritidae. E. A. ANDREWS. 1935. *Journ. Morph.*, vol. 57, no. 1, pp. 31-54, 3 pls.

Egg Capsules and Development of Some Marine Prosobranchs from Tropical West Africa. J. KNUDSEN. 1950. *Atlantide Report* (Copenhagen), no. 1, pp. 85-130. (Large bibliography.)

The Eggs and Larvae of Some Prosobranchs from Bermuda. M. V. LEBOUR. 1945. *Proc.* Zool. Soc. London, vol. 114, pp. 462-489, 43 text-figs.

Biologie et Ponte de Mollusques Gastéropodes Néo-Calédoniens. J. RISBEC. 1935. *Bull.* Soc. Zool. de France, vol. 60, pp. 387-417. Illus.

Spawning of the Whelk [Busycon canaliculata]. L. B. SPENCER. 1910. *Zool. Soc. Bull.*, N.Y., no. 38, pp. 637-638. (Two photos of process.)

Studies on the Egg-Capsules and Development of Arctic Marine Prosobranchs. GUNNAR THORSON. 1935. *Komm. Vidensk. Undersog. Gronland*, vol. 100, no. 5, pp. 5-71. 75 figs.

Studies on the Egg Masses and Larval Development of Gastropoda from the Iranian Gulf. GUNNAR THORSON. 1940. *Danish Sci. Invest. in Iran*, pt. 2, pp. 159-238, 32 figs.

Spawning of Fulgur perversus [Busycon contrarium]. J. WILLCOX. 1885. *Proc.* Acad. Nat. Sci., Phila., 1885, pp. 119-120.

CHAPTER III—LIFE OF THE CLAM

An Annotated Bibliography of Oysters, with Pertinent Material on Mussels and Other Shellfish. J. L. BAUGHMAN. 1948. Texas A. and M. Research Foundation. pp. 1-794. Over 2000 articles on the biology of mollusks are listed, with abstracts and a subject index. Very useful.

Habits, Life Histories and Ecology

The Habits of Life of Some West Coast Bivalves. FRITZ HAAS. 1942. *The Nautilus*, vol. 55, pp. 109-113 (Lithophaga, Diplodonta and Cooperella); *ibid.*, vol. 56, pp. 30-33 (Mytilus and Brachidontes).

The Life History and Growth of the Razor Clam. H. C. McMILLIN. 1924. Wash. State Dept. Fish., 52 pp., 5 pls., 3 graphs.

The Pismo Clam: Further Studies of Its Life History and Depletion. W. C. HERRINGTON. 1930. Calif. State Fish Lab. *Contrib. no. 81, Bull. 18*, 67 pp., 16 figs.

Notes on the Ecology of the Butter Clam, Saxidomus giganteus Deshays. C. M. FRASER. 1928. *Trans.* Royal Soc. Canada, ser. 3, vol. 22, pp. 271-277, 2 pls., 7 tables; also see *ibid.*, vol. 22, pp. 249-270 for Paphia staminea; also vol. 25, pp. 59-72 for "Cardium corbis."

The Edible Bivalves of California. PAUL BONNOT. 1940. Calif. Fish and Game, vol. 26, no. 3, pp. 212-239. Notes, habits, sketches, and laws pertaining to catches.

A Resurgent Population of the California Bay-Mussel (Mytilus edulis diegensis). W. R. COE. 1946. *Journ. Morph.*, vol. 78, no. 1, pp. 85-104.

The Marine Borers of the San Francisco Bay Region. C. A. KOFOID. 1921. *Report* San Francisco Bay Marine Piling Survey, pp. 23-61.

A Brief Study of the Succession of Clams on a Marine Terrace. PAUL T. WILSON. 1926. *Publ.* Puget Sound Biol. Sta., vol. 5, pp. 137-148.

The Habits and Movements of the Razor-Shell Clam, Ensis directus Con. G. A. DREW. 1907. *Biol. Bull.*, vol. 12, no. 3, pp. 127-140, 1 pl.

Locomotion in Solenomya and Its Relatives. G. A. DREW. 1900. *Anat. Anzeiger*, vol. 17, no. 15, pp. 257-266, 12 figs.

The Structure and Behaviour of "Hiatella gallicana" (Lamarck) and "H. arctica" (L.), with Special Reference to the Boring Habit. W. R. HUNTER. 1949. *Proc.* Royal Soc. Edinburgh, B, vol. 63, pt. 3, no. 19, pp. 271-289. 12 figs.

Growth and Feeding

The Life-History and Growth of the Pismo Clam (Tivela stultorum Mawe). F. W. WEYMOUTH. 1923. *Bull.* 7, Calif. State Fish and Game Comm., pp. 5-120.

The Age and Growth of the Pacific Cockle (Cardium corbis). F. W. WEYMOUTH and S. H. THOMPSON. 1931. Bureau of Fisheries *Doc. no. 1101* (Wash., D.C.), pp. 633-641. (also see *Doc. nos. 984, 1099* and *1100*).

Food Material as a Factor in Growth Rate of Some Pacific Clams. G. M. SMITH. 1928. *Trans.* Royal Soc. Canada, ser. 3, vol. 22, sec. 5, pp. 287-292. (see also Further Observations. . . . *ibid.*, 1933, vol. 27, sec. 5, pp. 229-245).

The Mechanism of Feeding, Digestion, and Assimilation in the Lamellibranch Mya. C. M. YONGE. 1923. *Brit. Jour. Exper. Biol.*, vol. 1, pp. 15-63, 27 figs.

On the Morphology, Feeding Mechanisms, and Digestion of Ensis siliqua (Schumacher). A. GRAHAM. 1931. *Trans.* Royal Soc. Edinburgh, vol. 56, pp. 725-751, 8 figs.

The Habitat and Food of the California Sea Mussel. D. L. Fox *et al.* 1936. *Bull.* Scripps Inst. Oceanogr. Univ. Calif., Tech. ser., vol. 4, no. 1, pp. 1-64.

The Systematic Value of a Study of Molluscan Faeces. H. B. MOORE. 1931. *Proc.* Malacological Soc. London, vol. 19, pp. 281-291, 4 pls.

Anatomy and Shell Structure

The Anatomy of Some Protobranch Mollusks. HAROLD HEATH. 1937. *Mem.* Mus. Royal d'Hist. Nat. Belgique, ser. 2, fasc. 10, pp. 3-25, 10 pls.

The Anatomy of the Pelecypod Family Arcidae. HAROLD HEATH. 1941. *Trans.* Amer. Philos. Soc. (n.s.), vol. 31, pt. 5, pp. 287-319, 22 pls.

Uber die Anatomie von Chama pellucida Broderip. E. GRIESER. 1913. *Zool. Jahrb. Suppl.*, vol. 13, pp. 207-280, 1 pl., 11 figs.

Literature on the Shell Structure of Pelecypods. H. G. SCHENCK. 1934. *Bull.* Mus. Royal d'hist. Nat. Belgique, vol. 10, no. 34, pp. 1-20.

The Structure and Composition of the Shell of Tellina tenuis. E. R. TRUEMAN. 1942. *Journ. Royal Micro. Soc.*, vol. 62, pp. 69-92.

The Shell Structure of the Mollusks. O. B. Boggild. 1930. D. Kgl. Danske Vidensk. Selsk. Skrifter, Naturv. Og Mathem., Afd. 9, Raekke II.2., pp. 233-326, 10 figs., 15 pls.

The Pigmentation of Molluscan Shells. ALEX COMFORT. 1951. *Biol. Reviews* Cambridge Philos. Soc., vol. 26, no. 3, pp. 285-301. 5 figs.

The Mystery of the Pearl. J. BOLMAN. 1941. *Internat. Archiv. Ethnographie*, Leiden, suppl. to vol. 39, pp. 1-170, 37 pls. [Excellent semi-popular account in English.]

Reproduction, Development and Larvae

Sexual Differentiation in Mollusks. W. R. COE. 1943. *Quart. Review of Biology*, vol. 18, pp. 154-164; also vol. 19, pp. 85-97. Large bibliography.

Development of the Oyster. W. K. BROOKS. 1880. *Studies Biol. Lab.*, Johns Hopkins Univ., vol. 4, pp. 1-93, 10 pls.

Spawning Habits of the Mussel, Mytilus californianus Conrad, with Notes on the Possible Relation to Mussel Poison. W. W. FOREST. 1936. Univ. Calif. Publ. Zool., vol. 41, no. 5, pp. 35-44, pl. 3.

Reproduction and Larval Ecology of Marine Bottom Invertebrates. G. THORSON. 1950. *Biol. Reviews*, London, vol. 25, pp. 1-45. Large bibliography.

Relations Between the Moon and Periodicity in the Breeding of Marine Animals. P. KORRINGA. 1947. *Ecol. Monogr.*, Durham, N.C., vol. 17, no. 3, pp. 347-381 (*Ostrea, Littorina, Mytilus* and *Pacten*).

The Identification and Classification of Lamellibranch Larvae. C. B. REES. 1950. *Hull Bull. Mar. Ecol.*, vol. 3, no. 19, pp. 73-104, figs.

Bivalve Larvae of Malpeque Bay, P. E. I. M. C. SULLIVAN. 1948. *Bull.* 77, Fish. Research Board Canada, pp. 1-36, 22 pls.

CHAPTER IV—LIVES OF THE OTHER MOLLUSKS

Cephalopoda

The Biology of Spirula spirula L. A. F. Brunn. 1943. *Dana Report* (Copenhagen), vol. 4, pt. 24, pp. 1-44, 13 figs., 2 pls.

Sexual Activities of the Squid, Loligo pealii (Les). I Copulation, Egglaying and Fertilization. G. A. Drew. 1911. *Journ. Morph.*, vol. 22, pp. 327-359, pls. 1-4; also 1919, vol. 32, pp. 379-435, 6 pls.

The Anatomy of the Common Squid, Loligo pealii Lesueur. L. W. Williams. 1909. E. J. Brill, Leiden. 92 pp., 16 figs., 3 pls. Excellent.

Report on the Cephalopods of the Northeastern Coast of America. A. E. Verrill. 1882. *Rep.* U.S. Comm. Fish, pt. 7, for 1879, 245 pp., 47 pls.

Amphineura

On the Presence of Eyes in the Shells of Certain Chitonidae. H. N. Moseley. 1885. *Quart. Journ. Micro. Sci.* (N.S.), vol. 25, pp. 2-26, 6 pls.

Notes on the Post-larval Development of the Giant Chiton, Cryptochiton stelleri (Midd.). S. Okuda. 1947. *Journ. Facul. Sci.*, Hokkaido Univ., ser. 6, Zool., vol. 9, no. 3, pp. 267-275, 18 figs.

Die Anatomie und Phylogenie der Chitonen. L. H. Plate. 1899. *Fauna Chilensis* (Jena), vol. 2, pp. 15-216, 9 pls. Excellent.

Scaphopoda

Scaphopoda. H. Simroth. 1892. In *Bronn's Thier-Reichs* (Leipzig), vol. 4, pp. 356-467, 10 figs., 5 pls. (In German; with long bibliography.)

A Monograph of the East American Scaphopod Mollusks. J. B. Henderson. 1920. *Bull.* U.S. Nat. Museum, 111, pp. 1-177, pls. 1-20. Out of print.

Generic and Subgeneric Names in the Molluscan Class Scaphopoda. W. K. Emerson. 1952. *Jour. Wash. Acad. Sci.*, vol. 42, no. 9, pp. 296-303.

CHAPTER V—COLLECTING AMERICAN SEASHELLS

Night Collecting; The Bar; Digging 'Em Out; and other articles. B. R. Bales. 1945-46. *The Nautilus*, vols. 58 and 59. Extremely interesting and informative.

Killing and Preservation of Bivalve Larvae in Fluids. M. R. Carriker. 1950. *The Nautilus*, vol. 64, pp. 14-17.

Instructions for Collecting Mollusks, and Other Useful Hints for the Conchologist. W. H. Dall. 1892. Part G, *Bull.* U.S. Nat. Mus., no. 39, 55 pp. Out of print.

The Collection and Preparation of Shells. T. C. Stephens. 1946-47. Turtox News, vol. 24, no. 9 and vol. 25, no. 1, 15 pp. A rather full annotated list of references.

Symposium on "Methods of Collecting and Preserving Mollusks." *Annual Report,* American Malacological Union, 1941. Several authors. Excellent. Write: Secretary, A. M. U., Buffalo Museum of Science, Buffalo 11, N.Y.

CHAPTER VI—HOW TO KNOW AMERICAN SEASHELLS

Malacology and the Official List of Generic Names. C. G. Aguayo. 1949. *The Nautilus,* vol. 63, pp. 17-19.

Illustrated Glossary of Gastropoda, Scaphopoda, Amphineura. *Beatrice Burch.* 1950. *Minutes 105,* Conch. Soc. Southern Calif. Mimeographed. $1.50. Pelecypoda glossary also available.

Systematics and the Origin of Species. E. MAYR. 1942. Columbia Univ. Press, N.Y. 334 pp. Good bibliography.

Procedure in Taxonomy. E. T. SCHENK and J. H. McMASTERS. 1948. Stanford Univ. Press, Calif. Revised ed. by A. M. Keen and S. W. Muller.

GEOGRAPHICAL GUIDE

ATLANTIC COAST

General

List of Marine Mollusca of the Atlantic Coast from Labrador to Texas. C. W. JOHNSON. 1934. *Proc.* Boston Soc. Nat. Hist., vol. 40, pp. 1-204. 2632 species listed with ranges. About 75% complete, some names now obsolete. Bibliography very good, with over 500 entries. No illustrations or descriptions.

Johnsonia. *Monographs of the Marine Mollusca of the Western Atlantic.* 1942- Vols. 1 and 2 (600 pp., quarto size now complete). Museum Comparative Zoology, Cambridge 38, Mass. Thorough treatment of various genera with excellent plates, descriptions, ranges, records and collecting localities. W. J. Clench, ed.

A Preliminary Catalogue of the Shell-bearing Marine Mollusks and Brachiopods of the Southeastern Coast of the United States. W. H. DALL. 1889 (and a 1903 reprint). *Bull. 37,* U. S. Nat. Mus., 232 pp., 45 pls. Illustrations very useful, ranges fairly good, but the names out-of-date. Out of print.

Blake Reports. Reports on the Results of the Dredging . . . in the Gulf of Mexico and the Caribbean Sea by the . . . Steamer 'Blake'. 1886. Pt. I. Brachiopoda and Pelecypoda. W. H. DALL. *Bull.* Mus. Comp. Zool., vol. 12, pp. 171-318, 9 pls.; 1889, Pt. II, Gastropoda and Scaphopoda. *Ibid.,* vol. 18, pp. 1-492, 30 pls. Numerous deep-sea species with excellent drawings.

Contributions to the Tertiary Fauna of Florida. W. H. DALL. 1890-1903. 6 parts. *Trans.* Wagner Inst., vol. 3, 1654 pp., 60 pls. A monumental work containing many important revisions. Contains many recent species, but the nomenclature is out-of-date.

Greenland, Labrador and Newfoundland

Mollusca of the Crocker Land Expedition to Northwest Greenland and Grinnel Land. F. C. BAKER. 1919. *Bull.* Amer. Mus. Nat. Hist., vol. 41, pp. 479-517.

Report on the Mollusks Collected by L. M. Turner at Ungava Bay, North Labrador. W. H. DALL. 1886. *Proc.* U.S. Nat. Mus., vol. 9, pp. 202-208, 3 figs.

Observations . . . Recent Invertebrate Fauna of Labrador. A. S. PACKARD. 1867. Mem. Boston Soc. Nat. Hist., vol. 1, pp. 210-303, pl. 7-8.

A List of the Mollusks Collected by O. Bryant along the Coasts of Labrador, Newfoundland and Nova Scotia. C. W. JOHNSON. 1926. *The Nautilus,* vol. 39, pp. 128-135, and vol. 40, pp. 21-25.

Eastern Canada

The Mollusca of Nova Scotia. J. M. JONES. 1877. *Proc.* Trans. Nova Scotian Inst. Nat. Hist., vol. 4, pp. 321-330.

Catalogue of the Marine Invertebrates of Eastern Canada. J. F. Whiteaves. 1901. 272 pp. Ottawa: Geol. Survey of Canada.

Bivalve Larvae of Malpeque Bay, Prince Edward Island. *Bull.* 77, *Fish.* Research Board Canada, pp. 1-36, 22 pls. M. C. SULLIVAN. 1948.

New England

Report on the Invertebrata of Massachusetts. Second ed. 1870. A. A. GOULD. W. G. BINNEY, ed. 524 pp., 755 fig., 27 pls. *Mollusca.* Boston: Wright and Potter, State Printers.

Fauna of New England. List of the Mollusca. C. W. JOHNSON. 1915. *Occas. Papers,* Boston Soc. Nat. Hist., vol. 7, no. 13, pp. 1-231.

Plankton of the Offshore Waters of the Gulf of Maine. H. B. BIGELOW. 1926. *Bull.* U.S. Bur. Fish., vol. 40, pp. 1-509, figs. 1-134.

List of Shell-bearing Mollusca of Frenchman's Bay, Maine. D. BLANEY. 1906. *Proc.* Boston Soc. Nat. Hist., vol. 32, pp. 23-41, pl. 1.

Preliminary Catalogue of the Marine Invertebrata of Casco Bay, Maine. J. S. KINGSLEY. 1901. *Proc.* Portland Soc. Nat. Hist., vol. 2, pp. 159-183.

A Bibliography of the Recent Mollusca of Maine—1605-1930. N. W. LERMOND and A. H. NORTON. 1930. *Maine Nat.,* vol. 10, pp. 49-73; 100-121.

Shells of Maine, A Catalogue. N. W. LERMOND. 1909. State Entomol. of Maine, *Ann. Report 4,* pp. 25-70.

Report Upon the Invertebrate Animals of Vineyard Sound . . . Mollusca. A. E. VERRILL. 1873. *Moll. Rep.* U.S. Fish Comm., vol. 1, pp. 634-698, pls. 20-32. (Also numerous similar reports by same author in *Trans.* Conn. Acad.)

Notes on the Marine Mollusks of Cape Ann, Mass. R. W. DEXTER. 1942. *The Nautilus,* vol. 56, pp. 57-61. Good list. See also *ibid.,* vol. 58, pp. 18-24; 135-142.

The Shell-bearing Mollusca of Rhode Island. H. F. CARPENTER. 1884-1902. Numerous articles by this title in *The Nautilus,* vols. 1-16.

Marine Mollusca of the Bridgeport, Conn. Region. A. P. JACOT. 1924. *The Nautilus,* vol. 38, pp. 49-51.

Molluscan Fauna of New Haven. G. W. PERKINS. 1869. *Proc.* Boston Soc. Nat. Hist., vol. 13, pp. 109-164.

New York

Check List of the Mollusca of New York. E. J. Letson. 1905. *Bull.* N.Y. State Museum, vol. 88, pp. 1-112.

Some Marine Mollusca about New York City. A. P. Jacot. 1919. *The Nautilus*, vol. 32, pp. 90-94; vol. 34, pp. 59-60.

List of Marine Mollusca of Coldspring Harbor, Long Island. F. N. Balch. 1899. *Proc.* Boston Soc. Nat. Hist., vol. 29, pp. 133-162, pl. 1.

On the Mollusca of Peconic and Gardiner's Bays, Long Island, N.Y. S. Smith. 1860. *Ann.* Lyceum Nat. Hist. N.Y., vol. 7, pp. 147-168; also vol. 9, pp. 377-407, 6 figs.

Maryland and Virginia

Recent Oyster Researches on Chesapeake Bay in Maryland. R. V. Truitt. 1931. *Report* Chesapeake Biol. Lab. 1931, pp. 1-28.

Littoral Marine Mollusks of Chincoteague Island, Virginia. J. B. Henderson and P. Bartsch. 1914. *Proc.* U.S. Nat. Mus., vol. 47, pp. 411-421, 2 pls. Out of print.

Oyster Bars of the Potomac River. D. G. Frey. 1946. Fish and Wildlife Service, *Special Sci. Report*, 32, pp. 1-93, mimeo. List of Mollusks.

North and South Carolina

Additions to the Shallow-water Mollusca of Cape Hatteras, N. C., dredged by the . . . Albatross in 1883-84. K. J. Bush. 1885. *Trans.* Conn. Acad. Arts Sci., vol. 6, pp. 453-480, illus.

List of Mollusca from around Beaufort, N. Carolina, with Notes on *Tethys*. A. G. Hackney. 1944. *The Nautilus*, vol. 58, pp. 56-64. 174 marine species listed with habitats.

Notes on the Natural History of Fort Macon, N.C., and Vicinity. E. Coues. 1871. *Proc.* Acad. Nat. Sci. Phila. for 1871, pp. 120-148; and *ibid.* for 1878, pp. 297-315.

Some Marine Molluscan Shells of Beaufort and Vicinity. A. P. Jacot. 1921. *Journ. Elisha Mitchell Sci. Soc.*, vol. 36, pp. 129-145, pls. 11-13.

Catalog of Mollusca of South Carolina. W. G. Mazÿck. 1913. *Contrib.* Charleston Mus., 2, xvi-39 pp.

Florida

Marine Shells of the Southwest Coast of Florida. Louise Perry. 1940. *Bull.* Paleontol. Research Inst., Ithaca, N.Y. 260 pp., 39 excellent plates.

Catalogue of the Marine Shells of Florida, with Notes and Descriptions of Several New Species. W. W. Calkins. 1878. *Proc.* Davenport Acad. Nat. Sci., vol. 2, pp. 232-252, 1 pl.

An Annotated List of the Shells of St. Augustine, Fla. C. W. Johnson. 1890. *The Nautilus*, vol. 3, pp. 103-105.

List of Mollusca Obtained in South Carolina and Florida in 1871-72. J. C. Melville. 1881. *Journal of Conchology* (Great Britain), vol. 3, pp. 155-173.

Mass Mortality of Marine Animals on the Lower West Coast of Florida, Nov. 1946 to Jan. 1947. GORDON GUNTER *et al. Science*, vol. 105, pp. 256-257.
Numerous articles on Florida mollusks appear in *The Nautilus.*

Gulf States

Recent Molluscs of the Gulf of Mexico, and Pleistocene and Pliocene Species from the Gulf States. C. J. MAURY. 1920-22. *Bull.* Amer. Paleont., vol. 8, no. 34, pp. 1-115; vol. 9, no. 38, pp. 34-142.
Some New and Interesting Mollusks from the Deeper Waters of the Gulf of Mexico. H. A. REHDER and R. T. ABBOTT. 1951. *Revista Soc. Malac.* "C. de la Torre," vol. 8, no. 2, pp. 53-66, 2 pls.
A Contribution to the Fauna of the Coast of Louisiana. L. R. CARY. 1906. Gulf Biol. Sta., *Louisiana Bull.*, vol. 6, pp. 50-59.
Pteropoda from Louisiana. M. D. BURKENROAD. 1933. *The Nautilus*, vol. 47, pp. 54-57.
Brackish-Water and Marine Assemblages of the Texas Coast, with Special Reference to Mollusks. H. S. LADD. 1951. *Publ.* Inst. Marine Sci. Texas, vol. 2, no. 1, pp. 129-163, tables, maps.
Notes on the Marine Shells of the Texas Coast. J. K. STRECKER. 1935. *Baylor Bull.*, vol. 38, no. 3, pp. 48-60.
Mollusks from Point Isabel in Texas. H. B. STENZEL. 1940. *The Nautilus*, vol. 54, pp. 20-21 (list only).
An Illustrated Check List of the Marine Mollusks of Texas. T. E. PULLEY. 1952. *Texas Jour. Science*, vol. 4, no. 2, pp. 167-199, 13 pls.

PACIFIC COAST

General

An Abridged Check List and Bibliography of West North American Marine Mollusca. A. MYRA KEEN. 1937. Stanford Univ. Press, 87 pp. Very valuable paper with a useful bibliography.
Catalogue of the Marine Pliocene and Pleistocene Mollusca of California and Adjacent Regions. U. S. GRANT, IV and H. R. GALE. 1931. Mem. San Diego Soc. Nat. Hist., vol. 1, 1036 pp., 32 pls.
The Marine Shells of the West Coast of North America. IDA S. OLDROYD. 1924-1927. Stanford Univ. Publ., Univ. Ser., *Geol. Sci.*, vol. 1 (1924), Pelecypoda and Brachiopoda, 248 pp., 57 pls.; vol. 2, Gastropoda, Scaphopoda and Amphineura. Pt. 1, 298 pp., 29 pls.; Pt. 2, 304 pp., 42 pls.; Pt. 3, 340 pp., 35 pls. A compilation of original descriptions. Illustrations recommended with the corrections given by A. Myra Keen; see above.
Summary of the Marine Shell-bearing Mollusks of the Northwest Coast of America, from San Diego, California, to the Polar Sea. . . . W. H. DALL. 1921. *Bull. 112*, U.S. Nat. Mus., 217 pp., 22 pls. Good bibliography and illustrations. Names out-of-date. Ranges since extended.
Illustrated Key to West North American Pelecypod Genera. A. M. KEEN and D. FRIZZELL. 1946 ed. Stanford Univ. Press. 28 pp., illus. Very helpful.

Illustrated Key to West North American Gastropod Genera. A. M. Keen and J. C. Pearson. 1952. Stanford Univ. Press. 39 pp., illus. Rather helpful, with useful pen drawings.

Mollusks from the West Coast of Mexico and Central America. L. G. Hertlein and A. M. Strong. 1940-50. Pts. 1 to 10. *Zoologica* (N.Y. Zool. Soc.), vols. 25 to 36. Very useful for more southern species.

Distributional List of the West American Marine Mollusks. By numerous authors; edited by John Q. Burch. *Proc.* Conch. Club Southern Calif. 1945-46. Mimeographed. Very valuable, with up-to-date names, locality records, habits, some identification hints and drawings.

Alaska

Marine Shells of Drier Bay, Knight Island, Prince William Sound, Alaska. W. J. Eyerdam. 1924. *The Nautilus*, vol. 38, pp. 22-28.

Shell Collecting in Puget Sound and Alaska. Fred Baker. 1910. *The Nautilus*, vol. 24, pp. 25-31.

Notes on the Mollusca of Forrester Island, Alaska. G. Willett. 1918. *The Nautilus*, vol. 32, pp. 65-69; *ibid.*, vol. 33, pp. 21-28.

British Columbia

Notes on the Marine Mollusca of the Pacific Coast of Canada. G. W. Taylor. 1899. *Trans.* Royal Soc. Canada, ser. 2, vol. 5, sec. 4, pp. 233-250; see also *ibid.*, 1895, vol. 1, pp. 17-100.

Report on the Marine Shells of British Columbia. C. F. Newcombe. 1893. *Bull.* Nat. Hist. Soc. Brit. Col., pp. 31-72.

On Some Marine Invertebrata from the Queen Charlotte Islands. J. F. Whiteaves. 1880. Geol. Survey Canada, *Report of Progress, 1878-79*, pp. 190B-205B.

Washington

Marine Shells of Puget Sound and Vicinity. Ida S. Oldroyd. 1924. Puget Sound Biol. Sta., Univ. Wash., vol. 4, pp. 1-272, 49 pls.

Oregon

Edible Mollusca of the Oregon Coast. C. H. Edmondson. 1920. *Occas. Papers*, B. P. Bishop Mus., Honolulu, vol. 7, no. 9, pp. 179-201, 6 figs.

The Pelecypoda of the Coos Bay Region, Oregon. Yocum, H. B. and E. R. Edge. *The Nautilus*, vol. 43, pp. 49-51. 1929.

An Index Method for Comparing Molluscan Faunules. H. G. Schenck and Myra Keen. 1937. *Proc.* Amer. Philos. Soc., vol. 77, pp. 161-182.

California

The Marine Molluscan Fauna from the Vicinity of Bolinas Bay, California. Bruce L. Clark. 1914. *The Nautilus*, vol. 28, pp. 25-28.

The Gastropod Fauna of the Intertidal Zone at Moss Beach, San Mateo Co., Calif. H. E. Vokes. 1936. *The Nautilus*, vol. 50, pp. 46-50.

Molluscan Fauna from San Francisco Bay. E. L. PACKARD. 1918. *Univ. Calif. Publ. Zool.*, vol. 14, no. 2, pp. 199-452, pls. 14-60.

Ecological Aspects of a California Marine Estuary. G. E. MacGINITIE. 1935. *Amer. Midland Naturalist*, vol. 16, no. 5, pp. 629-765.

Partial List of the Molluscan Fauna of Catalina Island. A. M. STRONG. 1923. *The Nautilus*, vol. 37, pp. 37-43.

Mollusks of Anaheim Bay, California. E. P. CHACE. 1916. *The Nautilus*, vol. 29, pp. 129-131.

The Marine Mollusks and Brachiopods of Monterey Bay, California, and Vicinity. A. G. SMITH and M. GORDON. 1948. *Proc.* Calif. Acad. Sci., 4th ser., vol. 26, no. 8, pp. 147-245. Excellent and with a full bibliography of the region.

Mollusks and Brachiopods Collected in San Diego, California. F. W. KELSEY. 1907. *Trans.* San Diego Soc. Nat. Hist., vol. 1, no. 2, pp. 31-55.

Common Marine Bivalves of California. J. E. FITCH. 1953. *Calif. Fish Bull.*, no. 90, 102 pp., 63 figs. Excellent and reliable.

ADDITIONAL REFERENCES

ABBOTT, R. TUCKER 1955: *Introducing Seashells.* vi + 64 pp., 10 pls. (6 in color), text figs. D. Van Nostrand, N.Y. $2.50. A guide and introduction for beginners.

AMERICAN MALACOLOGICAL SOCIETY 1955: *How to Collect Shells.* 50 pp. Buffalo Museum of Science, Buffalo 11, N.Y. $1.00. A very useful booklet compiled by numerous experts.

BARNARD, K. H. 1952: *A Beginner's Guide to South African Shells.* 215 pp., 5 color pls., numerous text figs. Maskew Miller Ltd., Cape Town. $3.00.

GUTSELL, J. S. 1931: *Natural History of the Bay Scallop.* 63 pp. Bulletin Bureau Fisheries, vol. 46 (Document no. 1100). Washington, D.C. Technical.

HORNELL, JAMES 1951: *Indian Molluscs.* iv + 96 pp., 1 color pl., 70 text figs. Bombay Natural History Society, Apollo St., Bombay. $1.50. Excellent and interesting reading.

HUTCHINSON, W. M. 1954: *A Child's Book of Sea Shells.* 28 pp., illustrated in color. Maxton Publishers Inc., N.Y. 75 cents. Excellent for 7 to 9 year olds.

JOHNSON, MYRTLE E. 1954: *West Coast Marine Shells.* 36 pp., 16 text figs. Natural History Museum, San Diego 1, Calif. 65 cents. A useful beginner's booklet.

OLSSON, A. A. and ANNE HARBISON 1953: *Pliocene Mollusca of Southern Florida.* viii + 457 pp., 65 pls. Academy Natural Sciences, Phila. 3, Pa. $8.00. Many living species included. Excellent plates.

PERRY, L. M. and J. S. SCHWENGEL 1955: *Marine Shells of the Western Coast of Florida.* 198 pp., 55 pls. Paleontological Research Institution, Ithaca, N.Y. $7.00. Very useful.

Index to
Subject Matter and Common Names

Common names not used in this book appear in *italics* and are followed by the modern usage.

Index to Scientific Names

This is a cross-index to subspecies, species, subgenera, genera and families. Names that are considered synonyms or homonyms have been entered in *italics*.

Index